ANCIENT INVENTIONS

ANCIENT INVENTIONS

PETER JAMES
&
NICK THORPE

Michael O'Mara Books Limited

First published in Great Britain in 1995 by
Michael O'Mara Books Limited
9 Lion Yard, Tremadoc Road, London SW4 7NQ

A CIP catalogue record of this book is available from the British Library

ISBN 1 85479 777 8

Design by Beth Tondreau Design

Printed and bound in Great Britain by Mackays of Chatham PLC, Chatham, Kent

10 9 8 7 6 5 4 3

FOR NORI AND ALLISON

CONTENTS

ACKNOWLEDGMENTS

A book of this scope would have been impossible without picking the brains of numerous friends and colleagues, whose kindness in answering our questions is gratefully acknowledged:

Richard Baker (architect)
Dr. Nick Barton (palaeolithic archaeologist)
Lucy Blue (marine archaeologist)
Dr. Lucilla Burn (classical archaeologist)
Andrij Cholij (film buff)
Dr. Victor Clube (astronomer)
Joyce Filer (palaeopathologist)
Dr. Joann Fletcher (Egyptologist)
John Frankish (Aegean archaeologist and nurse)
Dr. Keiji Imamura (Japanese prehistorian)
Dr. Nikos Kokkinos (Greco-Roman archaeologist)
Stanley Lee (librarian)
Brian Moore (librarian)
Dr. Robert Morkot (Egyptologist)
Dr. Jack Ogden (expert in ancient jewelry)
Sandra Parker (librarian)
Bob Porter (engineer)
Sue Rollin (ancient historian)
Adel Salameh (musician)
Dr. Delwyn Samuel (palaeobotanist)
Ken Tyler (geologist)
Dick Vygers (linguist)
Gareth Williams (Indologist)

Special thanks are due to our agent Leslie Gardner; to wizard proofreader and classicist Katherine Stott, who went through the text with a fine-toothed comb and suggested many valuable improvements; and to Tom Gayton (at *The Nightingale*) who over the years has acted as an invaluable "clippings agency" of new archaeological discoveries in the press. At Ballantine, we need to thank many people for their patience and generous help—in particular Lesley Malin Helm, Joëlle Delbourgo and Daia Gerson. We would also like to thank Belinda Barratt and the other librarians at the Institute of Archaeology (London) and the staff of Liverpool City Libraries for their generous assistance. Our thanks also to Stuart Laidlaw (Institute of Archaeology) and Leslie Primo for help with the reproductions.

Finally, our extra-special thanks to artist Peter Koenig, who appeared out of the blue and undertook the herculean task of providing the new illustrations. Peter has been immensely sanguine in the face of the massive demands we imposed on him to draw (and redraw) so many difficult reconstructions and interpretations of ancient gadgets and scenes, many never illustrated before. Without his help, the book as you see it would not have been possible. Many thanks also to Patricia Briggs, Rosemary Burnard and Franco Vartuca for additional illustrations, done mostly before Peter arrived to save the day.

PETER JAMES AND NICK THORPE,
London and Liverpool,
July 1994

Rather than attempting perfect consistency in our spelling of non-European names, we have used those forms which we feel will be most familiar to our readers. While we have expressed serious doubts about the conventional dates for the Egyptian New Kingdom (1550–1070 B.C.) and the Late Bronze Age in the Old World (see *Centuries of Darkness* in the bibliography), we have kept to generally accepted chronology for this book.

Working in the fields of ancient history and archaeology, we have been plagued for years by questions from friends and acquaintances about scientific, technological and practical know-how in the remote past. Is it true that the ancient Greeks discovered the steam engine? How good were doctors and surgeons in ancient times? When was brewing invented? Where was gunpowder developed—in Europe or in China? How did prehistoric people shave? What is the oldest city? When did people first take to the air? How did agriculture begin? Why do we live with cats and dogs? Who first devised a calendar? What did ancient people use as contraceptives? Who invented soccer or, for that matter, golf, badminton and tennis? The list was endless.

Trying to find the real answers led us time and again back to the libraries. In researching this material—recorded and attested to in the scientific and archaeological literature, though often ignored by textbook histories—we found we had amassed a collection of extraordinary facts and curiosities, the highlights of ancient and medieval ingenuity. The result is this book.

Our cutoff point is A.D. 1492, when one Christopher Columbus bumped into the West Indies and set into motion a train of events that were to transform (and largely destroy) the native civilizations of the New World. Only six years after Columbus's first voyage the Portuguese explorer Vasco da Gama found the sea route to India by circumnavigating Africa. The world suddenly opened up like a book, and a new chapter of history began.

At the same time, Europe was undergoing political, religious and scientific revolutions on an unprecedented scale. Byzantium, the last bastion of the ancient Greco-Roman civilization, had fallen to the Turks in 1453. Byzantine refugees, carrying with them knowledge and manuscripts, fled to Europe via Italy. There they played no small part in the Renaissance, inspired by a rediscovery of the arts and sciences of the ancient world. Many European countries then threw off the shackles long imposed by a corrupt Church in the startling intellectual revolution, mildly described as "the Reformation," which led to the birth of all the Protestant faiths.

It was in this intellectual climate that such a genius as Leonardo da Vinci (1452-1519) flourished. As the Spanish *conquistadores* systematically eradicated the knowledge accumulated over thousands of years by the native cultures of Central America, Leonardo was painting the Mona Lisa and producing futuristic designs for tanks, submarines and even flying machines. With Leonardo we enter the beginning of the "modern" era in terms of the history of science.

During the time of the extraordinary upheavals that were turning the New World, Africa and Europe upside down, the ancient civilizations of the Orient continued to develop: in China, Japan and Korea, for example, arts and sciences from cartography to acupuncture flourished and grew. These regions were to reach a cultural and scientific crisis point only with the first major interference of European culture in the seventeenth century. All the same, the events that were transforming the world in the 1490s inevitably caught up with them.

So this book is about the technological and scientific achievements of humankind from every period before the rise of the modern world in the late fifteenth century. It has one simple message: Our ancestors, however long ago they may have lived and whatever part of the globe they may have occupied, were no idiots.

Sad to say, over the last thirty years there have been many books on ancient technology that are not only ill researched but that also have an ax to grind. Most commonly this has been a desire to attribute the wonders of the ancient world to the efforts of "ancient astronauts." While the present authors would not dismiss the possibility of past extraterrestrial contact out of hand, we firmly reject the assumption that the technological advances of the ancient world were only achieved because of some kind of "outside help." To us this seems like a kind of racism, in which our ancestors are looked down on simply because they lived in the past.

The ancient-astronaut theories are only plausible if we denigrate the intelligence and abilities of ancient civilizations. A popular misconception exists that the builders of the pyramids or the cave painters of prehistory were somehow less intelligent than we are. This simply isn't true—there is no evidence that the human brain has evolved at all in the last fifty thousand years at least. Modern people are merely benefiting from thousands of years of accumulated knowledge and experimentation, not from increased intellect.

The real story of ancient ingenuity makes the spaceman theories redundant. These ideas are part of a mistaken view of history best described as temporocentrism—the belief that our own time is the most important and represents a "pinnacle" of achievement. The temporocentric view is a hangover from nineteenth-century ideas of progress. This crude version of Darwinian evolution has led to many misinterpretations of the archaeological evidence for ancient technological and cultural achievements.

A famous case concerns the cave paintings of Europe. In 1879 Don Marcelino Sanz de Sautuola was digging Stone Age remains out of the floor of a cave at Altamira on the north coast of Spain while his small daughter, Maria, played nearby. Suddenly she cried out, "*Mira, Papa, bueyes!* (Look, Papa, oxen!)." What she had spotted was a group of massive multicolored bison painted on the cave roof. At first Don Marcelino couldn't believe that the magnificent paintings were made by the same people who left the Stone Age relics, but he gradually had to accept that there was no real alternative. He wrote to Professor Juan Vilanova y Piera, Spain's greatest expert on early prehistory, who inspected the cave art and was convinced. The discovery became front-page news in Spain, and hordes of visitors, including King Alfonso XII, came to view the paintings for themselves.

However, it was a very different story when Professor Vilanova presented his conclusions to an international meeting at Lisbon in 1880, where the assembled experts dismissed the Altamira paintings out of hand. They refused to visit the site to judge for themselves, convinced that Don Marcelino had either been fooled by a forger or conspired with one. But why did the archaeological world react so blindly? The reason lay primarily in the great antiquity of the paintings. The evolutionists of the day saw history in terms of steady progress through the ages. The Altamira finds, which came from a time long before even agriculture, completely upset that notion—but rather than change the theory, the experts tried to discredit the evidence. Only in 1902, after a series of discoveries of French cave paintings, was the genuineness of these great works of art accepted. Today we recognize them as a flowering of artistic talent some fifteen thousand years ago.

Hand in hand with the belief that our time is the most advanced goes cultural arrogance and the assumption that other races are less inventive than one's own. A typically wrong-headed example is the view of China held by the British military in the nineteenth century. At the Battle of Wusung, in the Yangtze estuary, in 1842, during the Opium Wars, both navies used paddleboats: the British had a fleet of fourteen steam-powered ones, while the Chinese used five treadmill-powered vessels, all of which were captured or destroyed during the battle. The British officers, surprised at being confronted by Chinese paddle wheelers, fell to wondering where the Chinese had come by the idea. The memoirs of Commander W. H. Hall, captain of one of the British ships, include this entry:

> The most remarkable improvement of all, and which shewed the rapid stride towards a great change which they [the Chinese] were daily making, as well as the ingenuity of the Chinese character, was the construction of several large *wheeled* vessels, which were afterwards brought forward against us with great confidence at the engagement at Woosung. . . . each commanded by a mandarin of high rank, shewing the importance they attached to their new vessels. . . . The idea must have been suggested to them by the reports they received concerning the wonderful power of our steamers or wheeled vehicles.

In fact, far from copying the British, the Chinese had revived an invention of their own. A thousand years before, the Chinese navy had whole fleets of paddle wheelers.

The invention of the paddleboat provides a classic instance of how preconceptions have often distorted Western views of history. And some views, however incorrect or out-of-date, have a tendency to stick. The prehistoric cave painters of 15,000 B.C. had a lifestyle nothing remotely like the half-naked savages that popular accounts—reinforced by endless cartoons—would have us believe. Instead we should picture them living more like the Plains Indians of the recent past. They used a large repertoire of tools ranging from fine bone needles to superbly fashioned stone spears and were clad in neatly tailored garments of sewn leather. While they chose to decorate caves with vivid paintings and carvings, very few so-called cavemen lived in them. Far more common were tents and other structures, including solid dwellings made of mammoth bones.

By 7000 B.C. agriculture had been firmly established in many parts of the globe, and settled communities were blossoming into towns. Some, such as Çatal Hüyük in Turkey, were amazingly advanced. Covering some thirty acres, the town was a honeycomb of extremely regular brick-built houses, clustered together around small courtyards and streets. There were about one thousand of these dwellings, which had neatly plastered interiors, some decorated with elaborate murals. The inhabitants—about seven thousand in number—grew wheat, barley, and a dozen other edible plants, and raised herds of cattle. A wide range of arts and crafts were practiced by the townspeople: they made pottery and simple metal tools, wove linen garments, fashioned elaborate jewelry and used mirrors of polished volcanic stone. Looking at the remains from Çatal Hüyük, it is often easy to forget that they come from a time 7,500 to 8,500 years ago. In the words of British archaeologist Jacquetta Hawkes, the excavation of this "precociously advanced civilization" in the 1960s "transformed our whole conception of human life and behaviour in that period."

Thousands of years of further invention and discovery culminated in the fantastic civilization of Alexandria, a Greek city on the Egyptian coast that was heir to the knowledge of both cultures. By 200 B.C. this bustling metropolis of more than 500,000 people boasted the world's first lighthouse, a university, a library with over half a million volumes, luxurious multidecked liners, theaters with mechanical figures and moving scenery, temples with automatic sliding doors and slot machines, and engineers capable of devising every conceivable gadget, from executive toys to a simple steam engine—altogether a wonderland of human ingenuity.

The technological know-how developed in the Greco-Roman world was most impressively put to use in feats of engineering, many of which can still be seen today. Aqueducts and plumbing to bring water to the urban centers, defense systems along frontiers and around cities, roads for military and civilian use, magnificent temples, apartment buildings to house the massive urban populations, lighthouses and harbor installations—all were constructed around the Mediterranean.

The quality of ancient civil engineering often matches, and frequently exceeds, that of present-day work. One of the most unexpected discoveries of modern archaeology has been the great harbor of Caesarea Maritima in Israel, built by King Herod the Great, tyrant of biblical fame, between 22 and 9 B.C. According to a detailed description left by the ancient Jewish historian Josephus, Caesarea's harbor was bigger than the port of Athens and was surrounded by a glistening metropolis of white limestone, its streets laid out on a grid plan and graced with numerous freshwater fountains. The harbor itself had an artificial breakwater two hundred feet wide, big enough to hold arched shelters for sailors. Scholars thought that Josephus was exaggerating wildly, until in 1960 modern divers began to spot massive stone remains on the seabed. These have now been fully explored by marine archaeologists, who were staggered by their findings. In the words of Dr. Avner Raban, codirector of the Caesarea excavations,

> This Herodian port is an example of a 21st century harbour built two thousand years ago. In fact, if the modern harbours of Ashdod and Haifa had employed such systems of design and engineering, they would not have had the problems they face today.

We should not imagine that such engineering feats were achieved simply by bringing together enormous labor forces. The ancients also had at their disposal highly sophisticated technical devices, ranging from surveying equipment like modern theodolites to labor-saving machinery, such as cranes, used in the construction of great buildings such as the temples of Rome.

When the West declined with the fall of the Roman Empire, much of its technology was preserved and developed to ever-greater heights by the Arabs, while most of Europe slumbered in a "dark age." The following description of Baghdad early in the Middle Ages, written by the Lebanese historian Amin Maalouf, gives a flavor of the highly sophisticated urban civilization created by the Arabs:

(Opposite) A crane powered by slaves working a treadmill—as shown on the temple-tomb of the Haterii family, Rome (1st century A.D.).

> At the beginning of the ninth century, during the reign of . . . Harun al-Rashid, the caliphate had been the world's richest and most powerful state, its capital the centre of the planet's most ad-

vanced civilization. It had a thousand physicians, an enormous free hospital, a regular postal service, several banks (some of which had branches as far afield as China), an excellent water-supply system, a comprehensive sewage system, and a paper mill.

Great cities were not of course exclusive to the Old World. The Spanish invaders of Mexico were almost dumbstruck when they entered the ancient Aztec capital of Tenochtitlán (now Mexico City) in 1519. Nothing in Europe could compare to it—with its magnificent temples and palaces, thriving marketplace, drugstores, busy canals crossed by portable bridges, extraordinary floating market gardens, ball-game courts and unparalleled zoological and botanical collections. Two hundred thousand people lived in the five square miles covered by the city and its suburbs; at this time the population of Spain's largest city, Seville, was a mere forty-five thousand. Bernal Díaz, a *conquistador* who left a personal account of the Spanish conquest of Mexico, summed up the amazement of the invaders on arriving at Tenochtitlán: "Are not the things we see a dream?"

Even greater wonders were in store for modern Europeans when they began to penetrate China. Here for more than five thousand years the Chinese had steadily progressed along their own technological path, largely independent of developments in the West. The activities of countless scientists and technicians had completely transformed every aspect of life—the basic economy (from drilling for salt and natural gas to earthquake detection), military technology (from the crossbow to the cannon) and the development of luxury products (from perfume burners to mirrors) had all benefited from thousands of years of intensive and energetic experimentation.

The depth of scientific knowledge in ancient and medieval China remained unappreciated in the Western world until recently. One of the most famous statements in the history of inventions is that recorded in 1620 by the English philosopher Francis Bacon in his *Novum Organum*, as his "Aphorism Number 129": "We should note the force, effect and consequences of inventions, which are nowhere more conspicuous than in those which were unknown to the ancients, namely, printing, gunpowder and the compass. For these three have changed the appearance and state of the whole world." In fact all three had long histories before Bacon's time—but in China, not in the medieval West.

Bacon can be partly forgiven, however, as so many inventions have been developed only to be forgotten and reinvented much later. A classic example of this phenomenon is the reaping machine, invented in Celtic Gaul in the first century A.D. and then again in Australia in the nineteenth century. A less discussed case is the apparent use of aluminum in ancient China. In 1956 archaeologists excavating the tomb of a military commander who died in A.D. 297 found some twenty metal belt ornaments, which they sent for laboratory analysis. Several were found to be made of aluminum, with an admixture of up to 10 percent copper and 5 percent manganese. This astounded the archaeologists, as aluminum is extremely hard to smelt and was, as far as anyone knew, first isolated in 1827 and produced on a large scale only since 1889. Modern Chinese experts doubt that their ancestors could have produced such an aluminum-rich alloy and have even questioned the reliability of the excavation. But Joseph Needham, the leading historian of Chinese science, finds this "Altamira-type" conclusion unacceptable. Instead he argues that it is quite likely that a Chinese alchemist produced the alloy by chance and kept the secret of his lucky accident to himself, the knowledge being lost with his death.

Like Needham, we feel that it is a fatal mistake to underestimate the technological and intellectual achievements of an ancient people. Because over time most materials will simply rot, rust and crumble away, what does survive in the archaeological record is clearly only the tip of the iceberg. But every now and again an object or a text turns up that completely surpasses all previous estimates of an ancient culture's technical skill. The increasing number of such finds is now forcing us to reassess completely our view of ancient technological abilities.

The scientific exploration of oceans and space presses on, but we also have much to learn from the exploration of ancient science and technology. The goal of such exploration is unimaginably rich: a vast store of accumulated knowledge, arrived at through thousands of years of trial-and-error experience on the part of all our ancestors.

MEDICINE

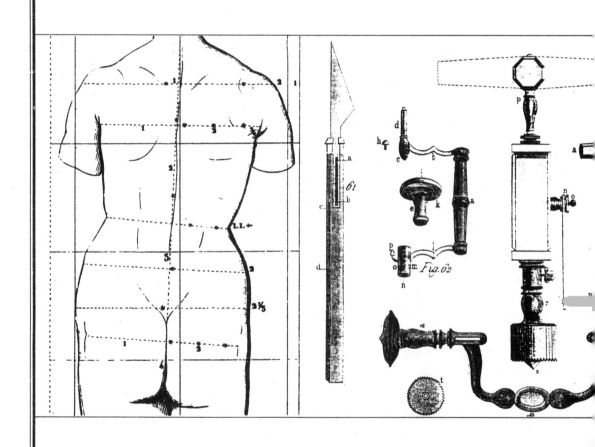

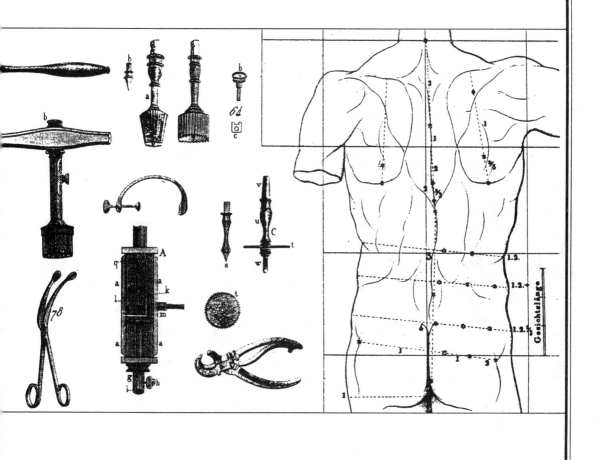

CHAPTER ONE

INTRODUCTION

From its remotest beginnings the human race has showed compassion for its fellows. Neanderthal Man, a primitive relative of *Homo sapiens sapiens*, was until quite recently thought of as a dull, unthinking brute, a view completely overturned in the late 1960s by the discovery of Neanderthal burials at Shanidar Cave, in northern Iraq. Dating from 60,000–30,000 B.C., several were in carefully dug graves, and two of these contained large amounts of pollen, possibly from flowers scattered by mourners.

One of the dead Neanderthals was a forty-year-old man with a withered arm and a leg crippled by arthritis, killed by a rockfall in the cave about fifty-thousand years ago. His teeth were abnormally worn, suggesting that since he was unable to hunt and gather with the rest of the group, he was given other work to do at home, such as softening skins by chewing. The very fact that he lived so long despite his disabilities indicates that he was carefully looked after. Even more remarkable is the strong likelihood that one of his tribe had actually performed surgery on him, for the skeleton's right arm is missing from just above the elbow. The end of the arm bone has a rounded surface, suggesting it may have healed up after the operation.

Many other Neanderthal skeletons are those of individuals who reached a relatively ripe old age despite becoming disabled. Richard Klein, professor of anthropology at the University of Chicago, sees this as a key stage in the evolution of human society: "The implicit group concern for the old and sick may have permitted Neanderthals to live longer than any of their predecessors, and it is the most recognizably human, nonmaterial aspect of their behavior that can be directly inferred from the archaeological record."

There is also a common belief that early humans killed off children born disabled. The earliest certain evidence that this was not always the case also comes from the Old Stone Age. At the cave site of Riparo del Romito, in southern Italy, archaeologists found the twelve-thousand-year-old burial of an adolescent dwarf. Despite his severe condition, which must have greatly limited his ability to contribute to either hunting or gathering, the young man survived to

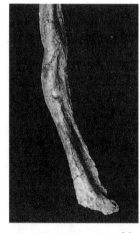

An amputation fifty thousand years ago? This arm bone from a Neanderthal skeleton found at the Shanidar Cave, in northern Iraq, shows signs of having been cut off just above the elbow.

the age of seventeen. He must have been supported by the rest of the community throughout his life.

The emergence of a specialist group of doctors is presumably a later development. Herbal medicines were undoubtedly used throughout prehistory, but our knowledge of medical practice begins only with the earliest literate civilizations of the Near East. From these we learn that care of the sick had already become an actual profession. The first individual doctor we know of was a man named Lulu, who practiced in the country of Sumer, in southern Iraq, around 2700 B.C. Doctors appear in the Egyptian records at about the same time: fifty or so are known from the Old Kingdom period of the pyramid builders (2600–2100 B.C.), mainly from the inscriptions in their tombs. Shortly afterward the first medical papyri were written, describing some of the operations these pioneer doctors carried out. They must also have developed a considerable body of herbal medicines over time, as the classical Greeks knew Egypt as a wonderland of pharmaceutical knowledge (see **Drugs** in **Food, Drink and Drugs**).

The Greeks were quite happy to admit that they owed much of their medical expertise to Egypt. Practically everything we know of ancient Greek medicine is associated with the name of the great Hippocrates (around 460–380 B.C.). He lived and worked on the Aegean Island of Kos during the peak of ancient Greek civilization that followed the successful conclusion of their long struggle with Persia. The medical school on Kos preserved and developed Hippocrates' techniques, and this is probably the source of the Hippocratic writings, the definitive summary of ancient Greek medicine. How much of it, if any, was actually the work of Hippocrates himself has always been a moot point; more importantly he founded a medical tradition that lasted from the fifth century B.C. until modern times. Hippocrates must be counted as the father of modern Western medicine.

His approach was a scientific one, based on careful observation of the patient and selection of the most suitable treatment. Hippocrates took a holistic view of treatment, and many of the writings attributed to him focus on the general health of the individual as opposed to specific illnesses. The doctor's role, as he understood it, was to assist nature in curing the patient. That prevention was better than cure Hippocrates' successors fully understood. The best way to ensure good health, they believed, was through a moderate diet accompanied by regular exercise. Walking was recommended as the best exercise for the sedentary person, a conclusion with which modern doctors would agree. They were particularly interested in the effects of climate on health, an area that present-day medicine tends to neglect. They also understood the importance of the psy-

Roman copy of a Greek bust of the medical genius Hippocrates, found at Ostia, the port of Rome. Hippocrates (460–380 B.C.) founded the tradition of Greek medicine from which modern medical science eventually grew.

chological state of the patient, as described in the *Hippocratic Precepts*: "For some patients, though conscious that their condition is perilous, recover their health simply through their contentment with the goodness of the physician."

Hippocrates is also famous for inspiring the "Hippocratic Oath" (to treat patients irrespective of payment) traditionally sworn by Western doctors. The *Precepts* advise, "Sometimes give your services for nothing, calling to mind a previous benefaction or present satisfaction. And if there be an opportunity of serving one who is a stranger in financial straits, give full assistance to all such."

We are indeed fortunate that Hippocratic ethics were followed by later doctors. By contrast, Persian physicians of the fifth century B.C. subscribing to the official state religion of Zoroastrianism had to carry out three successful operations on nonbelievers before being allowed to practice on the faithful.

The Roman rich were generally attended at home by resident or visiting doctors. The majority of doctors were, however, public physicians, paid by the local town council to treat anyone. This didn't necessarily attract the highest-quality doctors, and the first "public" physician at Rome—a Greek named Archagathus, appointed in 220 B.C.—relied exclusively on the knife and the cauterizing iron. His patients unkindly dubbed him the Butcher, and the irascible politician Cato the Elder, who detested all Greeks, was convinced that Hellenic doctors were part of an international conspiracy to murder the people of Rome. Other Roman doctors seem to have employed gentler methods. A shipwreck dating to c. 100 B.C., found off the coast of Tuscany in 1989, produced the remains of a doctor's chest; in it were 136 small wooden boxes containing ointments and herbs, a bronze cupping vessel for bloodletting, and a wooden statuette of Asclepius, the Greco-Roman god of healing.

The Chinese, too, had a system of publicly financed doctors, paid for—unlike the Romans—by the central government. Set up in the second century B.C., it was originally confined to the major cities, but was extended in the first century A.D. to cover all of China. This national medical coverage was unique to China until modern times. The Chinese backed up their state medical service with systematic training. There were Chinese professors of medicine in the Imperial University at Lo-yang by A.D. 493. About a century later the Imperial Medical College was established at Ch'ang-an, together with medical colleges in all the chief provincial cities, all of which had the power to award degrees. Similar institutions were established in Baghdad in A.D. 931 and Salerno, Sicily, in A.D. 1140.

We also have to look to China for the earliest hospital. In A.D. 2 Emperor Wang Mang set up a temporary hospital to deal with an emergency caused by the combination of a severe drought and a plague of locusts. The Romans built military hospitals in many of the larger forts constructed during the expansion of their empire. At the major base of Neuss on the Rhine, archaeologists found more than one hundred medical and pharmaceutical implements in one room alone. The hospital wards were well drained and lit and were situated in the quietest part of the fort. The military hospital included casualty reception centers and space for administration, staff and supplies. In many cases the courtyard was used to grow healing plants, and legionnaires would rest in the surrounding corridor during their convalescence.

Model of the military hospital from the legionary fortress built by the Romans in the 1st century A.D. at Xanten on the Rhine (Rome's frontier with the German tribes).

Many Roman doctors set up clinics and nursing homes in their own houses, but the earliest civilian hospitals in the Roman Empire were founded by leading Christians, from A.D. 350 onward. These were charitable institutions. The idea may have begun even earlier in

India, as thriving hospitals are described by the Chinese Buddhist pilgrim Fa-Hsien around A.D. 400. According to his account, these hospitals, providing their service free of charge, were places to which "the poor of all countries, the destitute, crippled and diseased may repair." The great hospital at Jundishapur (modern Shushtar), in Iran, was founded in A.D. 489 and was staffed, ironically, by heretical Christians expelled from Athens and Edessa (eastern Turkey). The first permanent hospital in China, no doubt copying the idea from India, was set up in A.D. 491 by Hsaio Tzu-Liang, a Buddhist prince of the southern Ch'i Dynasty. This was a private foundation; the earliest official government hospital was opened in A.D. 510 and was intended mainly for poor people.

One of the best equipped of ancient hospitals was that founded in Byzantium by Emperor John II (1118–1143). Men and women were housed in separate buildings, each containing ten wards of fifty beds, with one ward reserved for surgical cases and another for long-term patients. The staff was a team of twelve male doctors and one fully qualified woman physician as well as a female surgeon. The men had twelve qualified assistants and eight helpers each, but the women doctors were allowed only four qualified assistants and two helpers each. In addition two pathologists were attached to the hospital staff. Vegetarian meals were available for the patients. A dispensary provided treatment for outpatients, while next door to the hospital was a school in which the sons of the entire medical staff were trained as doctors.

The first drugstores were opened for business in the Muslim world during the ninth century A.D. Ordinary people as well as doctors could buy medicinal herbs and substances collected from all over the Islamic world, including alcohol, cassia, senna, manna, and arsenic. We make our pills taste nice by coating them with sugar; the Arabs did the same using rose water and perfumes. Chemists' shops also sold plasters and ointments. It was here that pharmacy first emerged as a calling distinct from medicine. Most of the pharmacies were located around Baghdad. There was a minority of well-educated practitioners who were examined and licensed by the authorities and whose shops were routinely inspected. Aside from these privately owned apothecaries' shops, there were dispensaries attached to hospitals, where outpatients could buy medicines.

S anguinaria herba.

After the Spanish conquest of Mexico some traditional Amerindian medical knowledge was committed to writing. This depiction of a nettle (*sanguinaria herba*— "blood herb") comes from an Aztec "herbal" of 1552. An accompanying text in Latin gives a cure for nosebleeds—nettle juice, mixed with salt in urine and milk, poured into the nostrils.

Similar drugstores were being operated by the Aztecs of Mexico by the time of the Spanish conquest. Hernán Cortés wrote that in the market at the capital Tenochtitlán "there are places like apothecaries' shops, where they sell medicines, ready to be taken, ointments and poultices." Plasters made of rubber could also be bought there.

The remedies sold over the counter in these ancient drugstores were of course herbal medicines. These have a long history, although archaeologists undoubtedly have much more to discover about them. A variety of plant-based painkillers, and even anesthetics, were developed, many of them taken before operations. Certain plants, such as opium and cannabis, which were used by some societies as intoxicants (see **Drugs** in **Food, Drink and Drugs**) were valued by others as medicines. Assyrian medical texts from ancient Iraq recommend cannabis as an antidepressant: its fumes were inhaled by patients to dispel sorrow or grief. Greek and Roman doctors prescribed cannabis to relieve earache. Some drugs produced by plants, such as belladonna, can be fatal yet are medically beneficial if taken in small, controlled doses. The custodians of the oracle centers of ancient Greece, where quite deadly narcotics were sometimes taken to induce visions (see **Drugs** in **Food, Drink and Drugs**), must have played an important role in establishing the properties and safe doses of such drugs. The Greek writer Philostratus explicitly states how the priests of Asclepius, god of medicine, learned their arts through divination. It is no coincidence that a major temple of Asclepius was founded in the fourth century B.C. on the island of Kos, the birthplace of Hippocrates.

Many spices also have medicinal values, long appreciated by the ancients but only recently accepted by modern science. For example, in the third century B.C. the Greek botanist Theophrastus recommended a mixture of burned resin, cassia, cinnamon and myrrh "to relieve the inflammation caused by any wound." Most of these ingredients have medicinal properties: it has been shown, for example, that oil of cinnamon is a powerful germicide.

The value of other ancient herbal knowledge, such as that preserved in traditional Chinese and Amerindian medicine, is now being reappraised and increasingly appreciated by Western doctors. One hundred and seventy plant-based drugs discovered by the native peoples of the Americas had been catalogued in the *Pharmacopeia of the United States of America* or the *National Formulary* by 1970. The list of officially recognized Amerindian herbal drugs includes diuretics, emetics, laxatives, astringents, painkillers, antiseptics, cathartics, stimulants, tonics and remedies for snake and insect bites. Native Americans even developed the first oral contraceptives, whose existence sparked the Western quest for the "pill" (see **Contraceptives** in **Sex Life**). The Amerindians' vast pharmacological knowledge undoubtedly goes back to pre-Columbian times.

Indeed many of the major discoveries of medical science we fondly believe to be modern were actually made hundreds, if not thousands, of years ago. Though they had no word for them, the ancient Egyptians used the antibiotics contained in moldy bread to heal wounds by applying it as a poultice. And while it is hard to imagine that the Romans, in an age before the modern microscope, were aware of bacteria, an intriguing passage in the writings of the Roman scientist Varro (116–27 B.C.) may suggest otherwise: "Precautions must also be taken in the neighborhood of swamps . . . because there are bred certain minute creatures which cannot be seen by the eyes, which float in the air and enter the body through the mouth and nose and there cause serious diseases."

The circulation of the blood is generally assumed to have been discovered by European physicians in the seventeenth century A.D., yet it was well known to Chinese medicine from at least the time of the Han Dynasty (202 B.C.–A.D. 220). Chinese doctors were also aware of hormones and were distilling them from urine for use in various sexual disorders by the second century A.D. (see **Introduction** to **Sex Life**).

The range of breakthroughs in medical knowledge achieved by ancient doctors is quite extraordinary. How they made them is another matter. The Egyptian use of antibiotics may have been an accidental discovery, while Varro's apparent description of microorganisms may just be a lucky guess. But most ancient medical advances were the result of traditions of practical medical science much like our own, in which observation, theory and even experimentation played their part.

It was experiments carried out in classical times that laid the foundations of our understanding of anatomy. The work was initiated in the third century B.C. at Alexandria, then the scientific capital of the ancient Mediterranean (see **Introductions** to **High Tech** and **Communications**). Its pioneer was the Greek physician Herophilus, hailed by Roman doctors as the first ever to perform dissections on the human body. While this may not be exactly true, Herophilus certainly invented the basic techniques of dissection. His investigation of the human body led him to many fundamental discoveries—he distinguished tendons from nerves, wrote the first detailed description of the brain and was the first to suggest a connection between nerves and the senses. The darker side of Herophilus's work is that he and his pupil Erasistratus (inventor of the catheter) apparently performed some of their work on live subjects—criminals awaiting the death sentence in state prisons. A defense of this practice written by the Roman writer Celsus, that the suffering of a few criminals is justifiable in order that "we should seek remedies for innocent people of all future ages," is disturbingly reminiscent of that used to defend the grotesque tests performed on live animals today.

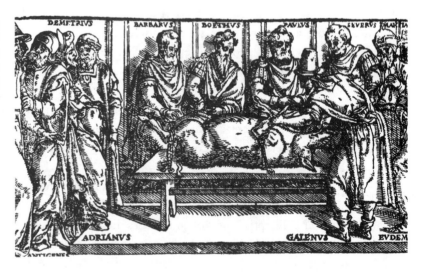

Renaissance engraving showing the famous Roman doctor Galen studying the workings of the nervous system by opening up a living pig.

The anatomical work of the Alexandrian doctors was brought to fruition by the Greek surgeon Galen (c. A.D. 130–200). By his time the dissection of human beings, whether dead or alive, was no longer considered ethical; still, Galen managed to acquire considerable familiarity with the insides of human beings by treating casualties at the gladiatorial games. And he continued the Alexandrian work on the dissection of animals, notably rhesus monkeys. Again the experiments were often grisly, but led to major advances, proving beyond any doubt, for example, that arteries contained blood, that urine came from the kidneys and that organs such as the larynx were controlled by nerves. Galen summarized his knowledge in an extensive sixteen-volume work entitled *On Anatomical Procedure*, the first systematic study of human anatomy ever written. It was probably also the most influential—Galen's conclusions, both correct and incorrect, dominated medical thinking in Europe during the Middle Ages and Renaissance. Advances on the knowledge accumulated by Galen only began to be made in Europe in the seventeenth century.

Ancient Chinese medicine must have also combined theory with practical experiments in order to achieve its extraordinary results. The foundations of immunology were laid in the tenth century A.D. by Chinese alchemists, who developed the first inoculation against smallpox. Plugs of cotton smeared with material containing the virus would be placed in the nose. During the sixteenth century the technique became widely used in China and spread from there to Turkey, where Westerners first became acquainted with inoculation.

Modern medicine could still learn much from the underestimated healers of the ancient world, whose achievements were quite astounding.

SURGICAL INSTRUMENTS

Human beings have been digging out thorns, splinters, arrows and other foreign bodies from their skin since time immemorial. The sharp stone blades used by our prehistoric ancestors for so many purposes would also have been the tools for the first surgical probings. Minor operations on animals became common during the agricultural revolution of the New Stone Age. Cattle were domesticated some nine thousand years ago, and the surest way of making the aggressive male of the species docile was to castrate it.

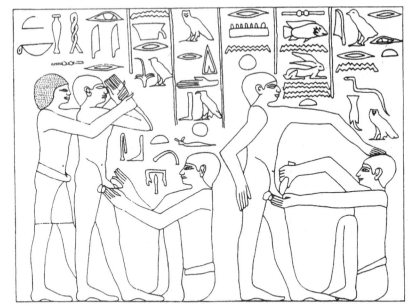

Egyptian relief, dating from around 2500 B.C., showing youths being circumcised with knives. The hieroglyphic inscription has the surgeon on the left saying to his assistant, "Hold on to him; don't let him faint!"

Knives made from flint or obsidian, a volcanic glass that can produce an extremely sharp edge (see **Razors** in **Personal Effects**), would have been used to slice off the unfortunate bull's testicles.

The first evidence of more sophisticated surgical instruments comes from Egypt, regarded by the later Greeks and Romans as the cradle of medicine. By the time of the pyramid builders (around 2600 B.C.), the Egyptians were making copper surgical blades sharp enough for simple operations, such as circumcision. Ancient Egyptian men were normally circumcised; the custom, prevalent in Africa, seems to have spread to the Jews, Arabs and other Near Eastern peoples from Egypt.

Copper needles must also have been among the first surgical instruments, as we can see from the extraordinary Smith Papyrus: entirely concerned with surgery, it describes a number of procedures for stitching up wounds to the head, shoulder and other parts of the body. A surprising factor is its date—the papyrus itself was written around 1700 B.C., but the archaic language used shows that it was copied from a text composed up to a thousand years earlier. Yet as Professor Harry Saggs, a leading British authority on ancient Near

Eastern civilization, noted, the Smith Papyrus "is the Egyptian medical document which comes nearest to the modern scientific approach." Indeed Egyptian surgery seems to have peaked very early and actually declined after the time of the Smith Papyrus.

Surgery also made an early start in Iraq, where Babylonian physicians were using surgical knives made of obsidian or, more commonly, bronze some four thousand years ago. They also had saws and trephines, drills for boring into the skull (see **Brain Surgery**). Medical tools from the preclassical age generally only occur as single, chance survivals, with one extraordinary exception—the kit belonging to a doctor of the Minoan civilization, discovered in a tomb at Nauplion, on Crete. Dating to the fifteenth century B.C., it includes forceps, drills, scalpels, a large dilator for internal examination and specially shaped pieces of stone and marble, presumably used for grinding medicinal ingredients. Even more extraordinary is the resemblance of the surgical instruments to classical Greek examples. Except for being made of copper, they are virtually identical to the bronze and iron instruments from Greek temples of healing of over a thousand years later.

One of the most complete sets of surgical instruments from the ancient world is that shown at the great temple of Kom Ombo, in Egypt, on a relief carved about 100 B.C., when Egyptian civilization had been "hellenized" by two centuries of Greek influence. The Kom Ombo relief depicts a probe, forceps, saws, a retractor, a cautery (burning iron), bandages, a flask, scales, medicinal plants, a pair of shears, a sponge, a variety of scalpels, an instrument case, and cupping vessels.

Cupping vessels—the only equipment unfamiliar today—were used for drawing blood. Made of glass, horn or bronze, they were among the most common instruments of any Greek or Roman surgeon. The doctor would cut a patient's vein with a scalpel, then apply to the wound the bleeding-cup, from which the air had been removed by suction or burning. The vacuum created would gently suck out a measure of blood from the patient; then the wound was bandaged. Bleeding was as popular in classical times as it was during the Middle Ages and Renaissance, when leeches were used to suck blood from patients suffering from almost every complaint. Indeed the Romans seem to have begun the trend toward bleeding as a

universal remedy. Celsus, whose work, *On Medicine*, was a major influence on medieval medicine, extolled the virtues of bleeding: "To
let blood by incising a vein is no novelty; what is novel is that there
should be scarcely any malady in which blood may not be let." He

Surgical instruments
depicted on a relief from
the Egyptian Temple of
Kom Ombo, about 100
B.C. At the bottom left are
two cupping vessels, used
for bleeding patients.

then lists the complaints for which bleeding is useful: they include severe fever, swollen blood vessels, diseases of the intestines, paralysis, muscular spasms, loss of breath and voice, intolerable pain and internal ruptures.

While the idea of draining blood from people for such a wide range of complaints seems rather horrific to us today, Celsus did not prescribe the treatment as casually as did physicians during the Middle Ages. He recommended that doctors check the color of the blood, and if it was healthy-looking, call a halt to the operation. And while he felt that the procedure was good for "all acute diseases," this was only "provided that . . . they are doing harm, not by weakness, but by overloading." Nor is bloodletting quite as daft as it may sound. A small amount of bleeding *can* help conditions associated with high blood pressure, and in the late 1980s some doctors even recommended a return to leeching for such complaints. The Roman cupping devices were also perfect for removing poison from snakebite. Sucking with the mouth is as effective, but cups are more hygienic, as they protect the surgeon from any diseases carried in the patient's blood.

Roman doctors were very particular about the quality of their scalpels. Wanting tools that could not be easily blunted, chipped or bent, they insisted on using the best-quality steel from Austria, if available. They would visit a blacksmith, specialist instrument maker or cutler when a scalpel blade wore out. As well as the *scalpellus*, Roman doctors had heavier knives for slicing through flesh for amputations and, for cutting through bone, metal-bladed saws identical to those used by surgeons today.

For genito-urinary disorders the Romans used catheters, first invented by Erasistratus, a Greek doctor of the third century B.C. Roman catheters were fine tubes very similar to the modern instruments except that they were made of bronze and were used to treat a number of disorders, such as strangury, in which the urinary tract clogs, making urination extremely difficult and painful. Catheters were used to clear the problem, and Celsus recommended that doctors should keep graded sets of different bores, three for men and two for women. They were also useful for removing the stones that sometimes form in the bladder, blocking its mouth and preventing urination. The Roman gynecologist Soranus describes how these

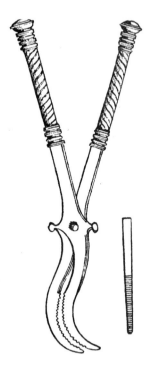

Curved forceps with finely serrated edges (interior also shown)—one of the highly sophisticated surgical instruments found at the Roman city of Pompeii, preserved by the volcanic ash of Vesuvius that covered the city in A.D. 79.

stones can be pushed out of the way with catheters and back into the cavity of the bladder. The earliest-known double-curved catheter (a tube with two bends in it to follow the contours of the urinary tract) comes from the "House of the Surgeon" in Pompeii, sealed by the volcanic eruption of A.D. 79. Such instruments were not made again until A.D. 1700.

Some of the instruments used by Roman doctors were of such superb quality that any modern surgeon would be proud to own them. The forceps found in Pompeii cannot be excelled today for the alignment of their fine-toothed jaws. Soranus and other Roman gynecological writers often refer to the use of a *speculum*, a specialized tool for dilating the vagina for inspection and treatment. The "House of the Surgeon" contained three *specula*, complex instruments of extraordinary precision. The surfaces inserted into the patient were immaculately smooth. Similar devices from Renaissance times could not match this standard of craftsmanship.

The ancient Hindus also excelled in the manufacture of surgical instruments. Surgery seems to have reached quite remarkable heights in India during the late first millennium B.C. (see **Plastic Surgery**), and the ancient Hindu surgeons had a vast array of instruments available to them, many of which show improvements in design over Roman ones. The *Sushruta Samhita*, most important of the ancient Hindu medical works, describes 20 sharp and 101 blunt medical instruments, each designed to perform a different task. They include scalpels, razors, saws, probes, needles, hooks, forceps, pincers, hammers, tubular appliances, syringes, hollow hemispheres, horn instruments and specula, all made of tempered, hardened iron, steel or other appropriate metals. The training of Indian surgeons was extensive, and the *Sushruta Samhita* stipulates a series of exercises to be carried out by the trainee on inanimate objects. Lancing abcesses was practiced on a bag of thin leather filled with slime, stitching on cloth, cauterizing on pieces of meat, the opening of veins on those of a dead animal and so on.

As a final illustration of the inventiveness of the ancient Hindu surgeons, the *Sushruta Samhita* also contains a remarkable solution to the tricky problem of stitching up the intestines, which are highly vulnerable to infection after surgery. The instruments used were large Bengali ants, placed side by side along the opening. The ants clamped the wound shut with their jaws; the surgeon then cut away their bodies, leaving only the heads behind. The intestines, complete with their macabre suture, were then pushed back into the stomach and the abdomen sewn up. The ants' heads would gradually dissolve as the wound healed.

The use of ants as surgical tools seems to have been independently discovered in South America. The indigenous people of Guyana still suture their wounds with the jaws of giant worker ants. Such is the staying power of their jaws that in 1921 one William Beebe reported that, a whole year after returning from the Guyanan jungle, he found the jaws of two such ants still clamped to his boots, "with a mechanical vise-like grip, wholly independent of life or death."

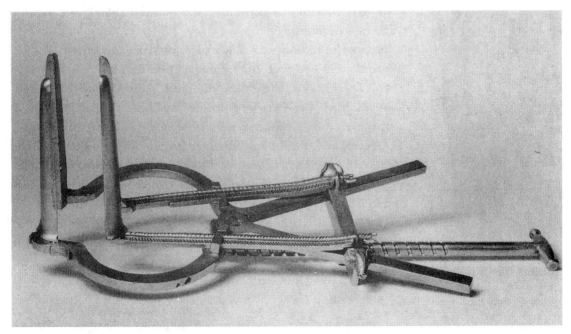

Modern reconstruction of a gynecologist's instrument found at Pompeii. The precision-made screw mechanism gradually parted the four prongs, gently dilating the vagina for inspection and treatment.

EYE OPERATIONS

A Roman doctor steadies a patient's head with his left hand while he uses a surgical instrument on her eye with the other. The patient's hands are bound to prevent them jerking around during the treatment. From a Roman relief found at Malmaison, France, and thought to belong to the shrine of the healing god Apollo-Grannus.

One of the most delicate parts of the body on which a surgeon can operate is the eye. Yet ophthalmic surgery was one of the most advanced areas of medicine in the ancient world. The extraordinary skills of ancient eye doctors developed in response to a desperate need: eye complaints were particularly common in the ancient world, as they are in the Third World today.

The Romans, for example, suffered particularly from trichiasis, ingrowing eyelashes that cause constant irritation. Removing them was a straightforward procedure: The eyelid was turned outward and the troublesome lashes plucked with forceps; to prevent the hair from growing again a fine iron needle was heated and inserted into the root to cauterize it.

Such an operation was child's play for the experienced Roman surgeon, compared with the techniques necessary for removing cataracts. Then as now these were the most frequent cause of near or total blindness. Cataracts form in the crystalline lens near the front of the eye when fluid accumulates between the fibers of the lens. These deposits build up into round blobs, the fibers of the lens itself decay, and large areas of pus fill the otherwise transparent lens. The lens will eventually become completely clouded and vision reduced to zero. Physical removal of the cataract is the only cure.

The medical writings of Cornelius Celsus, who lived under the Emperor Tiberius (A.D. 14–37), include a detailed description of a cataract operation. Great care was needed, particularly in the preparation, as Celsus stressed:

The patient . . . is to be seated opposite the surgeon in a light room facing the light, while the surgeon sits on a slightly higher seat; the assistant from behind holds the head so that the patient does not move; for vision can be destroyed permanently by slight movement. In order also that the eye to be treated may be held more still, wool is put over the other eye and bandaged on. Further, the left eye should be operated on with the right hand, and the right eye with the left hand.

With everything prepared, the Roman eye surgeon could set to work. Some used sophisticated instruments, such as the superb tools discovered at Montbellet, France, in 1975. In a copper case together with three more ordinary needles with handles were found two "needle syringes"—extremely fine retractable needles set in close-fitting, slender tubes. Their discovery confirmed the feasibility of a highly delicate procedure for cataract removal described by Galen, a Greek doctor of the second century A.D. The instrument was inserted into the lens and the needle pushed through the tube to break up the cataract. After withdrawing the needle, the surgeon used the tube to suck out the debris and clear the lens.

Through the Eye with a Needle

Instruments of such quality must have been rare, and the repeated probing needed to break up a cataract with a needle syringe would have made the operation risky, except in the hands of the most experienced surgeons. Celsus himself recommended a different operation, simpler but far bolder. Using a plain bronze needle, the surgeon would actually push the cataract lens completely out of the way:

Roman needle syringe from Montbellet, France (actual size).

> The needle used is to be sharp enough to penetrate, yet not too fine; and this to be inserted straight through . . . at a spot between the pupil of the eye and the angle adjacent to the temple, away from the middle of the cataract, in such a way that no vein is wounded. The needle, however, should not be inserted timidly. When the spot is reached, the needle is to be sloped against the colored area [lens] itself and rotated gently, guiding it little by little below the pupil; when the cataract has passed below the pupil, it is pressed upon more firmly in order that it may settle below. . . . After this the needle is drawn straight out; and soft wool soaked in white of egg is to be put on, and above this something to check inflammation; and then bandages.

The operation described by Celsus in squeamish detail is now known as cataract "couching" (from the French *coucher*, "to lay down"). The same procedure—pushing the infected lens down and

"Couching" restores sight by removal of the crystalline lens if clouded by a cataract. A needle is inserted at the edge of the cornea and behind the iris. The lens is then pushed down and out of the way. The patient feels only slight pain if the procedure is carried out with precision. The operation was well known to the Romans but was most likely an Indian or Babylonian invention.

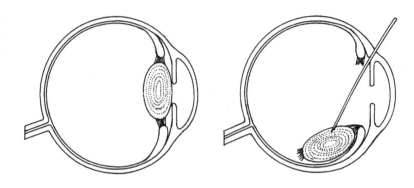

out of the way by using a lancet or needle—is still used today. Provided infection does not set in, the operation will restore a measure of vision. Results for people who are already extremely shortsighted can be excellent, as the procedure helps correct the focal distance between the back of the eye and the cornea.

Where did Roman surgeons learn this remarkable technique? It appears full-blown in Celsus, and nothing like it occurs in the work of Greek doctors, forerunners of Roman medicine in so many other respects. Unless they invented it themselves, it is tempting to see here a Roman borrowing from Indian medicine. The classical Hindu medical text *Sushruta Samhita* was probably compiled during the last few centuries B.C. (see **Plastic Surgery**). It contains a section on eye diseases four times the length of that in Celsus and describes the operation of cataract couching in broadly similar terms, including the advice that surgeons should pierce the right eye with the left hand and vice versa.

But the Babylonians of southern Iraq seem to have developed eye surgery long before either the Romans or the Indians, both of whom may have learned the technique from them. Unfortunately we lack Babylonian descriptions of eye operations, but they are mentioned in the famous legal code drawn up by King Hammurabi of Babylon in the eighteenth century B.C. The text refers to the "opening of a *nakkaptu*" with a bronze lancet as a way of curing blindness. There is some difficulty in translating the word *nakkaptu*; an ophthalmic specialist has argued it must mean "cataract," but the following is a more literal rendering: "If a physician performed a major operation on a nobleman with a bronze lancet and has saved the nobleman's life, or he opened up the eye-socket [*nakkaptu*] of a nobleman with a bronze lancet and has saved the nobleman's eye, he shall receive ten shekels of silver." For saving a commoner's eye, the reward was five shekels, for a slave two shekels. It is hard to imagine what the operation involved could have been, other than the cataract couching described by Celsus nearly two thousand years later.

The fact that such surgery required legislation shows that it must have been fairly common. There may even have been specialist eye surgeons in ancient Babylonia as there were in the Roman Empire. Still, while it may have been a lucrative profession, it was also a risky one. Hammurabi's code continues with the penalty for failure in such a hazardous operation: If a surgeon destroyed the sight of a gentleman's eye by piercing it with a lancet, the law demanded that his hand be chopped off.

PLASTIC SURGERY

One of the most startling achievements of ancient Indian medicine was the world's earliest plastic surgery. The procedures involved are described in detail in a Hindu medical work entitled the *Sushruta Samhita*. Like most ancient Indian classics, it is difficult to date with any precision, but modern scholarship places its supposed author, the famous surgeon Sushruta, sometime in the last few centuries B.C.

In any case Sushruta's techniques antedate modern plastic surgery by a good two thousand years. Their development at such an early date seems to have come about because of the Indian custom of earlobe stretching. All good parents in early Hindu society had the ears of their children pierced—it was thought to ward off malignant spirits as well as looking attractive. The piercing was carried out by trained physicians, who made a hole in the earlobe and plugged it with cotton lint. Provided no infection developed, the child returned to the surgery every three days, so that increasingly larger plugs of lint, and then circles of wood or lead weights, could be added to stretch the skin gradually. The longer the lobes, the more beautiful they were thought to be. But as well as creating stretched lobes, one of the surgeon's main jobs was to repair them. Accidents meant that patients frequently came to them with torn ears, and the different kinds of damage that could happen to stretched lobes had been classified into a system by Sushruta's time, as were the remedies, which he described as "innumerable." This experience with piercing flesh and stretching pieces of skin provided the perfect background for the crowning achievement of ancient Hindu surgery.

Sushruta describes how an earlobe that had been completely torn off could be restored by transplanting flesh from the cheek:

> A surgeon well versed in the knowledge of surgery should slice off a patch of living flesh from the cheek of a person devoid of earlobes in a manner so as to have one of its ends attached to its former seat [i.e., the cheek]. Then the part where the artificial earlobe is to be made should be slightly scarified [with a scalpel] and the living flesh, full of blood and sliced off as previously directed, should be stuck to it.

This ancient operation used a classic technique of modern surgery—the "pedicle flap"—in which a long, U-shaped piece of flesh is cut and looped over to another, where the tissue has been prepared for grafting. The blood congeals, holding the two surfaces together; they eventually grow together and begin exchanging blood, at which point the flesh can be separated from the "donor" area. Similar operations are performed today to make new earlobes, but using skin from behind the ear rather than the cheek.

The same techniques were used by the ancient Hindu surgeons to carry out an even greater wonder—rhinoplasty, the repair, or even

The manufacture of a new nose (rhinoplasty) as described by ancient Hindu medical texts. A leaf-shaped flap of skin is cut from the forehead, twisted over, lowered over two small tubes (usually bamboo) inserted to act as nostrils, and attached to the face. The most difficult part of the operation is to twist around the flap without constricting the blood vessels at the tip of the "nose."

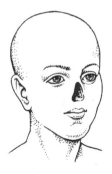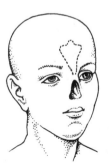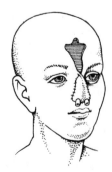

creation, of noses. Again, the method is described in the *Sushruta Samhita*. To create a nose, a leaf-shaped flap of skin was cut from the patient's forehead, leaving a "stalk" at the bridge of the nose. The flap was then folded down over two artificial tubes, which acted as nostrils. Scarring the tissue, as in the earlobe operation, allowed the sides of the leaf shape to grow onto the face.

The Romans also experimented with plastic surgery. The poet Martial (late 1st century A.D.) refers to a barber who specialized in removing brand marks from freed slaves. Slightly before Martial's time the Roman writer Celsus compiled a medical encyclopedia that describes two plastic-surgery operations. In one a wound or sore that refuses to heal is covered by stretching over flaps of skin and sewing them together. In the other Celsus describes how the lobe of someone who wanted to get rid of the holes created by wearing heavy earrings could be restored to normality. It was effective, but not up to the standards of Sushruta, who could fashion a new lobe from scratch.

 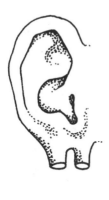 

Simple repair operation for closing a hole in a lobe stretched by wearing a heavy earring, as described by Celsus (1st century A.D.). The lobe is shortened, then the inner skin surfaces are scraped open and stitched together.

Indeed nothing as advanced as the ancient Indian techniques was practiced anywhere else in the world for almost two millennia. Some knowledge of Hindu surgery passed to the Arabs, and from there to the Mediterranean, where we read of Sicilian physicians making new noses about A.D. 1400. But modern Western plastic surgery began in the eighteenth century when British doctors working with the East Indian Company learned the secrets of rhinoplasty from Hindu surgeons.

BRAIN SURGERY

In 1865 an archaeologist investigating a Stone Age tomb in southern France uncovered a curious human skull. A large piece of bone had been sawed out of the cranium, and one edge of the hole appeared to have been polished. The archaeologist surmised that the polished area "had been expressly made for the application of the lips"; in other words the skull had been fashioned into a gruesome drinking vessel. His conclusion fitted well with nineteenth-century thinking. Early scholars usually saw our prehistoric ancestors as barbarous savages who would have enjoyed nothing better than quaffing some primitive brew from the skulls of their defeated enemies.

An Unthinkable Operation

The real nature of the skull came to light a few years later when it was sent to the laboratory of Professor Paul Broca in Paris. Broca was an outstanding pioneer in the fields of surgery, anatomy and anthropology, whose fundamental discoveries include the location of the speech centers in the left side of the brain. Reexamining the mysterious "drinking cup" skull, Broca made a staggering observation. The supposed "polish" was actually *regrowth of the bone tissue after cutting*. A cut bone surface reveals tiny pores that are easily visible to the naked eye. When the bone heals after cutting, tissue builds up, blocking the pores to give a smooth appearance. No other process but healing could have produced such a surface.

The conclusion reached by Broca was inescapable: During the Late Stone Age surgical operations were being performed on the human skull. In Broca's day brain surgery was a field of medicine that was still largely theoretical. Attempts at penetrating the skull had an appallingly high mortality rate, the result of damaged or infected brain tissue. Yet the skull that Broca held in his hands, bearing the unmistakable traces of successful surgery, was more than four thousand years old.

The notion that skull surgery, known as trepanation, or trephination, was performed in the remote past is one that archaeologists once found hard to swallow. As late as 1960 British archaeologist Dame Kathleen Kenyon wondered whether the small squares cut from some Palestinian skulls of the sixth century B.C. were actually the result of ghastly "experiments" carried out on prisoners "in the manner of the Nazis," rather than the traces of beneficial surgery. The evidence, however, persuaded her toward a medical explanation. While two of the cases had clearly not survived the trepanation, the third "had apparently lived long enough for the bone to heal," and she concluded that "trephining may thus have been a recognised surgical practice, employed . . . on battle casualties."

Somewhat reluctantly—through the sheer weight of the evidence, comprising literally hundreds of skulls showing signs of surgery— modern science has been forced to accept the fact that operations were performed on the skull in ancient, even prehistoric times. Trepanned skulls have now been discovered at burial sites of most of the great civilizations of the past, from the megalith builders of prehistoric Europe to the Incas of Peru. (One exception is ancient Egypt, where there was a religious taboo against interfering with the body during the lifetime.)

Four-thousand-year-old skull from Crichel Down, Dorset (England), bearing unmistakable traces of the surgical operation known as trepanation. A circle of bone had been carefully removed from the patient's head, using stone tools. The man recovered, as regrowth of the bone tissue clearly shows. The circle of bone had been saved by the patient, apparently for good luck, and was buried with him when he died.

A remarkable fact gleaned from the study of ancient trepanations is the high rate of survival. Between 1870 and 1877, 32 trepanations were performed at Saint George's and Guy's Hospitals in London; only 25 percent of the patients survived. This should be compared with a sample of 214 trepanned Inca skulls from ancient Peru, 55 percent of which show that *complete* healing had occurred—the owners must have survived the operation for several years. Only 28 percent died with no trace of healing. Again, from a sample of ancient German trepanations, only 23 percent showed no signs of recovery.

To assess properly the success rate shown by the figures from ancient skulls, we should also remember that many of the "failures" may not actually have died "under the knife" but simply as a result of the injury or illness that prompted the trepanation in the first place. An extraordinary tribute to the skill of the ancient surgeons is provided by a Peruvian skull with no less than seven healed trepanations, evidence of as many successful operations.

"Worms on the Brain"

But why did our ancestors undergo such horrendous treatment? A few traditional accounts describe the motives in terms of release of "evil spirits" or "winds" from the head. This might lead us to see ancient trepanning as a dangerous ritual of self-mutilation, performed *only* because of superstitious beliefs. Surviving historical records discredit this explanation. Together with archaeological evidence, such records can tell us not only how but why trepanations were usually performed.

We are lucky to have a number of descriptions of trepanation from the medieval and ancient worlds. We know that during the Middle Ages monks performed the operation to cure the effects of skull fractures, usually battle wounds, relieving the distress caused by splinters of bone touching the brain. Such wounds may have been the main reason for trepanation throughout the ancient past. Pia Bennicke and Joyce Filer (paleopathologists at the University of Copenhagen and the British Museum respectively) have observed that most trepanations on ancient skulls appear on the left side or front of the head. As most people are right-handed, their victims in battle would have a tendency to be clouted on the left side of the head if engaged in face-to-face combat. Bennicke has also noted that pre-historic Danish subjects of trepanation are overwhelmingly male, which may offer further support to the war-wound theory.

Still, there were many other motives in the ancient world for carrying out this dangerous operation. Medical treatises by the Roman writer Celsus (around A.D. 30) and the famous Greek doctor Hippocrates (late 5th century B.C.) prescribe trepanation for incurable headaches as well as for injuries. The ancients seem to have believed that trepanation helped to relieve pressure on the brain in cases associated with acute conditions such as hydrocephaly ("water on the brain"), which may have arisen after injury or illness. The Greco-Roman doctors carefully detail the reasons for operating and specify the different methods of penetrating the skull. These accounts, written in strikingly matter-of-fact terms, show conclusively that the practice was part of a serious medical tradition.

Other accounts seem to refer to the actual removal of tissue from the brain, although the terms in which these stories are couched may be strange to the modern reader. Yugoslavian folklore tells how, in the nineteenth century, a physician was secretly observed

Woodcut depicting a 16th-century attempt to operate on the brain.

extracting a beetle from the brain of the czar's daughter. Medieval Chinese accounts relate that foreigners from Ta-Chhin (Syria) knew the art of removing worms from people's heads to cure blindness. Visiting surgeons from Arabia were also known in China as specialists in this art. Ancient Indian tradition holds that at Taxila (in modern Pakistan) "Atreya, king of the physicians" taught the skill of opening the skull to extract dangerous "worms." An intriguing tale is told about Atreya's student Jivaka, who became a famous doctor in his own right and, tradition has it, personal physician to the Buddha (5th century B.C.). Buddhist folklore tells that Jivaka had a magical gem which "X-rayed" his patients: "when placed before an invalid, it illuminated his body as a lamp lights up all the objects in a house, and so revealed the nature of his malady." Whatever we make of the miraculous stone, Jivaka was said to have used it to locate a "centipede" inside a man's head. Cutting open the skull, he removed the intrusive object with heated tongs, and the patient recovered.

How should we understand these stories of "worms," "beetles" and "centipedes" extracted from people's brains? The answer may lie with the dog tapeworm known as *Taenia multiceps*. Sheep act as hosts to the larvae of these parasites before they enter dogs—for example they can form cysts in a sheep's liver that, when eaten by a dog, grow into a large tapeworm. The worst infestation occurs when the larvae settle in sheep's brains, causing them to wobble about with an unsteady, drunken gait. This condition is known as "the staggers." Farmers in eastern Europe traditionally treated it by scraping the unfortunate creature's skull with an iron knife to locate and remove the larvae—though few animals actually recovered. Regular worming of farm dogs has now considerably reduced the incidence of this horrible disease.

Do tapeworm larvae also lodge in the human brain? The grisly answer is that they can. Cases today are mercifully rare, again due to the widespread treatment of dogs. But it is a fair assumption that the condition was once common among prehistoric farmers, who coexisted closely with their animals. In fact it has been suggested by Michael Ryder, the leading authority on ancient sheep farming, that "man may originally have been the sole intermediate host (of the larvae) of this tapeworm of dogs," sheep only entering the picture later, after they were domesticated by man. The theory receives gruesome support from a pathologist's report on Egyptian mummies kept in the Manchester Museum, England, published in 1984. The brain of one mummy was carefully examined and found to contain several cysts containing embryo tapeworms from dogs. The report notes that as such cysts expand with the growth of the worm, they produce headaches, loss of vision and fits before the sufferer lapses into unconsciousness.

Drilling Through the Skull

Skull fractures and parasites were not the only motives for trepanation. Incurable headaches, as we have seen, were another common complaint that ancient surgeons believed could be cured by "relieving the pressure" on the brain. The experience gained through thousands of operations would also have made ancient doctors aware of other visible brain diseases. It is more than likely that they were familiar with brain tumors and attempted to remove them. Direct work on the brain itself is suggested by accounts such as the story of the Chinese physician Thai Tshang Kung (c. 150 B.C.), who "used to cut open the skulls of patients and arrange their brains in order."

Any doubts as to the seriousness of the ancient skull surgeons are dispelled when one looks at the intricacy and refinement of the instruments they employed. Specially designed saws were used by the Celts of ancient Europe. An example recovered from a Rumanian grave of the second century B.C. had a wedge-shaped edge that prevented the blade from penetrating the bone more than a few millimeters at a time and damaging the underlying tissue.

The burial of a Roman surgeon discovered in Germany included a full trepanning kit made of bronze. The tube-shaped drill had a removable center pin to guide it in the first stages of the operation, a bow to drive it and a cup to protect the surgeon's hand. The reconstruction drawing shows how the drill was used to remove a circular piece of bone from the head. Ancient medical handbooks by

Hippocrates, Celsus and others stress that these drills *(trephines)*, had to be plunged frequently into cold water to avoid overheating from friction between the drill bit and bone.

The very idea of drilling into someone's skull is almost too painful to consider. Though it is not usually known, anesthetics other than alcohol were available to the ancients (see **Anesthetics**), and these could have relieved the agony of trepanation. Yet, as we shall see, the operation can actually be performed *without* the use of anesthetics.

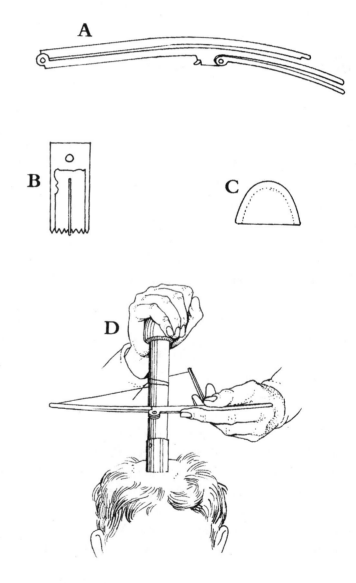

The bronze trepanning kit buried with a Roman physician at Bingen on the Rhine, Germany, in the early 2nd century A.D. It includes a complex folding bow to drive the drill (A); a shaft for the drill; a tube-shaped drill bit with serrated edge (B); and a cup (C) to protect the surgeon's hand as he pressed down on the patient's skull (D).

Trepanation Today

A fact almost as surprising as the fantastic antiquity of trepanation is that *it is still performed today* in some remote societies around the world. It was practiced in Europe as recently as the last century by the peasants of Yugoslavia. From other parts of the world, particularly East Africa, anthropologists have collected a number of cases from tribal cultures that have preserved this technique of prehistoric surgery as an integral part of their traditional medicine.

For a striking example we turn to a case investigated by psychiatrist Dr. Edward Margetts in 1958 that he described as "the most spectacular curiosity that one would ever hope to see." The subject is a gentleman known, for reasons that become immediately apparent, as "Hat On, Hat Off," a fifty-year-old man who was once a tribal policeman of the Kisii tribe in Kenya. Wearing his battered hat, he looked quite normal. When he removed it, "one was amazed to see the whole top of his head missing," like a half-eaten boiled egg. X-ray photographs showed an oval hole of about thirty square inches in his skull!

"Hat On, Hat Off" explained his story as follows: While pursuing inquiries, he entered a hut one day and struck his head on the door lintel. Severe headaches continued for four or five years, and in 1945 he decided to undergo trepanation to alleviate them. To relieve his continuing headaches, a number of further operations followed—his own estimates varied between five and thirty—in the course of which an enormous portion of his cranium was successively removed. Eventually he was fitted with a plastic skullcap to protect his exposed brain.

The operations, said "Hat On, Hat Off," were extremely painful, as we can well imagine—he apparently underwent them without the aid of any anesthetic. Incredible though it may seem, his experience is confirmed by many other modern trepanation subjects who were given no painkiller.

The tribal doctor who had operated on "Hat On, Hat Off" was also interviewed by Margetts, who described him as "an interesting old fellow" between seventy and eighty, calmly confident of his work and especially proud of the fact that he had never lost a pa-

Mr. "Hat On, Hat Off," as he is known to anthropologists, was a local policeman from Kenya. His story provides one of the most extraordinary cases of the survival of the prehistoric technique of skull surgery. X-ray photography showed that some thirty square inches of his skull were removed in a series of operations by a tribal doctor.

tient. He had lost count of the number of operations he had performed, having been trepanning people since he was about twenty. He had learned the art of skull surgery from his father, who in turn had been taught by his father, and his by his, and so on. In such a way the secrets of an almost unbelievably dangerous medical operation have been handed down since time immemorial. In some of the areas where it has been preserved, the level of technology is little different from that available to our remotest ancestors. Until recently the islanders of Melanesia performed the operation with tools made of obsidian (volcanic glass) or sharpened seashells.

Where trepanation was first invented is uncertain. As we have seen, the East Africans and Chinese may have learned the skill from the Arabs; they in turn may have learned it from the Greeks and Romans, or from India. But the ancient trepanners *par excellence* were the Celts of Europe, and they were most likely the teachers of the Greeks and Romans. Western and central Europe have provided the earliest examples of trepanning drills and saws and by far the earliest skull operations. The four-thousand-year-old example examined by Broca is not unique—in fact the Neolithic (or Late Stone)

Age has been seen by some as the high point of ancient skull surgery. From Europe comes by far the oldest trepanned skull, dating to around 6000 B.C..

These dates lead to some quite baffling conclusions. An operation too dangerous for nineteenth-century surgeons was being successfully performed by our ancestors at a time when agriculture was in its infancy. In fact the trepanned skulls of prehistoric Europe are the oldest indisputable traces of any surgical operation performed by the human race.

FALSE TEETH AND DENTISTRY

What do Winston Churchill, George Washington, Queen Elizabeth I, King Henry III of France and many ancient Etruscans have in common? The surprising answer is false teeth, which were invented almost three thousand years ago by the Etruscans of central Italy, fathers of much of what we now know as Roman civilization.

From 700 B.C. Etruscan craftsmen were producing partial dentures of bridgework, good enough to wear while eating. Some were re-

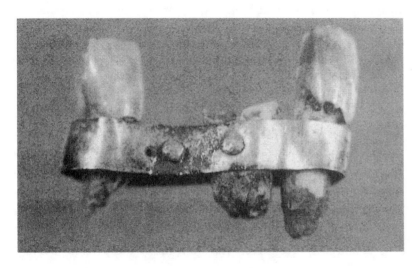

A fine example of Etruscan dental bridgework, 7th–6th centuries B.C.

movable for cleaning while others were permanently attached to the surviving original teeth. The procedure in the latter case was to solder together wide bands of gold, which fitted over the natural teeth with the substitute held in place by a pin in one of the bands. The gold bands were kept well above the line of the gums to avoid rubbing on them, while the false teeth themselves rested not on the gums but on the genuine teeth to either side. The replacements either came from the mouth of another person or were carved from ox teeth. Dentistry of this degree of skill was not practiced again until the nineteenth century A.D.

One explanation for the high quality of Etruscan work, suggested by Vincenzo Guerini in his 1909 study of *The History of Dentistry*, is that the Etruscans wore their false teeth with pride:

> Since the gold bands of which they were constructed covered a considerable part of the crowns, they certainly could not have had the pretension of escaping notice, being, on the contrary, most visible. It is thus to be surmised that in those times the wearing of false teeth and other kinds of dental appliance was not a thing to be ashamed of; indeed that it rather constituted a luxury, a sort of refinement only accessible to persons of means.

Decorative false teeth were also developed by the Maya of Central America: a jaw dating to A.D. 600 had three of its teeth replaced by pieces of shell. That this had been done during the life of the patient is shown by the fact that bone had begun to grow back around the false teeth.

Stopping the Rot

In a way it is strange that the idea of false teeth took so long to develop. One of the worst afflictions among the ancient Egyptians, for example, was bad teeth. The large amount of sand and grit in their stone-milled bread would erode away the enamel on their teeth, leaving them susceptible to decay. Dentistry, accordingly, may have been one of the earliest medical specializations in ancient Egypt. The famous sage Hesy-Re, who lived under the pyramid-building

Pharaoh Djoser (mid-3rd millennium B.C.), is one of the earliest known doctors from Egypt. On his tomb inscriptions the hieroglyphics for Hesy-Re's medical profession are accompanied by a sign for "tooth," suggesting that he specialized in dentistry. But it seems that Egyptian doctors did not provide their patients with false teeth. Religious factors apparently lay behind this reluctance to replace lost teeth—the Egyptians believed that it was wrong to interfere with the body during life. Ancient Egyptian dentistry rather concentrated on tooth extraction, general dental care and, as we know from the examination of some mummies, drilling abscesses in jaws.

The oldest example of preventive dentistry of this kind comes from Stone Age Denmark. A burial in the Hulbjerg tomb, dating to around 2500 B.C., had a drill hole made by a flint borer in one of his upper molars, to relieve an abscess. The operation had clearly been a success, as the tooth was still in place in the jaw.

One development of later times was the invention by the Syrian doctor Archigenese around A.D. 100 of a tiny drill to release diseased matter from inside decaying teeth. Otherwise, in Roman times, teeth were pulled. The recent excavation of a barbershop in front of the Forum in Rome suggests that most people went to the barbers for such relief, as more than a hundred human teeth came from the site. In medieval China fillings were available, made with a paste of mercury, silver and tin, which hardened in the mouth. By the same time in the West, however, things had declined to the point where candle wax, raven's dung and stale bread were all considered to be no less suitable for fillings than lead, gold or various resins.

There were some bizarre attempts to ward off toothache in the ancient world. For example the Egyptians would place a live mouse on the gums of a sufferer, because mice have such good teeth. Things were little better in Roman times: the encyclopedist Pliny the Elder noted without comment the theories that a frog tied to the jaws would make teeth firm, that painful gums could be banished by being scratched with a tooth from a man who had died violently, and that toothache responded to eardrops of olive oil in which earthworms had been boiled.

There were, however, some more sensible attempts at prevention, through oral hygiene. Toothpicks were known from 3000 B.C. in Sumer (southern Iraq) and were a common element in Roman "pocket sets" (see **Introduction** to **Personal Effects**). Simple sticklike toothbrushes were also used by the Romans. (Ones with bristles originated in China about A.D. 1000, where they were used with a tooth powder made from soap beans.) Unfortunately the positive effects of attempts at oral hygiene in the Roman world were ruined by the use of tooth powders with abrasive ingredients such as emery; these made teeth look nice and shiny but at the same time ground away the surface, exposing the pulp.

FALSE LIMBS

For most of human history those who lost limbs through accident or war could be made comfortable but could not be enabled to resume a normal life. In the first millennium B.C., however, several widely separated societies, from Italy to India, started to make replacement limbs.

The earliest definite case was recorded by Herodotus, the "Father of History." It concerns one Hegesistratus, a Greek diviner who read omens for the Persian army invading Greece before the Battle of Plataea in 479 B.C. He performed this service not only for a fat fee but also because of his hatred of the Spartans, the Persians' opponents. Sometime earlier the Spartans had arrested Hegesistratus, locked him up in prison with one foot in the stocks and told him that he was to be tortured before being executed. Hegesistratus managed to get hold of a knife, cut off the front part of his foot and escaped. He then had a wooden replacement made and it was on this that Hegesistratus took part in the battle. Although he survived the Persian defeat, the Spartans eventually caught up with Hegesistratus and put him to death.

The Romans also made false limbs, one of which has survived from antiquity. A skeleton with a remarkably well modeled artificial leg was discovered in a tomb of c. 300 B.C. at Capua in central Italy. The leg was made of wood sheathed in two thin bronze sheets, which were attached to the wooden core by an iron pin. It was concave at the top to hold the surviving stump of the thigh, and would have reached from the knee to the ankle. A separate wooden foot was probably attached at the base, but this has not survived. The scientist Pliny records in his *Natural History* that Sergius Silus, a veteran of the Second Punic War (218–201 B.C.) against Carthage, with a mere twenty-three wounds, had an "iron hand" made to replace the one he had lost in battle.

Ex-soldiers were not the only ones to benefit from replacement limbs, as the case of a young woman from Kazakhstan in Central Asia shows. Her body was found in a tomb dating from 300 B.C. Her left foot had been amputated and replaced by two bones taken from a ram; the leg healed, and she survived for several years after the surgery, eventually succumbing to illness. Not long afterward similar experiments seem to have been being performed in India. The great Indian doctor Sushruta, who probably lived sometime in the last centuries B.C. (see **Plastic Surgery**), was said to have conducted many amputations, substituting iron limbs for the diseased or damaged ones.

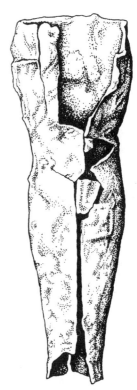

Front view of the Roman artificial leg from Capua. Only the bronze sheeting that had covered the wooden core survived when the photograph from which this drawing was made was taken.

Legs for the Afterlife

The strangest examples of false limbs are those found on Egyptian mummies. These were not provided for medical reasons, as Egyptian surgeons were extremely reluctant to perform amputations. They believed firmly that the physical body had to be preserved in its entirety for the immortal soul to reach the next world. However, the other side of this religious belief was that embalmers would add false limbs to restore bodies for the afterlife when necessary.

The three-thousand-year-old mummy of a fourteen-year-old girl now in the Manchester Museum, England, proved to have false legs when it was unwrapped in June 1975. It seems that her body had been badly damaged after her initial burial, so when the mummy was unwrapped hundreds of years later, artificial legs of wood were splinted onto the remaining bone and a pair of feet was modeled out of reeds and mud.

ANESTHETICS

Corridors of Blood, a classic 1958 horror movie starring Boris Karloff, took as its theme the unspeakable agonies suffered by surgeons' victims before the introduction of modern anesthetics in the 1840s. Much suffering there undoubtedly was, but we should not imagine that every operation before the advent of modern surgery was accompanied by bloodcurdling screams of pain. One of the earliest of mankind's wishes must have been for a magical substance that would kill pain. To this end ancient doctors made extensive investigations of the pain-deadening properties of certain plants, carefully building up their knowledge through practical experience.

We know that cocaine (from the coca plant) and opium (from the poppy) were familiar as psychoactive drugs by at least four thousand years ago—in South America and the Near East, respectively (see **Drugs** in **Food, Drink and Drugs**). That they were used in medicine at an equally early date seems likely.

Roman doctors had a number of drugs at their disposal that could be used as painkillers and sleeping drafts. Opium was one of the most commonly used: Celsus, a writer of the early first century A.D., recommends taking the juice of wild poppies to cope with "headache, ulceration, ophthalmia, toothache, difficulty in breathing, intestinal pains, inflammation of the womb and pain in the hips, liver, spleen or ribs." He also mentions the use of henbane as a sedative. A large quantity of henbane seeds was discovered in the Roman military hospital at Neuss in western Germany, along with other medicinal herbs. Henbane contains scopolamine, a drug still used today as a preanesthetic—in low doses it induces sleep and amnesia.

The most powerful anesthetic known to the Romans was the mandrake, a plant with an awesome reputation. Though it is difficult to be sure that all references to "mandrake" in ancient literature are actually to the same plant (see **Aphrodisiacs** in **Sex Life**), it is reasonably certain that the mandrake used in medicine was datura (also known as jimson weed or thorn apple). Datura contains atropine and hyoscine, drugs that slow down the heart rate and, in the right doses, can completely deaden pain as well as reduce the trauma experienced by a patient undergoing surgery. Datura tea was one of the substances fed to their sacrificial victims by the Aztecs to keep them drowsy and oblivious to their eventual end. Around A.D. 75 Pliny described how it was put to more constructive use by Roman doctors: "When the mandrake is used as a sleeping draught the quantity administered should be proportional to the strength of the patient, a moderate dose being one cyathus [about three tablespoonfuls]. It is also taken in drink for snakebite, and before surgical operations and punctures to produce anesthesia. For this purpose some find it enough to put themselves to sleep by the smell."

Clearly the perils of overdosing must have already been learned the hard way by Pliny's time. Perhaps because of the dangers of such drugs, they were not used on a routine basis but only when strictly necessary. Most minor surgery was carried out without anesthetics, and Celsus notes that one qualification of a good surgeon was the ability to concentrate on his work without being distracted by the cries of his patients.

The anesthetic drugs known to the Romans continued to be used in Europe until the Middle Ages. The most common way of ingesting them was by means of a "soporific sponge," referred to in numerous texts from the ninth to fifteenth centuries A.D. The drugs, including opium, mandrake, henbane and hemlock, were mixed together and soaked into a sponge, which was then dried. When anesthesia was needed, the sponge was moistened and placed over the mouth of the patient, who inhaled the fumes of the drugs. The cocktail of drugs would certainly have rendered anyone completely insensible, though the inclusion of the deadly poison hemlock (which depresses the motor and then the sensory centers of the nervous system) would have made the whole exercise extremely risky.

There was a similar problem with the various recipes for oral anesthetics described in a number of fifteenth-century manuscripts. The fact that there are usually only a few copies (often only one) for each recipe suggests that they were not that popular—presumably because they contained potentially deadly ingredients in dangerously high doses. But one English recipe, for a potion called *dwale*, seems to have hit upon the right formula. Nearly thirty copies are known, all of which give exactly the same ingredients and measures—to be mixed together, boiled for a short while and diluted in wine. As well as hemlock juice, opium and henbane, all known from other anesthetic recipes, it also contained vinegar, which might have counteracted the toxic effects of the hemlock, and "wild nept," or briony, which, like henbane, is a strong purgative. After drinking the potion, it seems a patient would rapidly have become unconscious and insensible (the wine helping considerably) for enough time to undertake simple surgery such as amputation, draining an abscess or stitching a wound. The powerful laxative properties of the briony and henbane would then have given the patient a fairly unpleasant time, but must have successfully expelled the other ingredients, which could have been fatal if allowed to linger in the system.

To judge from its popularity, *dwale* (which in Middle English means something like "unconsciousness") must have been fairly reliable. The manuscripts give a quaint description of its use: "When it is needed, let him that shall be carved sit against a good fire and make him drink thereof [the dwale] until he fall asleep and then you may safely carve him, and when you have done the cure and will have him awake, take vinegar and salt and wash well his temples and his cheekbones and he shall wake immediately."

The Power of Mandrake

The value of anesthetics was marked in ancient China by honoring their discoverer, a Doctor Hua T'o of the third century A.D., as the god of surgery. He used an effervescent mixture of *mafeisan* dissolved in wine as his "knockout drops" before operations. *Mafeisan* was a herbal drug the composition of which has long been lost. Cannabis has been suggested as an ingredient, but though it would have helped to put the patient to sleep, it would not have removed the agony of surgery. It is more likely that *mafeisan* contained mandrake (datura), known to have been the main ingredient of later Oriental anesthetics. But it seems its use may have been forgotten in China for some centuries after Hua T'o's time. Writing around A.D. 1300, the Chinese geographer Chou Mi was clearly unfamiliar with mandrake when he described its use by the nomads of Outer Mongolia: "As soon as a man takes a little amount of the herb with wine, he is paralyzed so completely that he becomes unconscious of the pain even though he would be wounded with a sword. And after three days he revives upon taking a certain drug."

Mandrake had returned to the mainstream of Chinese medicine by A.D. 1313, when Wei I-lin, professor of Medicine at Nan-fang University in China, developed the use of white datura in wine as an anesthetic.

The efficacy of ancient herbal anesthesia is shown by the success of the Japanese surgeon Hanaoka Seishu in 1805, when he carried out an operation for the removal of breast cancer using datura as the anesthetic. This was the first modern example of such a delicate procedure, a generation before William Thomas Morton cut a tumor from the neck of a patient anesthetized by ether at Massachusetts General Hospital on October 16, 1846.

A traditional print showing Hua T'o, Chinese god of medicine, operating on General Kuan Yü, Chinese god of war. Kuan Yü's bravery is made clear by his refusal of a draft of Hua T'o's anesthetic. In some versions of this scene Kuan Yü is shown taking his mind off the pain by playing *Go*, an ancient Chinese board game.

ACUPUNCTURE

In 1971 American scientists visiting China witnessed a spectacular demonstration of the ancient art of acupuncture. In a Peking hospital an operation was performed to remove an ovarian cyst entirely without the aid of modern drugs, except for a standard "premed" tranquilizer to steady the patient's nerves. The patient was fully conscious throughout the operation and showed no discomfort. The acute pain she would normally have experienced was alleviated simply by implanting needles into key points on her body and gently twirling them.

The use of acupuncture for anesthesia is one of the crowning glories of the revival of this ancient technique in modern China. It played a crucial role in convincing Western scientists beyond any doubt that acupuncture works. If it could be used to render someone insensible to pain even under the surgical knife, the acupuncturists' claim that it had relieved the suffering of patients with mundane complaints for more than two thousand years had to be taken seriously.

The renaissance of acupuncture was one of the more positive results of the Communist party takeover in 1949. Needing cheap medical services for the troops and peasants involved in his popular revolution, Chairman Mao had wisely encouraged a return to traditional practices officially out of favor for more than a century. During the Ching Dynasty (A.D. 1644–1911) prudishness increasingly restricted medical practice, as patients considered it immoral to bare their bodies for examination. Acupuncture, which involves close contact between doctor and patient, suffered particularly. In 1822 a royal decree eliminated it from the curriculum of the Imperial Medical College. As part of its program to Westernize the country, the republican government of Chiang Kai-shek tried to outlaw acupuncture in 1922.

While traditional herbal medicine continued to thrive as the trade of respectable Chinese doctors during these troubled times, acupuncture survived only as a popular art carried on by street practitioners who often had very little training. It has now been fully restored to the position that it has generally held in the broad sweep of Chinese history. In both ancient and modern China, and increasingly in the West, its efficacy in treating sciatica, lumbago, sprains, asthma, migraine, ulcers, the aftereffects of stroke, drug addiction, angina and neuralgia is now well established. Success has also been claimed for numerous other complaints, including hemorrhoids, depression and schizophrenia, but the results still await confirmation.

The successful revival of this ancient science was helped by the survival of medieval texts, particularly from the eleventh to sixteenth centuries, a golden age for acupuncture. One of the most important texts was that prepared by the great Wang Wei-I in A.D. 1026, the *Illustrated Manual Explaining Acupuncture and Moxibustion with the Aid of the Bronze Figure and Its Acu-points*. It was written to accompany two life-size bronze models of the human figure, made according to his directions, which showed the precise locations of needle points for training acupuncture students. Wang Wei-I defined with great precision the points of the body where acupuncture needles are to be inserted. The same points were used in moxibustion, a treatment closely related to acupuncture, in which dried leaves of the moxa herb (*Artemisia vulgaris*), usually shaped into small cones, are placed on the patient's skin and then lit. Wang Wei-I's definitive manual, particularly its illustrations, cleared up the confusion that had been growing in the field over previous centuries due to the mass circulation of books after the discovery of printing (see **Books and Printing** in **Communications**). The texts for these works were often prepared by calligraphers untrained in medicine, with the result that the location of the needle points, as well as the medical commentaries that accompanied them, were becoming garbled through copying.

Wang Wei-I refined and reformed the whole system and, in recognition of his success, his work was immortalized in stone. The entire book was copied onto two enormous slabs, some six feet high and twenty-two feet wide, which were displayed at Khaifeng, the capital of the Sung Dynasty (A.D. 960–1279). Here all could read the work or make a copy in the same way that brass rubbings are made. Under the Mongol (Yuan) Dynasty (A.D. 1260–1368) founded by Kublai Khan, the tablets and bronzes were transferred to the Imperial Medical College at the new capital of Peking. Remarkably

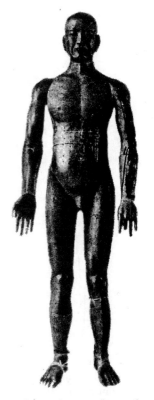

Life-size bronze figure for training acupuncturists. The acu-points are represented by small holes. During examinations the model would be filled with water and covered with a thin layer of wax. Medical students would be asked to located specific points by piercing them with a needle. If they were correct, drops of water would trickle out.

enough, five fragments of the original stone tablets were discovered in 1971, hidden inside the city wall. Though worn with time, much of the text was still legible.

An Ancient Art

Gold acupuncture needle from the tomb of Prince Liu Sheng, who was buried at Man-ch'eng in Hubei Province in 113 B.C.

Acupuncture was already ancient when Wang Wei-I wrote his classic in the eleventh century A.D. But exactly how old is it? Under the Tang Dynasty (A.D. 618–906) the Imperial Medical College had a professor of acupuncture, with ten assistant teachers, twenty students and twenty craftsmen to manufacture needles. The two main textbooks used were the *Classic of Acupuncture and Moxibustion*, written by one Huangfu Mi sometime between A.D. 256 and 282, and the *Nei Ching*, whose authorship was ascribed to Huang Ti, the "Yellow Emperor." Chinese tradition places the reign of this emperor as far back as 2600 B.C.

That the *Nei Ching* was really composed more than 4,500 years ago is highly doubtful. The Yellow Emperor himself is a shadowy figure, perhaps entirely legendary. It is generally thought that the text we now have was compiled over many generations, taking its present form sometime between the second and first centuries B.C. In any event it is one of the oldest Chinese medical tracts extant. Taking the form of a dialogue between the Yellow Emperor and his prime minister, Ch'i Po, it is a rambling discourse as much concerned with Taoist philosophy as it is with medicine. A striking fact, however, is that acupuncture and moxibustion dominate the medical discussion; the use of drugs is of only minor importance.

Acupuncture and moxibustion were therefore already well established by the first century B.C. Their earlier history is less clear, but a remarkable archaeological discovery in 1973 has pushed back our knowledge of Chinese medical writings. Excavation of a nobleman's tomb near the village of Mawangdui in Hunan Province revealed three medical texts written on silk, two of which are detailed treatises on moxibustion. The burial dates to 168 B.C., but the style of the texts suggests that they were written up to two centuries earlier. The earliest mention of acupuncture itself is in an entry for the year 580 B.C. in court records known as the *Spring and Autumn Annals*.

Other archaeological discoveries have confirmed the popularity of acupuncture during the last centuries B.C. Excavation of the tomb of the Prince Liu Sheng at Man-ch'eng in Hubei Province, dating to 113 B.C., produced a set of needles, four of gold and five of silver. Their shape and number (conforming to the "nine needles" of traditional acupuncture) leave no doubt as to their purpose. Possible acupuncture needles made of bronze date from the eighth century B.C., while the use of earlier, stone needles is known from the classic texts. The *Nei Ching* talks of stone probes. Indeed some Chinese scholars maintain that the origins of acupuncture lay in the Stone Age.

East Meets West

The rationale behind acupuncture is—at least to the Western mind—as mysterious as its origins. How was it invented? When we encounter it in the *Nei Ching*, it already appears as a fully developed system. Trial and error have undoubtedly refined the art, but conceiving a theoretical starting point for its invention is difficult. The most curious fact about acupuncture of course is the fact that the needles are generally inserted and twirled in a part of the body that is different from where the ailment is located.

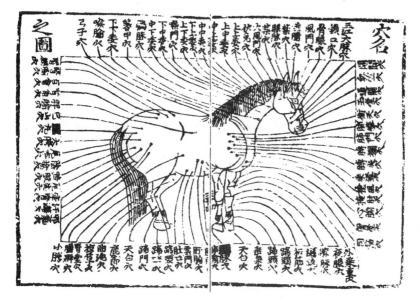

Acupuncture has been used on animals in China since at least the 14th century A.D. This diagram from a veterinary manual of A.D. 1399 shows the acu-points on a horse's body.

The traditional Chinese explanation is that most complaints are due to disharmony between the *yin* (spirit) and *yang* (blood) essences in the human body (partly equivalent to the female and male aspects of life). These are thought to flow in channels, comprising fourteen main "meridians" with 361 points on the surface of the body. Put very simply, the imbalance between *yin* and *yang* at one end of a channel can be restored by releasing or draining energy at the other end by opening it with a needle. The problem with this, as modern Chinese doctors admit, is that it involves intangible forces that cannot as yet be studied by "normal" scientific means. Nerve endings can be studied under the microscope, but as yet the only energy we know to pass through the nervous system is electricity. That acupuncture involves the stimulation of nerve endings and pain receptors can hardly be doubted, but exactly how remains a mystery.

One Western medical concept that has approached an explanation is the "gate theory of pain," the Nobel Prize–winning idea developed in 1965 by Ronald Melzack, professor of psychology at McGill University, Montreal, and Patrick Wall, professor of anatomy at University College, London. The theory states that while pain is transmitted to the central nervous system (and thence to the brain) via specific nerve fibers, other nerve fibers work to inhibit its transmission. The two meet at the spinal cord. Pain is transmitted from the bare nerve endings found in the skin, and without the action of the inhibiting nerves we would be in pain every time we touched something. If the pain receptors are overloaded, the "gate" at the interface in the spinal cord is opened and we experience the sensation of pain. Hence stimulation of the pain-inhibiting nerves, by various means such as massage, can relieve pain even in different areas of the body.

By 1977 Dr. Melzack and other researchers had identified numerous "trigger points" around the body—sensitive areas where the prodding of a finger causes a deep aching feeling; pressure on these areas can also relieve pain in other parts of the body. Further, Melzack found that every trigger point reported in Western medical literature corresponded to an acu-point. The "gate" theory must be along the right lines, but it still falls very short of an explanation of acupuncture. For example, stimulation of a trigger point may have some effect while pressure is applied, yet the twirling of an acupuncture needle can have effects that last for months, or even a lifetime.

So the work still continues to find a theoretical marriage between acupuncture and modern science. The day will surely come when Western medicine can understand, in its own terms, a technique developed by a totally different medical tradition. The breakthrough will undoubtedly change the nature of Western medicine completely. And it will be one we owe to an apparently irrational yet highly successful technique, invented more than 2,500 years ago, which has stubbornly refused to go away.

TRANSPORTATION

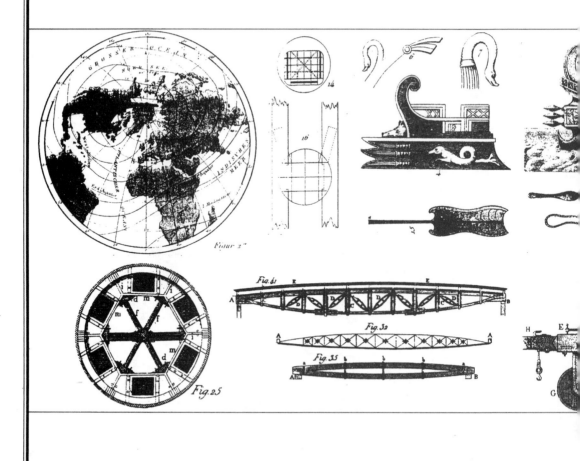

CHAPTER TWO

Fig 22
Fig.38
Fig.30
Fig.24
Figur 2
G R O S S E R O C E A N
INDISCHES MEER
Erd-Aequator
Madagascar

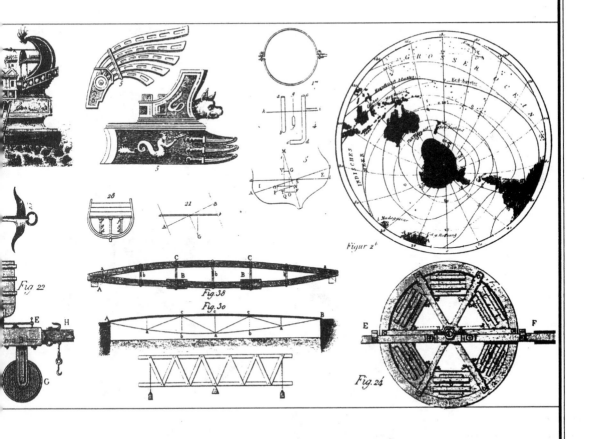

INTRODUCTION

Well equipped as our ancestors were with sturdy legs, they must have become dissatisfied at a very early stage with this sole means of transport. It wouldn't have taken long to appreciate that other, larger, animals were fleeter of foot and that, by clambering on them, a long-distance slog could be made into an easy ride. But this process would have involved the domestication of animals, which archaeological textbooks do not deem to have begun before about 7000 B.C. (see **Introduction** to **Out of Town**).

This leaves a good many years between the arrival of fully fledged *Homo sapiens sapiens* at least fifty thousand years ago and his first official exploitation of domesticated animals. It seems unlikely that in

Stone Age engraving on wall of cave at La Marche, France, from about 15,000 B.C., possibly showing a bridled horse's head.

the intervening period no experiments were made to hitch rides on horses, reindeer, llamas, camels and other likely beasts. Here we simply lack any tangible evidence, though an intriguing case has been presented over the last few years by British archaeologist Paul Bahn that cave paintings in France show that horses were bridled by Stone Age hunters as long ago as 15,000 B.C. Still the earliest hard evidence for the domestication of the horse dates back only to 4000 B.C.—teeth of horses from the Copper Age site of Dereivka, in southern Ukraine, examined under the microscope show traces of wear caused by having bits in their mouths. These were probably made of rope, as the earliest known metal horse bits date from about 1500 B.C.

The wheel, as far as present evidence suggests, was another relatively late arrival. But for a vast stretch of time, crucial in terms of human technological advance, large parts of the inhabited globe were in the grip of the ice sheets of the last glacial period (lasting from forty thousand to ten thousand years ago), resulting in conditions that would have made a wheel quite useless. Wheeled vehicles are also utterly impractical in jungles and deserts, and even on the right terrain they would have been of little real value without animals strong enough to pull them. The sledge undoubtedly preceded the wheel and must have been used by Stone Age hunters (with or without animals) long before the ice sheets retreated. There is some evidence that wheeled vehicles evolved from the sledge—vehicles of basically the same shape, some with wheels but some on sledge bases, are shown in the earliest picture-writing from Sumer, in southern Iraq, dating to about 3200 B.C. The first evidence of carts in Europe follows soon after, with some curious pottery models of wagons found in Hungary.

Wheeled vehicles by themselves are not enough to get around with comfort and speed. To make them effective, good roads are needed. The most famous ancient roads of course were those built by the Romans. Not only did all roads lead to Rome, but the empire itself was held together by its network of roads—the arteries of its tax system and the means by which troops could be moved swiftly to suppress revolts. One of the first acts of many rebels and barbarians as the empire disintegrated in the fifth century was to hack up the roads. Nevertheless some stretches of Roman road still

Signs found on inscribed tablets of the late 4th millennium B.C. from Uruk, in Iraq. It has been argued that they demonstrate the development of wheeled vehicles from sledges.

A pottery cup in the form of a wagon from a grave at Szigetszentmarton, in Hungary, dating to around 2900 B.C.

survive intact—a tribute to their excellent construction. The earliest Roman roads were made from wood and were often constructed to bridge marshy areas. Later ones, beginning with the heart of the network begun in Italy during the fourth century B.C., were built to last indefinitely. They consist of a flagstone foundation covered with successive layers of rubble and concrete and topped with more flagstones, rammed-down gravel or concrete. Like modern roads they were convex, allowing easy drainage.

By the second century A.D. Roman engineers had constructed some 48,500 miles of solidly built roads, a distance roughly equal to twice the circumference of the earth.

The Romans were not the only great road builders; indeed, other ancient civilizations surpassed them in certain respects. The first official roads were being built in China in the ninth century B.C., a good hundred years before the city of Rome was even founded. By the end of the third century B.C. the total length of the Chinese imperial highways was about 4,250 miles. Shih Huang Ti, the first emperor of all China (221–210 B.C.), standardized the width of chariots at five feet. He constructed numerous nine-lane highways, the central lanes of which were for the exclusive use of the Emperor and his family; even powerful nobles were executed for daring to ride on them. By the end of the Han Dynasty, around A.D. 200, there were some 20,000 miles of imperial roads. But the credit for the longest single road of the ancient world seems to go to the Persian Empire: the "King's Road," built in the sixth century B.C., ran 1,600 miles from the capital of Susa (in modern Iran) to Sardis, on the west coast of Turkey (see **Postal Systems** in **Communications**).

Between A.D. 1400 and 1520 the Inca emperors of South America built more than 15,500 miles of roads, running from sea level up to 17,500 feet in the Andes. Obstacles such as fierce mountain rivers were overcome by means of suspension bridges anchored by twin stone towers at each end. Made entirely of tightly twisted plant fiber, the cables were as thick as a man's body. The most famous is that spanning the great Apurímac River in the Peruvian Andes, with cables nearly 150 feet long. It was still being used in 1890, some five hundred years after its construction, when it was immortalized by Thornton Wilder in his great novel *The Bridge of San Luis Rey*. In the lowlands wide rivers were crossed on ingeniously built floating bridges made of balsa.

"However many new roads are built, there will be always be more traffic to fill them." This is one of the irritating "laws" of modern life, but traffic congestion was also a familiar problem in the ancient world. Under the Han Dynasty "traffic cops" were appointed to control the flow of vehicles through the great cities of China. In Rome, focal point of a ghastly confluence of traffic streams, Julius Caesar had to enact a law in 45 B.C. to regulate the wheeled vehicles allowed into the city: "In all streets which are, or come to be within, the built-up area of Rome, it is forbidden from 1 January next to lead or drive a cart between sunrise and the 10th hour [late in the afternoon], except for bringing materials necessary for building temples to the Gods or public works, or for removing rubble." VIPs of course were exempt: "Carts used for transporting Vestal Virgins, the "king" in charge of sacrifices and the flamines [a group of fifteen priests], carts used for triumphal processions or for public games held in Rome or within one mile of Rome shall be exempted from this law." It all sounds terribly familiar.

Rivers provide the easiest way, apart from roads, to move about inland, and where the natural world failed to provide navigable waterways in the right places, our ancestors made their own. Canal building on a large scale began with the earliest urban civilizations. In 2400 B.C. Uni, a governor of Upper Egypt under the Sixth Dynasty, built a shipping canal to bypass the waterfall that interrupts the river Nile at Aswan (the First Cataract). At much the same time a canal nearly one hundred miles long was being constructed by the Sumerian citizens of Lagash, in southern Iraq, the first of an enormous, and successful, network of canals that has continued for more than four millennia as the backbone of commerce as well as irrigation. In China a 260-mile link known as the Canal of the Wild Geese (or Hung Kou) was built between the Yellow River and the Pien and Ssu rivers as early as the fifth century B.C. Under the First Emperor, Shih Huang Ti, the digging of the 120-mile-long "Magic Canal" through a mountain range joined together north and south China, creating 1,250 miles of continuous navigable waterway. The "Grand Canal," which was begun in A.D. 70, reached a length of

1,060 miles by A.D. 1327, running for most of the length of eastern China. It was built on a massive scale, ten to thirty feet deep and, at points, one hundred feet wide.

The most surprising feat of ancient canal construction must be the Suez Canal, linking the Mediterranean and Red Seas, dug by the pharaohs and restored by the Persian emperor Darius. His megalomaniac son, Xerxes (485–465 B.C.), undertook another massive excavation project to facilitate his naval invasion of Greece. The quickest route for the Persian fleet along the coast of northern Greece was around the mountainous peninsula of Athos. But it was also dangerous—a previous Persian expedition had been caught by a storm and wrecked there. So in 480 B.C. Xerxes decided to dig a canal straight through the neck of the rocky peninsula. The Greek historian Herodotus describes how huge gangs of workmen sweated to complete the Athos Canal, a mile and half long, and wide enough for two warships to be rowed abreast, in as short a time as possible. Traces of the canal were still visible early this century.

In the same campaign Xerxes accomplished the even more outrageous feat of building a pontoon bridge across the Hellespont, the strip of water that divides Europe and Asia. Even at its narrowest point, where Xerxes built his bridge, it is still almost a mile wide. Six hundred and seventy-four ships were tied side by side with ropes, shunted into place as platforms for building two bridges and steadied with heavy anchors. Enormous rope cables were then stretched between the continents and drawn taut to steady the whole construction, while the ship "roadways" were covered with soil, brushwood and planks. For a short while Europe and Asia were joined; according to Herodotus, it took a week for Xerxes to drive his army under whiplash from one side to the other. The mad Roman emperor Caligula (A.D. 37–41) emulated this feat when he had a similar bridge of boats built across the three-mile-wide Gulf of Baiae (near the island of Capri) so that he could gallop across it on horseback.

With so much expertise available to satisfy the whims of unbalanced dictators, it is not surprising that ancient engineers also achieved many extraordinary practical advances. For example, the principle of the rail or tramway was discovered thousands of years before the invention of the steam locomotive. The most famous ancient railway was the Diolkos, which ran for nearly four miles across the Isthmus of Corinth in Greece. It was probably built by the tyrant Periander of Corinth in the early sixth century B.C. and remained in use until A.D. 900. The Diolkos consisted of a roadway paved with

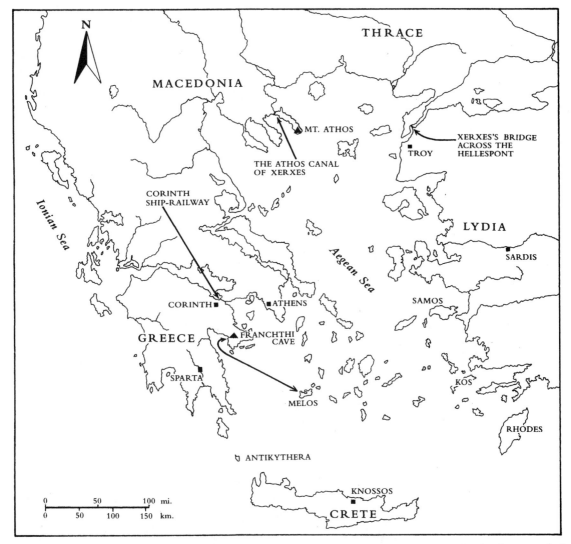

Map showing location of the Athos canal, Xerxes' bridge over the Hellespont, and the Diolkos (Corinth ship-railway), some of the most brilliant engineering feats of the ancient world. (The route taken by the obsidian traders between the island of Melos and the Franchthi Cave in the 11th millennium B.C. is also shown—see **Ships and Liners**).

limestone blocks, cut with two parallel grooves five feet apart to hold the wheels of transport vehicles. Small warships or empty cargo vessels were pushed along the track on large trolleys by their crews and slave gangs; in 1961 remains of a mechanical device used to load the ships onto the trolleys was discovered. After the Diolkos fell into disuse, the principle of railways seems to have been forgotten. Modern railways developed from medieval mining practices: Carts filled with ore were being pushed to the surface on rails from the fourteenth century onward.

The brilliance of the ancient transport engineers, in inventing new ways of overcoming almost any obstacle presented by nature is breathtaking. In the following pages we review the highlights of their ingenious efforts to produce efficient means of traveling by land, sea and even air.

MAPMAKING

The maps prepared by Greek and Roman geographers during the last few centuries B.C. provide the most familiar textbook examples of the "world's earliest maps." In fact, however, these maps represent the culmination of thousands of years of earlier experimentation. There are huge gaps in our knowledge of early cartography for one obvious reason: For practical purposes maps were usually drawn on portable, lightweight materials such as bone, wood, bark and papyrus. Unfortunately these materials are also easily perishable, so our record of the earliest history of mapmaking is frustratingly patchy.

All the same, enough tantalizing scraps of evidence survive to show that the history of cartography actually begins with the Stone Age. In 1966 a fragment of inscribed mammoth tusk was excavated at Mezhirich, in Ukraine. Some eleven to twelve thousand years old, it is believed to show the plan of a stream with a row of riverside dwellings (possibly mammoth-bone houses—see **House and**

Possible Stone Age map showing a line of dwellings beside a river, engraved on a mammoth tusk from Mezhirich in Ukraine, dating to between 12,000 and 11,000 B.C.

Home). From later Stone Age times archaeologists have discovered sporadic cases of artwork with maplike elements, such as enclosures containing houses, animals or people. Other prehistoric etchings have been interpreted as more complex maps showing, for example, rivers and their tributaries, but it is hard to be certain—even more so than in the Mezhirich case, where the pictorial elements allow some grasp of the artist's intentions. But some examples from the New Stone Age are perfectly clear. A wall painting from the town of Çatal Hüyük, in Turkey, dating to around 6200 B.C., shows the layout of houses and streets against the backdrop of an erupting volcano (see **Introduction** to **Urban Life**). While not a map in the full sense of the word, it is certainly the world's oldest urban plan.

The best claim to the title of "world's earliest map" is held by a beautifully engraved silver vase from Maikop, in Ukraine, found in a tomb dating from 3000 B.C. A scene on the vase shows in charming detail two rivers running down from a mountain range, meeting in a lake or sea, surrounded by well-executed figures of wild animals.

Picture map on a silver vase from a tomb at Maikop, Ukraine, dating to about 3000 B.C.

Clay and Papyrus Maps

The Maikop map is ornamental, but more practical examples begin to appear ·not long afterward among the civilizations of Mesopotamia (Iraq). At Yorgan Tepe, in northern Iraq, a small, battered, clay map dating to 2300 B.C. was found. Showing a district bounded by hills and divided by a canal or waterway, it charts the location of a landholding—cuneiform signs in the center of the map give the owner's name and the size of his claim (about thirty acres). Most of the other writing on the tablet is too damaged to read, but clearly visible on the edges are the words for *North, South, East* and *West*, making this the first known map to show cardinal points.

The next two thousand years saw a wide variety of maps and plans being used in Mesopotamia. An unusual example can be seen on a stone statue of Gudea, ruler of the Sumerian city of Lagash around 2100 B.C., shown seated with a tablet on his lap. The tablet is neatly engraved with the outline of a large building, while the top edge of the tablet shows a graduated measuring rod—the earliest use of a scale. The bulk of the Mesopotamian evidence comes from clay tablets, including a map of the great Babylonian city of Nippur from

about 1500 B.C. (perhaps the first town plan drawn to scale), showing the great temple, city walls and gateways, canals, warehouses and
a park, with captions in cuneiform writing.

Dozens of other maps are known from ancient Mesopotamia,
mainly of fields, walls and canals, evidently drawn up as plans for use
in land deals. But the ancient Mesopotamians also had a broader in-

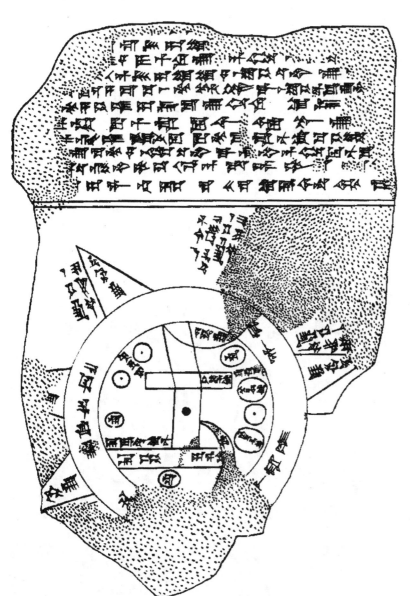

The Babylonian "world
map" from about 600 B.C.
The parallel lines running
down the center show
the Euphrates, crossed by
a rectangle marked, in
cuneiform writing,
Babylon. At the bottom
another rectangle shows the
marshland of southern Iraq,
next to a horn-shaped
Persian Gulf. The small
circles show the countries
surrounding Babylonia,
including Assyria
(northern Iraq), Urartu
(Armenia) and parts of
western Iran. The map is
highly schematic, its main
purpose being to show
the "four regions" at the
edges of the world in
relationship to Babylonia.
These are shown by the
triangles jutting out from
the circular ocean and have
captions describing the
fabulous beasts that lived
there. Overall it is very
reminiscent of the *mappae
mundi* produced two
thousand years later in
medieval Europe.

terest in geography—their imperial ambitions reached as far as Turkey, Iran and Egypt and their commercial contacts as far as India. They were also fascinated by cosmology and had firm ideas about the shape of the world and the universe surrounding it. Some of these are expressed on a clay tablet from Babylonia dating to around 600 B.C., the earliest surviving map of the world. It shows Babylon surrounded by immediately neighboring countries, while the "known" world is bounded by a circular ocean. Beyond this were remote regions, such as the "land where the sun is not seen"; captions name the legendary beasts that lived in these remote lands.

The ancient Egyptians seem to have had less interest in practical maps than the Babylonians. Most of their cartographic efforts seem to have been channeled into religious matters, producing symbolic maps of the universe or the routes to the Underworld. Only one artifact from ancient Egypt survives that can be recognized as a truly secular map, but it is quite outstanding. This is the map on the fragmentary "Turin Papyrus" (named after its present location). Dating to about 1150 B.C., it is the world's oldest mining map and shows the layout of, and routes to, gold mines near the coast of the Red Sea.

The main fragment of the Egyptian mining map on the "Turin Papyrus," dating to around 1150 B.C. Routes to the mines are shown running through a mountain range colored pink, except, as a caption explains, for "the mountains where gold is worked: they are colored red."

Maps of the Classical World

So when the Greeks began their mapmaking, they had considerable experience to lean on—particularly from the Babylonians, who had already developed scales, cardinal points and the concept of global maps. A debt to both Babylonia and Egypt was acknowledged by the Ionian Greeks of the western coast of Turkey, who led the Mediterranean world in scientific research during the seventh and sixth centuries B.C. It was an Ionian astronomer of this time, Anaximander of Miletus, who was hailed by later Greek encyclopedists as the inventor of maps. We no longer have a copy of Anaximander's world map, but that drawn around 500 B.C. by his successor, Hecataeus (also of Miletus), was described by classical writers in enough detail for modern reconstructions to be made.

While it is clear that the Greeks had already discovered that the world was spherical, the easiest way to represent it "on the flat" was by a circle. This led many geographers to idealize the shape of the continents to fit them neatly inside a perfectly circular sea. Around 430 B.C. the great historian and traveler Herodotus (yet another Ionian Greek) mocked this geometrical school of mapmaking: "It makes me laugh when I see some people drawing maps of the world without having any reason to guide them; they show the Ocean running like a river round the earth, and the earth itself to be an exact circle, as if drawn by a pair of compasses, with Europe and Asia just of the same size." Herodotus was clearly aware of better work; he describes a realistic map of the Persian Empire (reaching from the Balkans to India) used by a Milesian agent in an attempt to draw the king of Sparta into a war against Persia around 490 B.C.

The geographical knowledge of the fifth century B.C. was well ahead of that shown in the earliest Greek maps transmitted to us by one means or another. Most show Africa joined by land to India, but Herodotus knew from explorers' accounts (see **The First Suez Canal**) that Africa was a vast landmass attached at only one point to Asia. By the third century B.C. Greek maps could show with reasonable accuracy the relative positions of Arabia, India and Ceylon in the east, and Scandinavia, Britain and even Iceland in the west—all based on the empirical evidence of Greek maritime explorers. Yet many classical mapmakers continued to perpetuate major errors, such as joining Africa to Asia at two points, as late as the second century A.D. The problem typifies the jerky progress made by Western cartography until modern times: At any given period great advances in knowledge could be matched, or even eclipsed, by equally great lapses into ignorance.

We owe to the ancient Greeks the invention of latitude and longitude, the framework for all modern mapmaking. The major lines of latitude (such as the Equator and the Tropics of Capricorn and Cancer) are part of a system developed from the fourth century B.C. by Greek astronomers, who divided the globe mathematically into horizontal bands (latitudes). Vertical meridians or longitudes were added more randomly until about A.D. 100, when the geographer Marinus of Tyre developed a grid with latitudes and longitudes at regularly spaced intervals. He was also the first to tackle the problem of projection, the method of transferring the features of a spherical globe onto a flat surface, and his map was the first European example to stretch as far as China. His knowledge of Asian distances came from Syrian traders who traveled the silk route to China.

Under the Roman Empire (1st to 5th centuries A.D.) most of Europe, as well as large parts of north Africa and the Near East, was united under one authority. The experience of Greek mariners could now be blended with the practical know-how acquired by the Romans on land in building their massive network of roads with milestones. The Roman state encouraged the study of geography for military and commercial reasons, as well as out of sheer scientific curiosity, and reliable itineraries and maps were produced of home areas such as the central Mediterranean.

The greatest geographer of Roman times was a Greek-speaking Egyptian, Ptolemy of Alexandria (c. A.D. 90–168). His main contribution to cartography was the development of projection techniques to a far more sophisticated level than that used by Marinus. Six books of his eight-volume work *The Geography* are occupied with tables of longitude and latitude of specific places, given to within one-twelfth of a degree. While his latitudes are generally accurate, his longitudes were largely guesswork, but this was a problem that plagued every cartographer until the invention of modern navigational instruments in the eighteenth century. Ptolemy also made some major blunders. He ignored the evidence given in Herodotus and joined up southern Africa to eastern Asia by a strip of land called *Terra Incognita*. Worst of all, he seriously underestimated the size of the earth. In the late third century B.C. the Greek mathematician Eratosthenes had already calculated the world's circumference as 24,700 miles. Ptolemy, however, used an estimate of 17,800 miles. The real figure is 24,902 miles, meaning that whereas Eratosthenes was correct within an error of only one percent, Ptolemy was way off the mark.

Ptolemy's error meant that the landmass well known to him, from the west coast of Spain to India, occupied much too great a percentage of the globe's surface. This had far-reaching consequences, influencing Christopher Columbus in making his own underestimate of the earth's circumference—hence his belief when he arrived in the Americas that he had actually reached India. The "West Indies" ultimately owe their name to Ptolemy's blunder.

Had Columbus been an Arab, he might not have made such a mistake. During the Middle Ages Greek science was preserved in the Islamic world, and in the ninth century A.D. the Caliph of Baghdad commissioned two of the Banu Musa brothers, renowned Iraqi scientists (see **Automata** in **High Tech**), to check the circumference of the globe. Starting in the desert of northern Iraq, they measured the altitude of the Pole Star (North Star) from a base point and then traveled north in a straight line until the Pole Star had changed its altitude by one degree. They repeated the procedure by returning to the base point and walking due south until the same shift in the Pole Star occurred. Finally, they checked their results by conducting an identical experiment in southern Iraq. Knowing the distances they had traveled, they could use trigonometry to calculate the earth's circumference at 24,000 miles, very close to Eratosthenes' estimate and the real figure.

MAPMAKING MARVELS OF THE ANCIENT ORIENT

The evidence for mapmaking starts later in the East than it does in the Mediterranean world, but the Chinese soon outstripped Western cartographers in many respects. The first historical reference to a Chinese map comes from 227 B.C., when one Ching Ko attempted to assassinate the future Ch'in emperor, Shih Huang Ti, by concealing a dagger inside a map case. During the following Han Dynasty the emperors employed a flock of state officials to prepare maps. When the Emperor toured the provinces, the imperial geographer rode close to the royal chariot with maps to explain to his master the lay of the land and its produce.

In 1973 actual examples of Han Dynasty maps were discovered in the tomb of a nobleman at Mawangdui, in Hunan Province, dated to 168 B.C. Painted on silk, they are the earliest surviving maps from the ancient Orient. One is purely geographical, showing rivers, roads, towns and other features. The second is military and gives the disposition of troops during a campaign of 183 B.C. Garrisons are represented by rectangles containing the names of their commanders; villages, by circles with captions recording whether they were inhabited or deserted; and the fortress used as the campaign headquarters, by a triangle.

Maps produced under the early dynasties must have been of variable quality, however, as Phei Hsu, who was appointed minister of works in A.D. 267, was highly critical of them. Like the Han Dynasty examples recently excavated, they lacked grids and therefore precision. Scientific cartography began with the foundations laid around A.D. 100 by Chang Hêng, a polymath and inventor (see Earthquake Detectors in High Tech) who was astronomer royal to the imperial court. It is likely that he developed the first Chinese grid system, and Joseph Needham, the leading historian of Chinese science, has suggested that Chang Hêng may have been influenced by the work of the classical geographer Marinus, who had contacts with China. In the late third century Phei Hsu set up rigid official standards for the making of maps, including scales and rectangular grids of parallel lines.

At the very time that cartography declined in Europe during its Dark Ages, a golden age of mapmaking was beginning in the Far East. This culminated in two magnificent maps of China carved in stone at Sian in Shensi Province, northern China, in A.D. 1137. One seems to be from an older source and is less exact. The other has a rectangular grid system (with one hundred *li*, or Chinese miles, to each square) and is remarkably accurate. As Needham remarked: "Comparison of the network of river systems with a modern chart shows at once the extraordinary correctness of the pattern. Anyone who compares this map with the contemporary productions of European religious cosmography cannot but be amazed at the extent to which Chinese geography was at that time ahead of the West."

Korean cartographers, working within the Chinese tradition, also excelled in mapmaking in medieval times. A Korean "world map" produced in A.D. 1402 is outstanding. The shapes of Africa and Europe are roughly correct, and the map includes the Azores, islands in the mid-Atlantic. A charming detail is the representation of the Pharos at Alexandria (see Lighthouses) as a pagoda.

In addition to the sheer bulk and quality of their output, ancient Oriental cartographers must be credited with the invention of several novel forms, particularly with respect to three-dimensional maps. A reliable account of the (as yet unexcavated) tomb of the megalomaniac first emperor, Shih Huang Ti (221–210 B.C.), states that the inner chamber contained a marvelous map. The rivers of China and the "great sea" (Pacific) were imitated by streams of liquid mercury circulated by hidden pumps, making it the world's first mechanized map as well as its first relief map.

In succeeding periods more practical relief maps were modeled from wood, clay and even rice. Another novelty was the creation of a relief "jigsaw map" by Hsieh Chuang (A.D. 421–466), who, according to the official history of the Liu Sung Dynasty, "made a wooden map ten feet square, on which mountains, water-courses and the configuration of the earth were all well shown. When one separated [the pieces of the map], then all the districts were divided and the provinces isolated; when one put them together again, the whole empire then once more formed a unity."

Maps of the Ancient Sea Kings

Strangely, medieval mapmaking, so much closer in time to our own than that of the Greeks and Romans, presents some of the most baffling problems of cartography. After the fall of the Roman Empire, in the sixth century A.D., mapmaking in the West, like so many other sciences, went into serious decline. The art of cartography also remained static among the Arabs. In fact only the Chinese made major strides in cartography during the early Middle Ages. In Europe the tendency to produce idealized maps fitting geometrical shapes returned, and the most typical medieval world maps, or *mappae mundi*, while often beautifully illustrated, show a pitiful knowledge of geography. They usually placed Jerusalem at the center of the world and marked mysterious lands at the edges with captions such as "Here be dragons." Conceptually they were no more advanced than the Babylonian world map of 600 B.C.

The existence of extremely advanced maps from the Middle Ages therefore places historians of cartography in something of a quandary. The famous Vinland Map is a case in point. Discovered in 1957, bound into a book containing medieval manuscripts, it is drawn on fifteenth-century parchment but supposedly records much earlier information. A caption describes the western voyage in A.D. 1118 of Henricus, appointed by the pope as bishop of Greenland. Further, the map shows the northern shore of Greenland, which became invisible after A.D. 1200, when it disappeared under the polar ice sheet. Surprisingly the coast of Greenland is drawn far more accurately than it is on the North Atlantic map prepared by Icelander Sigurdur Stefansson in A.D. 1590. It also shows Vinland, the farthest point of Henricus's journey and the Viking name for New England, as we know from the *Vinland Sagas*, which record the discovery of North America by the adventurer Leif Eiriksson around A.D. 1000.

After initial doubts regarding its authenticity, the Vinland Map was hailed by experts at Yale University as "the most exciting cartographic discovery of the century" and as further authentication of the Vikings' familiarity with North America during the early Middle Ages. But in 1974 analysis of ink taken from the map appeared to show that it contained large amounts of titanium dioxide; ink pigments based on this substance were not manufactured until after

(Opposite) Oval *mappa mundi* from a manuscript by the English monk and historian Ranulph Higden, dating to around A.D. 1350. It shows east at the top and Jerusalem at the center.

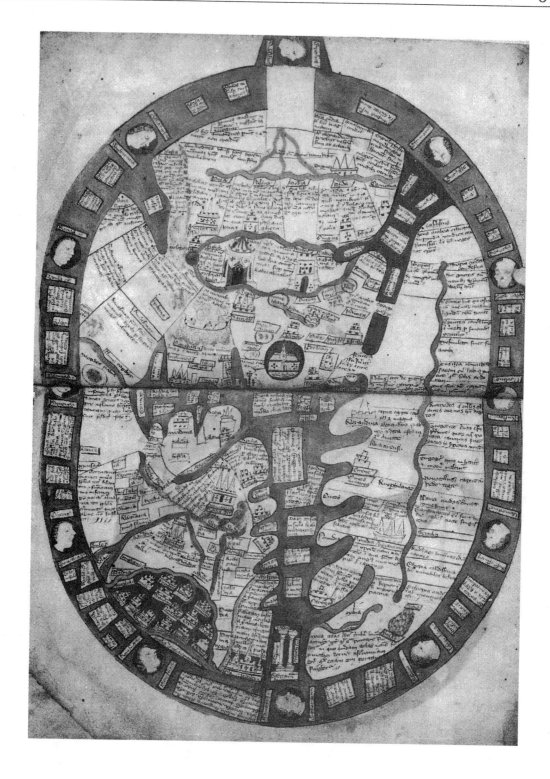

1920. The skeptics were satisfied, and the map was widely proclaimed a modern forgery. Yet a new study undertaken at the University of California in 1985 has shown that the original chemical analysis was quite inadequate. More thorough investigation demonstrates that the levels of titanium dioxide are actually extremely low and are consistent with amounts found in medieval inks. This result has reopened the controversy over the Vinland Map, which may after all prove to be the oldest pictorial record of part of the North American coastline.

Just as controversial are the sailors' maps from the late thirteenth to fifteenth centuries known as portolans (from an Italian word for "sailing directions"), even though their authenticity has never been in

The Cortona Chart, one of the earliest known Portolans, drawn around A.D. 1300. The surviving portion shows the eastern Mediterranean and Black seas.

doubt. They came to the attention of modern scholars at the turn of this century and were enthusiastically greeted as "the first true maps." The moniker is well deserved. The surviving portolans (mainly from Portugal, Spain and Italy) are meticulously drawn on sheets of parchment (sometimes bound into atlases) and exhibit remarkable quality and consistency of scale. Instead of latitude and longitude the portolans are crisscrossed with a curious system of superimposed lines, which radiate like spokes from focal points in the sea. The purpose of these lines is uncertain—were they used as an aid for drawing the maps, as straightforward mariner's directions, or do they relate to the use of the compass? In scope the portolans generally concentrate on the Mediterranean area, but some range as far as the Black Sea to the east and Britain and the Atlantic islands to the west.

The outlines of the Mediterranean coast shown on an early portolan dating from A.D. 1311 were not improved on until the eighteenth century. Viewed side by side, the portolans and the naive *mappae mundi* look like the products of totally different ages—yet our copies of both come from the Middle Ages. The portolans drew on an entirely separate tradition of medieval mapmaking, one fostered by practical seafarers rather than the Church. But where did this tradition come from? Unlike the *mappae mundi*, whose development can be clearly traced from late Roman times through the Dark Ages, the portolans appear, as if out of the blue, in a fully developed state around A.D. 1270. Many of the later versions are slightly updated copies of their predecessors, and it has been argued that the entire corpus derives from a few key originals.

Who, then, could have made the original portolan maps? A number of suggestions have been made, including the Knights Templars (a secretive order of Crusaders), the Chinese, the ancient Phoenicians and even the megalithic mariners of prehistoric Europe (see **Ships and Liners**). None of these more exotic solutions is backed by any sound evidence, though some impetus may have come indirectly from the Chinese: They invented the compass, which was introduced by Arab traders into Europe during the late twelfth century (see **The Compass**).

On the other hand, there are intriguing clues that draw the investigation back to the classical world. One of the finest of the portolans was drawn by one Ibn Ben Zara, a mapmaker from Alexandria.

In its corners are four faces that look like miniatures from the Christian art of Greece and Egypt of the seventh and eighth centuries A.D. Could Ibn Ben Zara have copied his map from an original made during the early Byzantine Empire? The Turkish admiral Piri, who also made portolan-style maps, stated clearly that among the sources he used were maps dating back to the time of Alexander the Great (336–323 B.C.).

Both clues suggest an input into the portolan tradition from the Greek mariners of the classical world, a possibility seriously contemplated by many specialists in the field. Though none has survived, many ancient Greek geographers refer to the sea charts used by navigators. Alexandria, the scientific capital of the Greek world, was a likely center for the production of complex sea charts. It is conceivable that some examples could have survived the destruction of its great library (see **Introduction** to **Communications**) and come into the hands of Mediterranean mariners during the thirteenth century A.D. During the same century a copy of Ptolemy's *Geography* was recovered by scholars of the late Byzantine Empire from the Arabic Near East, where it had been preserved throughout the Dark Ages.

We should not underestimate the contribution of the medieval navigators who prepared the surviving portolans. But as with many other aspects of medieval knowledge, perhaps it was the find of an ancient Greek mariner's chart that sparked a renaissance in Western cartography during the Middle Ages.

SKIS AND SKATES

In the frozen wastes of northern Europe and Asia, skis and skates have been the preferred ways of crossing land and ice for thousands of years. As the first skis were made of wood—which can easily rot away—the chances of finding examples in archaeological excavations are very slim. Despite this a remarkable discovery was made in a peat bog at Vis, near the Ural Mountains in northeastern Russia, in the 1960s. The acids in the peat had preserved several fragments of Stone Age skis, made no less than eight thousand years ago. One is the front end of a ski, superbly carved in the shape of an elk's head, which would have acted as a "brake" as well as symbolizing rapid movement.

For our earliest picture of a traveler using skis, we have to turn to the tiny island of Rødøy, in the far north of Norway. Here a rock carving, fashioned around 2500 B.C., shows a figure wearing skis about twice his own length, propelling himself along with a single pole. Depicted in an approved skiing stance—leaning forward with knees bent—the "Rødøy Man" seems to be wearing a costume with harelike ears, maybe to give him luck in hunting.

By the seventh century A.D. skiing had reached Central Asia and had even become known in China. In the year 640 the Lui-Kuei tribe of nomads, from the Lake Baikal area, sent representatives to the Chinese court, where official records said about them: "As their country is so quickly covered with frost and snow, they use wooden boards six inches wide and seven feet long with which to glide over the piled-up ice and go hunting the deer and other animals."

In Scandinavia skiing came to have a military value, and spies on skis were used by King Sverre, of Norway, at the battle of Isen, near Oslo, in A.D. 1200. Six years later one of the most dramatic incidents in Norwegian history took place when the infant king, Haakon Haakonson, was spirited away from a raging civil war by two body-guards on skis. Their loyalty is still commemorated today by an annual race following the thirty-five-mile mountain escape route they took.

Carved ski head in the shape of an elk found at Vis in northeastern Russia (side view). The head would have faced into the snow, helping to stabilize the skier; more importantly it would have hooked onto packed snow on slopes and acted as a brake.

The Rødøy Man, a Norwegian rock carving of a figure on enormous skis, dating to around 2500 B.C.

Skating on Bones

Early skates were mostly smaller versions of skis, still requiring poles for propulsion. They were made of bone, from 1000 B.C. up to medieval times. About A.D. 1120 the Arab traveler Sharaf Al-Zaman visited the Yura people, living north of the River Volga in southern Russia. He was greatly impressed by their strange but practical method of transport: "It is impossible for a man to go over these snows, unless he binds on to his feet the thigh-bones of oxen, and takes in his hands a pair of javelins which he thrusts backwards into the snow, so that his feet slide forwards over the surface of the ice; with a favorable wind he will travel a great distance by the day."

Skating had reached England as a sport by the late twelfth century. Its popularity with the youth of London is vividly described by the chronicler William FitzStephen:

> Others, more skilled at winter sports, put on their feet the shin-bones of animals, binding them firmly round their ankles, and, holding poles shod with iron in their hands, which they strike from time to time against the ice, they are propelled swift as a bird in flight or a bolt shot from an engine of war. Sometimes, by mutual consent, two of them run against each other in this way from a great distance, and lifting their poles, each tilts against the other. Either one or both fall, not without some bodily injury, for, as they fall, they are carried along a great way beyond each other by the impetus of their run, and wherever the ice comes in contact with their heads, it scrapes off the skin utterly. Often a leg or an arm is broken, if the victim falls with it underneath him; but theirs is an age greedy for glory, youth yearns for victory and exercises itself in mock combats in order to carry itself more bravely in real battles.

Even when modern skates were invented, using an iron blade slotted into the sole of a wooden clog, skating didn't become safe. A fifteen-year-old Dutch girl named Liedwi was knocked down and badly hurt by another skater in the winter of 1395. She retreated to a convent and dedicated herself to religious work until her death in 1443, as a result of which she came to be the patron saint of skaters.

The skating accident of Saint Liedwi, patron saint of skaters—as shown in a biography of her written in A.D. 1498, about a century after the incident.

ODOMETERS

An essential part of every automobile, the odometer must be one of the single most widely used mechanisms in the world. As we can see from their passion for maps, the ancients were just as curious as we are to know the exact distances between destinations. Perhaps not surprisingly, then, several designs for odometers were developed in ancient and medieval times.

A Chinese book from about A.D. 300 discusses a device for measuring the distance of journeys, stating that a description of its workings can be found in a scientific work written under the early Han Dynasty. Unfortunately this early work is now lost. Much later the *Sung Shih*, or court records of the Sung Dynasty (A.D. 960–1279) describe in some detail the manufacture of a vehicle known as "*li*-recording drum carriage" (a *li* being the Chinese equivalent of a mile).

The first of these carriages was built by engineer Lu Daolong in A.D. 1027. When the carriage started, the main wheels set in motion a system of gears made of toothed wheels arranged in a reduction chain. This scaled down the large distances covered by the wheels of the carriage into much smaller movements at the top of the device, so that the last axle in the system revolved only once when the whole engine had traveled one *li*. The sequence ended with a catch that activated the arms of a mechanical wooden figure to strike a drum as each *li* passed. The addition of further geared wheels meant that another wooden figure would strike a bell after every ten *li*. Despite some problems with the mathematics, Wang Zhenduo, a specialist in ancient Chinese science, succeeded in making a working replica during the 1950s by following the original texts.

The Roman Odometer

But the oldest surviving odometer design, from the works of the Roman architect Vitruvius, managed to defy interpretation until 1981. Writing in the early first century B.C., Vitruvius introduced the *hodometer* (from Greek *hodos*, "way" and *metron*, "measure") as a device "which enables us, while sitting in a carriage on the road . . . to know how many miles of a journey we have accomplished." Vitruvius went on to describe the hodometer's gear system and how the passage of miles was shown by a device at the top of the carriage: as each mile passed, a stone dropped into a box with a clang; at the end of the journey the box was opened and the stones counted to reveal the distance covered.

On paper Vitruvius's description seems to make sense. His main gear was a wheel with four hundred teeth turned by a single tooth attached to the main wheel of the vehicle. Thus four hundred actual rotations were scaled down to one movement of the recording mechanism. But problems arose when attempts were made to translate Vitruvius's description into a practical design. Even the great Renaissance engineer Leonardo da Vinci (A.D. 1452–1519) had difficulties. Leonardo left two sketches of the hodometer, adapting Vitruvius's account to the measurements of his own day. For simplicity's sake one of his drawings shows the main gearwheels in the machine with forty teeth as opposed to the four hundred given in the Roman text. As such, his reconstruction provided a plausible working model. But, as he seems to have realized, such a device could not possibly work with as many as four hundred teeth—even if the wheel were six feet in circumference, the teeth would still be so small that they would barely be able to mesh with the single-toothed wheel. Leonardo solved this part of the problem in his second hodometer drawing, by interpreting the "single tooth" as an endless screw or worm gear (which is rather like an elongated tooth wrapped around a cylinder).

The trouble with Leonardo's model, though it would have worked, is that after all his modifications it bore little resemblance to the gear system described by Vitruvius. In 1963 Professor A. G. Drachmann, a classical scholar and expert on ancient mechanics, was also stumped and declared that the problems were insoluble. Vitruvius's hodometer

The hodometer of Vitruvius, as reconstructed by engineer André Sleeswyk. The single-toothed gear attached to the hub of a carriage wheel advances the vertical gear by one step with each rotation. A similar arrangement moves the horizontal gear by one step for every four hundred rotations of the carriage wheel, equivalent to a distance of five thousand feet, or one Roman mile. (For clarity fewer than four hundred teeth are shown on the gearwheels.) The holes in the top hold round stones; with every mile one pebble drops through a slot into a box (behind the vertical gear), giving a record of the distance traveled.

was relegated to being an ancient Roman "armchair invention," which could never have been built or used.

This does seem rather unfair to old Vitruvius. The other complex mechanisms he described (see **Watermills** and **Windmills** in **Working the Land**, and **Keyboards** in **Sport and Leisure**) are known to have been manufactured by the hundred. Moreover another, though much simpler, hodometer design was described by the Greco-Egyptian engineer Heron of Alexandria, a century after Vitruvius. There is thus no good reason to doubt that the Roman hodometer existed as a working device. This was the approach taken by engineer André Sleeswyk, who published a new discussion of the problem in 1981. Vitruvius clearly stated that the device he was describing was "transmitted [to us] by our predecessors," suggesting that he was describing a device that he has aware of but had probably never examined at first hand. Bearing this in mind, Sleeswyk started again from scratch. The teeth of the ancient Greek computer from Antikythera (see **Computers** in **High Tech**) were not the square-shaped ones familiar to Leonardo, but triangular, pointed ones. By using teeth of this shape, adding a deep notch to the single-toothed wheel and placing it at a ninety-degree angle to the four-hundred-toothed wheel, Sleeswyk solved the problem. Adding a second single tooth to the large wheel enabled it to drive a third, horizontal, wheel on top of the device.

Sleeswyk then made a quarter-scale working model following the other details given by Vitruvius. He drilled a series of holes through the horizontal wheel on top to hold ball bearings. When a hole passed over a slot every quarter of a mile, a ball dropped through. Vitruvius's hodometer could be made to work, without taking any liberties with his design.

A final problem was to identify the original designer of the machine that Vitruvius stated was passed down from earlier times. Indeed what use would the hodometer have been in Italy by his own time, when the highways had already been marked with milestones for a couple of hundred years? The answer may lie in a medieval Arabic manuscript that describes a similar pebble-dropping mechanism in a water-clock design attributed to the great Archimedes (287–212 B.C.). The greatest engineer of the classical world, Archimedes was certainly capable of producing the gearing system involved. He was the leading scientific adviser to the king of Syracuse, in Sicily, a close ally of Rome for thirty-six years of Archimedes' working career (see **The "Claws" of Archimedes** in **Military Technology**). During this very period the network of Roman roads in Italy was being completed. It is possible that Rome commissioned this great scientific genius to help them in their systematic road-building program, replete with milestones, by devising a means of accurately measuring distances. Sleeswyk's case is built on circumstantial evidence but is certainly persuasive.

WIND CARS AND ROCKET CARS

The idea of harnessing a source of energy more reliable and powerful than a horse or donkey to propel vehicles is an attractive one, and medieval engineers devised various ways of achieving this end. The oldest and most practical was to harness the power of the wind. Not surprisingly this was the invention of the Chinese, presumably born of their long tradition of kite flying.

The oldest reference to "land sailing" is in the *Book of the Golden Hall Master*, by the noted scholar Emperor Luan, who reigned in A.D. 552–554. He wrote that Kaots'ang Wu-Shu, a famous philosopher, constructed a carriage driven by the wind that could carry

Vignettes of sail car on a map of China drawn in 1577 by the Portuguese cartographer Luis Jorge de Barbuda.

thirty men and was capable of traveling hundreds of miles in a day. A much larger version, which was pulled by horses but also had sails, was supposedly built for Emperor Yang, of the short-lived Sui Dynasty, in A.D. 610. This was claimed to have had room for several thousand passengers, which must be an exaggeration but does point to the existence of smaller versions. Much more common were sails added to wheelbarrows and even to plows to make them easier to push. In the far north of China wind power was adapted to a different environment by fixing sails onto ice yachts, which ran across the ice on little wheels.

Toward the end of the sixteenth century European travelers to China brought home reports of sail cars, which became an overnight sensation and were widely copied and talked about. The great English poet John Milton, for example, writes in his masterpiece *Paradise Lost* (1663) of "Sericana, where Chineses drive With Sails and Wind their canie Waggons light." Sail cars managed speeds of thirty miles an hour, faster by far than the first steam railways of the early nineteenth century. This appears to have given some people the idea of wind-powered railways, but because no alternative power source existed for use on those days when the wind failed or was in the wrong direction, this idea never caught on.

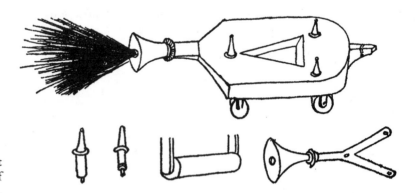

Giovanni di Fontana's rocket car, from a manuscript of c. A.D. 1425.

However, there were other notions of how to power a vehicle without the aid of animals. Giovanni di Fontana, the Venetian engineer, gave two alternatives in a manuscript of c. A.D. 1425. One was a "self-driving" carriage in which the driver pulled on a rope connected by cogwheels to the wheels of the car. This would hardly have been practical, as the rope would soon fray—if the arms of the driver didn't give out first. His other design, which bears a passing resemblance to the Batmobile, was for a jet-propelled car. The car was to be propelled along on its two rollers by a gunpowder-based fuel, whose exact recipe di Fontana kept a secret. We don't know whether any foolhardy engineer actually tried to build di Fontana's rocket car, but the driver would undoubtedly have wished he'd never agreed to have anything to do with it.

SHIPS AND LINERS

Oddly enough the first evidence of sea travel comes from a time tens of thousands of years before the earliest remains of boats. During the last Ice Age, some forty thousand to ten thousand years ago, Stone Age hunters colonized every continent except Antarctica. While lower sea levels meant that the Americas could be reached by a land bridge between Siberia and Alaska, Australasia was a different matter. Although the Australian continent extended to the north as far as New Guinea during the Ice Age, it had been separated by water from Southeast Asia for millions of years. Yet the latest archaeological evidence confirms that the first settlers reached New Guinea and Australia as long ago as 40,000 B.C. There is thus no alternative to accepting that, even this early, mankind had already taken its first steps toward mastering the sea.

This makes the boat one of the earliest of all human inventions. What kind of craft the first Australians used to reach their future home can only be guessed at, but it must have been quite sturdy. While part of the journey involved relatively easy "island hopping," stretches of fifty to sixty miles across open sea also had to be covered. Current thinking is that fishermen on the coast of Southeast Asia discovered that attaching a hollowed-out bamboo log to several spars could give both the stability and the maneuverability needed to tackle the open sea. The principle is exactly the same as that used in building modern catamarans.

There are a few other glimpses of prehistoric sea travel before we come to the actual remains of boats themselves. Regular contact by sea has been detected between sites around the Aegean Sea. At the Franchthi Cave, in southern Greece, archaeologists have discovered quantities of obsidian in several successive levels dating from the eleventh millennium B.C. onward. Obsidian is a volcanic glass that was highly valued in prehistoric times (see **Razors, Jewelry and Mirrors** in **Personal Effects**). Laboratory analysis of the Franchthi obsidian showed that it came from the island of Melos one hundred miles away (see map, p. 55). This sea trade, the earliest yet demonstrated anywhere in the world, continued and expanded over the next few millennia, as finds of Melian obsidian at sites in northern Greece show.

Megalithic Mariners

On the other side of Europe, deep-sea fishing was being developed by about 5000 B.C. Archaeologists working at Stone Age sites on the coasts of Ireland and Scotland, such as Oban, in Strathclyde, have discovered large amounts of fish bones belonging to cod and other species that rarely come close to the shore. Boats must have been used to reach these deepwater fish. These vessels were most likely made of animal skins tightly sewn together, since the excavation of Oban, for example, produced no sign of woodworking tools.

The first tangible remains of boats also come from western Europe. The earliest known paddle, dating from around 8500 B.C., was excavated at Star Carr in Yorkshire. From about a thousand years

The dugout canoe from Pesse, Netherlands, dating to c. 7400 B.C. (ten feet long).

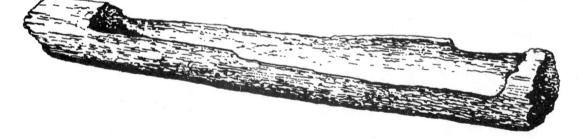

later comes the oldest surviving boat, a simple dugout carved from a pine log, discovered at Pesse, in Holland.

This early European evidence of boatbuilding—from the Aegean traders in obsidian to the Scottish fishermen—belongs to the Mesolithic, or Middle Stone Age. The New Stone Age which followed saw the beginnings of agriculture, accompanied by an explosion of technological and social organization. The most fascinating development was the mushrooming of the megalithic cultures along the western seaboard of Europe after 4500 B.C. In Portugal, Spain, France, Ireland, Britain and southern Scandinavia, similar societies began to erect large structures of rough-hewn stone for burial and ritual activities. This monument building culminated in the famous alignments and circles of standing stones—Stonehenge being the classic example—whose purpose was doubtless astronomical in part but whose full meaning still eludes archaeologists (see **Calendars** in **Communications**). The distribution of megalithic architecture, concentrating in coastal areas, has naturally encouraged speculations about a great maritime civilization that once dominated the Atlantic. The legend of lost Atlantis is attractive, but there is unfortunately not a shred of hard evidence to support the idea that such a landmass ever existed.

All the same the common elements of the megalithic cultures must have been spread by sea routes: the archaeological evidence demonstrates contact between areas as far apart as Portugal and Ireland. The boats used by the megalithic mariners were almost certainly made of leather. Later Celtic tribes of the region were renowned for their skin boats, such as the round one-man coracles used by river fishermen and seen by Julius Caesar when he invaded Britain in 55 B.C. The Romans were amazed that the Celts braved the stormy seas of the Atlantic in such apparently flimsy craft. But the fact is that stitched leather, stretched over a wooden frame, can make an extremely robust, seaworthy vessel. It also had the advantage of lightness—a leather boat forty feet long with a capacity of two tons can be easily carried by two men.

A longboat made of leather was the preferred sea transport of the Celts of Ireland. The *curragh* (Irish for "coracle"), as it was known, was the vessel used by the seafaring Irish monks of the early Middle Ages in their wide-ranging Atlantic voyages. They certainly reached the Shetland Islands and Iceland, while the legend of Saint Brendan

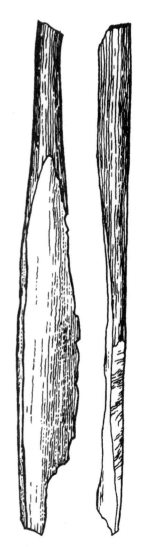

Blade of the paddle found at Star Carr, England, c. 8500 B.C. (seventeen inches long).

tells how he and an intrepid band of monks made a lengthy voyage in a *curragh* around the Atlantic, reaching mysterious lands far to the west. Some believe that the descriptions given in the medieval account of Saint Brendan's voyage suggest they may even have reached the New World during the seventh century A.D. The possibility was tested by explorer Tim Severin in 1978 in an extraordinary archaeological experiment. Using pre-industrial-age techniques, he and his team constructed a *curragh*, the *Brendan*, and used it to cross the Atlantic. When they touched down in America, both the crew and the ship were in good shape.

The skin boats of megalithic and Bronze Age times were much like these later *curraghs*, to judge from engravings and rock drawings found in Denmark (for an illustration, see **Razors** in **Personal Effects**). Other Danish rock drawings show what must be large dugouts. It is unclear when the first plank-built boats were made on the Atlantic coast, but the Celts, always master seafarers, excelled at their construction. The massive ships of the Veneti tribe made a lasting impression on Julius Caesar when he reached the coast of northwestern France in 56 B.C. Made of solid oak with huge leather sails, they could easily withstand the most violent Atlantic storms. The cross-timbers, Caesar noted, were beams "a foot wide, fastened with iron bolts as thick as a man's thumb," while their bows towered over the Roman galleys: "We could not injure them by ramming because they were so solidly built, and their height made it difficult to reach them with missiles or board them with grappling-irons."

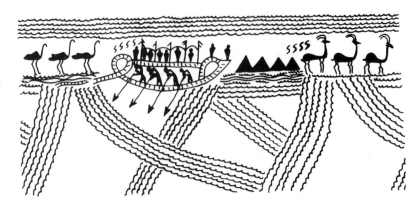

Early Egyptian ship painted on a pot, c. 3100 B.C. The triangular shapes in the background are hills, as the picture long predates the building of the pyramids.

The First Navies

The achievements of prehistoric mariners make some of the great civilizations of the ancient world seem like Johnny-come-latelies. The Egyptians, for example, were a basically landlocked people, most of their experience with boats being restricted to the Nile Valley. Even so, rock drawings from the Red Sea site of Wadi Hammamat show that they were exploring the coasts by the fourth millennium B.C. The boats depicted were probably made of papyrus or reed bundles, the standard materials for Egyptian river-craft construction over the next few thousand years.

One of the ships carved on the wall of the mortuary temple of Pharaoh Sahurê at Abusir, in Egypt, about 2450 B.C. The presence of non-Egyptian sailors (with long hair and beards) suggests this may have been a seagoing vessel, returning from a voyage to Palestine.

Though behind in some other respects, the civilizations of the ancient Mediterranean seem to have invented the art of building boats from planks. Egypt never disappoints when it comes to spectacular archaeological discoveries, and the world's earliest known plank-built ship was discovered right next to the Great Pyramid. Around the pyramid, built as the tomb of Pharaoh Cheops (about 2600 B.C.), five enormous pits were dug to house ceremonial riverboats for the dead king and then closed with stone covers. One was still completely airtight and watertight when it was opened in 1952, and the ship it contained was in an almost perfect state of preservation— dismantled into 1,224 component parts made from cedar and sycamore wood. Careful reconstruction restored it to its original splendor, a vessel 142 feet long and 19 feet wide, displacing some forty tons.

The Egyptians excelled at organization as well as craftsmanship and have left us the earliest evidence of state-organized navies. From about 2300 B.C. their texts speak of large cargo ships for traffic along the Nile, and seagoing vessels sent by the pharaohs to collect cedar and other products from the coast of Lebanon. Marine forces played an important role in the Egyptian conquest of Palestine and Syria during the Eighteenth Dynasty (beginning around 1550 B.C.). The pharaohs of this dynasty also launched large-scale expeditions for peaceful purposes. Queen Hatshepsut, one of the few women to rule Egypt as a pharaoh in her own right, sent a small fleet to bring back whole frankincense trees from the remote land of Punt on the east coast of Africa (see **The First Suez Canal**, and **Perfume** in **Personal Effects**). A relief at her great mortuary temple depicts how two giant obelisks of about 350 tons each were transported down the Nile from the stone quarries at Aswan on an enormous barge: it has been calculated that the barge must have been well over 200 feet long and 70 feet wide, requiring a fleet of over thirty oar-powered tugs to pull it.

The Chinese were also late starters in ship construction, but, like the Egyptians, they made up for lost time through their organization and ingenuity. As we know from court records, the southern Chinese state of Wu had developed a navy of no less than ten thousand men by the late sixth century B.C. The Guangzhou shipyard of the late Ch'in or early Han Dynasty (late 3rd to early 2nd century B.C.) had three building platforms capable of holding wooden ships 262 feet long and 98 feet wide, with a weight capacity of fifty-nine tons. (For comparison the *Mayflower*, which brought the Pilgrim Fathers to America in 1620, was about 60 feet long and 26 feet wide.) By A.D. 260 even larger oceangoing junks appeared.

This period also saw the invention of the rudder. There are a number of literary clues pointing to its early development in China, confirmed in 1958 by the discovery of a number of detailed pottery ship models in tombs at Canton. The models, which clearly show axial rudders, date from the first to second centuries A.D., a thousand years earlier than the earliest European representation of a rudder. The Chinese also introduced the compass to the Middle East and Europe (see **The Compass**).

By the Middle Ages the Chinese undoubtedly had the largest navy in the world. The great fleet commanded by Cheng Hô, grand eunuch of the imperial court under the Ming Dynasty, left China in the year 1405 carrying 28,000 men in 62 four-decked ships 440 feet long and 184 feet wide. It traveled Southeast Asia collecting tribute—from peoples overawed by the might of the Imperial fleet—for two years before returning to base. By 1420 the Ming navy boasted 250 long-distance galleons, 400 large warships, 400 grain transport freighters, 1,350 patrol vessels and 1,350 combat ships attached to coast-guard stations.

Ancient Superfreighters

The most extravagant ships ever built in the ancient world were those of the Hellenistic Age (3rd to 1st centuries B.C.), when Greek culture and technology spread throughout the Mediterranean and Middle East. Then, as in modern times, the Greeks were renowned shipbuilders.

In the last quarter of the third century B.C., Ptolemy IV, one of the Greek dynasty which ruled Egypt before the Romans, was famous for the magnificent ships he commissioned. Whereas the typical Greek war galley (or trireme) was never much longer than 120 feet, Ptolemy built a warship that was 420 feet long and 57 feet wide, with sides 72 feet high. On a trial voyage it took a crew of 4,000 to man the forty banks of oars, with 400 rowers in reserve. Manning the deck were 2,850 marines. While undoubtedly invincible, it was also so massive that it was extremely difficult to maneuver.

Detail of a Roman sailing vessel— from a mosaic in a house at Rome, c. A.D. 400.

After the enormous struggle of launching it for its trial run, it stayed as a showpiece at its moorings, except when it had to be taken in to a specially built dry dock for repairs. The ship never saw active service, though it must have served its purpose as a deterrent.

Ptolemy's contemporary, King Hiero II of Syracuse, in Sicily (270–215 B.C.), had an equally ambitious ship built under the direction of Archias, the leading Greek naval architect. Named the *Syracusa*, it took a year to construct, using materials imported from as far afield as Spain and Germany and consuming enough timber to build sixty normal ships. With twenty banks of oars and three masts, it was a curious mixture of armored luxury liner and superfreighter (for exporting wheat) and could carry a cargo of up to 1,800 tons. Its armaments included cranes to drop stones on, and grapple with, enemy vessels, and a giant catapult designed by Hiero's scientific adviser, the great Archimedes. He also undertook the difficult task of launching the ship (see **The "Claws" of Archimedes** in **Military Technology**). The weapons were manned by two hundred marines stationed in eight armored turrets, packed with stones and other missiles, to ward off any pirates who weren't simply overawed. The figure for the full complement of sailors and galley slaves is missing from our main ancient source, the later Greek writer Athenaeus (see **Cookbooks** in **Food, Drink and Drugs**)—but there were enough to justify its own on-board court, presided over by the captain and other officers.

Few ships can ever have been as lavishly equipped as the *Syracusa*, described in glorious detail by Athenaeus. Its freshwater tank held twenty thousand gallons, next to which was a seawater tank to keep fresh fish. There were twenty stalls for horses, huge kitchen areas provided with flour mills and ovens, and numerous storerooms for wood, food, animal fodder, tools and weapons. The officers' decks had a bathroom with three bronze tubs, a gymnasium, a shrine to the goddess Aphrodite with a floor of semiprecious stones and doors made of ivory and cedar, walkways with grapevines and flowerbeds and even a library with couches and walls inlaid with boxwood. The floors, walls, ceilings and furniture of the main gangways and "first class" accommodation were decorated with elaborate mosaics and carvings illustrating the entire story of Homer's epic poem of the Trojan War, the *Iliad*.

The dimensions of this monstrous liner are not actually given by Athenaeus, but it was so enormous that only one of the harbors of Hiero's trading partners throughout the Mediterranean was actually equipped to receive it. Hiero was said to have realized this only after the ship's completion. As it happened, his superliner made only one voyage—from Sicily straight to Alexandria, where, filled with grain, it was given as a present to King Ptolemy, who immediately grounded it. The *Syracusa* was a naval dinosaur. By the mid-second century B.C. the fashion for ships on such an absurd scale had run its course, and Mediterranean boat builders returned to more practical designs.

Paddleboats

The naval architects of the Greco-Roman world did, however, produce one last creative flourish before the Dark Ages descended on Europe. This was the invention, albeit on paper, of the paddleboat. Paddle-wheel devices were well known to the Romans, such as those powered by water to drive mills (see **Watermills and Windmills** in **Working the Land**) and a theoretical version of the odometer for naval use described by Vitruvius. If the principle were reversed, paddle wheels driven by muscle power could propel craft through water. An anonymous Roman manuscript written about A.D. 370 entitled *De Rebus Bellicus* ("On Military Matters") describes a number of fanciful devices, designed to save the Roman Empire from the growing tide of barbarian invasions. One was a plan for a ship with six paddle wheels, each powered by three oxen walking around a treadmill.

The thirteenth-century English inventor Roger Bacon developed the idea further, but it was not until much later that the idea was actually put into effect in Europe. In 1543 small tugs with paddle wheels were built in Barcelona, each powered by forty men working a treadmill. These eventually gave rise to the steam-powered version, used by the British navy against the Chinese in the 1840s. But the Chinese also had paddleboats. The British were surprised that they had apparently copied their design so quickly, but the fact is that the Chinese had been building paddleboats for at least a thousand years.

Reconstruction drawing of a Chinese military paddleboat, c. A.D. 1135. This would be the largest type, with twenty-two paddlewheels and a stern-wheel. It would have carried some three hundred marines and crew.

They were probably invented in the fifth or sixth century A.D. The earliest certain evidence is a record concerning Li Kao, governor of Hungchao, in A.D. 783: "Li Kao, always eager about ingenious machines, caused naval vessels to be constructed, each of which had two wheels attached to the side of the boat, and made to revolve by treadmills. These ships moved like the wind, raising waves as if sails were set."

In the early twelfth century a naval arms race set in during a protracted civil war. Much of the fighting centered on the control of strategic lakes and rivers, for which paddleboats, powered by the legs of their crew and not dependent on wind, were ideal. The size of wheel ships increased until they were two hundred to three hundred feet long. They carried seven hundred to eight hundred men apiece and were propelled by more than twenty wheels. By the end of the war both sides had thousands of these giants, but with the end of hostilities paddleboats went out of favor.

THE FIRST SUEZ CANAL

When Napoleon conquered Egypt in 1798, his dream was to build a canal connecting the Mediterranean and the Red Sea. This would have enabled him to undermine the economy of the British Empire by forging a direct route to the riches of India and the Far East.

Even if he hadn't been driven out of Egypt by the British navy under Nelson, it is unlikely that Napoleon's plan would have been realized. His scientific advisers insisted that the undertaking would require so many man-hours that it could never be completed. What is more, they said, the plan was incredibly dangerous. If such a project were undertaken, they argued, the Egyptian Delta would be drowned by seawater. Napoleon's surveyors had miscalculated the height of the two bodies of water, concluding that the Red Sea was several meters higher than the Mediterranean.

These fears about the viability of the project continued well into the nineteenth century. They were only dispelled when the brilliant French engineer Ferdinand de Lesseps, who had spent twenty years bringing the project to fruition, completed the Suez Canal in 1869.

A Triumph of Ancient Engineering

The ancient Egyptians don't seem to have been unduly bothered by such problems. As it happened, the Persian emperor, Darius, who ruled Egypt between 522 and 486 B.C., was warned by some of his advisers of exactly the same danger as Napoleon. Nevertheless Darius's engineers went ahead and built the canal, as we know from several inscriptions that proudly record its construction. One of these reads, "I commanded this canal to be dug from the river Nile which flows in Egypt, to the sea which goes from Persia. This canal was afterwards dug as I had commanded, and ships passed from Egypt through this canal to Persia as was my will."

We can only conclude that, in this respect at least, the geographical and engineering knowledge of the ancient Egyptians was more

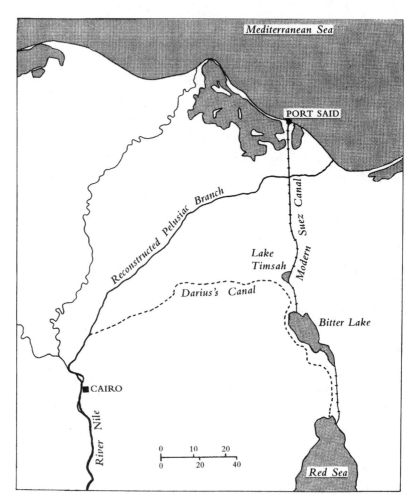

Route (broken line) of the Suez Canal as it was in the time of the Persian ruler Darius (522–486 B.C.). First built by the pharaohs at an unknown date, the ancient canal utilized a natural waterway, the Wadi Tumilat, which flowed eastward from the Nile. Reaching Lake Timsah, it then took the same route to the Red Sea as the modern canal.

sound than that of Napoleon's time. The canal of Darius was wide enough for two large galleys to pass each other. For the first half of its route, going from the Red Sea northward, it followed much the same course as the modern Suez Canal. For the second half it forked westward, using a natural waterway and then the river Nile itself to complete the journey to the Mediterranean.

Not all the credit should go to Darius, however. It seems that his workmen were really only completing and repairing the work of much earlier kings. The Egyptians themselves said that the first canal was built by the great conqueror Sesostris, a legendary figure whose identity and date are uncertain. But we do know that by 1470 B.C. such a waterway already existed. It was at that time that Queen

Hatshepsut sent her expedition to the mysterious land of Punt on the east coast of Africa (see **Ships and Liners**). The magnificent reliefs carved on her temple at Deir el-Bahri (Thebes) depict the voyage and make it clear that her navy was able to sail all the way from the Nile to the Red Sea without breaking its journey. Whoever began this ancient canal, it was an engineering achievement predating the modern Suez Canal by at least three and a half thousand years.

The ships of the Egyptian queen, Hatshepsut, on their way to the Land of Punt, in Africa. The fish, depicted in great detail on her reliefs, are typical of the Red Sea. This shows that Hatshepsut's navy sailed directly from the capital of Thebes on the Nile to the sea and that the ancient Suez canal already existed by this date, about 1470 B.C.

Over time the shifting sands of the desert encroached on the artificial waterway. Repairs seem to have been carried out by the great builder Pharaoh Ramesses II around 1250 B.C. The next to undertake the project was Necho (610–595 B.C.), one of the last great native rulers of Egypt. Necho was a remarkably ambitious ruler, who dragged Egypt out of the doldrums and transformed her into a great international power once more. He was also the first to send a navy to sail around the African continent—a feat of navigation not repeated for another two thousand years, until Vasco da Gama's voyage in 1498. But it is doubtful whether Necho actually completed his canal. According to the Greek historian Herodotus, 120,000 people lost their lives during the construction work, and the task had to be abandoned.

Necho's project was eventually finished by Darius, and for many years ships plied their way between the Mediterranean and the Orient. By the time of the Greek dynasty of the Ptolemies, it had silted up again and become unnavigable. The second Ptolemy (285–246 B.C.) was responsible for the construction of the spectacular Pharos lighthouse (see **Lighthouses**), as well as the transformation of Alexandria into the scientific and technological capital of the ancient world (see **Introduction** to **High Tech**). To revive Egypt's trade, he ordered the complete refurbishment of the ancient canal.

One intriguing possibility is that Cleopatra, a descendant of Ptolemy and the last queen of ancient Egypt, may have taken one of her pleasure cruises along a canal that the "experts" of early nineteenth-century Europe had denounced as a scientific and economic impossibility.

The Prophecy of Disaster

A lasting irony concerns the only worry that the ancient Egyptians *did* have about linking the Mediterranean and Red seas. Another reason why Pharaoh Necho is said to have abandoned his work on the canal is that the gods had told him of some dire consequences. According to the Greek historian Herodotus, "an oracle warned him that his labor was all for the advantage of the 'barbarian'—as the Egyptians call anyone who does not speak their language."

This ancient prophecy came true—more than once. Ptolemy's canal helped Egypt develop commercially into the rich prize that drew the Roman invaders in 30 B.C. The Romans fully exploited the sea route to Arabia and India, and the old canal was rebuilt by the Emperor Trajan in A.D. 98. A special police force was established to patrol the canal, and a navy was founded to protect the lucrative Red Sea trade against piracy.

But the prophecy had not yet run its course. It was fulfilled again—with a vengeance—when the modern Suez Canal was built. Immediately after the building of de Lesseps's canal, the astute western politicians Disraeli and Napoleon III bought the bulk of the shares in the project for their countries. Foreign interests—first commercial, then military—rapidly staked their claims on the canal, and Egypt was turned into a virtual colony of Britain and France. It only achieved real independence from Britain in 1936. The greedy colonialist scramble for control of the canal was finally ended when President Nasser expelled the British army from the canal zone in 1956.

THE COMPASS

The great sea voyages of Europe's "Age of Discovery," such as Christopher Columbus's trip to the West Indies in 1492, would have been inconceivable without the use of a compass. This supreme aid to navigation was introduced to the West from China, via the Arab traders of the Middle East, during the thirteenth century A.D. So much is generally acknowledged. What is less commonly appreciated is the great antiquity of the Chinese acquaintance with the compass.

Two thousand years ago the Chinese had already developed a primitive working compass. A chunk of lodestone (a naturally occurring magnetic iron ore) would be carved into the shape of a ladle. Placed on a stone board with a smooth, polished surface, it would swivel until the "handle" pointed south (the bulk of the ladle being

Reconstruction of a *sinan*, the earliest form of the Chinese compass, from the Ontario Science Center, Toronto. The south-pointing ladle is carved from lodestone and swivels on a polished plate of bronze.

attracted to magnetic north). A ladle shape was chosen in order to represent the constellation of the Great Bear (or Big Dipper)—as is well known, the two stars on the left of this constellation are aligned with the North Star, the main astronomical aid to navigators. This curiously shaped device, called a *sinan*, is mentioned in a book from about A.D. 80, with other references possibly dating as far back as the fourth century B.C. Jade miners traveling enormous distances in search of their raw material (see **Jewelry** in **Personal Effects**) would take a *sinan* with them in order to keep their bearings.

The jade connection is significant. Lodestone is a hard material—though not as hard as some jade—and the manufacture of a magnetic "ladle," as well as a highly polished stone board, would have required considerable skill. This suggests that the compass was invented by the jade craftsmen of ancient China. Nevertheless the considerable grinding and pounding needed to shape a *sinan* from lodestone would have deprived it of much of its natural magnetism. The care needed meant that few could be produced, while friction with the board on which it was placed, however carefully this was made, would also have reduced the efficiency of the device. These problems, and the growing need for compasses in navigation, led to a search for ways to mass-produce smaller, more reliable, compasses.

Navigation by Needle

Since so much of ancient science was inextricably bound to the study of the occult, it is not surprising that the next major advance in compass technology was attributed by the Chinese to magicians. A work written in A.D. 1086 by Shen Kua, greatest of the medieval Chinese scientists, describes how "magicians rub the point of a needle with a lodestone, making it capable of pointing to the south."

Today most magnets are made by placing a piece of iron inside the field produced by passing an electric current through a wire coil (electromagnetic induction). But until the late nineteenth century the main method remained that used by the ancient Chinese magicians—stroking a piece of iron with a lodestone or another piece of iron that had already been magnetized. The main alternative, without the use of electric current, is to hammer a heated strip of iron pointed toward north; the hammering jiggles the molecules which, as they settle, align themselves in the earth's magnetic field into a north-south direction. Another method is to superheat an iron bar, loosening the alignment of the molecules, and then, holding it with one end pointed due north, rapidly cool it by plunging it in water.

In medieval China all three methods were used to manufacture magnets. Small factories were devoted to their mass production for use in a variety of compass designs (see **Magnets and Magnetism in High Tech**). Among the simplest was the "south-pointing fish" described in an early eleventh-century text—a fish shape was cut from a leaf of iron flat and thin enough to float by surface tension in a bowl of water, in which it turned until its head pointed toward south. Alternatively a larger magnet could be fixed onto a wooden fish and floated in water, or a magnetic needle could be suspended from a thread or even balanced on the thumbnail.

The *Shi Lin Kuang Ji* ("Guide Through the Forest of Affairs"), written in the time of the Southern Sung Dynasty (A.D. 1127–1279), describes the first real prototype of the modern compass—the magnetic turtle. This was a "dry" compass with a fixed pivot, just like

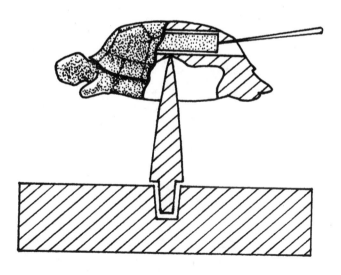

Reconstruction of the earliest known "dry" compass, as described in a Chinese text of the 12th to 13th centuries A.D. North was indicated by the direction of the needle, attached to a magnet concealed inside a model turtle. The model pivoted on a bamboo pin.

a modern compass, the only difference being that the place of a wobbling needle was rather colorfully taken by a model turtle concealing a small magnet mounted on a bamboo pin. Another refinement, developed at about the same time, was to add to the available compass designs (fish, needle or turtle) an underlying board, developed from that of the old ladle type (*sinan*), marked with twenty-four direction points. The result was an easy-reference mariner's compass.

Increasing reliance on the compass for navigation can be seen in Chinese texts from the early twelfth century onward. The earliest description comes from a text written about 1113 concerning events in the port of Kuang-chow (Canton) that took place in the 1090s: "The pilots steer by the stars at night, and in the day-time by the sun; in dark weather they look at the south-pointing needle." Under the Southern Sung Dynasty, the use of compasses became routine on all long voyages. By the thirteenth century the Chinese had explored and charted the waters of East and Southeast Asia, and sailing manuals, known as "needle guides," included precise compass directions.

The sophistication of the medieval Chinese navigators' knowledge can be judged from the fact that they clearly understood that magnetic north is not quite the same as geographical north. This phenomenon, known as magnetic deviation, was only realized in the West in 1492, during Columbus's voyage to the "New World." As early as 1086 Shen Kua, in his description of how magicians made magnetic needles, noted that "the needle often points slightly east of south and not strictly southward." By the time of the Yuan Dynasty (A.D. 1271–1368), the deviation of magnetic north from true north was being clearly marked on compass boards.

China or America?

The earliest Chinese references to the use of the needle as a navigational aid predate by nearly a century the first European mention, found in the writings of the English scholar Alexander Neckam about A.D. 1190. Chinese primacy in the development of the compass at every stage—from the lodestone "ladles" used on land to needles used at sea—would therefore seem to be assured, but for the surprising discovery of evidence that such instruments were actually used many centuries earlier in Central America.

During the mid-1960s excavations were undertaken at the Olmec site of San Lorenzo in the Veracruz region of Mexico. The Olmecs, like their much later successors, the Aztecs, did not know how to cast iron or even smelt it from ores. All the same, the finds from San Lorenzo and other sites of the "Early Formative Olmec" period, dating no later than 1400–1000 B.C., include a number that show a surprising familiarity with iron in its native form. In a premetallurgical age the Olmecs were carving and polishing iron ore into jewelry and even mirrors of high optic quality (see **Mirrors** in **Personal Effects**).

The most enigmatic object from San Lorenzo is a fragment of a neatly carved bar of hematite (a kind of lodestone), about 1¼ inches long, with a groove down the middle. Michael Coe, director of the archaeological project, became intrigued: The well-known magnetic properties of lodestone suggested to him that it might be part of a compass. To test this possibility, he floated it on a piece of cork mat in a bowl of water and discovered that "it consistently oriented itself to the same direction, which was slightly west of magnetic north. Turned over, the pointer always aligned itself to a consistent orientation slightly east of magnetic north."

The case was taken up by physicist John B. Carlson, whose examination revealed the extraordinary skill needed to shape and polish the brittle material into a grooved bar: "To my knowledge, [it] is unique in morphology among all known examples of worked Mesoamerican iron ore." His own efforts at polishing such ores met "with no outstanding success." The mysterious object was still strongly magnetic and, when floated on mercury or water (on a cork base), it consistently aligned itself 35.5 degrees west of magnetic north, with only half a degree of variation. During Carlson's experiments the object fractured into two pieces, but the accident proved to be a fortunate one. Tests showed that the magnetic orientation of the two parts differed considerably from that of the whole piece, enabling Carlson to make a further deduction: If the Olmec fragment was originally twice as long (about 2½ inches), then "there is a strong likelihood that it would have aligned itself close to magnetic north-south." The groove along the top of the piece could have acted as a sighting mark. Floated in water or mercury (with which the Olmecs were familiar), it would have made an effective compass. The motive for its manufacture may have been the Olmecs' interest in the orientation of their buildings, burials and ritual sites—many were carefully aligned eight degrees west of north.

Although there is no final proof, the evidence is compelling enough. There is no reason to believe that the mysterious Olmec object was anything other than a compass. Made no later than 1000 B.C., it predates the earliest Chinese examples by almost a millennium. The history of the compass provides an extraordinary example of how a similar device can be independently invented by two cultures widely separated in time and space.

LIGHTHOUSES

Off the coast of northern Egypt lies a small island called Pharos. It faces the city of Alexandria, seat of the Greek Ptolemaic Dynasty, which ruled Egypt from the late fourth century until it was aborbed into the Roman Empire in 30 B.C. The island of Pharos was to play an important role in the Ptolemies' plan to revive Egypt economically. An enormous mole was built connecting the island to the mainland, forming two safe harbors for the merchant ships of the Mediterranean. Then about 270 B.C. Ptolemy II commissioned the Greek architect Sostratus of Cnidos to build the world's first lighthouse on the eastern end of the island. Its purpose was twofold—it guided ships to the harbor of Alexandria and also served as a colossal prestige symbol of the revived Egyptian monarchy.

The lighthouse on Pharos was later remembered as one of the Seven Wonders of the Ancient World. It was a massive structure, 350 feet high, built in three sections: The lowest, standing on a huge stone platform, was square in shape; the middle one was octagonal and the uppermost circular. Near the top of the tower was kept a fire, the light from which could be seen over thirty miles out to sea. Sailors came to refer to the structure, rather than the island, as "Pharos," and the name entered a number of languages as the generic word for lighthouse (Greek *pharos*, Italian *faro* and French *phare*).

It was the ever-efficient Romans, however, who were the first to create a network of lighthouses. The earliest in the system was built in A.D. 50 at Ostia, the port of Rome, by order of the Emperor Claudius. The four-story lighthouse stood on an artificial island created by filling with stone and concrete the hulk of a wooden vessel measuring 320 by 65 feet, which had originally been used to bring an enormous obelisk from Alexandria to Rome. The Ostia lighthouse survived until the fifteenth century A.D.

The best known of the Roman lighthouses outside Italy were those of La Coruña, Boulogne and Dover. The La Coruña lighthouse, on the northwest tip of Spain, has been claimed as Phoenician, but it was really built by the Emperor Trajan in the early second century. It had a very similar design to the Pharos at Alexandria, with a rectangular base and a dome on the top. The dome was occupied by watchmen with a store of fuel for the braziers that burned on top of the whole structure. That at Boulogne in France was originally erected in A.D. 46 by the mad emperor Caligula as a monument to himself, but in A.D. 191 it was converted to useful purposes. It was allegedly some 200 feet tall, with twelve or thirteen stories that narrowed as one climbed up. On the other side of the channel at Dover there were originally two octagonal-shaped lighthouses, built around A.D. 100, one of which still survives to just over half its estimated original height of eighty feet.

By A.D. 400 there was a network of some thirty lighthouses around the Roman Empire. In the major lighthouses the bright light guiding ships was produced by a large metal mirror reflecting the light given off by a massive fire. With the end of the empire the skills and organization required for the upkeep of the lighthouses were lost: many of them fell into disrepair, while others were converted into military watchtowers.

The Arabs and Indians took up the cause of safety at sea from the Romans, and by A.D. 800 they had built a string of lighthouses around the coastlines of the Indian Ocean, according to the Chinese

The Pharos of Alexandria, the world's first known lighthouse, as reconstructed from ancient descriptions.

This 16th-century drawing of the siege of Boulogne, France, by the British in 1544 shows the ancient Roman lighthouse.

geographer Cha Tan. Despite their traditional disdain for sailing, even the Chinese built a few lighthouses at ports with large numbers of foreign merchants.

In the meantime maritime travel in northern Europe had become increasingly dangerous. Most of the coastal lights captains could see were lit by wreckers hoping to lure them onto the rocks. King Richard the Lion-Heart, himself the victim of a shipwreck when returning from the Holy Land in 1190, introduced heavy punishments for wreckers, particularly those working as pilots who took bribes and enticed ships to run aground:

> All false pilots and any other persons wrecking ships shall suffer a rigorous and merciless death and be hung on high gibbets, while the wicked lords [who pay them] will be tied to a stake in the middle of their own houses which shall then be set on fire at all four corners and be burned to the ground with all that shall be therein.

Even these threats did not work. The only measure that did prove successful was the twelfth-century revival of lighthouses proper. These used great mirrors to cast the light out to sea and were built by returning Crusaders copying Arab practices.

DIVING GEAR

The idea of being able to move freely under water has long exercised a fascination over humanity. In medieval Europe a common theme of the chivalrous tales (or "romances") that starred Alexander the Great was his descent to the ocean bed in a glass diving bell to talk to the king of the fishes. The story is, of course, a piece of medieval science fiction; the historical Alexander (336–323 B.C.), however megalomaniac, had no illusions of extending his conquests to the marine kingdom.

But oddly enough the first real evidence of technical aids to protracted underwater exploration comes from the writings of Alexander's personal tutor, the great philosopher and scientist Aristotle. His work *The Parts of Animals* mentions, in passing, the breathing tubes used by divers of his time: "Just as divers are sometimes provided with instruments for respiration, through which they can draw air from above the water, and thus remain for a long time under the sea, so also have elephants been furnished by Nature with their lengthened nostril; and, whenever they have to traverse the water they lift this up above the surface and breathe through it."

Subsequent references to breathing tubes and masks occur in German ballads of the twelfth century A.D. and later on in Chinese ac-

Alexander The Great's descent to the deep, as depicted in a late 13th-century French romance. In this particular version Alexander is rather mysteriously accompanied by a dog, a cat and a cockerel.

counts of pearl divers from the island of Hainan off the southern coast. Indeed the most detailed of the medieval accounts of diving apparatus concerns new equipment being used by pearl divers in the Arabian Gulf. Early in the eleventh century the scientist al-Biruni described how the divers fixed around their chests a leather hood that was filled with air before descent. He suggested that the apparatus could be improved by linking it to the surface and a permanent source of air:

> A more suitable arrangement would be to attach to the upper end of this gear opposite the forehead a leather tube similar to a sleeve, sealed at its seams by wax and bitumen [asphalt], and its length equal to the depth of diving. The upper end of the tube will be fitted to a large dish at a hole in its bottom. To this dish are attached one or more inflated bags to keep it floating. The breath of the diver will flow in and out through the tube as long as he desires to stay in the water, even for days.

The Hussite Wars manuscript of c. A.D. 1425 shows a diver in a suit with flippers, his leather helmet connected to the surface by a breathing-tube. The barrel and box are probably gunpowder mines.

Unfortunately no later writer confirms whether al-Biruni's practical-sounding design suggestion was adopted by the pearl divers.

Inevitably the military possibilities of prolonged underwater travel also came to be exploited. The German military expert Konrad Keyser illustrated two divers fighting underwater in his manual of warfare around A.D. 1400. The nature of their diving gear is alas unclear. A generation later another illustrated manuscript, written by an unknown military genius who fought in the Czechoslovakian Hussite Wars, clearly shows a diver in a suit and helmet.

Near the end of the fifteenth century the great inventor Leonardo da Vinci took the idea of underwater travel farther with a design for a submarine that was, however, never built. The first practical blueprint for a truly submersible vessel was presented by the Englishman William Bourne, a carpenter, gunmaker and writer, in his splendidly titled *Inventions and Devices Very Necessary for Generalles, Captaines or Leaders of Men* (1578).

MAN-BEARING KITES AND PARACHUTES

The invention of kites strong enough to bear the weight of a person came about in a particularly unpleasant way during the fortunately short-lived Ch'i Dynasty, which ruled northern China from A.D. 550 to 577. The first Ch'i emperor, Kao Yang, carried out a systematic extermination of the families who had controlled the previous dynasty. In one year he murdered no less than 721 of them by a most bizarre method. At the time, Kao Yang was living in a 100-foot-high tower, while receiving instruction in the Buddhist faith. One of the good deeds of Buddhism is to release trapped animals and not to kill them. According to the official history, the Emperor devised his own perverted version of this act of charity: "He caused many prisoners condemned to death to be brought forward, had them harnessed with great bamboo mats as wings, and ordered them to fly to the ground from the top of the tower. This was called a 'liberation of living creatures.' . . . All the prisoners died, but the Emperor contemplated the spectacle with enjoyment and much laughter."

By 559 he was regularly using the condemned as test pilots for man-bearing kites controlled from the ground. One prisoner, Prince Yuan Huang-T'ou, managed a flight of nearly two miles. The abominable Kao Yang did not reward this feat of skill with freedom; rather he starved the unlucky prince to death.

Kite manufacture in China goes back at least to the third century B.C., when the philosopher Mo Ti wrote of how he spent three years making one. They became a common sight in Asian skies and are described in a variety of roles in surviving texts. Kites were children's toys, but they also played a part in sporting contests as "fighting kites," with the winner being the owner of the kite that cut through the line of the loser. They were frequently employed by the military, sometimes to drop propaganda leaflets on enemy armies. But they also developed their use as temporary observation platforms, carrying a man up to spy on the enemy.

The Japanese also had a military use for kites, apparently em-

ploying them to carry men in or out of besieged cities. The most famous story of Japanese kite flying is told of the twelfth-century samurai hero Minamoto no Tametamo. He and his son had been exiled to an island off Japan. According to the tale, Tametamo strapped his son to a kite and succeeded in flying him back to the mainland. Less extraordinary, but more practical, was the Japanese use of kites as a means of transporting bricks to workmen at the tops of high buildings.

The oddest role for the man-bearing kite, however, was as an aid to divination. The Venetian merchant Marco Polo recorded the practice during his travels in China at the end of the thirteenth century:

> When any ship must go on a voyage, they prove whether its business will go well or ill. The crew make a hurdle [rectangular frame] . . . of withies [flexible twigs] and at each corner and side of the hurdle tie a cord, so that there are eight cords, and all tied at the other end with a long rope. They find someone stupid or drunken and bind him on the hurdle; for no wise man nor

This 19th-century print by the famous Japanese artist Hokusai shows the 12th-century samurai hero Tametamo helping his son escape their island exile by lashing him to an enormous kite.

undepraved would expose himself to that danger. And this is done when a strong wind prevails. They set up the hurdle opposite the wind, and the wind lifts the hurdle and carries it into the sky, and the men hold on by the long rope. If, while it is in the air, the hurdle leans toward the way of the wind, they pull the rope to them a little and then the hurdle is set upright, and they let out some rope and the hurdle rises. If it leans down again, they pull the rope by so much till the hurdle is set up and rises, and they let out some rope; so that in this way it would rise so much that it could not be seen, if only the rope should be so long. This proof is made in this way, namely that if the hurdle going straight up makes for the sky, they say that the ship for which that proof has been made will make a quick and prosperous voyage, and all the merchants run together to her for the sake of sailing and going with her. And if the hurdle has not been able to go up, no merchant will be willing to enter the ship for which the proof was made, because they say that she could not finish her voyage and many ills would oppress her. And so that ship stays in port that year.

What goes up must come down of course, and it is no surprise that ancient peoples were also interested in parachutes. The legendary Chinese emperor Shun supposedly used straw hats tied together to jump from a burning tower. More plausible is the account left by a daring Chinese robber and local celebrity of his getaway from the great Arab merchants' mosque at Canton in A.D. 1180 with the leg of its greatest treasure, a golden cockerel. Trapped at the top of the minaret with his loot, he escaped "by holding on to two umbrellas without handles. After I jumped into the air the high wind kept them fully open, making them like wings for me, and so I reached the ground without injury."

For the earliest actual parachute we must turn to the West. A pyramid-shaped parachute was drawn by the great Leonardo da Vinci in his notebooks around 1485, and the idea has long been credited to him. However, dating from a few years earlier, the notebook of an unknown Italian engineer shows a parachute of conical—and therefore more practical—shape, probably made of cloth. The jumper has a sponge strapped into his mouth to protect his jaws from the shock of landing; unfortunately we have no idea whether this helped the intrepid pioneer.

A conical parachute of cloth depicted in the Italian engineer's sketchbook of c. A.D. 1480.

GLIDERS

Some of the most bizarre stories from ancient India concern flying machines. For example, a collection of tales compiled in the tenth century A.D., but including stories composed many centuries earlier, tells how one hero made a trip through the air on a mechanical cockerel. The secret of making such flying machines, the story stresses, was known only to the *Yavanas*, or Greeks.

What lay behind this ancient belief? It seems that the ancient Greek engineers did make experiments with heavier-than-air flight—though only on a tiny scale—and that exaggerated reports of their success may have reached India. The clearest evidence concerns Archytas, a scientist from Tarentum, in southern Sicily, who lived about 400 B.C. He was probably the greatest Greek engineer before Archimedes and was famed for creating a flying dove, hailed by later Greek and Roman writers as one of the wonders of antiquity. Made of wood, the dove was said to have been propelled across the sky by a current of air "hidden and enclosed" within. What mechanism, if any, drove Archytas's flying dove has been the subject of much discussion, though the reference to air stored within it suggests some kind of pneumatic propulsion. Since there is no reference to the dove flapping its wings, we can assume that it was basically a kind of wooden glider.

Archytas's experiment would have been just the kind to be followed up by the engineers of Alexandria in Egypt (see Introduction to High Tech). A curious archaeological find from northern Egypt seems to confirm this. A collection of relics, dating to the fourth or third century B.C., were excavated at Sakkara in 1898 and deposited in the Cairo Museum. One object, a small wooden glider with a wingspan of about seven inches, was displayed there in a case together with model birds until 1969, when it caught the attention of Dr. Khalil Messiha, a medical doctor and aeronautics enthusiast. Messiha observed that though the object is vaguely birdlike, the tail is placed vertically and the wings are flat just like a model airplane—in short it shows a fair degree of aerodynamic sophistication.

But although it would fly, there is no evidence that the Sakkara glider was anything other than a toy. While there may have been a fair degree of experimentation by the ancient Greek engineers into heavier-than-air flight, it was something, despite the Indian legends, that they never managed to scale up to the point where it could transport human beings.

The Sakkara glider.

BALLOONING

As in the case of kites, the Chinese were pioneers of hot-air ballooning. More than two thousand years ago they were making miniature hot-air balloons from empty eggshells. Chinese children
would place dry plant stems in the eggshells, set them alight to heat
the air inside and send the shells flying away. No great advance on
this game seems to have taken place until medieval times, when the
military possibilities of balloons began to be exploited. The Mongols
learned of balloons around A.D. 1200 from their new Chinese subjects, and there are several European descriptions of their use at the
Battle of Legnica, fought by the Poles against the Mongol invaders
in 1241. The balloon used by the Mongols was a dragon-shaped
windsock with a lighted torch in the mouth to throw out flames and
smoke and so appear threatening; it acted either as a signaling device
or as a standard for the Mongol troops to rally around.

One European thinker, the Venetian engineer Giovanni di
Fontana, writing around 1425 (see **Wind Cars and Rocket Cars**),
was inspired by these Oriental experiments to suggest the idea of
taking advantage of the lifting power of hot air to put people aloft,
but he dismissed the notion as impractical and dangerous.

This fear of flying does not appear to have affected the far earlier
engineers of the Nazca culture of Peru. The lines on the Nazca desert plain of coastal Peru are world famous, largely due to the extraordinary dedication of Maria Reiche, a German mathematician,
who has mapped them as a labor of love since 1946. They run for
miles across the plateau, generally in straight lines but sometimes
forming giant figures such as spiders, birds and fish. The lines truly
merit the title of the largest work of art in the world. While their
creation can be dated by archaeological evidence to between 500
B.C. and A.D. 900, their purpose remains a mystery. They cannot be
seen properly from the ground and are in many cases invisible except
from the air. This has led those adherents of the "ancient astronaut"
theory, who believe that the earth was once colonized by extraterrestrial beings, to seize on the site as proof. Yet nothing has ever

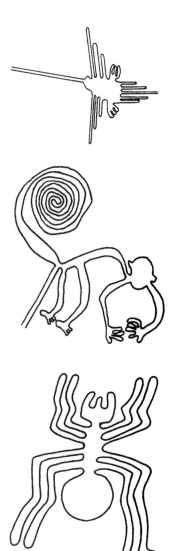

Some of the gigantic animal
figures of the Nazca Plain
as seen from the air—a
hummingbird (wingspan
200 ft.), a monkey (260 ft.
head to tail) and a spider
(150 ft. long).

turned up from archaeological excavations or the activities of tomb robbers that is even remotely "out of this world."

An Archaeological Experiment

A far more plausible explanation of the figures' visibility from the air has been put forward by members of the International Explorers Society of Miami. Legends of the Incas, who conquered the area a thousand years after the Nazca culture disappeared, speak of a boy named Antarqui, who helped the Incas in their battles by flying above enemy lines and reporting on their positions. One technologically feasible way in which this could have been done was in a cotton balloon. Nazca pottery is sometimes decorated with paintings of what could be balloons and kites with trailing streamers and lines, while Nazca textiles often depict flying men. The idea of flight was clearly important to the ancient Nazcans.

More interesting to the Explorers Society than the decorated textiles were the plain pieces of fabric found in looted Nazca tombs. When tested by a modern balloon-manufacturing company, they proved to have a tighter weave than the fabrics the company made themselves. Indeed the prehistoric Peruvians made some of the finest textiles of the ancient world. The Society tested the feasibility of Nazca ballooning by constructing their own balloon, which they named *Condor I*. It was made from cotton and wood, with a gondola of reeds from Lake Titicaca to hold the passengers. The balloon was to be filled with hot smoke produced from burning extremely dry wood. Similar fires may have been the cause of the "burn pits"— circular areas of rock thirty to fifty feet in diameter scorched by intense fires—found at the end of many of the straight Nazca lines.

On November 23, 1975, American Jim Woodman of the Explorers Society and English balloonist Julian Nott ascended to 380 feet in *Condor I*, stayed in the air for several minutes and then descended again safely, thus completing a remarkable piece of experimental archaeology. Of course this does not prove that the ancient Nazcans did fly, but it certainly shows that it is a very real possibility. Maria Reiche, the "Lady of the Lines," was convinced. Asked to comment on the flight of *Condor I*, she said, "I am certain the early Peruvians flew. I wept with emotion. It is the culmination of my work."

One final interesting but little known fact is that the Montgolfier brothers of Paris were probably not the first modern people to fly in a balloon. Instead the honor may well belong to a Jesuit priest from Brazil, Father Bartolomeu de Gusmão, who demonstrated his hot-air balloon to the Portuguese court at Lisbon in 1709. Could it be that he was inspired by South American legends or memories of successful flights nearly two thousand years before his day?

The *Condor I* balloon in flight over the Nazca Plain.

HIGH TECH

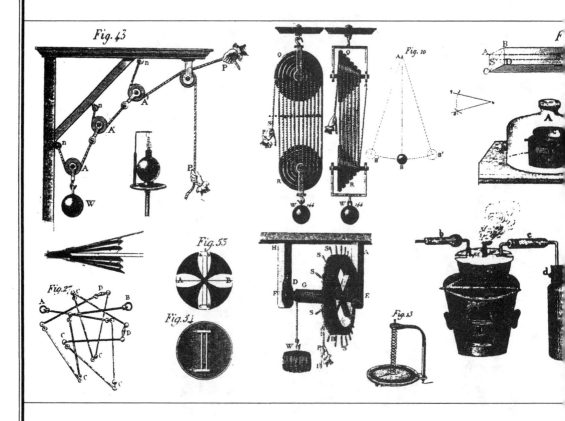

Fig. 43

Fig. 10

Fig. 53

Fig. 51

Fig. 27

Fig. 13

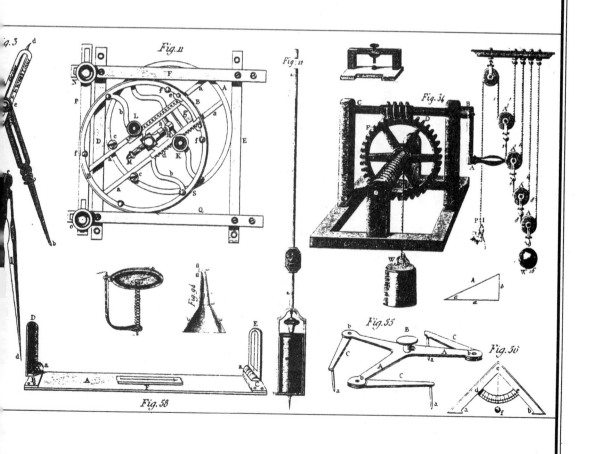

CHAPTER THREE

Fig. 3
Fig. 11
Fig. 12
Fig. 34
Fig. 94
Fig. 55
Fig. 50
Fig. 58

INTRODUCTION

very few years another article appears arguing that, despite
what we have been told, ancient Greece *did* produce "real
scientists." It may be surprising to readers of this book that
such a point needs to be made at all. The background to the problem is one of those beliefs that refuses to die gracefully, despite the
number of times it has been knocked down. It runs roughly as follows: The Greeks were not proper scientists because they did not
base their conclusions on observations from experiments; preoccupied with "pure" science (i.e., mathematics) they disdained any
practical application of their brilliance. Both ideas are complete nonsense, as shown by two favorite examples of the "Greeks were real
scientists after all" school of writing: Pythagoras and Strato.

Pythagoras, the greatest of the Greek sages of the sixth century
B.C., is the most frequently cited. He was a gifted mathematician,
best remembered for the geometrical theorems rightly or wrongly
attributed to him. As a popular philosopher Pythagoras also taught
the doctrine of reincarnation, a respect for all living creatures and
strict vegetarianism (see **Introduction** to **Food, Drink and
Drugs**), teachings curiously similar to those of his Indian contemporary, the Buddha. The mystical edge to Pythagoras's teachings
make him an easy target for the label "soft scientist." In fact, however, his mathematical theory of the musical scale was based directly
on experiments with varying the weights of hammers and the
lengths of strings (see **Written Music** in **Sport and Leisure**).

By the third century B.C., philosophy and experimentation had
been blended into an elegantly balanced system by Strato, the physicist who was head of the Lyceum at Athens between 287 and 269
B.C. His original writings are lost, but the clarity and startlingly
modern character of his thinking can be judged from a passage on
pneumatic experiments preserved in the writings of the later engineer Heron:

We must first correct a popular illusion. It must be clearly
grasped that vessels which are generally believed to be empty are
not really empty, but full of air. Now air, in the opinion of the

natural philosophers, consists of minute particles of matter for the most part invisible to us. Accordingly if one pours water into an apparently empty vessel, a volume of air comes out equal to the volume of the water poured in.

To prove this make the following experiment. Take a seemingly empty vessel. Turn it upside down, taking care to keep it vertical, and plunge it into a dish of water. Even if you depress it until it is completely covered no water will enter. This proves that air is a material thing which prevents the water entering the vessel. . . .

Now bore a hole in the bottom of the vessel. The water will then enter at the mouth while the air escapes by the hole. . . . This constitutes proof that air is a bodily substance.

This is just the simplest of the numerous experiments designed by Strato to illustrate the properties of gases, and the same experiments for elementary-grade physics are still carried out in classrooms the world over. All things considered, the concept that the ancient world somehow lacked "real scientists" seems absolutely bizarre.

Before he took over as head of the Lyceum at Athens, Strato was summoned to Egypt to be tutor to the future King Ptolemy II in the palace at Alexandria. This city, founded by Alexander the Great in 332 B.C. after his conquest of Egypt, was to become the scientific center of the ancient Greek world. As a result of Alexander's extraordinary conquests, this was now a much wider world, stretching from the Balkans to north Africa and India; Greek culture fused with Persian, Babylonian and Egyptian and thrived as a new hybrid, the Hellenistic civilization, under the patronage of the dynasties founded by Alexander's generals on his death in 323 B.C. In Egypt the ruling family started by Ptolemy I continued until the death of his descendant, Cleopatra, in 31 B.C.: during these three centuries the Ptolemies reigned as enlightened despots, investing vast sums of money in the development of the arts and sciences.

Strato did a good job of instilling scientific curiosity in his royal pupil. As King Ptolemy II (285–246 B.C.) he exhibited a passion verging on mania for engineering projects and scientific knowledge. He built the Pharos lighthouse, redug the ancient Suez Canal (see **Lighthouses** and **The First Suez Canal** in **Transportation**) and founded the prestigious university known as the "Museum" (Greek *museion*) of Alexandria, and a magnificent library. His agents were

— The water only mounts to A, the space B is filled up with air, which is a gas.

Simple classroom experiment demonstrating that air is a substance that occupies space, as depicted in a Victorian school primer (1880). The same experiment, together with numerous others, was already being used in Alexandria around 200 B.C. to demonstrate the laws of pneumatics.

Pharaoh Ptolemy II (285–246 B.C.) making an offering to the goddess Isis. Through his patronage of science, Ptolemy was responsible for the extraordinary advances in engineering made in the Greco-Egyptian city of Alexandria over the next three hundred years.

sent far and wide in pursuit of an audacious educational policy—to gather into the Library of Alexandria every scrap of available knowledge on every known subject (see **Introduction** to **Communications**).

Under Ptolemaic rule Alexandria remained unrivaled as a center of learning, and even after its conquest by the Romans in 31 B.C. it continued to be the "university of the Mediterranean." From the fertile intellectual soil of Alexandria sprang some giants of ancient engineering. The first was Ctesibius, the son of a barber, who began his career as an inventor with a gadget for adjusting mirrors in his father's salon. He developed a flair for mechanical novelties, and in 270 B.C. was said to have made a singing cornucopia for the statue of Ptolemy II's queen. His book describing his mechanical inventions and experiments is now lost, but we know from later authors who used it that Ctesibius made successful designs for water pumps, water clocks and a water organ, the world's first keyboard instrument (see **Keyboards** in **Sport and Leisure**).

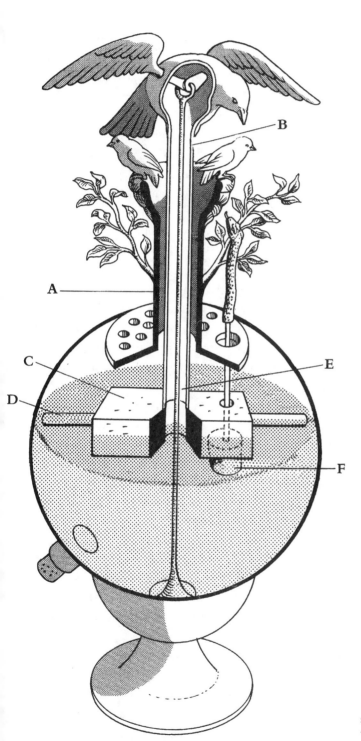

A charming gadget from the golden age of Hellenistic engineering—invented by Philon (2nd century B.C.). A nest containing model fledglings is attached to the tube-shaped "tree" (A), which is soldered to the perforated lid of the vessel. The mother-bird perches on an inner tube (B), which runs through the tree to a float (C), attached to the bottom of which are supports (D). Tube (B) contains a long wire (E), which runs through the float—one end is fixed to the bottom of the jar, the other to the bird's wings. A thin model snake is attached to a smaller float (F).

As water or wine is poured in through the perforations, the small float (F) raises the snake to menace the fledglings in the nest. But when the liquid has risen far enough, float (C) is lifted up; tube (B) rises, pushing up the bird. The sides of the bird lift up her wings into an aggressive posture to ward off the snake. When the vessel is drained (through a tap at the bottom), the snake descends along with its float (F). Float (C) also descends until it rests on its supports (D). The mother-bird folds her wings and recovers her composure.

The next great Hellenistic engineer was Philon of Byzantium (Istanbul), a rather more hard-nosed variety of scientist. A military engineer by profession, he worked in both Alexandria and Rhodes during the second century B.C. His main work was the application of science to warfare, described in a nine-volume treatise, much of which still survives (see **Catapults and Crossbows** in **Military Technology**). But he also shared the Alexandrian fascination with gadgets. One of his inventions was a novelty eight-sided ink pot with an opening on each side. One can turn the octagon so that any face is on top, dip in a pen and ink it—yet the ink never runs out through the holes in the other sides. The secret lies in the suspension of the inkwell at the center, which is mounted on a cunningly arranged series of concentric metal rings—known as gimbals—and remains stationary no matter which way one turns the pot itself. Gimbals were reinvented (or inspired by an imported inkwell?) in Han Dynasty China around 100 B.C. (see **Perfume** in **Personal Effects**) and discovered anew (or borrowed from China?) in thirteenth-century Europe—in both cases for use in incense burners. In 1546 the Spanish improved the Chinese invention of the compass (see **The Compass** in **Transportation**) by suspending it within gimbals, allowing it to maintain a constant position despite outside motion. It is a sobering thought that Europe's main contribution to the development of the compass was merely a reinvention of a Hellenistic executive toy designed eighteen centuries earlier.

Finally, there was Heron, "the Machine Man" *(mechanikos)*. Living in Alexandria in the late first century A.D., he was heir to a long tradition of advanced engineering, particularly in pneumatics and related fields. The list of his inventions is quite staggering, ranging from handy devices such as a self-trimming oil lamp to a sophisticated surveying instrument that was the prototype of the modern theodolite (see **Tunneling** in **Working the Land**). Other inventions included a variety of automata and devices designed for use in the theater and as gimmicks for Egyptian temples.

In recent years the question of the ethnic makeup of Alexandria's high-tech pioneers has been raised: Were they of Greek or Egyptian stock? Many black scholars today have rightly argued that a native-born Alexandrian such as Heron may well have been a pure Egyptian and that we should picture him with African rather than

European features. He had a Greek name, but so did many non-Greek inhabitants of Alexandria, and he often wrote from an Egyptian perspective when describing to a Greco-Roman audience the customs of the temples for which he tailored some of his inventions. But the real answer is that neither Europe nor Africa has an exclusive claim to the genius of Alexandria. Rather the extraordinary achievements of its scientists should be seen as a tribute to the wonderful cultural and intellectual melting pot that was Alexandria.

As Benjamin Farrington wrote in his classic study *Greek Science: Its Meaning for Us*, "With the science of Alexandria and of Rome we are in very truth on the threshold of the modern world." Yet even Farrington had some blind spots when it came to appreciating the truly advanced nature of ancient science. Apart from one grudging reference to the use of magnetism in the production of miraculous temple effects, Farrington completely avoids the subjects of magnetism and electricity. His index does not even include these words, a scandalous omission given the keen interest the ancients showed in understanding and harnessing these forces. It is difficult to see how ancient acquaintance with electricity can have no "meaning for us"—the very word derives from ancient Greek experiments with amber (*electron* in Greek). Farrington is also strangely silent about

Renaissance reconstruction of a toy designed by Heron in the 1st century A.D. Air heated by the fire at the top flowed through the bent tubes, making the platform revolve and the dancers circle around.

one of the most important discoveries in the history of science—the ancient Greek computer, discovered in a first-century B.C. shipwreck off the island of Antikythera in 1901. The chance survival of this remarkable example of ancient high technology has completely revolutionized our understanding of Greek and Roman engineering.

Such advanced achievements in the Old World were not limited to the area of the Roman Empire. To the east, Babylonian scientists made an extraordinary leap in physics by inventing the first electric batteries. And a curious number of major advances in technology seem to have taken place simultaneously in Europe and the Far East (see **Odometers** in **Transportation** and **Watermills and Windmills** in **Working the Land**). Indeed the great Chinese inventor Chang Hêng (early 2nd century A.D.) used mechanisms that have been described by Joseph Needham, the historian of Chinese science, as having "a distinctly Alexandrian air." About 100 A.D. China was beginning to appear on Roman maps, while Syrian merchants had long been familiar with the silk routes between the Mediterranean and Far East (see **Mapmaking** in **Transportation**). We must seriously consider the possibility that, even at this early date, news of scientific discoveries was passing between Europe and China and that some reciprocal influence between the two regions played a part in the scientific golden age that lasted from the third century B.C. to the second century A.D.

The explosion of scientific activity during these centuries brought the Old World to a stage of "high technology" that closely mirrors our own, both in its aspirations and in its achievements. All the same it failed to save the civilizations that nurtured it. As Professor Derek de Solla Price, the leading expert on the Antikythera computer, remarked, it is "a bit frightening to know that just before the fall of their great civilization the ancient Greeks had come so close to our age, not only in their thought but also in their scientific technology."

COMPUTERS

During Easter 1900 a party of Greek sponge fishermen were on their way home from their traditional fishing grounds off North Africa to the island of Syme, near Rhodes, when a storm struck. Blown off course, they finally reached shelter on the almost uninhabited rocky islet of Antikythera, northwest of Crete. Trying their hand at sponge fishing, they were amazed to discover the wreck of a huge ship. The most interesting item of the ancient cargo, as far as the fishermen were concerned, was a large heap of bronze and marble statues. They reported their discovery to the authorities and returned to Antikythera with archaeological advisers in November, working on the wreck until September 1901.

Nearly eight months after the excavation—as the finds were being painstakingly cleaned of the concretion that had built up around them during their long years under water—some tiny bronze fragments with an inscription in Greek were spotted. Further pieces soon came to light, and eventually a whole series of gearwheels, several with writing on them, was revealed.

From the outset these finds were controversial, with some archaeologists insisting that the mechanism was far too complicated to belong to the wreck, which, judging by the pottery in the cargo, belonged to the first century B.C. The experts were also hopelessly divided into two camps on the purpose of the object—one arguing that the remains came from an astrolabe, an instrument for calculating altitudes; the other equally convinced that they were from a planetarium, a device used to show the motions and orbits of the planets. With neither side giving way, the debate reached a stalemate, and the function of the Antikythera mechanism was left as an obscure, unsolved puzzle.

A Calendrical Computer

The situation changed in 1951, when Professor Derek de Solla Price, of Yale University, took an interest in the Antikythera mystery. Over the next twenty years he devoted many long hours to a

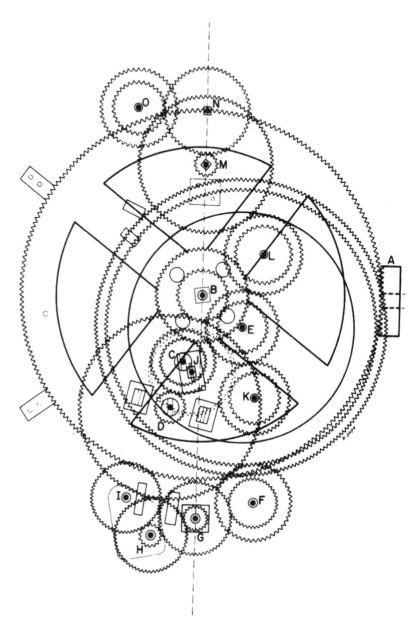

A reconstruction plan of the complex gearing of the Antikythera computer, by Professor Derek de Solla Price.

minute examination of the objects, with the aid of X-ray photography. His research eventually enabled him to reassemble the surviving pieces and so discover the device's true purpose.

The Antikythera mechanism turned out to be a complex computer for calculating the calendars of the sun and moon. One rotation of the main wheel represented a solar year, while the smaller wheels showed the positions of the sun and moon and of the rising of the most important stars. The wheels were housed in a wooden box, the doors of which would have been opened to reveal the mechanical marvels inside. Rather than being a navigational aid used by the captain of the ship, the device was probably part of the cargo, along with the statues.

The Antikythera discovery provides striking confirmation of some tantalizing literary clues that suggest that Greek scientists of this time had already been experimenting with such complex machines for astronomical purposes. Only a few years after the Antikythera ship went down, the Roman lawyer Cicero (106–43 B.C.) wrote that his friend and teacher, the philosopher Poseidonius, had "recently made a globe which in its revolutions shows the movements of the sun and stars and planets, by day and night, just as they appear in the sky." He also noted that the great Archimedes (see **The "Claws" of Archimedes** in **Military Technology**, and **Odometers** in **Transportation**) had devised an earlier model "imitating the motions of the heavenly bodies." It has even been suggested that it was the very machine made by Archimedes that was found in the shipwreck.

The implications of the Antikythera mechanism for our understanding of ancient Greek technology are enormous. Similar, though less complex, geared calendars are known from the Islamic world from around A.D. 1050. One such calendar, designed by the astronomer Abu Said al-Sijzi, recorded the moon's phases and the movements of the sun against the signs of the zodiac. Such devices were the ancestors of the astronomical clocks of medieval Europe. The development of such complicated mechanisms was therefore once seen as the product of a much later age. Now, as Professor de Solla Price has concluded, the Antikythera discovery "requires us completely to rethink our attitudes toward ancient Greek technology. Men who could build this could have built almost anything they wanted to. The technology was there, and it has just not survived like the great marble buildings, statuary and the constantly recopied literary works of high culture."

CLOCKS

We often like to think of the past as being relatively carefree, lacking our modern obsession with clock watching and not wasting time. However, clocks of varying degrees of accuracy have been around for at least four thousand years. The earliest attempts were those of the Egyptians, who made star-clock charts from which the time of night could be calculated by seeing which stars had already risen. For the daytime they later developed shadow clocks: The shadow of a crossbar gradually crossed a series of marks as the sun rose and then set. A set of instructions for making one was found in the tomb of Pharaoh Seti I, who reigned around 1300 B.C. Such simple shadow clocks were the ancestors of the sundial. The Romans perfected the kind of sundial we are familiar with today and even made portable ones for use when traveling.

The Water Clock

About 1500 B.C. the Egyptian court official Amenemhet, according to an inscription in his tomb, invented the water clock, or *clepsydra*. As the water flowed out, the drop in the water level in the vessel gave a measure of the passage of time. Such clocks were particularly

Interior of the Egyptian water clock from Karnak (14th century B.C.). The device was filled with water each evening to be used as a way of telling the time during the night. Water gradually flowed through a reed bung in a hole near the base. The passage of time was measured by the descent of the water past the "hour" marks. As the night was always divided into twelve parts—irrespective of the season—the "hours" shown by the clock varied in length through the year. Thus the twelve lines of "hour" marks (only parts of seven are visible from the front) are of slightly different heights.

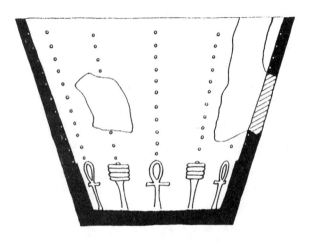

useful to priests, who needed to know the time during the night so that the rituals and sacrifices of the temple could be carried out at the correct hour. The oldest surviving water clock is from the reign of Amenhotep III (early 14th century B.C.)—it was found in fragments in the temple of Amun-Re, at Karnak, in 1905.

Water clocks became the standard way of telling time throughout the ancient world, being introduced into China in the sixth century B.C. They were common sights in towns such as Athens, where remains of the "town clock" built around 350 B.C. have been found. The movement of this clock was controlled by a float, which sank as a tiny outlet at the bottom allowed the water to trickle out. The float was probably connected to a shaft, which moved a pointing hand as it fell. The degree of wear on the steps leading down to the well suggests that the water reservoir had to be refilled every day.

The Greek world had more elaborate water clocks as well, such as the one manufactured by the inventor Ctesibius of Alexandria about 270 B.C. Its water outflow, precisely controlled by stopcocks, drove all sorts of automata, from ringing bells and moving puppets to singing birds—perhaps the earliest cuckoo clock! In Athens the "Tower of the Winds," built by the astronomer Andronichos sometime in the early first century B.C., contained sundials on the top and a complicated water clock inside, which gave the time on a dial, while a rotating disk showed the movements of the stars and the annual course of the sun through the constellations.

The impact of these water clocks is amply demonstrated by ancient authors, who recall their use in several different contexts. Writing around 360 B.C., Plato described lawyers as "driven by the *clepsydra* . . . never at leisure." The clock even began to affect literature. "The length of a tragedy," Aristotle complained, "should not be judged by the *clepsydra* . . . but by what is suitable for the plot." Clearly clock watching was already firmly in control. A more valuable role for the water clock was in Greek and Roman courts, where it was used to make sure speakers kept within their time; if the proceedings were interrupted for a while, for example to examine documents, the outflow pipe was stopped with wax until the speech started again. In sport it was used to time the races in Roman games.

Magnificent water clocks were later manufactured in the Islamic world. A particularly elaborate example was sent from Baghdad to Charlemagne, the first Holy Roman emperor (A.D. 742–814) by emissaries from Caliph Haroun al-Rashid. A pair of massive water clocks, consisting of two vessels that gradually filled as the moon was waxing, then emptied as it waned, was constructed at Toledo, Spain, by Arab engineers during the eleventh century. These were so cleverly made that they ran for more than one hundred years without needing any adjustment.

Mechanical Marvels

The next great step forward in clock making—the invention of the mechanical clock—has traditionally been claimed by western Europe. There is a reference to a machine for playing hymn tunes on bells in an abbey near Rouen, France, in 1321, which was probably a clock. The great Italian poet Dante clearly describes a striking clock in his *Divine Comedy*, penned around the same time, while the first reference to a specific clock, that at the church of Saint Gothard, in Milan, comes from a chronicle entry for the year 1335.

However, several hundred years before these earliest European examples, the mechanical clock had already been invented in China. The motivation was the need to record the precise time of conception of the various heirs to the throne, so that the court astrologers could then determine the cosmological influences at work on them and thereby determine the best candidate. Although over the centuries the Chinese had developed more accurate forms of water clocks, including a stopwatch form that used mercury rather than water, these were still not good enough for the specialized needs of the astrologers.

Accordingly, in A.D. 723 Buddhist monk and mathematician I-Hsing developed a clockwork astronomical instrument that he called the Water-Driven Spherical Bird's-Eye-View Map of the Heavens. As the name suggests, water provided the power—but machinery regulated the movements. Unfortunately within only a few years the bronze and iron mechanism started to corrode; in addition the clock had already run into problems during cold weather, when the water inside it froze. In 976 Chang Ssu-Hsün built a clock using mercury instead, but few details of this survive.

The greatest of the medieval Chinese clocks was the "Cosmic Engine," designed by the astronomer Su Sung at the command of Emperor Ying Zong and completed in A.D. 1090. His device was an astronomical clock tower over thirty feet high. On the top was a massive spherical astronomical instrument for observing the stars, constructed from bronze and driven by water power. Inside the tower was a celestial globe, whose movements were synchronized with those of the sphere above so that the two could be compared constantly. At the front of the tower was a pagodalike structure of five floors, each with a door through which wooden puppets appeared at regular intervals throughout the day and night. They beat

Reconstruction drawing of Su Sung's "Cosmic Engine," built at Khaifeng in A.D. 1090. The building was over thirty feet high. The clock was powered by the giant waterwheel, which in turn was fed by tanks. The clockwork mechanism drove a celestial globe on the upper story, an observational sphere below, and a vast array of miniature figures that announced the time at various hours of the day and night.

drums, rang bells and gongs, played stringed instruments and displayed tablets showing the time. All these figures were operated by the giant clock machinery, powered by a huge waterwheel with scoops on the end of blades, into which water dripped from a water clock, causing the machinery to advance by one scoop every hour.

Su Sung's great clock ran from 1090 until 1126; it was then dismantled and moved by the conquering Chin Tartars to Peking, where it ran for several more years. Su Sung's "Cosmic Engine" was the high point of medieval Chinese clock making; unfortunately the unsettled conditions of the next century seem to have resulted in the skill falling out of use.

COIN-OPERATED SLOT MACHINES

One of the strangest wonders of the ancient city of Alexandria was the world's first slot machine. It was designed by Heron, inventor extraordinaire and native of the city, whose passion for gadgetry and automated devices was unsurpassed. As well as creating the first steam engine, Heron designed mechanical puppet theaters, a fire engine, an odometer, a self-trimming oil lamp, a new kind of syringe, a surveying instrument similar to a modern theodolite, mechanical birds that sang, a solar-powered fountain, a water-powered organ, a windmill-driven organ and so on almost ad infinitum. The range of Heron's cunning gadgetry, described by him in a series of detailed manuals during the first century A.D., is staggering.

His coin-operated slot machine was, like many of his other marvels, designed for use in temples. The idea was that a worshiper would put a five-drachma bronze coin in the slot and receive in return a small amount of water for the ritual washing of face and hands required before entering the temple. At the end of the day the machine would be emptied and the priests would put away the takings, just as they do today in some modern Roman Catholic churches where people put their small change into coin-operated electric candles.

The ancient slot machine worked as follows: The coin fell into a small pan that hung from one end of a delicately balanced beam. Its weight depressed one end of the beam, thus raising the other end, which opened a valve and allowed the holy water to flow out. As the pan fell, the coin would slide off; the beam end with the pan then swung up and the other end down, closing the valve and shutting off the flow of water.

Heron's ingenious device may have been inspired by a party novelty designed some three hundred years earlier by Philon of Byzantium. This was a vessel with a built-in and rather spooky special effect, which served guests with water for washing their hands. Above the waterspout was a carved hand holding a ball of pumice stone. When a guest took the ball of pumice to scrub clean before dinner, the hand vanished inside the device, and water started to flow from the spout. After a time the water stopped flowing and the hand reappeared with a fresh piece of pumice ready for the next guest. Unfortunately Philon does not give a detailed description of how this particular mechanical marvel worked, but it must have relied on principles similar to those used in Heron's slot machine.

The slot machine designed for use in Egyptian temples by Heron. By dropping a coin into the machine, visitors would receive holy water from the tap for a ritual wash before worshiping.

AUTOMATIC DOORS

When you next step through the doors of a supermarket, spare a thought for Heron, the undoubted genius of ancient "high tech" engineering. Nearly two thousand years ago he designed automatically opening doors for the temples of the Egyptian city of Alexandria.

As much as anything, Heron was a showman, with a brilliant flair for designing mechanical wonders to tease, baffle and amaze. His design for automatic temple doors was a gift to the Egyptian priests, who for centuries had used marvels, mechanical or otherwise, as a way of reinforcing their authority and prestige.

Temple Wizardry

Employing relatively simple mechanical principles, Heron devised a means whereby the doors of a small temple would open—as if by unseen hands—when the priest lit a fire on the altar outside the temple. The fire heated the air in a metal globe concealed beneath the altar, forcing the water in it through a syphon into an enormous bucket. The bucket was suspended by chains from a system of weights and pulleys, which turned the doors on their pivots as the bucket became heavier.

A second surprise took place when the altar fire was extinguished: As a result of the sudden cooling of the air in the globe, the water was sucked the other way through the syphon. When the bucket emptied, it swung upward, activating the pulley system in reverse, and the doors solemnly closed again.

Another design included in Heron's writings could make a trumpet blow when the temple doors opened—a combination of musical doorbell and burglar alarm.

The automatic doors designed by Heron, as conceived by a Renaissance artist. Apart from the altar all the necessary apparatus was hidden in a chamber under the temple doors.

There need be little doubt that the automatic-door system described by Heron was actually used in Egyptian temples and possibly elsewhere in the Greco-Roman world. Heron himself referred in passing to an alternative system used by other engineers: "Some instead of water use quicksilver [mercury] as it is heavier than water and easily disunited by fire." What Heron meant by the Greek word translated as "disunited" is not really known, but using mercury instead of water in a machine similar to Heron's design would certainly have made it more efficient.

Much as we would like to examine such ancient temple wizardry, it is unlikely that the remains of one of Heron's door-opening machines will ever be discovered. In Egypt later Christian and Muslim settlers picked clean the old pagan temples, removing anything that was potentially useful—in the same way that the lead disappears from the roofs of old churches as soon as no one is minding the premises. The metal parts of Heron's automatic doors (chains, pulleys, containers and the fittings for the finely pivoted doors) will have been robbed centuries ago. It is only through the chance finds of smaller objects, such as the extraordinary device from the Antikythera shipwreck (see **Computers**), and the remarkable survival of the works written by Heron and other Hellenistic scientists, that we can still have a few glimpses of a past golden age of mechanical engineering.

THE STEAM ENGINE

The revolution of transport and industry brought about by the invention of the steam engine in the nineteenth century completely changed the world. But like so many key discoveries, the modern realization of the potential of steam power was really a *re*discovery of something learned two thousand years earlier.

The man responsible was the engineering genius Heron of Alexandria, who described in detail the first working steam engine, called by him an *aeolopile* ("wind ball"). His design was elegantly

The *aeolipile* ("wind ball") designed by Heron was a forerunner of both the steam engine and the jet engine.

simple: A large, sealed caldron of water was placed over a source of heat, such as a charcoal fire. As the water boiled, steam rose into two pipes, in the middle of which was pivoted a sphere. Jets of steam escaped from the sphere through two outlets, sending it spinning around at great speed. The principle involved is the same as that of modern jet propulsion.

The preface to Heron's treatise on pneumatics describes some of his inventions as having "useful everyday applications," and others "quite remarkable effects." It seems that the steam engine fell into Heron's second category, to judge from the description he gives, and had only a novelty value. But Heron was not the kind of man to ignore a potential source of energy that could have practical applications—for example he designed a small windmill to power a musical organ (see **Keyboards** in **Sport and Leisure**). Could his steam engine have been harnessed for more practical purposes?

To find an answer to this question, classicist Dr. J. G. Landels, of Reading University, made a working replica of Heron's device with the aid of the Engineering Department. He discovered that the model produced a remarkable speed of rotation, no less than 1,500 revolutions per minute: "The ball on Heron's machine may well have been the most rapidly rotating object in the world of his time."

Nevertheless Landels had difficulties in adjusting the joint between the revolving sphere and the steam pipe in order to make the device efficient. A loose joint allowed the sphere to spin more freely, but it also allowed steam to escape; on the other hand, a tight joint meant that energy was wasted in overcoming friction. Using a compromise, Landels estimated that the efficiency of Heron's machine could have been as low as one percent. Thus, for example, a large-scale version big enough to generate a tenth of a horsepower (the strength of one man) would have consumed an enormous amount of fuel, requiring far more man-hours of work to collect than would be produced by the device.

Could the Greeks Have Invented the Steam Locomotive?

Still it was well within Heron's grasp to have developed a more efficient way of harnessing steam power. As Landels noted, all the elements required for a useful steam engine are found in other devices described by Heron. The engineers of his time made extremely efficient cylinders and pistons, such as those used in Heron's design for a fire-fighting water pump (see **Fire Engines** in **Urban Life**). A suitable valve mechanism for a steam engine is found in his design for a water fountain powered by compressed air, a device similar to a modern insecticide sprayer. It also had a spherical bronze chamber—the best shape to withstand high pressures and one that would have been a great improvement on the sealed caldron used in his prototype steam engine.

It wouldn't have been too difficult for Heron or one of his contemporaries to combine all these elements (boiler, valves, piston and cylinder) to make a steam engine that could do real work. It has even been claimed that Heron *did* go farther in his experiments and combined the necessary elements for an effective steam engine but that he was killed in the attempt or frightened off by the idea. Nei-

Heron also employed steam power in constructing toys. In this example a small sphere is suspended in air by a jet of steam (Renaissance reconstruction).

ther story has any foundation. It is more likely that, if the idea occurred to him, he was simply too busy to pursue it. But there were plenty of other competent and inventive engineers in Alexandria and the Greco-Roman world. So why didn't any of them take the idea farther? The answer seems to be an economic one. The potential of many inventions was never fully realized in the ancient world because of its dependence on a slave-based economy (see **The Reaping Machine** in **Working the Land**). Even if some bright spark had managed to produce a steam engine that could do the work of a hundred men, the new-fangled device would have been of little interest to industrialists, who had a ready supply of man power in the slave markets.

History might well have taken a different turn, however, as the great historian Arnold Toynbee pointed out in a brilliant speculative essay. He considered what might have happened if Alexander the Great had not drunk himself to death in 323 B.C. at the age of thirty-three. Alexander was only partway through his plans of world conquest, having united the Greek world with the Persian Empire and taken his armies as far as Egypt and northern India. Had Alexander lived longer, Toynbee argues, there is no reason why his enormous and ever-growing army would not have gone on to subdue the Romans, the Carthaginian Empire of North Africa, Ethiopia and even China. If he had succeeded in uniting this vast area under one authority, the motivation would have arisen to develop swifter means of transport to connect the far reaches of the empire. As we know, the Greeks had already invented the railway (see **Introduction** to **Transportation**). In Toynbee's model they combined this with an improved version of Heron's steam engine and had steam locomotives running across Asia only a few generations after Alexander.

One of the by-products of Toynbee's speculative foray is too irresistible to overlook. The world religion he envisages under a global empire ruled by a continuous succession of Alexanders is a hellenized version of Buddhism, highly plausible given the extraordinary similarity between the teachings of the popular Greek philosopher Pythagoras and those of the Buddha (see **Introduction** to **Food, Drink and Drugs**). Christianity might never have gotten off the ground. Toynbee refers in passing to a failed prophet whose words fell on stony ground and who lived by the railway cuttings at Nazareth.

AUTOMATA

The concept of robots, machines so skillfully crafted that they can perform the tasks of human beings and serve their creators' every whim, was no stranger to the ancient Greeks. It can be found in their earliest surviving poetry, composed by Homer in the eighth century B.C. In his great epic the *Iliad* he describes some of the technological wonders produced by Hephaistos, god of metallurgy and craftsmanship. As helpers he manufactured mechanical golden maidservants, endowed with the gift of speech and intelligence. Hephaistos also owned a set of three-legged tables, which stood on golden wheels around his palace; they could move by themselves and follow him, doglike, to a meeting of the gods, or run back home at his command.

Though fictional, Hephaistos's creations typify the ideals toward which some ancient Greek engineers aspired. The great inventors of Alexandria, such as Philon and Heron, were particularly fascinated by automata. (An automaton, defined as "an organized machine which imitates the movement of a living body" is not quite a robot, which specifically mimics the behavior of human beings.) Working in Egypt, they were heir not only to the engineering genius of Greek scientists such as Archimedes but also to native Egyptian traditions. For some two thousand years the Egyptians had been making simple mechanical toys (see **Introduction** to **Sport and Leisure**), as well as arranging "marvels" in their temples such as statues with nodding heads and concealed speaking tubes that gave the illusion of talking gods. While Heron designed various special effects for the temples, he shifted the emphasis toward a particularly Greek medium—the theater. Some of his creations were sophisticated precursors of our own stage effects, such as thunder and lightning machines (see **Theaters** in **Sport and Leisure**). Other devices, however, created the illusion of mechanisms magically endowed with a life of their own.

Heron's Automatic Theaters

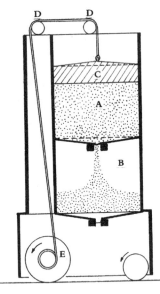

The basic design of Heron's self-moving stands. At the center of the device was a compartment (A) filled with grain. When the stopper at the bottom was removed, the grain trickled into a lower compartment (B). As the upper compartment emptied, the large weight (C) resting on the grain descended, pulling the rope over the pulleys (D) and turning wheel (E). Thus as the weight fell, the whole stand moved along.

Some of Heron's devices bear an uncanny resemblance to Hephaistos's self-propelled tripods. These were stands on wheels that ran around the theater as if by their own volition. In fact they were powered by gravity. A large weight connected by ropes and pulleys to the wheels rested on top of a container of grain. As the grain flowed through a hole in the bottom of its container into a recess below, the weight sank, drawing the ropes taut and turning the wheels of the automaton. The device was like a combination windup toy and giant egg timer, and the genius of Heron's design was the successful combination of the two elements.

The moving stand was merely the basic element in some of Heron's more complex arrangements. He designed variations that could trace circular, rectangular and figure-eight paths or even change direction and go into reverse—all managed by systems of weights and pulleys within the stands that altered the position of the wheels or lowered extra sets of wheels during their progress. The stands not only moved to and fro without human assistance, they could also carry mechanical puppets, powered, like the movement of the entire device, by falling weights. One of Heron's most complex designs was a miniature theater on wheels that rolled itself into place in front of the audience before putting on a show. Doors opened and closed, tiny altars lit up and mechanical figures revolved and moved about. At the end of the performance the whole device rolled off again.

Heron also designed stationary automatic theaters, which allowed for even more complex puppet movements as no power or gearings was needed to shift the thing about. Instead of grain, sand was used as the timing; slower running, it allowed a longer performance.

One miniature theater was programmed to enact an entire play called the *Nauplius*, a tale of tragedy and revenge set in the aftermath of the Trojan War. The son of King Nauplius has been falsely accused of treason by his comrade-in-arms Ajax and stoned to death. Nauplius plots revenge with the aid of the goddess Athena. When the curtain lifted, mechanical figures of nymphs were seen at work, repairing the ship of Ajax with hammers and saws. "There is a great noise," wrote Heron, "as of the sound of actual working." After a while doors closed and opened again on the second scene, the launching of the ship at sea. The third scene opened with an empty sea, and then the Greek fleet was seen sailing past in a line, while dolphins leaped in and out of the water; the calm sea became stormy and the ships drew in their sails. The door closed and opened again, and the ships had gone; the whole stage was lit up by a raised beacon carried by Nauplius to lure the Greeks to their death on the rocks, while the goddess Athena stood by approvingly. In the final scene the audience saw a shipwrecked boat. Ajax struggled in the water as Athena ascended into heaven. Thunder crashed as she hurled a bolt of lightning directly onto the swimming Ajax, who disappeared beneath the waves. Nauplius's revenge was complete.

Heron recounted his special effects with some relish, going on to explain in detail how each movement, from the leaping dolphins to the hammering nymphs, was produced by hidden mechanisms. The whole device was powered by weights. There were no microchips

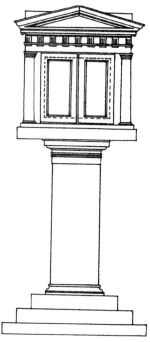

The exterior of Heron's miniature automatic theater (stationary version). The doors opened and a tragedy called the *Nauplius* was performed by mechanical figures. All the driving mechanisms were hidden in the pedestal.

The mechanism used by Heron in his miniature play *the Nauplius* to make a model nymph do hammering work. Weights attached to the large wheel swung it around, so that the pegs caught and depressed one end of the level. This raised the other end, to which the hammering arm was attached. When the peg clicked past the lever, the arm came crashing down, only to be lifted again by the next peg.

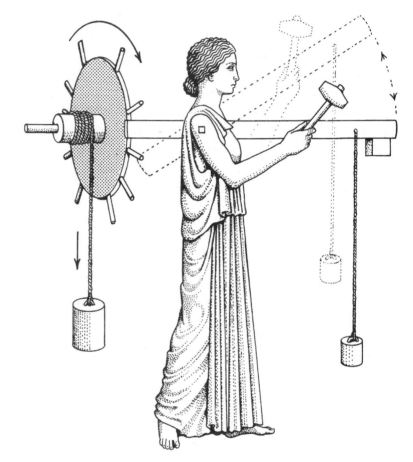

involved, but within the technical limitations of his day, Heron had achieved the effective programming of simple automata. His audiences must have been amazed.

The Book of Ingenious Devices

Alexandria had gone into decline by the third century A.D. Over the following centuries its museum, library, temples and theaters— centers of pagan philosophy and culture—suffered at the hands of the Christians. They were also the centers of learning and science and, by the time the library was burned during the Arab invasion of Egypt in A.D. 640, the golden age of Greek science had long finished. But some manuscripts survived to be treasured by the Arab conquerors. During the Dark Age of Europe that followed the collapse of the Roman Empire, the advanced engineering skills pioneered in Alexandria were preserved and developed in the Islamic world.

In the ninth century A.D. Heron's main successors were three Iraqis, the Banu Musa brothers of Baghdad. They were brilliant all-around scientists; two of the brothers came very close to a realistic estimate of the earth's circumference (see **Mapmaking** in **Transportation**). Their work on technological gadgets, with the wonderfully explicit title *The Book of Ingenious Devices*, still survives. It was described with awe by a later medieval Arab writer: "There exists a book that mentions every astonishing, remarkable and nice mechanical contrivance. It is often difficult to understand, because the geometrical proofs occurring in it are difficult."

Some of the gadgets described in the book are eminently practical, such as an oil lamp that raises its own wick and feeds itself more oil when needed. Others share Heron's delight in baffling an audience. The bulk of the devices are novelty vessels of various kinds that developed the pneumatic and hydraulic skills of Alexandrian science to an extreme of cunning. These were also automata in the sense that once the human element had intervened (for example, by pouring some liquid into the vessel), something else seemed to take over to produce a miraculous effect. In one of the Banu Musa designs for trick vessels, three colored liquids are poured in succession into a hole at the top; after a while they emerge in the same order from a spout at the bottom. To achieve this relatively simple effect, the vessel was stuffed with an incredibly complex system of tubes, tanks, valves, floats and syphons. Donald Hill, the leading expert on medieval Arab engineering, was moved to write, "Nothing like it is known to have been attempted before or since, until the advent of modern pneumatic instrumentation. Indeed, they had exhausted the subject, and it would have been impossible to emulate them in this kind of construction."

The Banu Musa brothers were pioneers in an extraordinary renaissance of high-tech engineering in the medieval Islamic world. The caliphs of Baghdad exploited it to create private playgrounds that must have rivaled Disneyland. Contemporary reports of the palace built by the Caliph in the early tenth century A.D. tell of a pond flanked by moving statues of mounted warriors. In the lake was a tree made of silver, with mechanical whistling birds of silver and gold. Another pond was made of mercury, on which gold boats floated. The gardens around the ponds were also decorated with automata—singing birds, roaring lions and other moving creatures.

The idea of Sinbad the Sailor encountering mechanical monsters doesn't come from the feverish imaginations of modern film directors. It comes straight from the traditional tales in *The Arabian Nights*—and these reflect the brilliant technological achievements of the golden age of Islamic engineering during the Middle Ages.

Miniature automata, reminiscent of Heron's automatic puppet theaters, also came back into fashion in the Islamic world. Three hundred years after their death the work of the Banu Musa brothers was still being carried on, notably by al-Jazari, another Iraqi scientist. His book on machines, completed in A.D. 1206, has numerous clock designs, all with automata that indicated the passage of time in different ways, including puppet musicians that struck drums or blew trumpets, and birds that dropped stones onto cymbals.

The "Cosmic Engine" clock completed by Su Sung in China in A.D. 1090 has strikingly similar automata: It had twenty-four wooden figures, each holding a tablet displaying an hour of the day, which popped in and out of a little door. Others struck drums and bells, while one puppet even played a miniature stringed instrument at regular intervals during the day and night (see **Clocks**). Clearly there must have been some exchange of ideas between Arab and Chinese engineers during the Middle Ages—the only uncertainty is who influenced whom.

The South-Pointing Carriage

Ancient China also produced a magnificent automaton known as the south-pointing carriage, which long baffled modern scientists. Many Chinese legends refer to the mysterious carriage. It was a chariot surmounted by a figure with an outstretched arm that continuously pointed south no matter which way the chariot turned—and thus could have been an invaluable guide to travelers. Some traditions ascribed the building of the south-pointing carriage to the twenty-third century B.C., a date once accepted by Western scholars, who also long believed that the device worked by magnetism.

Both these misunderstandings were finally cleared up in the 1960s by Dr. Joseph Needham and his Chinese collaborators. Examining the literary evidence closely, they discovered that the earliest certain reference to such a carriage came from the third century A.D. Nonetheless it is still amazingly ancient for such a complex device, which employed no magnetism but was instead a highly sophisticated automaton.

The human figure on top always "knew" where south was because it was connected to the wheels by a complex train of gears—similar to that in a modern automobile. Known today as differential gears,

A reconstruction of the "south-pointing carriage," on display in the Science Museum, London.

such trains allow the wheels of a car turning a corner to move at marginally different speeds (one having farther to go than the other), though powered by the same engine. The south-pointing carriage employed the same principle, with gears that translated the differences in the speed of the wheels as the vehicle turned into a correcting motion for the pointing figure. But as Needham stressed, such a device can only work if every component is manufactured with the highest degree of precision. An error of only one percent between the circumferences of the wheels would lead to the south-pointing figure being wildly off after only a few miles. Not only the wheels but also each of the gears must have been constructed with an extraordinary degree of accuracy to achieve the sensitivity required for what Needham has described as the world's "first cybernetic machine."

EARTHQUAKE DETECTORS

China has always been plagued by earthquakes. The dynastic histories detail numerous earthquakes over the centuries, such as the great disaster of February 2, 1556, in which more than 800,000 people died. Chinese emperors were deeply concerned about major earthquakes, since they frequently sparked social unrest in the form of food riots or even rebellions. To maintain control, the government needed to send both food and troops to the suffering region as quickly as possible. Some form of advance warning would therefore have been of great value.

In the early second century A.D. a remarkable device was presented to the Chinese court by Chang Hêng, an extraordinary polymath who was not only the astronomer royal but also a distinguished cartographer, mathematician, poet, painter and inventor (see **Mapmaking** in **Transportation**). He had produced a mechanism that he called an "earthquake weathercock," which he claimed could detect earthquakes at a distance—in other words a seismograph. The device was housed in a decorated barrel-shaped case, around the top of which were eight dragon's heads, each holding a bronze ball in its mouth. When an earthquake occurred, a ball would fall from a dragon's mouth into the open mouth of a bronze toad below. The clang would signify that there had been an earth tremor, while its direction was taken as being opposite to that in which the ball fell.

The amazement Chang Hêng's invention occasioned in the court is reflected by the state history for the year A.D. 132: "Nothing like this had ever been heard of since the earliest records began." Indeed, like many inventors across the ages, Chang Hêng at first came up against a skeptical audience. The court functionaries refused to believe that his device would really work, until a dramatic demonstration converted them. The official historian recounts the incident:

> On one occasion one of the dragons let fall a ball from its mouth though no perceptible shock could be felt. All the scholars at the capital were astonished at this strange effect occurring without any evidence [of an earthquake having occurred]. But several days later a messenger arrived bringing news of an earthquake in Lung-Hsi [four hundred miles northwest of the capital]. Upon this, everyone admitted the mysterious power of the instrument. Thenceforward it became the duty of the officials of the Bureau of Astronomy and Calendar to record the directions from which the earthquakes came.

But how did the device work? The ancient sources give few practical details, except for noting that inside there was "a central column capable of lateral displacement along tracks in eight directions, and so arranged that it would operate a closing and opening mechanism." Various reconstructions have been offered by twentieth-century scientists, but it is now generally agreed that the "central column," the key to the mechanism, must have been an inverted pendulum with a weighted bob at the top that would react to the shock waves released by an earthquake.

This is the principle used in the best reconstruction, built by Imamura Akitsune, of the Seismological Observatory of Tokyo University in 1939. This reconstruction—the diagram follows the slightly modified version of British historian of science Robert Temple—worked as follows: A shock wave from the earth tremor tilted the pendulum so that the spike at the top would swing into one of eight surrounding slots. These contained sliders, the ends of

A modern reconstruction of Chang Hêng's earthquake detector of A.D. 132. It uses an inverted pendulum with a weighted bob near the top, ending in a spike that can slide along any one of the eight different channels cut in the surrounding plates. As it enters a channel, it pushes the slider within that channel farther into the dragon's throat, ejecting the ball from the dragon's jaws. The ball falls into the open mouth of the toad waiting below. (Above left) The two plates with the eight channels cut into them; the tip of the pendulum is represented as a central dot. (Above right) Three of the sliders and balls are shown in their starting position.

which led into the dragons' mouths. When the spike swung into one of the slots, it would dislodge the slider, which in turn would eject a ball from the dragon's mouth. Trials on an actual model built by Akitsune showed that the device is indeed effective, though in some circumstances the direction of an earthquake's epicenter was found to be at right angles to the dropped ball.

Like many other Chinese inventions, the seismograph was forgotten but reinvented there a few hundred years later, and then lost again by A.D. 1300. The first modern seismograph was invented only in 1703 by the French scientist Abbé de Hautefeuille, who knew nothing of his great Chinese predecessor.

There is, however, an intriguing piece of evidence that suggests that news of Chang Hêng's invention may have reached Europe many centuries earlier. During the Middle Ages the Roman poet

Virgil, who composed Rome's epic the *Aeneid* in the last century B.C., acquired a (wholly unwarranted) reputation as a magician. According to one tradition, which may date back to the eighth century A.D., he installed in his palace a series of wooden statues representing the provinces of the Roman Empire. When trouble struck one of the provinces, its representative would ring a bell, while a bronze horseman on top of the palace would swivel around and point his spear in the direction of the trouble.

This tradition seems to be a curious mixture, blending impressions of the automata constructed by Greco-Roman and Arab engineers (see **Automata**) with a story about a system providing advance warning of disasters in remote provinces. If a garbled account of a scientific instrument is behind the story, it could only concern something like the "earthquake weathercock" invented by Chang Hêng—and it would be only one of many instances in which reports of such advanced discoveries traveled between China and the Mediterranean in ancient times.

ELECTRIC BATTERIES

Scattered through the writings of the Greeks and Romans are a number of clues suggesting that knowledge of electricity in ancient times may have been surprisingly advanced.

Natural electrical phenomena, such as the shock power carried by some fish, were keenly observed. The Roman poet Claudian described how a torpedo fish (or electrical ray) caught on a fishing hook could send its "effluence" along the line and rod to stun the fisherman. Ancient Babylonian doctors applied electric fish as a local anesthetic.

Lightning, the most powerful evidence of electricity visible to the naked eye, particularly intrigued the ancients. While most people believed that lightning was supernatural, the weapon of sky gods such as Zeus or Jupiter, the philosophers knew better. They understood lightning as the manifestation of a fundamental force. Some even held that lightning was the motive power behind the universe, an idea curiously close to our modern understanding that electricity binds together all matter.

We are usually told that the Greeks and Romans, because they had no specific word for electricity, knew nothing about it. Yet were they really so ignorant? Take the Roman writer Lucretius, whose philosophy was based on the theory that matter is composed of atoms—another idea that is often wrongly thought to be a modern invention. Lucretius described lightning, in the vocabulary available to him, as a kind of "rarified fire . . . composed of minute and mobile particles to which absolutely nothing can bar the way." He also made the acute observation that thunder and lightning occur simultaneously, even though lightning is usually seen before thunder is heard. Lucretius could even explain why—he knew and stated that sound travels more slowly than light.

Static Electricity

On a smaller scale, electricity could be studied firsthand through the mysterious properties of amber, the fossil resin of pine trees. Rubbed against a material such as fur, amber easily becomes charged with static electricity. It can then be used to attract small pieces of lightweight material, such as straw or paper. Chinese scientists made the same observation as the Greeks and Romans—a work from A.D. 83 describes how amber can pick up mustard seeds.

This "magical" property, as well as its beautiful color and ultrasmooth surface, made amber one of the most sought-after commodities of the ancient world. The Chinese imported it from Burma, and it was traded throughout Europe from the coasts of the North Sea and Baltic since Stone Age times. Amber disks, pendants and other ornaments were great status symbols, prestigious objects placed in the graves of the rich and powerful along with metalwork and precious stones. The Greeks were particularly fond of amber. As early as 1600 B.C. it was being imported in large quantities by the warrior kings of Mycenae to be buried in their magnificent Shaft Graves.

Greek scientists experimented with amber and tried to explain how it could draw certain materials "just in the same way as a magnet attracts iron." Yet despite the similar behavior of magnets and charged amber, they seemed to appreciate that a different force was involved. Lodestones—the naturally occurring magnets known to the Greeks—affect iron particles by magnetism, while amber picks up scraps of material through static electricity. Many centuries later the Elizabethan physicist William Gilbert started again at the point where the Greeks left off by studying the properties of amber. He called its attractive power *vis electrica*, from *electron*, the Greek for amber. So our word *electricity*, first attested in A.D. 1646, ultimately derives from the work of the ancient Greek thinkers.

One detail of the Greek stories about the origin of amber is particularly intriguing. They said that the spot where most amber came from on the North Sea coast was where a mythical character called Phaethon had been struck down by lightning. This suggests that the Greeks saw some connection between lightning and the properties of amber. In other words they may have been very close to understanding electricity as a distinct form of energy that manifests itself in many different ways.

Fifty-million-year-old insects entombed in amber, a fossil resin prized in ancient times for its color and mysterious electrical properties. The piece was found on the coast of Denmark, washed up from a prehistoric forest now under the sea.

Textbook versions of scientific history may insist that electricity as such was not "discovered" until the seventeenth century, but this can no longer be held to be true. Archaeological evidence has shown that electricity was actually being used some two thousand years ago.

The "Baghdad Battery"

The collection of the Iraq Museum, Baghdad, includes a small, plain clay jar that, in spite of its unprepossessing appearance, has been hailed as the most amazing find in the archaeology of science. For, although it is some two thousand years old, the jar appears to be, to all intents and purposes, the casing of an electric battery.

The enigmatic object was found in June 1936, after workmen moving earth for a new railway near Baghdad accidentally uncovered an ancient tomb. Archaeologists moved in and found that the tomb was part of a settlement of the Parthian period (roughly 250 B.C. to A.D. 250). Their excavation produced a wealth of material, including engraved bricks, pottery, glass and metal objects, plus the oval-shaped clay jar with its curious contents—a copper tube with one closed end, an iron rod, and some crumbling pieces of bitumen (asphalt).

Intrigued by reports of the discovery, physicist Walter Winton of the Science Museum in London examined the jar when visiting Baghdad. He was impressed: "Put some acid in the copper vessel—any acid, vinegar will do—and—hey presto!—you have a simple cell which will generate a voltage and give a current of electricity. Several such cells connected together in series would make a battery of cells which would give enough current to ring a bell, light up a bulb or drive a small electric motor."

That the object was an electric battery was, Winton stated, "completely obvious and completely credible." His only doubts sprang from its unique nature. Archaeological "one offs" are always the most difficult kind of find to interpret. In fact, though unknown to Winton, other jars had been discovered at the Parthian city of Ctesiphon, near Baghdad. They were found with various magical objects such as amulets, which suggests that alchemists used the jars

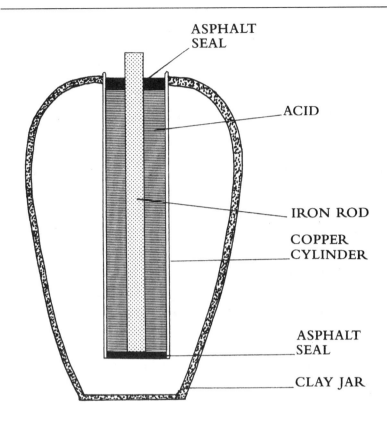

ASPHALT
SEAL

ACID

IRON ROD

COPPER
CYLINDER

ASPHALT
SEAL

CLAY JAR

A reconstruction of the ancient "Baghdad battery." The clay jar (just over seven inches in height) contained a copper cylinder sealed with asphalt (bitumen). Inside that was an iron rod. If the cylinder was filled with acid to act as an electrolyte (a conducting solution), an electric current would flow between the iron rod and the copper casing.

but still gives us no clue to their purpose. Ideally, as Winton noted, the pot should have been found with connecting wires, or even better a series of such jars joined together, which would have left no trace of doubt. Yet as Winton argued in 1967, if it wasn't a battery, what else could it be? "Not being an archaeologist, I jumped straight to the easiest scientific solution. I still can't see what else it would have been used for, and if there has been a better solution, I haven't been told it."

Twenty-five years later nobody has been able to suggest any other plausible explanation for the mysterious jar. And the overriding fact remains that it *does* work perfectly well as an electric cell. Replicas of the jar and its contents have been tested in two separate experiments in the United States. Topping up the copper tube with acetic acid (vinegar or wine) as an electrolyte—sulfuric or citric acid would also work—the model produced a current of one-half volt for eighteen days.

A Secret Art

Yet what would anyone have been doing with batteries two thousand years ago? Paul Keyser, of the University of Colorado, has recently suggested that they were used by Babylonian doctors as a substitute for electric fish in local anesthetics. But the best suggestion remains that made by German archaeologist Wilhelm König, director of the Iraq Museum Laboratory, who examined the "Baghdad battery" in 1938. He felt that the current from a number of these cells, joined together in series, could have been used for electroplating metals. The voltage produced by the experimental replicas is sufficient for the job.

A primitive method of electroplating is in fact still used by local craftsmen in Iraq to cover copper jewelry with a thin layer of silver. It may be that the technique has been handed down through the centuries from the Parthian period or even earlier. The Parthians were heirs to a Near Eastern tradition of scientific investigation going back more than three thousand years—and to the more recent ingenuity of the Greeks, who invaded the region with Alexander the Great in 330 B.C. As we have seen, the Greeks had rudimentary electrical theories and experimented with static electricity and magnetism. Magnetism was actually used by the Greeks to work some of the mechanical marvels displayed in their temples (see **Magnets and Magnetism**).

Exactly how far the ancient experiments with electricity went we may never know. The craftsmen of ancient Iraq carefully guarded their technical knowledge; Babylonian tablets giving the formulae for making colored glass (see **Jewelry** in **Personal Effects**) were often shrouded in jargon so that they were only accessible to the cognoscenti. The secrets of electroplating would certainly have been jealously guarded and perhaps never committed to writing in a straightforward way. Fortunately there are literally hundreds of

ANCIENT LIGHTNING RODS

The knowledge acquired by the ancients through observing lightning seems to have been put to practical use. There is evidence that some of the masts attached by the Egyptians to the pylons (gateways) at the front of their temples were intended to be lightning conductors. Inscriptions from the temple at Edfu, completed by Ptolemy IV in 212 B.C., record how copper-covered masts were built into the pylons: "This is the high pylon of the god of Edfu at the throne of Horus, the light-bringer; masts are arranged in pairs in order to cleave the thunderstorm in the heights of the heavens."

More than a thousand years earlier the Minoans of ancient Crete may also have had lightning rods. Many of their small temples were located on the peaks of mountain ranges, and Greek archaeologist Chryssoula Kardara has argued that they were associated with a cult of thunder worship whose aim was to bring down rain by inducing lightning. Kardara has produced some intriguing evidence from Minoan drawings of peak sanctuaries, which usually have spearlike masts at the front that are not unlike the Egyptian ones. Her argument is that these masts, if made of metal, would attract lightning from passing storm clouds and produce rain. The idea is attractive and would explain why the later Cretans told how the smith gods of olden times, the Telchines, could draw down lightning and rain from the sky. Interestingly the ancient Cretan masts are pointed, the shape for lightning conductors recommended by Benjamin Franklin when he reinvented them after his famous kite experiment in 1752.

Depiction of a Cretan "peak sanctuary," from a Minoan vessel of about 1500 B.C. It has been suggested that the spear-shaped masts at the front of the sanctuary were lightning rods.

mounds still unexcavated in Iraq and thousands of scientific tablets in museums awaiting translation. Perhaps the safest thing to say about the extent of ancient knowledge of electricity is that further surprises are likely.

MAGNETS AND MAGNETISM

In the fifth century A.D. Saint Augustine cited the invisible power of magnets as proof that phenomena can exist that defy rational explanation. With his own eyes he had seen iron rings jump into the air and latch onto each other; and, on the authority of a fellow bishop who had witnessed it at a dinner party, he described how a piece of iron could be moved around a silver dish by waving a magnet underneath. While it may seem surprising that magnets were so familiar to one of the founding fathers of the Church, knowledge of magnetism is extremely ancient—indeed it was far better understood in early Greek and Roman times than it was in Augustine's day.

Greek tradition has it that magnetism was originally discovered by a shepherd strolling through the mountains of western Turkey. His feet suddenly became stuck to the ground, the nails in his sandals caught by the natural magnetic power of an outcrop of lodestone. The account is legendary, but the lodestone-rich country near the town of Magnesia, where the story is set, has ever since given its name to the mysterious, and useful, force by which certain substances (usually iron but also cobalt and nickel) can attract or repel other pieces of metal.

The "official" discovery of magnetism belongs to the Greek scientist Thales of Miletus, who, in the sixth century B.C., attempted the earliest known explanation of the invisible power of lodestone: "The magnet has life in it because it moves the iron." It is unlikely that he believed a magnet was really "alive" in the same sense as a living creature, only that it shared the power of movement. Later Greek philosophers continued to be fascinated by the properties of lodestone and, through endless experimentation, became familiar with many of the basic laws of magnetism. The principle of magnetic induction—the ability of a magnet to transfer its properties to other pieces of iron—was clearly understood. A passage in the works of Plato (4th century B.C.) gives the following description: "The magnet . . . has power to attract iron rings. More, it invests those rings with a power like its own, so they can attract other rings, and the result is often a long chain of rings hanging one from another, though in all of them the magnetic power is derived from the original stone."

The Greeks also distinguished the attractive power of the magnet from the similar property induced by rubbing amber (see **Electric Batteries**). And, far from Saint Augustine's miraculous view of the phenomenon, ancient classical scientists produced some very down-to-earth theories not dissimilar to our own. The closest call was made by the Roman philosopher and poet Lucretius, whose explanation was couched in terms of a stream of invisible particles. In the works of scholars like Lucretius we find nearly all the basic pieces of the puzzle for the modern understanding of electromagnetism: the theory of atoms, the knowledge that they could emit invisible particles, and the understanding that electricity and magnetism were related but different phenomena.

Magnetism and Medicine

While on the verge of drawing all these strands together into a unified theory, Greco-Roman scientists never quite arrived at a satisfactory explanation of magnetism. Yet this didn't prevent them from exploiting it on a practical level. Lodestone was highly prized as a mineral throughout the ancient world and was extensively mined. But, apart from its inherent scientific interest and its novelty value for party tricks, what use did the ancients put it to?

The Assyrians had an aphrodisiac use for powdered lodestone that is naively comical: By placing it on his penis a man could supposedly attract a woman whose genitals were dusted with iron filings (see **Aphrodisiacs** in **Sex Life**)! By contrast we must take seriously the extensive use to which powdered lodestone was put by the ancient Greeks in medicines, particularly in the form of ointments for healing burns and salves for irritated eyes. The efficacy of these preparations is uncertain—indeed it never seems to have been examined by twentieth-century medical science. However, given the increasing experimental use of electromagnetism in modern healing processes and pain alleviation, a renewed study of the medicinal properties of lodestone as used by ancient Greek doctors would surely be worthwhile.

More straightforward was the use to which magnetism was put in ancient Chinese medicine. As early as the Sung Dynasty (A.D. 960–1279), lodestones were used to extract small iron objects or splinters that had settled in the eye or throat. The same practice is used today by modern surgeons.

Statues Suspended in Air

On a grander scale, lodestone was employed to produce some spectacular effects in ancient temples. The Roman poet Claudian left a fascinating description of some famous effigies of Venus and Mars, most likely housed at a temple in his home town of Alexandria. The beautiful statue of Venus was carved from a large piece of lodestone, while that of the warrior god Mars was made of pure iron. During festivals the two figures were introduced to each other on a nuptial couch spread with roses; to the accompaniment of music they were brought steadily closer until Mars leaped forward to clasp his lover in a magnetic embrace.

The use of magnetism in such temple wizardry was developed further. In the reign of Ptolemy II of Egypt (3rd century B.C.), the Greek architect Timochares planned a statue of the king's sister (and wife) Arsinoë, which would hang hovering in the air—suspended by the force of magnets strategically placed in the roof and walls. Both ruler and architect died before the project could be completed, but the idea was not forgotten. It was said to have been put into effect in the Temple of Serapis, in Alexandria, where an image of the Sun god hung—suspended by magnetism—until that temple was destroyed by the Christians in A.D. 391. Another floating magnet wonder, a statue of the god Mercury, was displayed in a Roman temple at Tréves, in France.

A Magnetic Security System

The ingenuity of the ancient Greeks and Romans in exploiting the power of magnetism was surpassed only by the Chinese. Unlike the scientists of the classical world, who never seem to have discovered the secret of making artificial magnets, the Chinese manufactured them on an industrial scale for use as compass needles—perhaps the most practical use to which the magnet was put in ancient times. There is some doubt, however, as to whether the Chinese or the Olmecs of South America were the first to invent the compass (see **The Compass** in **Transportation**).

The most ambitious use to which the ancient Chinese put lodestone is described in a traditional story about Shih Huang Ti, first emperor of China (221–210 B.C.). His tomb, near the city of Xian, is of megalomaniac proportions. Covered by an enormous mound of earth, it includes a complete subterranean palace. It was on the boundary of this burial complex that the deposit containing the now world-famous army of eight thousand terra-cotta warriors was discovered in the 1970s. The burial chamber of the emperor was said to have been filled with unrivaled treasures, including a model of the empire, with mercury rivers (see **Mapmaking** in **Transportation**). The tomb's door, it was said, could not be cut through with iron weapons, because the material of which it was made attracted them and held them fast. This tradition is paralleled by an account contained in some recently discovered fragments of the encyclopedia known as the *Yung-Lo Ta Tien* (compiled in A.D. 1406). This describes a similar security arrangement at another of Shih Huang Ti's buildings, the Ah Fang Palace, in the city of Hsienyang: "The gate was built of magnetic stone. Warriors wearing iron armor were detained or attracted and could not pass through."

These stories were examined by L. C. Tai, of the Research Institute of Iron and Steel, Beijing, who found them quite plausible. Strong circumstantial evidence supporting them has been provided by the discovery that the sands of the Wei River basin, near the cradle of ancient Chinese civilization, are extremely rich in magnetic iron oxides. Tai analyzed the oxides in the sands and found them highly magnetic—in fact of the same quality as laboratory-manufactured oxides. Further, the oxides can be fused together by gentle heating (sintering) to form powerful magnetic rocks of any size required.

The world was amazed when the terra-cotta army was unearthed

In Medieval China magnets for compasses were manufactured in small factories. This scene comes from a book of A.D. 1637 but shows techniques that had remained unchanged for hundreds of years. Needles, after being heated till red-hot, are pointed north and hammered by a seated craftsman. (Hammering shakes the molecules up and leaves them aligned in a north-south direction, forming an effective magnet.) The second craftsman makes holes in the needles with a bow drill.

on the fringes of Shih Huang's burial complex. The massive task of excavating the tomb itself has yet to be undertaken—but it is guaranteed to uncover a wonderland of ancient scientific marvels, one of which may well be the actual remains of the world's earliest magnetic security system.

MAGNIFYING GLASSES

In 1853 Sir Austen Henry Layard returned from his excavations at Nimrud, one of the capitals of the ancient kingdom of Assyria in northern Iraq. Of the many treasures he submitted to the British Museum, one particularly intrigued him. It was a small oval piece of polished rock crystal, about one-quarter inch thick, in the shape of a lens with one flat surface and one convex, which he had found among a collection of glassware of the ninth to seventh centuries B.C. Layard consulted Sir David Brewster, a famous physicist and specialist in optics, who pronounced that the mysterious object could have been used "either for magnifying or for concentrating the rays of the sun."

Cross-section of the Layard lens—actual size

Layard was clearly astonished. "Its properties," he noted, "would scarcely have been unknown to the Assyrians and consequently we have *the earliest known specimen of the burning and magnifying glass.*"

As Brewster noted, the lens works as a burning glass only "very imperfectly," so that its most likely purpose was magnification. The use to which the ancient Assyrians would have put such optical aids is easy enough to guess. Their craftsmen followed a long tradition in Mesopotamia of manufacturing intricately carved seals, most commonly in the shape of cylinders that were rolled on clay to leave an impression. As a matter of course archaeologists study these impressions with photographic enlargements or by using a magnifying glass—simply because the details on many seals are not clearly visible to the naked eye. It seems reasonable enough that the craftsmen themselves employed some optical aid.

Mesopotamian scribes could also produce breathtakingly miniscule texts on clay tablets using their wedge-shaped script (see **Intro-**

duction to **Communications**). The characters on one example written about 2000 B.C.—from the early Mesopotamian civilization of Sumer—are so tiny that they prompted Samuel Noah Kramer, a senior authority on the Sumerians, to write, "We wonder how the ancient scribe succeeded in writing them and how, once written, he could read them without magnifying glass or microscope."

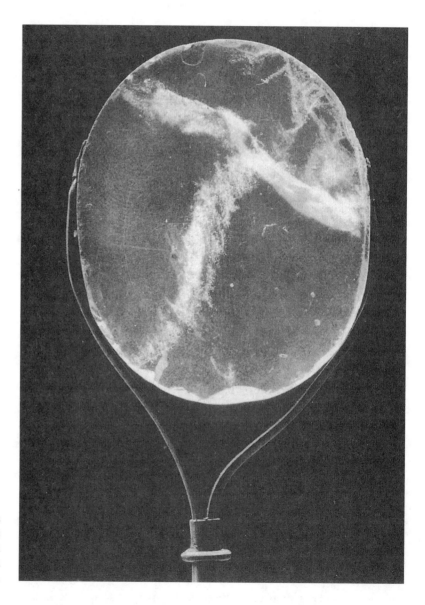

The ancient Assyrian lens found by Layard, as mounted for display in the British Museum (London).

Myopes or Microscopes?

The lens from Nimrud is not an isolated example. In the century and a half that have elapsed since Layard's discovery, similar rock-crystal lenses have turned up in archaeological excavations throughout the Mediterranean and Near East. Two were found at Gordion, the ancient capital of King Midas, in central Turkey. Twenty-three lenses of optical quality are on display at the Heraklion Museum of ancient Cretan civilization. As many as fifty were reported as having been found in the excavations of Troy, though only a handful have been properly published.

The rock-crystal lenses from these sites belong to a huge period of time, spanning the late third millennium to the second century A.D., when the Roman Empire was in decline. Some of them have impressive magnifying powers. One lens, probably of the fifth century B.C., found in the sacred cave on Mount Ida in Crete, can magnify with perfect clarity up to seven times. If you hold it farther away from the object viewed, it will actually magnify up to twenty times, though with considerable distortion.

By comparison the lens found by Layard is extremely feeble, being just strong enough to "demonstrate magnification." (It enlarges things by 0.5 times.) Nevertheless this was sufficient to start a controversy that has been simmering for well over a century. Though there is a mass of archaeological evidence that the ancients could manufacture perfectly good magnifying glasses, there have always been skeptics. Despite the fact that such lenses *work* as magnifying glasses, and although there was a clear need for such instruments in the manufacture of minutely decorated seals and gemstones, some scholars have gone to extraordinary lengths to explain away the archaeological evidence.

Ornaments or jewelry have frequently been suggested as the "real" function of ancient lenses. In 1984 Professor Cyril Smith, a noted metallurgist and historian of science at MIT, went as far as dismissing all the excavated examples as "mere baubles." Three years previously Leonard Gorelick and John Gwinnett, medical scientists at the State University of New York (Stony Brook), had advanced the hypothesis that there was no need for magnifying glasses in the ancient world. When they hold small objects near their eyes, people with myopia (nearsightedness) can see them much more clearly than normal-sighted people. Gwinnett and Gorelick argue that the miniature work of the ancient world was done by myopic craftsmen. As a tendency toward myopia is inherited, they feel that countless generations of interbreeding between craftsmen would have increased this tendency.

Many ancient craftsmen employed in close work were probably myopic—this would only be natural. But it does not mean that they would have had no use for magnifying glasses. Unfortunately Gorelick and Gwinnett take it for granted "that magnification, using eyeglasses, was not invented until the thirteenth century A.D." While spectacles were not invented until this time (see **Spectacles** in **Personal Effects**), classical sources show that the principle of magnification was understood more than a millennium earlier.

For example Seneca, Roman philosopher and tutor to the Emperor Nero (A.D. 54–68), noted that tiny letters could be magnified and made readable by looking at them through a water-filled ball of glass (which would work in exactly the same way as a convex crystal lens).

There is also a statement in Pliny that Nero would watch the Roman Games through a precious stone called *smaragdus*, most probably an emerald. A single magnifying glass would not help in improving long-distance vision. But two convex lenses together could act as a telescope, and it has been suggested that Nero's visual aid was a precursor of this seventeenth-century discovery—a concept that historians of science generally find hard to swallow. Alternatively Nero may have used a monocle made of a concave lens. It is lenses of this shape that are used to correct nearsightedness today. Such lenses were known in the Greco-Roman world—a number of concave rock-crystal examples were discovered at the Greek Temple of Artemis at Ephesus in western Turkey, and deemed by their excavators to be optical instruments. Interestingly Pliny notes that most *smaragdi* gemstones are "concave in shape, so that they concentrate the vision," a statement that, in itself, is enough to debunk the notion that the Romans had no knowledge of the optical properties of lenses.

Gorelick and Gwinnett do accept that lenses were used in classical times as burning glasses. The evidence for this is indisputable (see

Box on page 163). But a lens that can focus the sun's rays must also magnify. They therefore accept that "the use of magnification during Greek times was recorded, but only for use as a burning glass." This amounts to admitting that the Greeks had magnifying glasses but claiming they never used them! Gorelick and Gwinnett's arguments also gloss over the archaeological finds. Presumably they would dismiss all the powerful lenses now excavated from the ancient world as burning glasses. But will this explanation do for the glass lens found in the "House of the Engraver" at Pompeii? While this single find does not prove the case, it provides extra circumstantial evidence that ancient artists used magnifying glasses.

Secret Writing

All we really lack is a direct description in an ancient source of how lenses were manufactured to help with the close work of craftsmen. But there may be good reasons why such an explicit statement will never be found. Craftsmen in the ancient world were often very jealous of their technical secrets (see **Jewelry** in **Personal Effects**).

The detail on the cylinder seals of ancient Mesopotamia had a practical as well as aesthetic purpose. They were used for a variety of purposes, for example to stamp business contracts or the clay sealings placed on mouths of jars to show who owned the contents (wine for example). A businessman or official wanting to authenticate a stamp would check the tiny details in the seal impression. A magnifying lens would be as essential to him as to the seal engraver. It seems unlikely, then, that the art of manufacturing lenses would have been widely broadcast—rather it would have been carefully guarded by a small circle of highly placed individuals and the craftsmen in their employ.

Centuries later a similar way of hiding extra information on coins was developed. A silver Islamic coin of the tenth century A.D. includes a tiny inscription, almost invisible to the naked eye. When magnified four times, it can be read: "the work of al-Hasan, son of Muhammad," the craftsman who cut the die from which the coin was struck. The practice of using "microletters" on coins can be traced back to the classical period. It was first detected around 1918 by British numismatist Munroe Endicott. He was examining a Cypriot coin that bears the head of the Macedonian conqueror Alexander the Great (336–323 B.C.) wearing a lion-skin headdress like the god Heracles, when he noticed tiny letters concealed among the locks of the lion's mane. The letters, in Greek, spelled out the name Nikokles, a local king who reigned at Paphos in Cyprus during this period. This name, engraved on the die from which the coin was struck, was not meant to be easily visible to the naked eye—it was intended by Nikokles as a subtle proclamation of independence from the Macedonian Empire. A later numismatist, J.M.F. May, noted that on five other coins of the same design, the name of Nikokles "can be deciphered beyond doubt with the aid of a powerful glass." If it takes a powerful magnifying glass to read these tiny letters, it may be reasonable to assume that the engraver of the coin die had a similar optical aid.

Coin of King Nikokles of Cyprus, drawn at twice actual size. In the locks of hair one can see tiny Greek letters spelling out his name.

Since Endicott's discovery, many other Greco-Roman coins have been claimed to conceal microletters. Some are just visible to the naked eye. Others are so minuscule that they can only be seen clearly with the aid of microscopes. Dr. Nikos Kokkinos, a specialist in Roman archaeology from Oxford University, has made an extensive study of the case for microscopic writing in the classical world. While he feels that the detection of such tiny inscriptions can be a difficult and sometimes subjective field of research—the smaller the assumed letters, the more difficult their authentication becomes— Kokkinos has identified a few examples where the existence of microscopic lettering seems beyond reasonable doubt. His own theory is that these tiny letters were added by the die cutters as a means by which officials of the royal mint could detect forgeries. Fake coins, of course, are almost as old as coinage itself (see **Coins and Paper Money** in **Urban Life**). Microscopic letters on a coin would have been the ancient equivalent of watermarks on a modern banknote— except that unlike the latter, a magnifying glass, rather than just a strong light, was needed to detect them.

BURNING LENSES AND MIRRORS

Because they can concentrate the sun's rays on a single point, convex lenses make handy fire lighters—a fact well appreciated by the ancients. A joke involving a burning glass appears in one of the world's earliest comedies, *The Clouds* of Aristophanes, first performed in Athens in 423 B.C.:

> Strepsiades: **You have seen at the druggist's that fine transparent stone with which fires are kindled?**
>
> Socrates: **You mean glass?**
>
> Strepsiades: **Just so.**
>
> Socrates: **Well, what will you do with that?**
>
> Strepsiades: **When a summons is sent to me, I will get that stone and, holding it up to the sun, I will at long distance melt all the writings of the summons.**

The court summons in the story would have been written on a wax writing board.

Burning lenses also had a medical use in the ancient world—the encyclopedist Pliny mentions the use of lenses for cauterizing wounds.

By the first century A.D. literary references to burning lenses begin to appear in Chinese literature. Even earlier the Chinese appreciated that concave mirrors could be used to focus the sun's rays to a burning intensity. The properties of such mirrors were investigated in the fifth century B.C. by the scholar Mo Zi and his disciples and recorded in the famous scientific work *Mo Jing* ("Mohist Canon"). They called the focus of a concave mirror the *zhongsui* ("central fire"). Strangely there is no reference to the use of burning mirrors in the West. The story that Archimedes developed giant burning mirrors to destroy the Roman fleet at Syracuse in 212 B.C. is, however attractive, a medieval tradition with no foundation (see The "Claws" of Archimedes **in** Military Technology**).**

SEX LIFE

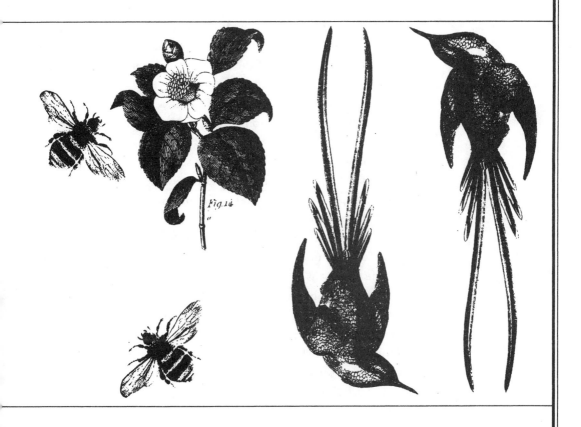

CHAPTER FOUR

Fig. 14

INTRODUCTION

In purely physical terms the human race has probably added little to the variety of sexual behavior enjoyed by the animal kingdom. Most of the relationships human beings fondly imagine are exclusive to themselves have long been explored by other creatures. There is an infinite range of behavior, reflected in both physiology and social life—many creatures are asexual (the best known being the humble amoeba), others hermaphroditic (such as many worms), while the variety among animals that do reproduce by the mating of two distinct sexes is equally enormous. As well as the heterosexual pair-bonding that we consider to be more usual, the natural world provides us with examples of single-parent families (among cats, for example) and even gay families. In 1977 14 percent of the gulls on an island off the coast of California were found to be lesbians. Females were conducting courtship rituals together, pairing off to build nests and incubate eggs jointly (although unfruitfully), and remaining with each other faithfully through several breeding seasons.

Even marriage is not sacred, in the sense that monogamy cannot be claimed as a human invention. Our distant cousins the gibbons, tree-living apes in the rain forests of Southeast Asia, pick one partner and strictly adhere to her or him for life.

What the human race *has* invented is the swathing of sexuality in a mass of institutions, rituals and other paraphernalia. Obsession with the subject is certainly peculiar to us—so much of social and religious history has been concerned with the "rights and wrongs" of sexuality and the elusive quest for what is the "natural" or "normal" way for us to behave.

Homosexuality, for example, was considered far less unusual in many ancient societies than it often is today. Indeed institutionalized homosexual behavior was not uncommon in the ancient world. In Athens and Sparta in the fifth century B.C. and later, homosexual relations between an older and a younger lover were considered to be educationally beneficial. The Spartans apparently believed that receiving the semen from a tried and tested warrior would increase one's own manliness. In Athens the moral and aesthetic aspects of such relationships were stressed.

Theban homosexuality, like that of Sparta, had a military edge. The Sacred Band of Thebes was an infantry formation composed entirely of pairs of male lovers. It was victorious for some forty years before being annihilated by Philip II of Macedon at the battle of Chaeronea in 338 B.C. During the fighting, all three hundred remaining members of the Band fell dead or were mortally injured. Spartan demographics were not helped by a similar tradition, and the separate education of boys and girls encouraged relationships between women, too, as well as late marriages. The Spartans were an elite class suppressing a massive peasant population; eventually, through a low birth rate and high losses in continual warfare, they disappeared completely.

The Cretans, on the other hand, consciously encouraged homosexuality as a means of population control. In classical times it was completely normal for teenage Cretan boys to be courted by older men and, after the required exchange of gifts and formal greetings with his parents, for the boy to go off and live with his lover. The latter would then be responsible for the boy's upkeep and education. Only in his late twenties or thirties were a man's thoughts supposed to turn to marrying, settling down and having children.

Of course much of human society has considered heterosexual marriage to be a useful norm (as well as the best way to proliferate), but even here the form it has taken in different civilizations has varied enormously. Modern society would heartily disapprove of ritualized "sex by numbers," but this was the expected form of marriage for ancient Chinese emperors. During the Han Dynasty (202 B.C.–A.D. 220) of China the emperor was expected to have one queen, three other consorts, nine second-rank wives, twenty-seven third-rank wives and eighty-one concubines—numbers considered in Chinese numerology to be of magical significance. Several court ladies were employed as "sex secretaries," their job being to ensure that the emperor had intercourse with the correct partner on the right day. The emperor worked his way up from the lowest to the highest-ranking partner over the course of a month. The day and hour of these consummations were noted in a diary and, as a cross-check, each woman had to wear a silver ring that she transferred from the right to the left hand after performing her duty. In this way the emperor was supposed to strengthen his vital sexual essences through the month by systematically absorbing those of his partners, a theme that dominated the sex manuals that began to appear in China at this time. The Han Dynasty also saw a liberal attitude toward homosexuality, and sexuality in general, while the concoction of aphrodisiacs to increase the pleasures of lovemaking was common.

Under its Mongol rulers (A.D. 1260–1368) China became a very different, increasingly puritan, society. One expression of this new mood was a curious scheme known as the Table of Merits and Demerits, listing good and bad deeds and giving points for various offenses. Among the items covered were the following: "Keeping an indecently large number of wives and concubines"—50 demerits; "touching the hands of one's womenfolks," either "by accident"—1 demerit, or "with lustful intent"—10 demerits, "unless when helping them in an emergency"—0 demerits, "but if such help arouses lust"—10 demerits; "telling one's women dirty stories"—20 demerits, "unless they are told to excite the women's sense of shame," in which case—0 demerits.

China had changed from what is known as a sex-positive culture to an increasingly negative one; in later centuries Chinese prudishness reached such heights that it even prevented women taking their clothes off for medical examination and treatment (see **Acupuncture** in **Medicine**). Today there are increasingly few sex-positive societies left in the world. (Thailand is one, and because of this it is now becoming the unfortunate victim of an AIDS epidemic through sex tourism.) The reasons behind these changes in attitude are hard to define. In the West, Christianity (particularly Roman Catholicism), with its fundamentally negative view of sexuality, is largely responsible. Judaism and Islam share somewhat less of the blame, having been (at least originally) less obsessed by the "guilt" with which Christianity has imbued the subject of sex.

Bizarre examples of the medieval Christian attitude toward sex are all too easy to find. In A.D. 585 the Church Council of Mâcon ruled that no male corpse should be laid to rest next to a female corpse until after the female body had decomposed. Around this time the Church forbade relations between spouses during the forty days before Christmas; the forty days before Easter and the eight after Pentecost; on the eve of great feast days; Sundays, Wednesdays and Fridays; during the wife's pregnancy, and for thirty days afterward if the child was a boy and for forty days if it was a girl; and, finally, for five days before Communion. Medieval Christians can't possibly have followed these rules, which, given the high death rate, would have seriously endangered the chances of communities reproducing themselves.

Generalizations are difficult to make, but many of the great civi-

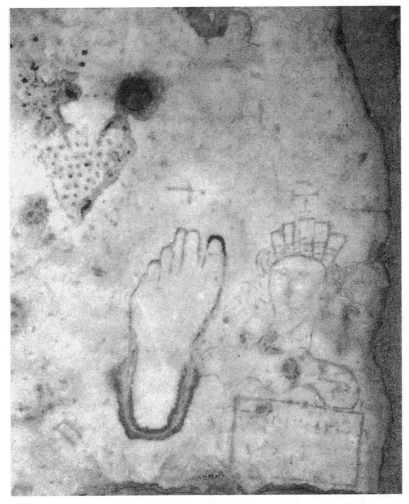

A Roman street sign from the city of Ephesus, on the western coast of Turkey, meaning "this way to the brothel."

lizations of the past highlighted in this book tended to have a more positive, and therefore more curious and inventive, attitude toward sexual matters than modern Western society. This certainly applies to prostitution, which, if not the oldest profession in the world, must be the second oldest—after that of witch doctor or shaman, which probably goes back to the Old Stone Age. In many ancient societies prostitution was an explicit, institutionalized, and even sacred, activity. The earliest evidence for prostitution comes from ancient Iraq, where texts such as the famous *Epic of Gilgamesh* show that prostitutes were an integral (and perfectly respectable) part of Babylonian temple life during the second millennium B.C.; the extremely graphic pictorial evidence of clay plaques excavated from temples takes us back another thousand years. According to the Greek historian Herodotus, who visited Babylon around 450 B.C., prostitution was a sacred duty for all its female citizens:

> Every Babylonian woman must once in her life go and sit in the temple of Aphrodite and there have sex with a stranger. . . . Once a woman has taken her seat, she is not allowed to return home until a man has thrown a silver coin into her lap and taken her outside the temple to lie with her. As he throws the coin, the man has to say, "In the name of the goddess Mylitta [Aphrodite]." . . . The value of the coin does not matter. . . . The woman has no choice and must go with the first man who throws her money. Once she has surrendered herself, her duty to the goddess has been rendered and she can go home.

Prostitution helped make the Babylonian temples rich, and from them the first great financial centers evolved (see **Banks** in **Urban Life**). There were also of course purely commercial brothels, such as the Porneion opened in Athens by Solon around 570 B.C. so that young men could work off their lusts. A temple to Aphrodite, goddess of love, was built with part of the profits. Later Greek and Roman brothels were licensed and supervised by the authorities. In the third century A.D. there were forty-five brothels in Rome, which were allowed to open for business only after 3 P.M., so that they didn't interfere with the working day. The prostitutes, mainly from the Near East, sat outside on benches wearing conspicuously gaudy clothes and jewelry, while phallic symbols were painted or carved

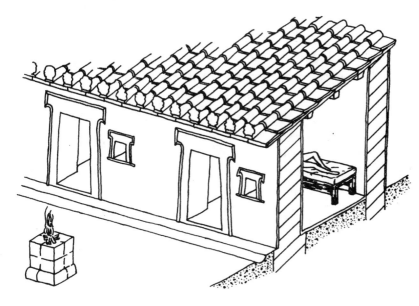

At the port of Pyrgi, in Italy, archaeologists discovered a row of small cells alongside the Roman Temple of Astarte (the Syrian version of Aphrodite). Despite commercialization under the Roman Empire, prostitution never completely lost its religious side; the excavators of Pyrgi interpret these rooms as units of a sacred brothel attached to the Temple.

above the door. Metal admission tokens, with equally explicit motifs, were minted especially for customers. The brothels of ancient Rome certainly lived up to its big-city reputation; there were 32,000 prostitutes on Rome's official police register early in the first century A.D. This compares with 35,000 in New York during the 1890s.

The "red-light district" seems to have been invented in ancient China, where, under the Sung Dynasty (A.D. 960–1279), "wine houses" doubling as brothels (many of which were run by the Imperial Board of Revenue), displayed a red light to inform the public. As a contemporary writer noted, "Such special wine houses have bamboo lamps of red [silk] suspended on their front doors; they are displayed both in dry and rainy weather, being protected by covers of plaited bamboo leaves; for it is by those lamps that such special wine houses can be recognized."

On a larger scale, a little-known religious cult that developed in Syria around A.D. 250 used live sex shows to pull in the public. Its ceremonies, which took place in shallow open-air pools, involved large numbers of naked young women frolicking in the water. Prudish Roman emperors banned the shows, while the Christian Fathers condemned them. At the end of the fourth century Saint John

Chrysostom ("golden-mouthed," in Greek) of Antioch despaired of his flock, who preferred "to see women swimming and revealing all their physical attributes" to attending church. Despite frequent official disapproval such spectacles lasted into the sixth century A.D.

The ancient world also saw the beginnings of sexual surgery. The ancient Chinese custom of opening the testicle sac to sew in small ringing bells can be passed over briefly as a rarity. However, castration was a run-of-the-mill operation in civilizations such as Rome and China, where there were once so many eunuchs that through sheer weight of numbers they have to be counted as a "third sex." Attractive boys were taken into slavery and emasculated, to become sexual playthings of the rich, or beardless household ornaments kept for their youthful beauty.

But the major role of the eunuch in ancient societies was a political one. Eunuchs were the perfect guardians of harems and provided safe companions and secretaries for royal ladies. They could also be entrusted with the very highest offices of state with no fear that they would want to muscle in and start their own dynasties. Less susceptible than other men to corruption and persuasion by sexual means, they were the ideal politicians and civil servants. Their reputations could not be sullied by the accusations of rape, paternity suits and other scandals that so often blight the careers of public figures.

The first civilization deliberately to select eunuchs as officers of state was the Assyrian Empire, which dominated the Near East during the early first millennium B.C. The practice was continued by its successors, including the Persian Empire, founded by Cyrus the Great (559–529 B.C.) who, according to the Greek writer Xenophon, "selected eunuchs for every post of personal service to him, from the doorkeepers up." Slightly earlier the King of Lydia (in western Turkey) favored female eunuchs, who had their ovaries removed before entering his service.

Eunuchs were becoming powerful in China during the same period. A sixth-to-fifth-century B.C. text complains about eunuchs as well as women meddling too much in the affairs of state. Eunuchs were particularly influential under the Han Dynasty (202 B.C.–A.D. 220), when some held tremendous power simply because of their looks, and it was normal for emperors to have as many male favorites as the recommended magical number of wives. But most were of the more professional variety, trained for a career in government.

The Roman civil service also employed eunuchs, despite the bans of castration imposed by various emperors. And, although the custom was condemned by the Church, the zenith of "eunuch power" in the Roman world actually came after it was Christianized, under the Eastern Roman (Byzantine) Empire, which ruled from Constantinople (Istanbul) between A.D. 395 and 1453. Thousands of young men entered public service by being castrated, providing the empire with some of its most distinguished state secretaries, generals and even Church leaders. By the tenth century, eunuchs actually took precedence over noneunuchs at the Byzantine court, aptly described by medieval historian Sir Steven Runciman as a "eunuch's paradise."

While most eunuchs were forcibly castrated or persuaded into it by their families to increase their career prospects—as not so long ago youngest sons were pressed into the Church—one suspects that a reasonable percentage were willing volunteers, perhaps the slice of society that is equivalent to modern-day transsexuals. A medieval Arabic text of the ninth century A.D. discusses cross-dressing eunuchs, "men who wish to be women," who worked as singers, dancers and prostitutes. There were also ancient religious cults that encouraged castration. The devotees of Cybele (a goddess from Phrygia, in modern Turkey), for example, horrified the citizens of Rome by castrating themselves in public. They roamed the streets chanting and clashing cymbals—not unlike members of the Hari Krishna movement—and, when the mood took them, produced a jagged piece of broken pottery, sliced off their genitals and flung them through the window of an unsuspecting Roman household.

A devotee of another Eastern cult, that of the Syrian Sun god, succeeded to the Roman throne in A.D. 218 at the age of fourteen. Known as Elagabalus, he made a triumphal entry into the capital wearing full drag, carried in a gaudy vehicle with dancing eunuchs in attendance (see **Makeup** in **Personal Effects**). The Romans were scandalized by his extravagance (see **Refrigeration** in **Food, Drink and Drugs**), outrageously effeminate behavior, appetite for male lovers and outspoken desire to be treated as a woman. Two contemporary Roman historians say that he requested doctors to create a vagina for him surgically. Unfortunately neither account gives more detail. Was the operation ever carried out? It seems un-

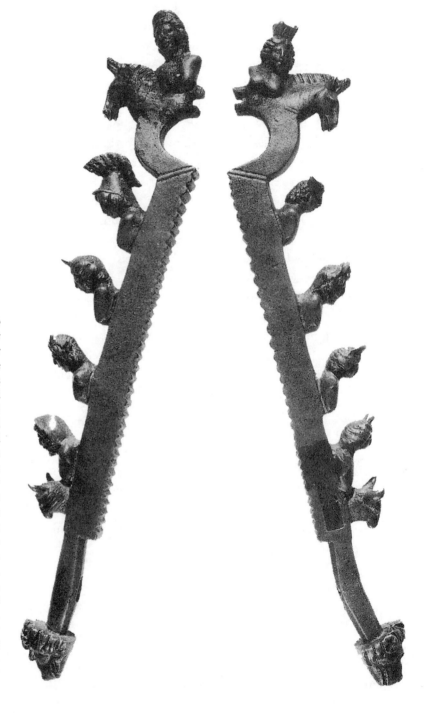

A Roman castration clamp of the 2nd or 3rd century A.D., found in the river Thames, London. Made of bronze, the two arms (originally hinged and closed by a screw nut) form an oval ring and two long arms with serrated edges. To perform the operation, the penis was kept safely out of the way by placing it through the ring, while the teeth gripped the skin attaching the scrotum to the body. The scrotum could then be cut off with a quick stroke of a surgical knife. This invention enabled much safer castration than earlier methods, avoiding damage to the penis through a swift, tidy operation.

likely, though the anecdote does show that the concept of sex change by surgical means is not a twentieth-century invention, but one that goes back nearly two thousand years.

Understanding of sexuality in the ancient world extended far beyond the obvious physiological differences between men and women. Pregnancy tests used by the ancient Egyptians as early as the thirteenth century B.C. show that they already had a practical grasp of some rudimentary biochemical laws.

In ancient China such knowledge was incredibly advanced. We don't know how they came by their knowledge, but by the second century A.D. the Chinese seem to have been extracting sex and pituitary-gland hormones from human urine for use in various medical preparations. A text dating to A.D. 1025 clearly sets out the methods for extracting them, first by evaporation in pans (in the same way that salt was extracted from brine—see **Drilling and Mining** in **Working the Land**). Further heating and chemical processing enabled the salts and urea to be removed, leaving a crystalline substance that the Chinese called the "autumn mineral"—this consisted of sex and pituitary hormones. The crystals were then made into medicines to treat various sexual ailments. By topping up and encouraging the production of the patient's own hormones, these preparations were useful for a wide range of problems, from impotence to hermaphroditism, and were even taken to stimulate beard growth. This remarkable ancient Chinese technique was not discovered in the West until 1927; the production of androgens and estrogens (male and female sex hormones) from urine is now routine. Nothing, it seems, is new under the sun, and sexual matters are no exception.

APHRODISIACS

A potion that can prolong the joys of lovemaking or stimulate the desire of an otherwise uninterested party has been one of humanity's endless quests. As a spur to inventiveness the idea of the aphrodisiac is probably unsurpassed, and the range of substances and methods recommended by the sages of ancient cultures is mind-boggling. Enormous success was sometimes claimed for the simplest of recipes.

Ibn Battuta, the great Arab geographer of the fourteenth century A.D., made a close study of these matters during his extensive travels through Asia. He became a great believer in the fish-and-coconut diet eaten by the inhabitants of the Maldive Islands in the Indian Ocean. As far as Ibn Battuta was concerned, they had clearly hit the jackpot. On his visit to the Maldives he observed how the inhabitants had "a striking and unequaled power in the practice of sexual intercourse. The islanders achieve astounding feats." He stayed for a while, ate the local food and was clearly impressed with the results: "I myself had four legitimate wives in this country, apart from concubines. I was potent for them all every day and besides that spent the whole night with whichever of them whose turn it was; I lived like this for a year and a half."

Allowing for exaggeration, was Ibn Battuta on an extended (as well as multiple) holiday romance, invigorated by the fresh sea air, or did the coconut-and-fish diet really have an effect on his libido? Coconut and seafood are still recommended today as aphrodisiacs in various societies. Whether they work or not has never really been tested seriously. While the idea that oysters make you amorous is commonplace, the effects are rather difficult to assess scientifically. Certain things are not amenable to laboratory testing—the classic case is how, why and when cats purr. Human beings are no different, and testing the aphrodisiac properties of foods can't be done simply by feeding a subject two ounces of coconut and then measuring his sperm count.

In many cases the only available evidence is what scientists term anecdotal—personal accounts and tradition. In the Western world oysters and other seafood still have a great aphrodisiac reputation, one only strengthened by the fact that this belief is at least two thousand years old. It was so well-established by the second century A.D. that the Roman novelist Apuleius was actually taken to court by the other suitors of a rich widow for inveigling her into marriage by giving her dishes of cuttlefish, spiced oysters, sea urchins and lobsters.

The aphrodisiac powers of "Spanish fly," a variety of blister beetle (*cantharides*), are still discussed frequently in the tabloid press. Like oysters, Spanish fly has a long history—the Roman empress Livia, scheming wife of Augustus (31 B.C.–A.D. 14), purportedly slipped it into the food of other members of the imperial family to stimulate them into committing sexual indiscretions that could later be used against them.

The Romans also used a vast range of herbs to increase the pleasures of lovemaking. They made an erotic ointment called *foliatum* from spikenard, an aromatic Oriental plant of the valerian family. They also made aphrodisiac wine from the rare herb gentian. Another brew, called *hippomane*, contained herbs supposedly mixed with secretions from the reproductive organs of colts and mares and was believed to give the man who drank it a stallionlike performance. Both the Greeks and the Romans used a plant called *satyrion*, which has red leaves and a double root with an erect fleshy stem. They usually added it to wine but believed it was so powerful that it would act as a love potion simply by being held in the hand. Likewise an herb known as southernwood (a type of wormwood) was thought to work simply by placing it under the bed—superstition or an early example of aromatherapy? Other favorites were skirret, a plant with edible tubers that the Emperor Tiberius (A.D. 14–37) imported from Germany, and nettle oil, used as a lubricant. Some Romans—presumably only really desperate cases—even whipped their genitals with stinging nettles to get them in the mood.

For those who might wish to experiment the Roman way, ancient Latin writings also recommend a number of more common herbs and vegetables. Martial, a poet of the first century A.D., wrote that "if your wife is old and your member is exhausted, eat onions in plenty." The Roman physician Galen (2nd century A.D.) also believed in onions; he prescribed powdered onion seed mixed with honey, to be taken while fasting, as a way of whipping up one's sexual appetite. The great gourmet Apicius (see **Cookbooks** in **Food, Drink and Drugs**) also recommended onions—stewed with pine kernels in water, or with pepper in cress juice. Cress also scores high in the number of mentions given it by Roman writers, including the doctor Marcellus Empericus, who prescribed a mixture of equal measures of cress, red onion, pine kernels and Indian nard (a kind of balsam) as a cure for impotence. The aromatic herb savory, still commonly used in cooking, was also thought to have aphrodisiac properties, as was chervil. The natural historian Pliny wrote that, in males, chervil "reinvigorates the body exhausted by sexual desire to an infinite degree." He also recommended parsnips, preferably wild ones grown in stony soil. Cabbage steeped in goat's milk, and even plain old carrots, were highly rated by other Romans.

Pills and Potions

Aphrodisiac recipes from the ancient East were usually more elaborate than the Roman. Wang Tao, an herbal physician of the T'ang Dynasty (A.D. 618–906), included in his *Collection of Secret Prescriptions* four recipes for the different seasons of the year, made from herbal mixtures ground into powder and bound together with honey into tablets. Some men would take dozens a day to maintain their potency. Different aphrodisiacs are referred to in another book of the same period entitled *Important Guidelines of the Jade Room*. They are mainly powders for applying to the penis or vagina and are often given charming names. Three for use by women are called "The Beauty Holds a Man Upside Down," "Happy Powder," and "A Beauty's Smile." One of the powders for male use supposedly enabled a man to satisfy more than ten women every night.

An Arabian manuscript of the ninth century A.D. (purporting to give a text written by the Greek physician Galen) has sections entitled "On the Secrets of Women" and "On the Secrets of Men," which make some staggering claims. For example, it lists ingredients for a product to "excite the desire of women so that they go wandering around, leaving their homes, looking for sexual satisfaction, throwing themselves before men and searching for a good time:

aged olive oil, orchid, garden carrot seed, turnip seed, ash of oleander leaf, pulverized burned nasturtium, dry alum, magpie excrement, powdered willow leaves and fine date pith." These were to be
ground together with coconut milk, dried into tablets, then pulverized again and mixed into a syrup with rosehip juice. The recipe
ends by saying that the desired effect, naturally enough, will happen
only if God wills it. The same document also includes recipes for
preparations that will turn women into lesbians and "increase men's
desire for women so that they forget about boys and eunuchs," another which "raises men's impatient desire for boys" and even
"drugs which bring about addiction to masturbation."

With charms like these we have clearly entered a fantasy world.
We can also discount the value of other magical recipes, such as an
Assyrian love spell from ancient Iraq. In this, iron filings placed on
a woman's genitals were supposed to attract her to a man who had
similarly anointed himself with a mixture of powdered lodestone
and oil. While it shows a grasp of the basics of electromagnetism
(see **Magnets and Magnetism** in **High Tech**), as an aphrodisiac,
the method is obviously hopeless. Iron filings maintained a reputation in the East, however; two thousand years later our friend Ibn
Battuta, the traveler, reported the death of an Indian prince from
eating iron filings as an aphrodisiac.

The claims made for the mandrake plant must be taken more seriously because of its almost global reputation. During medieval
times it was widely revered as a plant of amazing magical potency,
so much so that strict rituals had to be observed when collecting it.
The belief in mandrake's awesome power dates back to classical
times, as we can see from the writings of the Greek botanist
Theophrastus (4th century B.C.):

> It is said that one should draw three circles round mandrake
> with a sword and cut it with one's face towards the west; and at
> the cutting of the second piece one should dance round the plant
> and say as many things as possible about the mysteries of love. . . .
> The leaf of this mandrake, used with meal, is useful for wounds,
> and the root for erysipelas [facial inflammation], when scraped
> and steeped in vinegar, and also for gout, for sleeplessness, and for
> love potions.

But what exactly was this plant? Some confusion has arisen because the term *mandrake* has also been applied to the similarly named mandragora; to add to the muddle, ginseng, which, as Marco Polo noted in the thirteenth century A.D., was commonly used in China as an aphrodisiac, has also been mistakenly referred to as mandragora. The real mandrake of Near Eastern and European writings is probably a plant of the potato family, quite possibly datura, or "thorn apple," known in the United States as jimsonweed. Datura contains atropine and hyoscine (a form of scopolamine), both potent alkaloids that, when ingested, induce an initial phase of excitement followed by sedation and hallucination. One can see how mandrake, taken in the right dose, could enhance lovemaking. It was also used, by both the Romans and the Chinese, for its painkilling powers (see **Anesthetics** in **Medicine**).

As an aphrodisiac opium is as effective as any other known substance. It does not stimulate sexual urges directly: rather it makes

The mandrake was thought to be a powerful aphrodisiac charm. A relative of the potato, it has roots that sometimes look vaguely like a human figure, a characteristic strongly exaggerated in this illustration from a Byzantine manuscript of A.D. 512. The mandrake was thought to be so potent that gathering it by hand would be fatal; one method of collecting it was to dig down deeply to expose part of the root, tie it to a dog and let him pull it from the ground. According to tradition the mandrake shrieked as it was pulled up and the dog would die from horror. Here the nymph "Discovery" presents a mandrake, still attached to the dog, to the famous Greek physician Dioscorides.

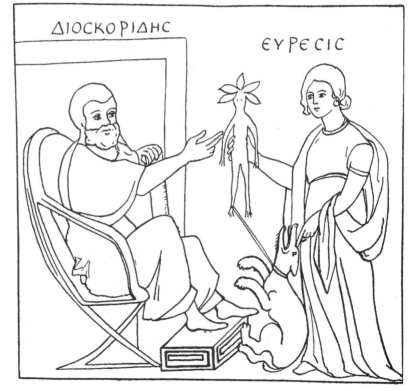

one extremely relaxed and at the same time heightens the enjoyment of sensual pleasures. It was heavily used in medieval Islamic cultures, where female circumcision (involving removal of the clitoris) was popular (though obviously not among the women on whom it was inflicted): smoking large amounts of opium helped make up for the woman's lack of sensual stimulation. It also seems to have been used by the ancient Greeks. At the suitably named site of Aphrodisias, in eastern Turkey, a sculpture of Eros, the god of sexual love, shows him clutching an opium poppy (see **Drugs** in **Food, Drink and Drugs**).

It is often stated that there is no such thing as a real aphrodisiac—perhaps not in the sense that the word is often interpreted. But a number of substances with a narcotic or invigorating effect can be effective in enhancing the delights of lovemaking. The most successful ancient recipe seems to be one contained in the *Kama Sutra* (see **Sex Manuals**), made from a mixture of powdered white thorn apple (datura), "long pepper" (chilis?), black pepper and honey. This is applied to the penis before intercourse and is supposed to make the woman "subject to the man's will." It wouldn't quite do that, but would certainly have had a significant effect on both partners, as the chemicals in the mixture would be quickly absorbed through the mucous membranes of the penis and the vagina. The pepper extracts (which must have been used in the *minutest* of quantities) would have a rubefacient effect—by irritating the skin they would draw blood into the penis, helping to increase and maintain an erection. The effect of the peppers would be quickly transmitted to the woman, the mild irritation of the clitoris increasing sexual arousal. Honey acts as a useful lubricant, while the drugs in the thorn apple (datura), as we have seen, would induce excitement as well as a pleasant, dreamy feeling. This recipe as a whole, therefore, has a completely rational basis for its action and was more than likely quite effective in practice.

DILDOS

Sex aids are not a twentieth-century invention. Far from it: our remote ancestors had a number of ingenious devices. Dildos, for example, were a speciality of the ancient Greeks. A short comedy written by Herondas of Alexandria in the third century B.C. gives some insight into their use in ancient Greek society.

Called *A Private Chat*, it consists of a dialogue between two women. Metro has dropped in to see her friend Corrito and asks her where she had her dildo made. Somewhat taken aback, Corrito asks where her friend saw it. Metro tells her in confidence that a woman called Nossis has it. Corrito is extremely annoyed—she had in fact lent it to a friend called Eubule and is furious to learn that the dildo, which she hasn't even used yet, has been passed on to Nossis, someone she particularly dislikes. After calming her down, Metro returns to the question of the manufacturer. Corrito tells Metro that it was a leather worker named Cerdon, who keeps his business secret to avoid the tax man, but that an introduction to him can be arranged via Artemis, a tanner's wife. The play ends with Metro hurrying off to see her.

The dildo (*olisbos*) in this ribald little play is made of softest leather. There is also reference to leather or woolen straps used to attach the dildo—but it is not clear whether the ladies in the dialogue want to use the device on themselves or to strap it on and employ it on someone else. Other Greek writers provide a little more detail: there is even an entry in a Byzantine encyclopedia of the tenth century A.D., according to which an *olisbos* was a leather phallus used by widows to satisfy themselves and by lesbians to have "immodest" relations. Pictures painted on Greek pottery from the fifth century B.C. onward leave little to the imagination; some, for example, show a double-headed *olisbos*, which could be used to satisfy two people at once. From the number of depictions that survive, dildos seem to have been fairly popular in ancient Greece.

The Greek evidence of dildos is the earliest known from anywhere in the world. However, it is unlikely that the Greeks invented the dildo as such, since simple ones can be made from almost anything, such as a smooth piece of wood or a cucumber. But they can probably claim the invention of the bread dildo. References to bread dildos in ancient Greek texts were only identified in 1986 by Professor Alexander Oikonomides, who noticed the occurrence of the word for a stick-shaped loaf of bread, *kollix*, in a joking Greek inscription. He then found the world *olisbokollix*, meaning "loaf-of-bread dildo," in a lexicon of classical Greek written in the fifth century A.D. Pictorial evidence confirmed his discovery: vase paintings show women carrying baskets full of phallus-shaped loaves.

The existence of ancient Greek bread dildos is now well established. They must have been much cheaper versions of the kind of expensive toy that Corrito and Metro were discussing. They were easily made at home, though perhaps an order was put in at the bakers if you were holding an orgy. Their edibility suggests variations on their use, and of course if accidents happened and something got stuck, bread has the great advantage over modern plastics in being soluble.

Few ancient societies produced the kind of explicit vase paintings

An Athenian lady entertains herself at home— from a vase of the 5th century B.C.

Ancient Greek dildos were often made to measure from bread; the size of the example shown on this 5th-century B.C. Athenian vase, however, suggests that it was a ceremonial object being carried to a religious festival.

that the ancient Greeks did. Textual evidence from other cultures is rare, with the exception of India and China. The *Kama Sutra*, a Hindu text composed sometime between the third and fifth centuries A.D. (see **Sex Manuals**), recommends hollow dildos (or penis extensions) as a way for a man to satisfy a well-endowed lady ("elephant woman") if his natural equipment is too small. They could be made of gold, silver, copper, iron, ivory, buffalo horn, various kinds of wood, tin or lead, and were supposed to be "soft, cool, provocative of sexual vigor and well fitted to serve the intended purpose." They could be made in one piece or from rings arranged like bangles on an arm. If these expensive materials were unavailable, then the *Kama Sutra* recommends alternatives: "a tube made of wood apple [*Feronia elephantum* or "elephant apple," an Indian gum-bearing tree], or tubular stalk of the bottle gourd, or a reed made soft with oil and extracts of plants, and tied to the waist with strings. . . . The above are the things that can be used in connection with or in place of the lingam."

In medieval China dildos (usually made of ivory or wood) were popular among lesbians, sometimes attached to one partner by a silk band around the waist. Double dildos were also used. Various writings cautioned against their excessive use, which might damage delicate internal tissue. There was also a natural dildo, a kind of phallic mushroom that supposedly swelled during use! This extraordinary plant was described in colorful terms by Ming Dynasty writer T'ao Tsung-i (14th century A.D.):

> In the pastures of the Tartars [Mongolia] wild horses often copulate with dragons. Drops of the semen will fall down and enter the earth and after some time put forth shoots resembling bamboo-shoots, of pointed shape and covered with small scales close together like the teeth of a comb, and with a network of veins, making them similar to the male member . . . lewd country-women insert these things into their vaginas; as soon as they meet the yin-essence they will suddenly swell and grow longer.

CONTRACEPTIVES

"Obvious quackery!" This was the opinion of Francis Llewellyn Griffith, one of the most eminent Egyptologists of his time, regarding the contraceptive advice given in an ancient gynecological papyrus found in April 1889 at Kahun, in northern Egypt. Living in a time when contraception was a taboo subject, it was impossible for Victorian scholars to imagine that the ancients had been successful in inventing technical means to avoid pregnancy. Modern medical science has proved them wrong.

We now know that many ancient recipes for preventing contraception, although not 100 percent effective, would actually have worked. The *Kahun Papyrus* was giving useful contraceptive advice—the earliest known—almost four thousand years ago. The prescriptions involved plugging the vagina with gum, a mixture of honey and sodium carbonate, or a paste of crocodile dung mixed

Acacia tree, from an Egyptian tomb painting. Acacia gum, which contains lactic acid, a natural spermicide, was used by the ancient Egyptians in contraceptive recipes from 1550 B.C.

with sour milk. Honey or gum would certainly have reduced the motility of the sperm. Even the use of crocodile dung is not as daft as it sounds, since it could have had absorbent qualities or, if more compact, might have acted as a plug. If nothing else, it would have dampened the ardor of the most passionate lover! However, elephant dung would have been better, as it is more acidic, and this unusual contraceptive is in fact recommended in thirteenth-century A.D. Islamic writings.

Another Egyptian document, the *Ebers Papyrus*, which dates from about 1525 B.C., gave the following method: "To make a woman not become pregnant for one year, two years or three years, acacia leaves are ground fine with honey, lint is moistened therewith and placed in her vulva." Such tampons would have functioned as contraceptives, since acacia gum contains lactic acid, a highly effective spermicide.

Even these very early contraceptive devices must have followed the adoption of other, simpler, methods of avoiding pregnancy. Texts of the third millennium B.C. from Sumer, in Iraq, refer to priestesses favoring anal intercourse during their ritual sexual activities, specifically explained as a way of preventing conception. Of course *coitus interruptus*, anal intercourse and attempts to expel sperm from the womb by violent movements have always been available as methods of contraception (though the last is of doubtful effectiveness). The Egyptian approach differed in its emphasis on equipment intended to allow ordinary sexual intercourse to take place without the usual consequences.

Olive Oil

Greek and Roman medical writers give a variety of different recipes for contraception. Predictably enough, all of these methods, like the Egyptian, put the onus of prevention entirely on the woman.

In the fourth century B.C. the great philosopher Aristotle noted that women of his time who did not wish to conceive would "anoint that part of the womb on which the seed falls with oil of cedar, ointment of lead, or frankincense commingled with olive oil." One of Aristotle's recipes unfortunately includes lead, a poisonous substance widely used by the Greeks and Romans in both medicines and cosmetics (see **Makeup** in **Personal Effects**). The application of oil, however, and particularly olive oil, makes perfect sense: the motility of sperm would have been reduced considerably. In the Western world this contraceptive property of olive oil was rediscovered by Marie Stopes, a modern pioneer of contraception who, in 1921, founded the Mothers' Clinic for Birth Control, in London. She made a study of the contraceptive properties of olive oil (applied to the vagina), and in 1938 she published the results of a series of controlled tests, reporting a zero failure rate for this method. More than two thousand years after Aristotle the already ancient practice he recorded was once again being taken seriously.

In Roman times the physicians Dioscorides (c. A.D. 40–80) and Galen (A.D. 129–199) listed around a dozen plants that acted as oral contraceptives. These include asafetida, juniper, pennyroyal, "squirting cucumber" and Queen Anne's lace (wild carrot). The effectiveness of these herbal potions has been amply confirmed by medical research within the last thirty years. Many of them, however, had potentially dangerous side effects, as some ancient doctors must have discovered.

This is probably why Soranus of Ephesus, in the first century A.D., revived the Aristotelian and ancient Egyptian traditions in his *Gynecology*, particularly favoring the use of vaginal plugs of wool anointed with various substances:

> Conception is prevented by smearing the orifice of the uterus all over with old olive oil, honey, cedar resin or the juice of the balsam tree, alone or together with white lead; or with an ointment containing myrtle oil and white lead; or before the act with moist alum [a mineral salt], or with gum resin together with wine; or put a lock of fine wool into the orifice of the uterus; or, before sexual relations use vaginal suppositories which have the power to contract and to condense. For such things as are astringent, clogging, and cooling cause the orifice of the uterus to shut before the time of coitus and do not let the seed into the womb.

The dangerous use of white lead aside, modern medical research has shown that every one of Soranus's methods could have worked—indeed some of them recur in the folk medicine of Western countries well into the present century. The use of sticky substances such as olive oil would, as we have seen, been highly effective, and alum, recommended in one of Soranus's recipes, is now known to be spermicidal. While the work of Soranus represents the best of Roman thinking on contraception, there were also many bogus claims made by quacks and charlatans. Some fooled even the great encyclopedist Pliny, who recommended (c. A.D. 75) the use of an amulet made of the insides of hairy spiders wrapped in deerskin or "mouse dung applied in the form of a liniment."

An alternative approach to contraception was tried by the Jews. Rabbis of the third to fifth centuries A.D. refer to the insertion of a sponge in the vagina. This would have absorbed the semen and so prevented contraception. The use of sponges was not advocated again until 1832, by Charles Knowlton, the American pioneer of contraception, and has only very recently become a standard method, used in conjunction with a spermicide.

The history of contraception in the West after the fall of the Roman Empire is difficult to follow because of its prohibition by the Church. With occasional exceptions, knowledge of contraception went underground, becoming the province of the wise women later accused of being witches.

The situation was less severe, however, in the Eastern Roman Empire. In his medical encyclopedia the sixth-century Byzantine court physician Aëtio of Amida adapted from Soranus the use of astringent, fatty or cooling ointments to close the opening of the womb and prevent sperm from entering. He also advised men to wash their genitals with vinegar or brine before intercourse. (Vinegar is one of the most effective of all spermicides.) However, Aëtio was influenced by Church opinion in particularly recommending the rhythm method—that is, having intercourse during the "safe" period at the beginning or end of menstruation.

The earliest definite evidence of Indian contraceptive practices comes in a collection of medical recipes from the eighth century A.D. This suggests that women should either smear the vagina with a mixture of honey and *ghee* (clarified butter) or use plugs of rock salt ground with oil. Either method would have been at least partly effective, as the sticky substances would slow down the sperm, while rock salt is an excellent spermicide.

A variety of contraceptive methods were in use in the medieval Islamic world, which was not affected by Christian condemnation of the practice. Islamic medical writers drew on Indian, Greek and Roman knowledge of the subject. The most thorough account occurs in the work of Ibn-Sina, known to the Western world as Avicenna (A.D. 980–1037). His enormous medical encyclopedia, containing some one million words, gives twenty different contraceptive methods, along with brief comments on their effectiveness. This reasoned discussion of the reliability of different contraceptive methods was not to be surpassed until the twentieth century. Norman Himes, the pioneer historian of contraception, noted in 1936 that Ibn-Sina was far in advance of many modern physicians in realizing that traditional contraceptives could actually work. The skepticism of scientists in Himes's day was the product of ignorance and a bias against the idea of contraception; they automatically assumed that natural and herbal—as opposed to chemical—remedies must be ineffective.

Oral Contraceptives

The mass of material on ancient contraceptives collected by Himes played an important part in undermining the then orthodox view of contraception as an unnatural practice. But Himes himself was a victim of a particular blind spot: a refusal to believe that such a thing as an oral contraceptive existed. In his discussion of the contraceptives used by the Indian tribes of America, Himes expressed a low opinion of the method traditionally used by Cherokee women, who believed that chewing and swallowing the roots of the spotted cowbane plant over four days would make them sterile. Himes's view was that this method was probably useless, "for no drug has yet been discovered which, when taken by mouth, will induce temporary sterility."

As it happens, research in the decades following Himes's work has shown that some of the oral contraceptives discovered by Amerindians *are* effective. Laboratory tests on rats have demonstrated, for example, that both stoneseed, which was taken as an infusion of "desert tea" by Shoshone women, and the powdered Paraguayan weed that was drunk with water by the women of the Matto Grosso reduce fertility considerably.

Investigation of the traditional herbal remedies of the Amerindians played a key role in inspiring the modern search for reliable oral contraceptives.

ROMAN CONDOMS?

There are various theories about the origin of the word *condom*, but they agree in placing their supposed inventor, a Dr. Condom, in the second half of the seventeenth century A.D. He was either a physician or a courtier in the service of Charles II of England (1660–1685), a notorious royal philanderer who was said to be bothered by the number of his illegitimate children. But painstaking efforts to trace any contemporary evidence of such an individual as Dr. Condom have failed. It is also clear that the condom dates back far earlier than the seventeenth century.

The first certain description of a condom is found a hundred years earlier, in a work of the great Italian anatomist Fallopius published in 1564. He claims to have invented the use of small linen sheaths, to be placed behind the foreskin, as a way of avoiding syphilis. The general feeling of experts on the subject is that Fallopius's invention was a refinement of an earlier device, but the whole matter remains shrouded in mystery. One theory has it that condoms were already in use in the Middle Ages and that their name was jokingly borrowed by a learned Latin scholar from a Persian word, *kondu* or *kendu*, for a long vessel made from the intestines of animals and used for storing corn. It is just as likely that the condom was invented in Persia and that the name came with it.

It is also often claimed that the ancient Romans were the real inventors of the condom. The evidence can hardly be described as decisive, but it is intriguing. It comes from a story recorded by the Roman writer Antoninus Liberalis (probably 2nd century A.D.) about Minos. Minos was a legendary king, supposed to have reigned about 1400 B.C., who gave his name to the Minoan civilization of prehistoric Crete. According to Liberalis, Minos produced semen containing serpents and scorpions; to sleep with him safely, women had to insert the bladder of a goat inside themselves. The myth actually suggests a female, rather than a male, sheath, though the basic principle is the same. Norman Himes, the great twentieth-century historian of contraception, thinks that it demonstrates the occasional use of protective sheaths, made from animal gut, in Roman times.

The evidence is suggestive, as the most common material for condoms before the invention of vulcanized rubber in the mid-nineteenth century was animal gut, like the goat bladder used by Minos. But if the legend described by Antoninus Liberalis reflects a memory of ancient condoms, it is an extremely garbled one. Could this mean that the story is far older than the Romans and actually descends from the Minoan civilization fifteen hundred years before?

PREGNANCY TESTS

Baffling as it may seem to find such an apparently modern technique in the ancient Near East, pregnancy tests were once commonplace there, centuries before the time of Christ.

A Babylonian tablet dating from about 700 B.C. sets out the method used in ancient Iraq. The test was carried out by midwives, whose expertise clearly extended beyond assistance during childbirth. It involved placing a woolen tampon impregnated with the juice of particular plants into the woman's vagina. The tampon was left in position for up to three days, during which time sexual intercourse was avoided. It was then removed by the midwife and the color examined. Depending on which plant extract was used, the midwife could decide if the woman was pregnant. For example, if a certain "white plant" dissolved in a solution of alum (a mineral salt) was used and the tampon turned red, then the woman was pregnant. The test worked because the pH value (degree of acidity or alkalinity) of a woman's secretions changes when she becomes pregnant. The plants were chosen for their ability to react to acids and alkalis, and the tampons worked in the same way that litmus paper does in a laboratory.

The ancient Egyptians invented a different method of testing for pregnancy. It is described in a papyrus of the thirteenth century B.C., now in the Berlin Museum. According to this, a woman wanting to know if she was pregnant would be asked to urinate daily on bags containing grains of wheat and barley. The papyrus states that the urine would accelerate the growth of the grain if the subject was pregnant.

Remarkably enough the method works, due to the action of a particular hormone present in the urine of pregnant women, of which modern science only became aware in 1927. A series of tests carried out from the 1930s onward has shown that in 40 percent of cases the hormones in the urine of a pregnant woman accelerate the growth of cereals, while the urine of a woman who is not pregnant (or a man) will retard growth. The *Berlin Papyrus* thus represents the earliest observation of these hormonal effects.

The ancient Egyptians also believed that a woman's urine would foster the growth of wheat grains if her child was to be a boy, or barley if it was a girl. The same belief was later held by the Greeks, as we can see from the writings left by the followers of the great doctor Hippocrates (see **Introduction** to **Medicine**). The Greeks clearly borrowed the idea from Egypt, whose doctors were particularly respected in the ancient world for their gynecological knowledge. Via the Greeks and Romans this sex-determination test became part of the medieval European medical tradition and was still current in Renaissance times. It provides an extraordinary survival of an ancient Egyptian gynecological practice over nearly three thousand years. Unlike the basic pregnancy test, however, modern science has as yet found no support for the claim that the sex of the unborn child can also be determined by the same method.

SEX MANUALS

In 1883 an English-language edition of the *Kama Sutra*, a manual of sexual love from ancient India, began to infiltrate the Western world. Translated by the famous British Orientalist Sir Richard Burton (who first brought *The Arabian Nights* to an English-speaking public) and F. F. Arbuthnot, it had to be printed in a strictly limited edition, and even anonymously. So the straightlaced values of Victorian times dictated. Published by the nonexistent "Kama Shastra Society of London and Benares," it was marked "for private circulation only." The very mention of the book continued to cause an enormous fuss in England until 1963, when Allen and Unwin risked prosecution by releasing a public edition of Burton's translation.

Looking back from the vantage point of the 1990s, it is hard to see what all the bother was about—the *Kama Sutra* is not actually a pornographic work. It was written to inform the leisured class of ancient India, in a very practical and detailed way, how to conduct their love affairs. Ancient Hindu society saw no shame whatsoever in the pleasures of sex; indeed sex, particularly within marriage, was considered a religious duty, an attitude reflected in the erotic art adorning so many Indian temples. The ancient Hindus sincerely believed that their literature on love and sex was ultimately of divine origin, derived from a definitive work in 100,000 chapters written by Prajapati, the creator of the universe. The *Kama Sutra*, which preserved some of this material, was not written to titillate but to educate; as it states in an introductory section, "Even young maids should study the *Kama Sutra* along with its arts and sciences before marriage, and after it they could continue to do so with the consent of their husbands."

Indeed the most offensive aspect of the *Kama Sutra*, from a contemporary perspective, is the sexist attitude toward women that pervades the writing. One passage recommends that a man, if his intended bride is dithering about getting married, should simply abduct her and conduct a forced marriage in the presence of fire taken from the house of a priest (Brahmin), which was thought to make the contract indissoluble. Still it also gives shameless advice on twenty-seven different ways for courtesans to wheedle money from their lovers. Another passage tells a courtesan what to do if she feels her lover is losing interest: get hold of his most valuable possessions and arrange for an accomplice to come around and seize them forcibly to pay off a phony debt. She should then stay with him if

A *Kama Sutra* in stone. Detail from the erotic sculptures from the Kandariya Mahaden temple at Khajuraho (western India), carved around A.D. 1000. Flanked by figures of gods, these scenes depict groups of two to four people involved in a variety of sexual acrobatics.

he is rich and behaves well to her, but drop him "as if she had never been acquainted with him" if he is poor!

Compiled from traditional material by one Vatsyayana sometime between the third and fifth centuries A.D., the *Kama Sutra* comes from a society very different from our own—and indeed nothing like that of modern India. It is a handbook for sophisticated town dwellers, a class of idle rich who had little better to do than indulge endlessly in leisure. Hence the idea of a book telling them how to enjoy themselves to the full, something they believed was their function in life. It was not, however, devoid of morality—the cases cited above are exceptions rather than the rule. Much of the work is devoted to the best ways to make and keep a marriage working. But in the society for which it was written, it was the done thing for a man not only to have a wife but also to keep as many courtesans as he could afford.

The Art of Love

Overall the text is fairly pedantic and not very erotic. It begins with general definitions, including that of *kama*, a term in Hindu philosophy meaning the enjoyment of sensual pleasures. It then gives a list of the arts and sciences that are useful adjuncts to lovemaking, such as music, dancing, flower arranging, poetry, conversation and coloring "teeth, garments, hair, nails and bodies" with dyes and paints, along with some surprising entries, such as "playing on musical glasses filled with water"!

The *Kama Sutra* then advises how to behave in society and defines the different kinds of men and women within it. There follows a quite obsessive classification of foreplay, with sixteen kinds of embrace, eight different types of kiss and numerous ways of scratching and biting. Then come the positions for intercourse, some of which involve such amazing contortions that it is strange that the study of yoga was not included in the recommended arts and sciences. The later chapters give classifications for oral sex (apparently a sideline for eunuchs who worked as hair shampooers), marriage, courtship and relationships with courtesans. The final section gives advice on improving one's appearance, sex aids (see **Dildos**) and love potions (see **Aphrodisiacs**).

For compiling the *Kama Sutra*, still the most famous sex manual in the world, Vatsyayana (of whom practically nothing else is known) came to be hailed as a *rishi*, or divinely inspired teacher. It was the first in a long line of similar manuals written in India during the Middle Ages. Some were eventually translated into Arabic and spread throughout the Islamic world.

But the Indians do not seem to have invented this genre of literature. An official state list of the most important books in circulation during the early Han Dynasty (202 B.C.–A.D. 9) includes no less than eight sex manuals, seven of them running to more than twenty chapters. These Chinese works predate the *Kama Sutra* by a good three centuries. Unfortunately, though they were endlessly copied and used as late as the Sui Dynasty (A.D. 581–618), they are now all lost. Yet much of their substance has been preserved in a tenth-century Japanese medical work entitled *Ishimpo*, written at a time when Chinese culture was beginning to dominate Japan. Long after the Chinese themselves had adopted prudish attitudes, the Japanese maintained the free appreciation of sex once enjoyed by the Chinese, and scholars accept that the *Ishimpo* contains material originating as far back as Han Dynasty times.

Much like the *Kama Sutra*, the *Ishimpo* classifies thirty positions for sexual intercourse, all with highly colorful names borrowed from the animal kingdom. Three involve more than one partner, such as the "Paired Dance of the Female Phoenixes": "A man and two women, one lying on her back and one lying on her stomach. The one on her back raises her legs, and the one on her stomach rides on top. Their two vaginas face one another. The man sits cross-legged, displaying his jade thing, and attacks [the jade gates] above and below." Some were even more complicated, and one—the "Dog of Early Autumn," in which a man and a woman connect rear to rear—seems to be downright impossible.

Self Control

The main character in the Chinese sex manuals was the Yellow Emperor, a mythical ruler who was supposed to have laid the foundations for Chinese medical knowledge around 2600 B.C. (see

Acupuncture in **Medicine**). The Yellow Emperor complained that he was feeling weary, and a "sexpert" called the "Plain Woman" explained to him that he was not indulging enough in the right kind of sexual intercourse. While it was thought that the sexual essence of women (*yin*) was stronger, men would lose their male essence (*yang*) through indiscriminate loss of semen. So the main purpose of the manuals was to teach men to enjoy sex without loss of *yang*, and even how to build it up, through certain practices, by absorbing and converting *yin*. The texts exhibit an almost paranoid attitude toward male ejaculation—a beneficial side effect for women whose partners suffered from premature ejaculation, especially as it was recommended that women must always be brought to orgasm.

Techniques for male self-control were therefore taught, such as this one given in the handbook written in the seventh century A.D. by Master Tung-hsuan, and preserved in a Japanese copy:

A man closes his eyes and concentrates his thoughts, he presses his tongue against the roof of his mouth, bends his back and stretches his neck. He opens his nostrils wide and squares his shoulders, closes his mouth and sucks in his breath. Then [he will not ejaculate and] the semen will ascend inwards on its own account. A man can completely regulate his ejaculations. When having intercourse with women he should only emit semen two or three times in ten.

In 1976 modern science caught up with this ancient Chinese discovery, when sexologist Dr. Mina Robins explained how the moment of ejaculation could be delayed by keeping still and breathing slowly. Thus in more respects than one the ancient sex counselors were far ahead of early twentieth-century knowledge.

Ancient texts also played an unwitting role in the gradual liberalizing of benighted Victorian attitudes toward sex generally. By the 1960s, Burton's translation of the *Kama Sutra* had at last become freely available and other Oriental works soon followed. The Western world was at last allowed to know and understand the sexual attitudes of different, but equally civilized, cultures. The results have only been healthy.

MILITARY
TECHNOLOGY

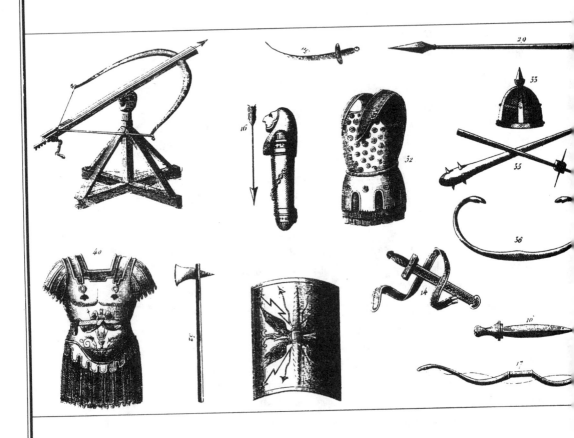

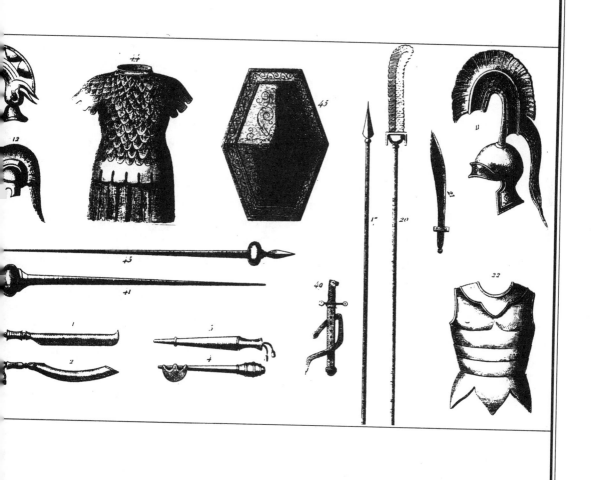

INTRODUCTION

Warfare, unfortunately, is as old as the human race itself. Despite modern fantasies to the contrary, there is not a shred of evidence that there was once a prehistoric "golden age" of peace and plenty, ruled over by peace-loving matriarchs who kept men's aggressive instincts in check. The sad truth is that human ingenuity has been feverishly applied to military technology since the beginnings of civilization.

Jericho, near the Dead Sea in Palestine, may well be the oldest town in the world, its origins going back some ten thousand years. It also has by far the earliest known fortifications. By 7000 B.C. the settlement was surrounded by a stone wall thirteen feet thick and ten feet high, enclosing some ten acres. At the center was a masterpiece of prehistoric construction—a solidly built stone tower, enclosing a central spiral stairway, the lower thirty feet of which were still standing when Jericho was excavated by a team of British archaeologists in the 1950s. Kathleen Kenyon, director of the excavation, clearly stood in awe of this discovery: "The whole comprises an amazing bit of architecture." The staggeringly early date of these constructions is underlined by the fact that they were built at Jericho *even before pottery came into use there.* At other prehistoric sites in the Near East ceramic skills were developed long before the growth of ordinary urban centers, let alone those with elaborate defensive works.

The eastern Mediterranean continued to be a region of mighty defensive works throughout ancient times—a reflection of the warlike nature of the civilizations that flourished there. The most spectacular walls are those built during the fourteenth to thirteenth centuries B.C. by the Mycenaean warrior-kings of Greece. These Bronze Age monarchs, the heroes of Homer's epic poetry, constructed elaborate defensive walls around their palaces and towns. The circuit walls were built of roughly hewn blocks of stone, the sheer size of which prompted the Greeks of classical times to believe that they were built by the cyclopes, the mythical race of one-eyed giants. As well as grand monumental entrances—the best known be-

(Opposite) Nine-thousand-year-old tower excavated at Jericho, Palestine, constructed in the "Pre-Pottery Neolithic" Period.

ing the Lion Gate at Mycenae—the citadels usually had small "postern" gates that could easily be concealed by brushwood, enabling surprise forays to be made against aggressors. Great attention was also paid to water supply—either the citadel would be situated to enclose a spring or secret tunnels would be cut under the walls to reach outside underground springs during times of siege.

There are nearly a hundred Mycenaean citadels in Greece—the product of a fragmented society in which city-states were constantly at war with their neighbors—yet their defense systems have a strange uniformity. A common origin is suggested, most likely Anatolia, where Greek legend placed the home of the cyclopes. Indeed, on the central Anatolian plateau the Hittites were building strikingly similar defenses at the same time as the Mycenaeans. But if the Hittites invented the "cyclopean" citadel style, the Mycenaean Greeks developed it to the full. Some structures are enormous, such as the walls of Gla in central Greece, which cover nearly two miles and enclose an area of half a square mile.

The construction of lengthy defensive systems was turned into a fine art by the Romans, who, during the second century A.D., decided that the best way of dealing with the troublesome barbarians on their frontiers was simply to erect physical barriers. The most impressive is the wall built across southern Scotland by the Emperor Hadrian (A.D. 117–138). Stretching from sea to sea, Hadrian's Wall is some 120 miles long. Built of stone for most of its course, it was dotted every mile with stations that acted as guardhouses, customs posts and signal stations, which communicated by a system of light telegraphy (see **Telegraphy** in **Communications**). The wall was like a raised road, some eighteen feet high and wide enough (about ten feet) for two soldiers to march abreast. On each side of the wall was a thirty-foot-wide ditch, and to the south a road running parallel that connected together various fortresses incorporated in the system.

But even the best efforts of the Romans pale by comparison with the Great Wall of China, at its greatest extent 2,150 miles long with another 1,780 miles of branches and outliers. It is the only human construction on earth made before the twentieth century that is visible from space.

The Great Wall began as a series of earthen banks thrown up by the northern Chinese kingdoms of the fourth century B.C. to keep out the nomads of the Central Asian steppe country. The megalomaniac first emperor of China, Shih Huang Ti (see **Magnets and Magnetism** in **High Tech**), built some 500 miles of wall to join up existing sections and create a continuous barrier some 1,300 miles long. Under the Han Dynasty (202 B.C.–A.D. 220) the Great Wall was strengthened by the addition of brick towers to the earthen base. Later Chinese dynasties lost control of the territory around the wall and it fell into disrepair, but in the late fifteenth century emperors of the Ming Dynasty took on the task of rebuilding it in stone. Their new wall was an imposing barrier up to forty feet high and thirty-two feet thick, incorporating as many as 25,000 watchtowers placed two arrow-shots apart. It is this final attempt to keep the northern steppe dwellers at bay that can be seen today—despite it the Manchus seized control of China in 1644.

Castles were developed later than defensive walls, with the earliest dating to the first centuries A.D. The original castle may be that in the Enda Mika'el palace at Axum, in Ethiopia. At the center of the

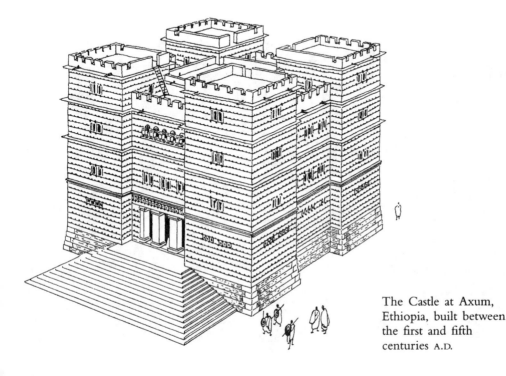

The Castle at Axum, Ethiopia, built between the first and fifth centuries A.D.

palace was a castlelike building of four square towers, standing up to four stories high on a stepped foundation. The castles of western Europe, however, seem more likely to be descended from those of the Byzantines. During his North African campaigns of the sixth century, the inventive general Belisarius (see **Watermills and Windmills** in **Working the Land**) constructed several square forts with a taller and stronger corner tower designed to be the last refuge of the garrison. The Muslim conquerors of North Africa a hundred years later copied the idea of the tower defense and introduced it to Spain. From here it spread to western France, where the castle as a private fortress appeared in the tenth century. These early castles of earth and timber were built to defend the nobility (together with their families, personal armies and servants) against Viking raiders.

Ironically it was William the Conqueror, descendant of the Viking invaders that captured Normandy, who constructed the most remarkable castle of the medieval world. Among the equipment carried by his navy when it sailed to England in 1066 was a prefabricated fort. The timbers had been cut, shaped, framed and pinned together in France before being loaded into massive barrels for the sea voyage. Landing at Pevensey, the Normans had reassembled the

Scene from the "Standard of Ur," c. 2500 B.C., showing the earliest known army. The Sumerian warriors are clad in leather armor and armed with pikes. The so-called Standard is a box decorated with a shell inlay on a lapis lazuli background.

fort by evening, giving themselves a secure base from which to launch their successful campaign.

We know much less about the very earliest offensive efforts. Compared with fortifications, they leave few obvious traces in the archaeological record, and here we have to rely mainly on the chance finds of weapons. However, from the third millennium B.C. written sources become available that throw light on the story of warfare. Around 2300 B.C. Sargon the Great of Babylonia created the world's first empire, stretching from the Mediterranean to the Persian Gulf, with the aid of the earliest known professional army. His inscriptions speak of 5,400 soldiers, who daily took their meals in his presence. This was a sudden and remarkable development— only a matter of generations earlier the casualties in battles between the rival states of Sumer (in southern Babylonia) were usually counted in single figures, according to their texts.

By Sargon's time the Bronze Age was well under way. The addition of tin to copper to make bronze meant that durable weapons such as swords, daggers, axes, spearheads and arrowheads could now be mass-produced by casting. Bronze was much harder than copper and far quicker to shape than flint, which required highly skilled and time-consuming techniques to make large weapons. Bronze was also less brittle than flint and could be formed into an almost infinite variety of forms. Mobility was also increased, since horses could now be shod.

It is no coincidence that the great age of warfare in the ancient Near East, initiated by Sargon's conquests, followed the invention of the chariot, the earliest war engine. Chariotry was almost the equivalent of an air force in the ancient world: it provided speedy transportation to the field of combat, where missiles (mostly spears and arrows) could be rained on a standing enemy while the charioteers passed by in relative safety because of their speed. The first known depiction of a chariot, pulled by four onagers (wild asses) is a bronze statuette found at Tell Agrab, in Iraq, dating to around 3000 B.C. Five hundred years later a decorated box from Ur, in Iraq (known as the "Standard of Ur"), shows chariots accompanied by the earliest known army.

Over the next two thousand years improvements (such as spoked wheels and metal axles) were continually made to chariots to reduce their weight and increase their speed, strength and deadliness. Different finishing touches were added by local inventors. A chariot from the royal cemetery at Salamis on Cyprus, eighth century B.C., has large iron rattles attached to its chassis to augment the terrifying thundering sound produced by a chariot charge. Scythed chariots, with long blades poking out from the axles to slice through the legs of enemy horses and men, seem to have been a Persian invention: the Greek historian Xenophon says they were introduced by the Persian emperor Cyrus (559–529 B.C.). The notion was somehow picked up by the Celts of ancient Britain, who were feared by the Romans for their scythed chariots.

Chariot production, as we shall see, was the focus of just one of the arms races of the ancient world. Ever-increasing swarms of chariots were used in the colossal battles of the second and first millennia B.C. At the battle of Qadesh, in Syria, around 1270 B.C., the 20,000 Egyptians led by Pharaoh Ramesses II faced an army of 17,000 Hittites from Turkey with 2,500 two-horsed chariots, each carrying at least two men. Both sides later recorded the battle as a great victory; the numbers, which we should perhaps take with a pinch of salt, are those given by Ramesses in a poem commissioned to celebrate his "triumph."

According to the records of the Assyrian king Shalmaneser III, at the battle of Qarqar (853 B.C.) his invading army faced a coalition of Syrian, Lebanese and Palestinian kings (including Ahab of Israel) with a combined muster of 52,900 foot soldiers, 3,940 chariots, 1,900 cavalrymen and 1,000 camel riders. For comparison, the crucial battle of New Orleans in 1815 was decided when 4,000 men under General Andrew Jackson defeated 9,000 British troops under Sir Edward Pakenham.

Past empires poured vast resources into their armies, churning out enormous quantities of arms. The Sung Dynasty of China, facing a constant threat from the nomadic Liao Empire, which had overrun the Great Wall, put their economy onto a war-footing in the 1040s A.D. Much of the iron production of Hebei Province (some eighty thousand tons a year) was dedicated to military production. Huge factories were created to supply the army of over a million men, making every year thirty-two thousand suits of armor, in three standard sizes, and sixteen million iron arrowheads for the crossbowmen who formed the backbone of the army. Despite their fearsome crossbows and their invention of incendiary weapons using gunpowder, the Sung armies were forced back by further nomad invaders, abandoning North China in A.D. 1125.

Naturally, in the ancient world as today, there were unsung geniuses in the back room keeping their side ahead in the arms race. An unknown Assyrian of the ninth century B.C. thought of equipping soldiers with inflatable animal skins so that they could cross rivers fully armed—he may have invented them or, perhaps more likely, adapted for military use something already developed by the marsh Arabs of southern Iraq. They became an essential part of the kit of Assyrian commandos.

The most ingenious of the back-room inventors were surely the military engineers of medieval China, whose early experiments with explosives even led to the invention of the mine. First used against the Mongol invaders in 1277, mines were manufactured in many shapes and sizes. One of the most fiendish was a land-mine booby trap, "The Underground Sky-Soaring Thunder." For this cunning device a group of spears and banners was stuck into the ground above the hidden mine to provide a tempting trophy for the enemy, who, as he approached to seize them, stepped on the mechanism, which lit the fuse and. . . . By the fourteenth century the Chinese were even making marine mines, which floated toward the enemy enclosed in inflated ox bladders. When the fuse, an incense stick, burned down, the mine went off.

In the late first century A.D. Julius Frontinus, Rome's leading mil-

(Left) Assyrian commandos inflating animal skins for use in crossing a river (right)—from a relief of the 8th century B.C.

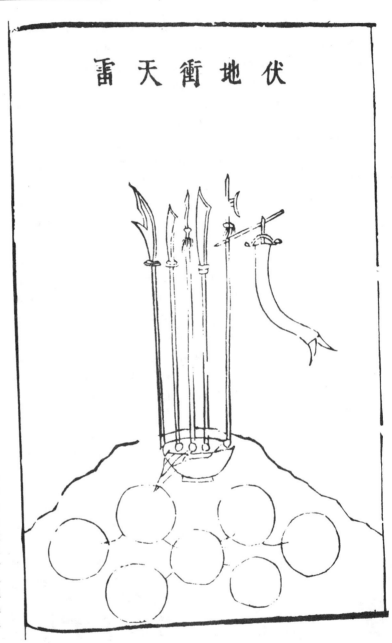

雷天衝地伏

Fourteenth-century depiction of the deadly Chinese booby trap known as the "Underground Sky-Soaring Thunder." A mechanism hidden beneath a cluster of banners and spears detonated the mines buried in the ground when trodden on.

itary scientist and Governor of Britain A.D. 74–78, wrote in his book *The Strategems*, "I shall ignore all ideas for new works and engines of war, the invention of which has reached its limits and for whose improvements I see no further hope." He could hardly have been more wrong, as we shall see in the following sections.

HUMAN AND ANIMAL ARMOR

As the human mind has devised ever more fiendish ways of killing people, so there has always been a need for increasing defense. The decorated box of the mid-third millenium B.C. known as the "Standard of Ur" shows the world's earliest known army—from the Sumerian city of Ur, in southern Iraq—wearing armor on its way to battle. Their gear, consisting of helmets, tunics and heavy cloaks, appears to have been made of leather, the most easily available material for defensive clothing in the ancient world, but one that was relatively easily penetrated. This problem was eventually tackled by the ancient Chinese, who made their leather armor far more robust by coating it with lacquer. Twelve suits of lacquered leather plates, reaching from the forearm to the thighs, were found in the tomb of the Marquis Yi of Zeng, dating to 433 B.C., along with a dozen helmets, also of lacquered leather.

The first item of armor to be made from metal, a vast improvement on leather, seems to have been the helmet. At the Royal Cemetery of Ur (see **Jewelry** in **Personal Effects**), where the "Standard of Ur" was discovered, archaeologists uncovered the grave of a prince buried around 2500 B.C. On his head was a magnificent helmet, beaten from a single sheet of electrum (a natural alloy of gold and silver). Beautifully contoured and decorated with chased lines to represent locks of hair, the helmet covered the owner's face down to the jawline; the ears were finely modeled, with holes enabling him to hear. This was clearly a "parade" helmet, but we can assume that simpler versions, of copper and bronze, were available to the ranks.

The world's earliest surviving body armor may be that from a tomb at Dendra, not far from Mycenae, in southern Greece. Here, in the late fifteenth century B.C., a Mycenaean nobleman was buried with his treasures, including an extraordinary suit of armor. Until this discovery, made in 1960, scholars had always been puzzled by Homer's description of the armor worn by the Mycenaean heroes of the Trojan War. Homer depicted his heroes donning corselets before going into battle, but what these were made of was hotly disputed; the poet called the warriors "bronze-shirted," but the absence of archaeological finds made scholars skeptical of the idea that anything like a solid metal breastplate was worn at such an early date. The Dendra find has completely revised our understanding. It is a complete set of bronze body armor, made from a series of overlapping plates—there is a round neckpiece, vertical plates to protect the shoulders, and three horizontal plates that encircled the body down to the thighs. The whole bears an uncanny resemblance to the suits of armor worn by medieval knights nearly three thousand years later.

Scale and Mail

The Dendra armor would have provided ample protection, but it had the disadvantage of being rather heavy and inflexible. Most likely it was worn by a charioteer, who drove around the battlefield in his chariot as he shot arrows at his enemies—traces of a quiver are visible on the armor.

The problem of making lighter, more flexible armor seems to have been tackled simultaneously in ancient China and Egypt. Scale armor, made of bronze plates mounted on leather clothing, has been found in Chinese tombs dating to around 1400 B.C. At almost the same date bronze scale armor was being produced in Egypt—some has been recovered from Pharaoh Amenhotep III's palace at Thebes. Whether the two inventions were genuinely independent is impossible to tell.

With the replacement of bronze by iron after 1000 B.C. came the first iron scale armor, of a kind used for the next twenty-five hundred years across the Old World. Iron armor scales of the mid-ninth century B.C., probably the oldest examples known, were uncovered by archaeologists at the Assyrian capital of Nimrud, in Iraq. Reliefs of the same period show Assyrian bowmen covered from head to foot in fine scale armor, looking rather like medieval Saracen knights.

Ninth-century B.C. Assyrian bowmen wearing full-length iron scale armor with the head protected by a conical helmet and the neck by a scale armor curtain.

As with scale armor, chain mail (made of small metal rings, each joined to four others—two in the row above and two in that below) seems to have appeared in widely separated regions at almost the same time. As we know only too well from recent history, military technology can spread with alarming rapidity—the ancient world was no different. Chain mail allowed the wearer even greater maneuverability than scale armor. The earliest evidence for chain mail in the West is on a triumphal sculpture of the second century B.C. at Pergamum, in Turkey, while the terra-cotta models of warriors guarding the tomb of the First Emperor (221–210 B.C.), Shih Huang Ti (see **Magnets and Magnetism** in **High Tech**), show seven different types of mail coat. We do know that the nomadic Scythian warriors of southern Russia wore iron scale armor, and it may be that they were the first to manufacture chain mail, carrying the idea eastward and westward as they migrated.

Armored cavalryman and horse depicted on Trajan's Column, in Rome, about A.D. 115. He was probably from Sarmatia (southern Russia) and is shown here in the army of the Dacians (from Romania) fighting against the Romans.

Paper Armor

The most surprising innovation in the history of armor must be the ancient Chinese use of paper. Since the sixth century B.C. they had been developing the technique of manufacturing clothes from mulberry paper (see **Clothing and Shoes** in **Personal Effects**), which they eventually applied to military uniforms. By the ninth century A.D. Governor Hsü Shang, of Ho-tung in Shensi Province, maintained a standing army of one thousand soldiers, clothed in thick pleated paper armor that could not be pierced by strong arrows—the cunningly designed layers obstructed the arrows' path.

After this, paper armor became common on both land and sea. In the twelfth century the local magistrate Chen Te-Hsiu requested permission from the central authorities to trade in 100 sets of iron armor for 50 of the superior paper type. At much the same time the captains of two pirate ships that surrendered during an official amnesty handed over 110 suits of paper armor. In the end, however, the increasing penetrative power of crossbows firing iron arrows meant that paper armor became useless.

The Aztecs of Mexico made similar use of thick cotton clothing as armor. Their army fighting the Spanish *conquistadores* wore quilted, tight-fitting suits of layered cotton about two fingers in thickness. The Aztecs had themselves learned the technique from the Mayans to the south, who, invaded by a Mexican army with powerful bows and arrows in the tenth century, had invented a kind of armor by soaking their quilted cotton tunics in salt brine.

Armor-Plated Elephants

Animals, too, were frequently armored in the ancient world. Undoubtedly the most awe-inspiring creatures in the Roman world were elephants with iron armor on their forehead and body. The idea came from the Greeks, who had encountered leather-armored elephants in India during Alexander the Great's campaigns there in 326 B.C. The sheer expense involved, however, meant that this was a rare practice.

Much more common was the armored horse. The Greek mercenary Xenophon, writing at the beginning of the fourth century B.C., said that the elite Persian cavalry force, both men and horses, was

Reconstruction of the elaborate headguard for a horse from the Roman fort of Vindolanda, close to Hadrian's Wall, in northern Britain— c. A.D. 100.

armored. The Greek successors of Alexander the Great in the East enthusiastically developed their own armored heavy cavalry, and in 189 B.C. Antiochus III of Syria had three thousand horses with iron breastplates. Only much later did the Romans adopt the idea. In the third century A.D. their heavy cavalry, the "shock troops" of the army, was finally fitted out with armor. One tower at the Roman fort of Dura-Europos, in Syria, collapsed during a Persian siege about A.D. 250, burying much of the military equipment that had been stored in it, including two horse covers of iron and copper-alloy scales to protect the animals against arrows.

TANKS

The most ruthless army of the ancient world was without doubt that of the Assyrians, masters of the Near East in the early first millennium B.C. All rebellions against their authority were systematically crushed, any cities that resisted being reduced by siege to piles of rubble. The siege-engines shown on Assyrian memorials are impressive structures, sheathed in wickerwork to protect their crews and mounted on either four or six wheels. They are depicted with two kinds of head: either a flat-ended battering-ram for cracking open masonry or a pair of massive lances for gouging out chunks from the walls after being driven into the cracks created by the ram.

The biggest danger to the crews engaged in battering down a wall was the rain of incendiary missiles hurled at them by the defenders above. To counter this threat, the Assyrians developed a prototype armored car with a tower topped by a platform from which archers could pick off defenders. The tower sometimes even incorporated a water tank for putting out the incendiaries before they could set the whole structure ablaze. The effort needed to move these giants up to the wall was of course enormous, even when the approach route had been leveled in preparation. This resulted in a gradual decrease in size over time, and the emphasis switched to making siege engines more mobile and fireproof.

A massive Assyrian siege engine of the 9th century B.C. While the battering ram smashes a hole in the wall, the archers on the tower of the armored car fight off the city's defenders.

Similar wheeled siege engines continued to be made up to modern times. Out on the battlefield the nearest thing in the ancient world to a tank was the elephant, a resemblance increased by the heavy armor in which they were often encased (see **Human and Animal Armor**). Elephants with troops mounted on their backs featured in some of the greatest campaigns of military history, including Alexander the Great's conquest of the Persian Empire and Hannibal's crossing of the Alps. But, important though elephants were, they were not of course on the line of development leading to the modern tank.

The Bohemian Tank

The most significant advance in the prehistory of the tank took place in medieval times. Contemporary texts record that the Sung Dynasty army of China developed an iron-plated armored car to use against the cavalry charges of the Tartars, who had pushed the Sung out of northern China. This invention helped to bring the Tartar advance to a halt in A.D. 1127, resulting in a border that survived for a century.

Little more is then heard of armored cars in China, and they were not used against the Mongols under Kublai Khan, who conquered the whole of China by 1279. The Mongols adapted the idea of the battle car to their traditional lifestyle, setting up large tents on wagons, which were drawn by oxen behind their advancing armies. While the Mongols in the West did not seem to build armored cars along Chinese lines, it is likely that they were responsible for transmitting the concept to eastern Europe during the fourteenth century, when their invading hordes swept as far as Poland and Hungry.

It was here, in Bohemia (the modern Czech Republic), that the tank came into its own, during the religious conflicts preceding the Protestant breakaway led by Martin Luther. Many theologians within the Church had expressed their desire for reform during this time. One of them, John Huss, the rector of Prague University, was burned at the stake as a heretic at the Council of Constance in 1415 for doing just that. This sparked off a peasant rebellion, led by Jan Zizka, against the mighty Imperial German Army.

Because his was a peasant army, lacking support in the cities of Bohemia, Zizka took up the idea of fortified wagons. He built wagon forts covered with sheets of iron and drawn by teams of horses. The crew, armed with spiked iron bars and large axes, fired crossbows and small handguns at any attackers through loopholes pierced in the sides of the wagons.

Between 1420 and 1431 these tanks enabled Zizka's force of 25,000 men to repeatedly defeat imperial German armies up to 200,000 strong. The Bohemian wagon forts were usually drawn up in defensive formations before the battle, with the gaps between them filled by folding ramparts of oak—they could then cut down the enemy with overlapping fields of fire. On one occasion, at Kutna

An encampment of Hussite tanks and carts resembling a town on the move. The vehicles in the outer ring are tanks, while those in the inner ring are for carrying supplies.

Hora in 1422, Zizka's army was caught in the open, and the tanks fought on the move in close formation. In fact Zizka's men were never defeated by the German invaders; they lost their long struggle through increasing war weariness and internal divisions.

As the accuracy and penetration power of firearms improved toward the end of the Middle Ages, the idea of tanks went out of favor—their armor would not stand up to a direct hit, and the horses pulling them could easily be shot. Only with the development of the internal combustion engine in the late nineteenth century did the tank again become a viable means of defense.

CATAPULTS AND CROSSBOWS

In 399 B.C. Dionysius the Elder, ruler of the Greek colony of Syracuse, in Sicily, prepared for war with the Carthaginian Empire by financing a massive research-and-development effort to produce new weapons. By far the greatest success of this crash program was the invention of the catapult, key weapon of the ancient world.

Even in these early days catapults fired arrows much farther than the strongest human could ever manage. Continued state-sponsored research, especially that funded by King Philip II of Macedon (359–336 B.C.), led to the development of large catapults using ropes made of animal sinews that could fire thirteen-foot-long arrows—more than twice the length of those hurled by Dionysius' machines. His son, Alexander the Great, used them to maximum effect in reducing the most strongly defended cities of the Persian Empire, such as Tyre, in Lebanon. Monster catapults were produced, like that designed by Archimedes and installed on the superfreighter *Syracusa*, built for Hiero II of Syracuse toward the end of the third century B.C. (see **Ships and Liners** in **Transportation**), which could hurl stones weighing 173 pounds distances of up to 200 yards. These would undoubtedly have smashed through the hull of any enemy ship the gunners managed to hit.

Such enormous catapults shifted the balance of power in favor of the besiegers. Philo of Byzantium, in his artillery manual written about 200 B.C., said that a wall had to be at least fifteen feet thick to withstand catapult stones weighing up to 350 pounds, while deep ditches had to be dug five hundred feet out from the city walls to keep the enemy catapults out of range.

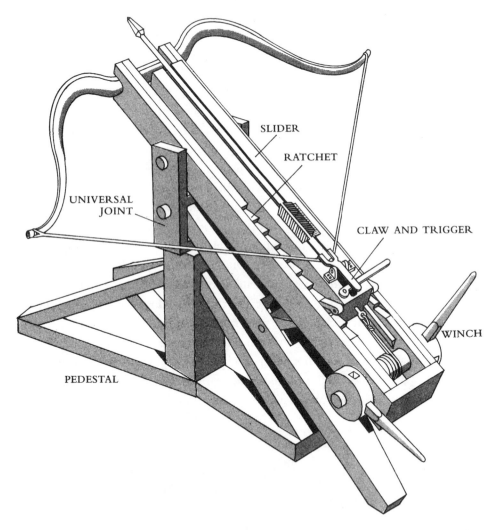

SLIDER

RATCHET

UNIVERSAL JOINT

CLAW AND TRIGGER

WINCH

PEDESTAL

A bow-firing catapult of the type devised for Dionysius the Elder of Syracuse in 399 B.C. The arrow, some six feet long, was loaded into a groove in the slider and the bowstring locked in place with the claw. The slider was then drawn back by the winch against the tension of the bow and held in place by a ratchet-and-hook arrangement. When the claw was lifted, by moving the trigger the arrow was released. A "universal joint" between the pedestal and the catapult enabled it to be aimed in any direction.

The main advance made by the Romans was to set smaller catapults on iron frames with wheels so that they could be shifted about on the battlefield. Rather than stones, these shot mainly bolts, firing a standard type twenty-seven inches long for some seven hundred yards. The awesome power of these Roman weapons is amply demonstrated by a find from the hill fort of Maiden Castle, captured by the future emperor Vespasian early in the conquest of Britain. This is a skull with a square hole in it made by a catapult bolt, which must have been moving at great velocity to pierce the bone so cleanly.

While the Romans were developing catapults, the Chinese were inventing the crossbow. The earliest reference to crossbows comes from *The Art of War* by Sun-Tzu, a remarkable manual of strategy and tactics written sometime in the fourth century B.C. but still studied by the Chinese People's Liberation Army today. Pictures of handheld crossbows are also found engraved on bronze vessels of the fourth century B.C. By the time of the First Emperor of China, Shih Huang Ti (221–210 B.C.), the crossbow had become the main weapon of the imperial army. Heavy crossbows were mounted on carts, and in the Number One Pit at the First Emperor's tomb, at Mount Li, there is a vanguard of two hundred unarmored crossbowmen among the terra-cotta warriors (see **Magnets and Magnetism** in **High Tech**). These were fast-moving long-range fighters, whose weapons could hit an enemy at a distance of 650 feet.

By 209 B.C. the imperial army had fifty thousand crossbowmen, and in 157 B.C. it is recorded that the imperial arsenals held half a million crossbows. This was a remarkable feat of mass production of

Skull with ballista-bolt hole found at Maiden Castle, England. The Celtic warriors of Maiden Castle stood no chance against the Roman legions under the future emperor Vespasian in A.D. 43. Their massively defended hill fort was soon overwhelmed by an enemy with vastly superior military technology.

highly complicated mechanisms, which were first cast in bronze, then machine-finished to a high standard and finally assembled. Used against the nomadic Huns of Mongolia, they were the ultimate "capture-proof" weapon. The Huns were incapable of keeping these extremely complex mechanisms in working order without workshops, while the shorter crossbow arrows were useless in an ordinary bow. Around this time the accuracy of crossbows was improved by the use of the world's first grid-sights for aiming. These were made of fine horizontal and vertical wires crossing each other to form grids, similar to those used in the cameras and antiaircraft guns of today.

Ancient Machine Guns

The ideal, with both the catapult and the crossbow, was to be able to shoot off a constant stream of arrows without having to stop to reload. Greco-Roman engineers struggled with this problem, and around A.D. 100 Dionysius of Alexandria invented a rapid-firing small catapult, a kind of ancient machine gun. A magazine of bolts was fitted above the groove in the catapult, into which each new arrow dropped as the previous one was fired off. Its range (six hundred feet) and accuracy were excellent, but it had one major drawback: The sighting could not be altered once it was set up. It was thus only really effective against attackers who had to cross a particular piece of ground.

The Chinese cracked this problem by applying the principle of a magazine of bolts to the crossbow, which could, of course, be aimed in any direction. The first machine-gun crossbows appeared in the eleventh or twelfth century A.D. Test firings of modern replicas have shown that a force of a hundred men armed with these lethal weapons could fire two thousand arrows in only fifteen seconds, over a range of 80 yards. At this time the range of large, winched crossbows was 1,160 yards, and that of the hand-held standard weapon 500 yards. Since the arrows were often dipped in poison, even a scratch from one of them could be fatal. It is no wonder that the Chinese crossbowmen were the most feared military corps of the medieval world.

Roman "machine-gun nest" made of trimmed saplings— from Trajan's Column, Rome, c. A.D. 115.

THE "CLAWS" OF ARCHIMEDES

The Romans were confident of an easy victory when, in 215 B.C., they sent a large expeditionary force to Sicily to capture the city of Syracuse. What greeted them was totally unexpected. The Greek scientist Archimedes had fortified the city with a range of military contraptions of such incredible power and accuracy that the mighty Roman war machine was stopped dead in its tracks.

The punishment dealt out by this weaponry is vividly described by the ancient sources. When the Roman legions attacked, they were met with a rain of missiles and immense stones launched from giant catapults. As they struggled to reach the wall, the relentless barrage was kept up with continuous volleys of arrows. Trying to protect themselves under a cover of shields, the helpless Roman infantry was crushed by boulders and large timbers dropped from cranes that swung out over the battlements. Most horrific of all were the enormous clawlike devices that wrecked the Roman fleet as it tried to enter the harbor, shaking the ships about and even plucking them clean out of the water. In the words of the Greek historian Plutarch,

> The ships, drawn by engines within and whirled about, were dashed against steep rocks that stood jutting out under the walls, with great destruction of the soldiers that were aboard them. A ship was frequently lifted up to a great height in the air—a dreadful thing to behold—and was rolled to and fro, and kept swinging, until the mariners were all thrown out, when at length it was dashed against the rocks, or was dropped.

And so the Roman army was kept at bay by the genius of one seventy-five-year-old man. Livy, who wrote the most detailed account from the Roman standpoint, commented,

> An operation launched with such strength might well have proved successful, had it not been for the presence in Syracuse at that time of one particular individual—Archimedes, unrivalled in his knowledge of astronomy, was even more remarkable as the inventor and constructor of types of artillery and military devices of various kinds, by the aid of which he was able with one finger, as it were, to frustrate the most laborious operations of the enemy.

The Amazing Archimedes

Who was this man whose mechanical wizardry could withstand the military might of Rome itself? First and foremost Archimedes was a mathematician, in fact the greatest ever produced by the Greeks. The son of an astronomer, he finished his education at the "Museum" of Alexandria in Egypt (see **Introduction** to **Communications**). Here he won fame with his theorems for calculating the volume of spheres, cones and any conceivable geometrical shape. An early practical application was the spiral-shaped water-lift known as Archimedes' Screw, still used today to draw water from the Nile. It has also been suggested that he was commissioned by the Romans to build the first odometer to assist them in their massive road-building program during the late third century B.C. (see **Odometers** in **Transportation**).

The rich kingdom of Syracuse was strategically situated between the rival powers of Rome and Carthage. Hannibal's epic crossing of the Alps in 218 B.C. began the second war between these two giants for control of the Mediterranean—after crossing the Alps he battled his way through Italy for another fourteen years before returning home. The capture of Syracuse in 212 B.C. by the Romans, after a long siege, was a key factor in their eventual victory.

But Archimedes spent most of his life in his native Syracuse, as scientific adviser to its ruler, King Hiero. It was while he was working on a problem for Hiero that Archimedes supposedly leaped from his bath and ran naked through the streets, screaming *Heyeureka, heyeureka* ("I've found it!"). Another story concerns Archimedes' genius for using levers to shift the most enormous of objects. He realized that, given a long enough lever and a suitable fixed point or fulcrum, literally any weight could be moved by a relatively small force. He demonstrated this by launching the massive royal ship *Syracusa* (see **Ships and Liners** in **Transportation**). According to the ancient accounts, Archimedes launched the fully loaded ship from its dry dock *single-handed*, using a system of levers and pulleys. "Give me somewhere to stand," Archimedes was said to have remarked on this occasion, "and I will move the earth!" He used the same principle to deadly effect when the Roman navy arrived.

The most important task that Archimedes undertook for Hiero was to design an integrated system of defenses for the city. He replanned the walls from scratch, tailoring them to hold an elaborate

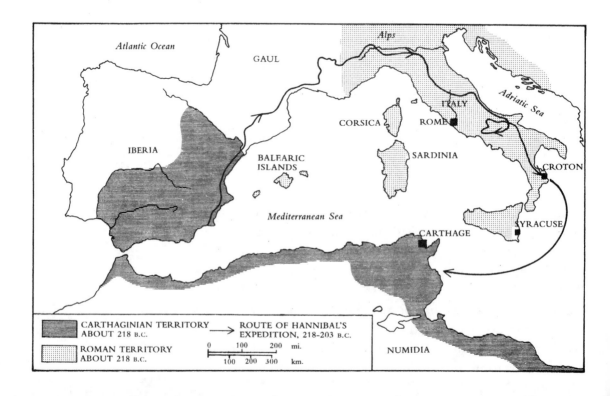

series of powerful ballistic devices, as well as new weapons such as the fiendish "claws." Both men had the foresight to see that Syracuse would soon be caught up in the titanic struggle brewing between the Roman and Carthaginian empires for control of the Mediterranean.

War broke out in 218 B.C. when the Carthaginian general Hannibal led an enormous army (including the famous elephants) from North Africa across the Straits of Gibraltar into Europe, in a desperate bid to break the power of Rome. But soon after Hannibal's spectacular crossing of the Alps, the bitter contest between Rome and Carthage reached deadlock. The losses incurred during the trek over the mountains left Hannibal's army too weak for a direct assault on Rome. The Romans, for their part, were unable to drive the invader from Italy, so concentrated instead on preventing reinforcements arriving from Carthage.

In 215 B.C., as the fate of the Western world hung in the balance, Archimedes' old friend King Hiero died. Hiero had been pro-Roman in his sympathies, but his successor made overtures to Carthage, throwing the Roman Senate into alarm. Sicily lay directly on the lines of communication between Carthage and Italy. As Syracuse wavered, a huge Roman force led by Consul Marcellus was dispatched to ensure its cooperation. He moved in with a fleet of sixty war galleys and an army of three legions (about fifteen thousand men), only to be stalled for two and a half years by Archimedes' weaponry.

A Ship-Shaking Machine

Before he arrived at Syracuse, Marcellus made sure that he was equipped with the best available military technology, including a catapult so enormous that it had to be housed on the decks of eight ships lashed together. Other ships were tied in pairs to carry siege towers high enough to dwarf the walls of the city. Plutarch wrote of the "abundance and magnificence" of Marcellus's preparations, "but all, it seems, were trifles for Archimedes and his machines."

Tradition has it that among the surprises prepared by Archimedes were gigantic concave mirrors that could focus the sun's rays to laserlike strength and set fire to the sails of enemy ships. Unfortunately the story is not substantiated by the earliest sources, and recent analyses suggest that such an idea is technically impossible. Far better documented are the ship-shaking "claws," mentioned in all the accounts of the siege of Syracuse. Exactly *how* they worked is less certain. It is clear that the Romans were baffled as to what had hit them, while Archimedes himself left no descriptions of his military devices. Some historians, including Plutarch, argue that Archimedes wrote nothing about them because he disdained mechanics as being "ignoble" compared with the more elevated subject of pure mathematics. It is just as likely that Archimedes committed nothing to paper simply for security reasons.

Plutarch's account refers to the claws working from "within," conceivably meaning the harbor. Indeed a reconstruction offered by Swedish engineer Sigvard Strandh shows how such a machine could have worked by grabbing the ships from *underneath* the waters. This seems rather fanciful, however: such underwater claws would have only been useful when a ship happened to pass directly above them. The accounts left by Livy and Polybius clearly describe something far more maneuverable, and controlled from above—namely, giant iron hands suspended by chains from long beams swung out over the harbor. A claw would clutch at the prow of a ship—like the miniature fairground cranes that grab at toys or sweets. Depressing the weighted end of the beam inside the city would lift the ship's prow, or even raise the vessel completely upright. Then the chain and hand would suddenly drop: "The result was that some of the vessels fell on their sides, some entirely capsized, while the greater number, when their prows were thus dropped from a height, went under water and filled, throwing all into confusion."

We can only wonder, as the Romans did, at the sheer scale and power of machines that could actually lift warships in the air. Few things could throw their highly trained army into panic. Yet the diabolical inventions of Archimedes made such a deep impression on the Roman troops that, after the initial assault on Syracuse, the mere sight of a piece of wood or rope appearing over the wall was enough to make them break rank.

(Opposite) A simplified reconstruction of Archimedes' ship-shaking device.

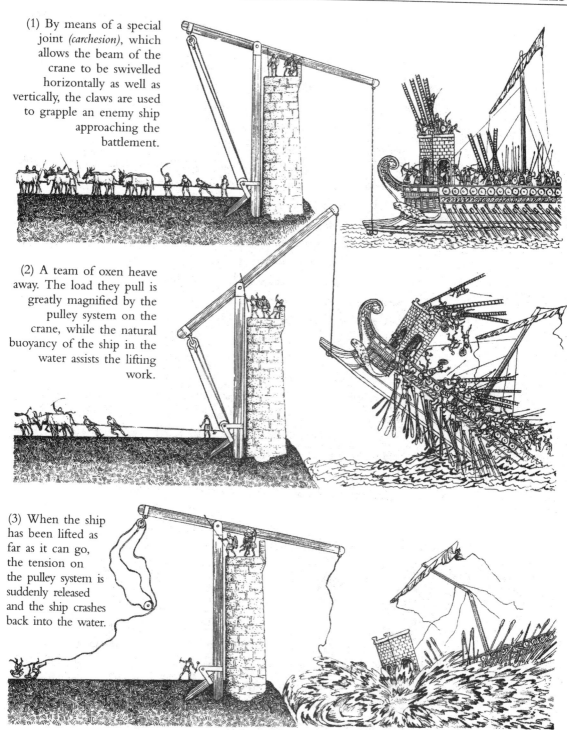

(1) By means of a special joint *(carchesion)*, which allows the beam of the crane to be swivelled horizontally as well as vertically, the claws are used to grapple an enemy ship approaching the battlement.

(2) A team of oxen heave away. The load they pull is greatly magnified by the pulley system on the crane, while the natural buoyancy of the ship in the water assists the lifting work.

(3) When the ship has been lifted as far as it can go, the tension on the pulley system is suddenly released and the ship crashes back into the water.

The Death of Archimedes

Syracuse only fell because of its inhabitants' negligence. Through his blockade Marcellus had ensured that food was scarce, but spies informed him that wine was still plentiful and was being greedily consumed during a lengthy religious festival. His troops scaled a small tower at one end of the city and seized it while the sentries were drunk. With a foothold inside the city the Romans poured in, the Syracusans being no match for them in hand-to-hand combat.

The citizens were blissfully unaware that their city was being taken. Archimedes himself was completely absorbed in his work when a Roman soldier broke into his quarters and killed him. Accounts of his death vary. Some say that he refused to be escorted to Marcellus until he had solved the mathematical puzzle he was contemplating. A different version says that Archimedes, worried about his precious instruments, was hurrying to hand them over to Marcellus for safekeeping when the invaders cut him down, thinking that he was trying to escape with treasure.

Marcellus was dismayed. He had seized one of the richest cities in the world and secured Roman mastery of the Mediterranean. Yet the most valuable prize of all had eluded him—the mind of Archimedes. Marcellus buried him with full honors and inscribed the tomb, as the old man had requested, with a design showing the ratio between the volume of a sphere and a cylinder, one of his favorite geometrical discoveries.

FLAMETHROWERS

In A.D. 674 the city of Byzantium (modern Istanbul) was in desperate straits, surrounded by the Islamic armies of the caliphs of Damascus. Fortunately a savior was at hand in the unlikely form of a Syrian architect and chemist named Callinicus. He was the inventor of "Greek Fire"—the secret weapon of the Byzantine Empire, whose formula was so closely guarded that its precise ingredients are still unknown. The besieging fleet was utterly destroyed in a massive onslaught with Greek Fire, and for another eight centuries Byzantium was safe from eastern invaders.

Later military writers, such as the thirteenth-century Mark the Greek, do let slip that Greek Fire contained sulfur, saltpeter, gasoline, pine resin and gum resin. It was either squirted over enemy ships through a pump, which had to be recharged between each burst, or poured into shells and fired off by catapult. In either case the effects were nearly always horrific. The flaming liquid, being oil-based, would float on the surface of the sea, setting fire to ships' hulls and frying any unfortunate sailors who hoped to save themselves by jumping overboard.

It was not only Islamic invaders that were to feel the power of Greek Fire. By the tenth century another threat to the Byzantine Empire had grown on its northern borders, where Vikings from southern Sweden known as the Rus had founded a powerful state at Kiev, the nucleus of the future kingdom of Russia. In A.D. 941 Prince

The Byzantines even employed their deadly flamethrowers against each other. This manuscript depicts an encounter in the civil war between Michael II (A.D. 826–829) and the rebel Joannes.

Igor of Kiev led the Rus fleet on a raid across the Black Sea to Byzantium. The outcome is recorded in the Russian *Primary Chronicle*:

> The Rus then came against the Greeks. There was fierce battle between them, and although the Greeks barely won, the Rus returned to their boats and fled. The Greeks met them in their boats and began to shoot fire through pipes onto the Russian boats. And a fearsome wonder was to be seen. The Rus, seeing the flame, threw themselves into the sea, wishing to swim away; and thus the rest returned home. When they came to their own land each one told his own people about what had happened. "The Greeks," they said, "possess something like the lightning in the heavens, and they released it and burned us. For this reason we did not conquer them."

Observers on the Byzantine side claimed that a mere fifteen ships equipped with Greek Fire routed a Russian fleet of thousands.

The psychological impact of Greek Fire was almost as devastating as the actual destruction it caused. Writing early in the twelfth century, Princess Anna Commena described how carefully her father, the Emperor Alexius (A.D. 1081–1118), prepared his fleet for battle with the Pisans of Italy:

> As the Emperor knew that the Pisans were skilled in sea warfare, and was apprehensive about having a battle with them, on the prow of each ship that he had built he had a head fixed of a lion, or other land-animal, open-mouthed, made in bronze or iron and gilded all over, so that just the sight of them was terrifying. And the fire that was to be directed against the enemy through tubes he made to pass through the mouths of the beasts, so that it looked as if they were vomiting fire.

Callinicus's devastating invention was the culmination of a thousand years of the development of incendiary devices in Greece. The ancient historian Thucydides records the use of an early fire weapon in the great Peloponnesian War between Athens and Sparta for mastery of the Aegean. In 424 B.C. the long siege of Delion was brought to an end when its partly wooden walls were burned down. The attackers achieved this through the use of a long tube on wheels containing burning charcoal, sulfur and pitch, behind which were bellows to blow the flame forward. This primitive burner could only work against extremely ill-organized defenders, since the absence of gasoline made it very slow-working.

It was actually the Chinese who developed this fiendish weapon to the full. They seem to have come across the idea of Greek Fire relatively late, around A.D. 900, but their much earlier invention of the double-acting piston bellows provided a ready-made technical breakthrough. This device had two inlet valves, so that, whether being pulled or pushed, it would suck air in on one side and compress the air on the other, thereby producing a continuous jet of flame.

It soon became a fearsomely destructive weapon, although not always of the enemy, as we see from an account of a naval battle on the Yangtze River in A.D. 975 given in Shih Hsu-Pai's *Talks at Fisherman's Rock*:

> Chu Ling-Pin as Admiral was attacked by the Sung Emperor's forces in strength. Chu was in command of a large warship more than ten decks high, with flags flying and drums beating. The Imperial ships were smaller but they came down the river attacking fiercely, and the arrows flew so fast that the ships under Admiral Chu were like porcupines. Chu hardly knew what to do. So he quickly projected petrol from flame-throwers to destroy the enemy. The Sung forces could not have withstood this, but all of a sudden a north wind sprang up and swept the smoke and flames over the sky towards his own ships and men. As many as 150,000 soldiers and sailors were caught in this and overwhelmed, whereupon Chu, being overcome with grief, flung himself into the flames and died.

The Chinese army certainly didn't lack ingenuity, and by the end of the tenth century the "fire lance"—a portable flamethrower—was in use. This was essentially a bamboo (later also cast-iron) tube filled with petroleum-based rocket fuel and tied firmly to the end of a spear so that it couldn't fly off. These flamethrowers, lasting about five minutes, were a key element in the defense of northern cities against invading nomads.

The ultimate development of these weapons was the wheeled flamethrower battery, created in the fourteenth century. These bat-

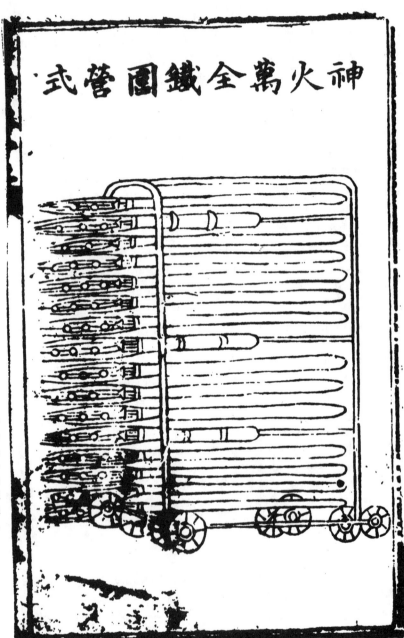

A mobile 14th-century Chinese flamethrower battery. The weapons could be fired all at once in a devastating broadside against the enemy or used to resupply guards on city walls.

teries, known as fierce-flame-sprouting shields, were like huge wine racks, covered with oxhide and mounted on wheels. In each were placed three dozen fire lances containing different mixtures of gunpowder, which could be lit as required by slow-burning matches. The military author of the *Hung Lung Chin* ("Fire-Drake Manual"), writing in 1412, describes the shields in action:

> When two opposing armies are confronting one another, at the sound of a rocket signal, the shields are rolled forward into action, and when they spout fire, the flames shoot twenty or thirty feet forward. One group of men in armor on the left hand work the shields, while another group on the right wield their cutlasses. They aim to decapitate the enemy soldiers, and to cut off the legs of their horses (during the confusion caused by the flame-throwers). A single one of these shields is in itself worth ten brave soldiers.

No enemy could withstand the wall of fire created by the fire-lance batteries. But with the increasing use of cannon they became extremely vulnerable to being blown up before being wheeled into position, with the result that they were never decisive on the battlefield.

HAND GRENADES

The grenade is one of the key weapons of modern infantry warfare. It therefore comes as quite a surprise to find that it was commonly used in battles between Christians and Muslims during the time of the Crusades. The use of flammable liquids in war originated in the Byzantine Empire (see **Flamethrowers**), but by A.D. 1000, according to written accounts of the day, grenades made of glass or earthenware filled with gasoline were used by defenders of besieged cities all over the Islamic world. These were flung at the enemy or at their wooden siege towers from catapults or by hand.

Until recently, however, there was little archaeological support for the literary evidence, with one possible exception. All over the Middle East, from Egypt to Samarkand, archaeologists this century have been picking up egg-shaped pottery vessels, some decorated with heraldic motifs and others inscribed with battle cries such as "Glory" or "Allah." While they are clearly Islamic objects from the Middle Ages, the function of these mysterious finds has been much debated, with one school claiming they are containers for mercury while others believe they are unused grenades.

A study published by Peter Pentz of the National Museum of Denmark in 1988 has completely changed this situation. In the 1930s Danish excavations at the Syrian town of Hama had revealed a strangely equipped workroom of the thirteenth century, where asphalt was manufactured for an unknown purpose. Pentz's reinterpretation of the site now makes this clear. If it was a workshop for the production of grenades, then all of its unusual features are explained. There was a fireplace for distilling gasoline, holes in the wall for ventilation, and a large jar with a lid set into the floor to protect the gasoline from stray sparks; outside was a large pit containing shells needed for making lime to mix with the gas. The scale of the operation is considerable, but it was all to no avail, as the workshop,

Examples of the elaborate Islamic grenades (made of pottery) used against the Christian Crusaders.

along with the whole town, was destroyed by the Mongol hordes in 1259.

The Chinese picked up the idea of grenades from Arab merchants and were making them in pottery containers by the early twelfth century. Sculptures dated to A.D. 1128 in the Buddhist cave temples at Ta-tsu in Szechuan include two extraordinary figures of demons

A Chinese demon prepares to throw a smoking grenade—as depicted in a cave temple at Ta-tsu, China, A.D. 1128.

holding weapons. One grasps a handgun (see **From Gunpowder to the Cannon**), while the other, carved on the opposite side of the cave, holds an egg-shaped bomb. The fuse and a trail of smoke disappearing over the demon's right shoulder are clearly visible.

In A.D. 1187 the Chinese chronicler Yuan Hao-Wen recorded an unusual use of the grenade in an anecdote about a hunter named T'ieh Li:

> One evening he found a great number of foxes in a certain place. So, knowing the path that they followed, he set a trap, and at the second watch of the night he climbed up into a tree carrying at his waist a vessel of gunpowder. The coven of foxes duly came under the tree, whereupon he lit the fuse and threw the vessel down; it burst with a great report, and scared all the foxes. They were so confused that with one accord they rushed into the net which he had prepared for them. Then he climbed down the tree and killed them all for their fur.

FROM GUNPOWDER TO THE CANNON

It is truly remarkable how little was known of the real history of gunpowder until recently, given its enormous impact on the world. Even twenty years ago it was still being touted as the greatest original invention of feudal Europe, the prime example of Western ingenuity. The proof of this claim apparently lay in the English soldier Walter de Milamete's manuscript "On the Majesty, Wisdom and Prudence of Kings," dating to A.D. 1327, which contained the earliest known depiction of a cannon.

In traditional accounts the inventor was usually claimed to be either the English philosopher Roger Bacon or a German monk called Black Berthold. The monk seems to be a wholly mythical figure, though Roger Bacon (1217–1292) did exist and did *describe* gunpowder. His account appears to refer to a Chinese firecracker, which led to an alternative version of the story admitting the priority of the Chinese in inventing gunpowder, but arguing that the only use they ever made of it was to create marvelous fireworks. The Chinese were certainly the first to make fireworks (see **Fireworks** in **Sport and Leisure**), but they also played the major role in the development of the cannon.

Gunpowder seems to have been discovered in China by accident—ironically, by alchemists seeking the elixir of immortality. Among the many mixtures they concocted was one of saltpeter, sulfur and carbon of charcoal. An alchemical text written about A.D. 850 warns against experimenting with this combination: "Some have heated together [this mixture] with honey; smoke and flames result, so that their hands and faces have been burnt, and even the whole house burnt down."

The earliest use of gunpowder by the Chinese military seems to

The oldest illustration of a cannon in Europe, a page from Walter de Milamete's manuscript of 1327. A figure in armor carefully applies a red-hot iron to the touch-hole of a vase-shaped cannon set on a carpenter's table; in the muzzle is a large arrow.

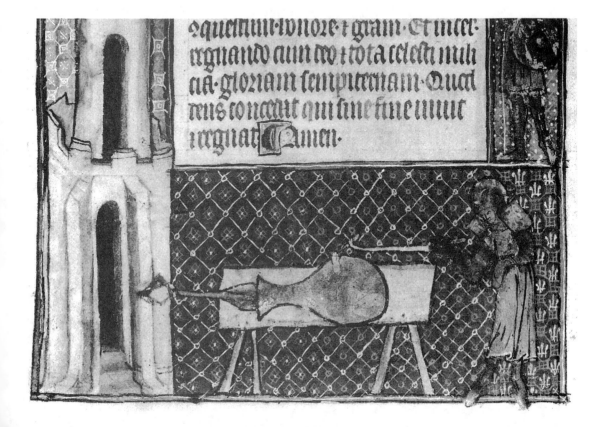

have been in A.D. 919, when it was used to impregnate the fumes of "fire lances" (see **Flamethrowers**). Incendiary arrows using gunpowder replaced those with oil and sulfur around A.D. 1000, and in the eleventh century explosive bombs filled with gunpowder and fired from catapults were introduced.

The First Guns

At about the same time, the Chinese adapted their "fire lances" to shoot out projectiles as the stream of flame came to an end. These could be anything from scrap metal to broken porcelain, though poisoned arrows were preferred. The inclusion of solid objects in the punishment dealt out by flamethrowers must have provided the inspiration behind the development of guns and cannons.

The true gun, however, did not appear until the twelfth century A.D. For this three things were needed: a metal barrel, gunpowder with a high nitrate content, and a projectile that neatly filled the barrel so that the powder charge could exert its full propellant effect. Until very recently it was believed that the first Chinese gun was that found in the early 1970s during an archaeological excavation at Pan-la-ch'eng-tzu village, Manchuria, and dated to around A.D. 1290. Continuing exploration of the Buddhist caves of western China has now revealed a remarkable temple at Ta-tsu in Szechuan Province, containing evidence that revolutionizes the history of the gun. Among the many reliefs in the cave are two showing gunpowder weapons. One depicts a small demon with two horns cradling a handgun while a second devil holds a grenade. The date of this extraordinary find, according to the inscriptions in the caves, is A.D. 1128.

The Chinese developed mobile battlefield artillery in the early part of the fourteenth century, at the time when gunpowder and simple cannon were first being adopted in the West. After this the development of such weaponry was rapid: In 1453 the Ottoman Turks deployed some enormous guns in their successful siege of Constantinople, last remnant of the once-great Byzantine Empire. The largest of these had a bore of some thirty-five inches and fired a ball weighing more than six hundred pounds. One shell, fired from the enormous distance of one and a half miles, cut a Venetian ship in two with its impact. Each pair of these massive guns was said to need an army of up to one thousand men and seventy oxen to shift them.

The earliest depiction of a gun, discovered in 1985. Flames and a ball shoot from a bulbous handheld cannon carried by a demon. The scene, from a cave temple at Ta-tsu, China, is dated A.D. 1128.

In parallel to their development of guns and cannon, the Chinese army also invested great efforts in producing military rockets. The idea of fixing a fire lance backward on a pike or arrow and shooting it off at the enemy occurred to some unknown military genius around A.D. 1180. A variety of rockets soon became common, including multiple rocket launchers mounted on wheelbarrows. The rockets of the fourteenth century were birdlike in shape, with wings made of wood and a range of 1,500 to 3,500 feet. The Chinese rockets that were in use by A.D. 1400 were clearly the forerunners of modern satellite launchers. They had large two-stage rockets with propulsion motors that ignited in successive stages; with a range of over one mile, they automatically released a swarm of rocket arrows toward the end of their trajectory. Such fearsome weapons were not created in the West until the end of the eighteenth century.

POISON GAS

When the Portuguese *conquistadores* reached Brazil at the beginning of the sixteenth century, they were met by natives attempting to defend their homeland by burning red peppers on pans of glowing charcoal to create toxic smoke. Despite this ingenious means of defense, the Brazilians eventually lost their land to invaders with superior firepower. Originally, however, poison gas was used against a far smaller enemy: the flea. In both ancient China and Egypt houses were regularly fumigated to rid them of this pest.

It was the Chinese who first developed this household invention into a deadly scourge of humankind. Texts of the fourth century B.C. record the deployment of chemical weapons against armies besieging cities. As the attackers built tunnels to undermine the city walls, the defenders countered by secretly tapping into the tunnels with terra-cotta pipes. At the end of the pipes were oxhide bellows connected to furnaces in which toxic substances such as mustard balls or the artemisia plant were burned. The bellows pumped the poison gases brewed up in the furnaces into the tunnels, causing fits or even death in the confined space.

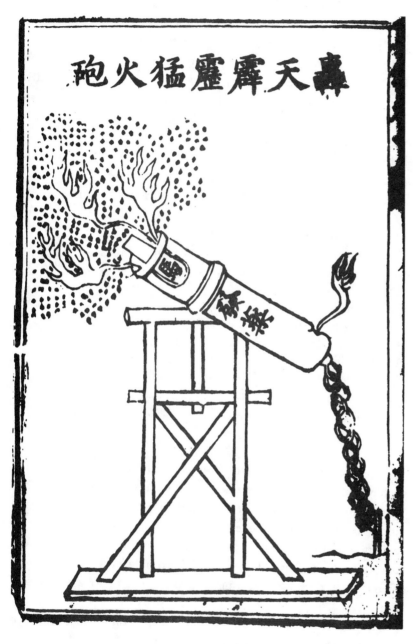

轟天霹靂猛火砲

A medieval Chinese mortar known as the Heaven-Rumbling Thunderclap Fierce Fire Erupter, which fired poison gas shells at the enemy.

By medieval times poisons had made their appearance on the battlefield, mixed with gunpowder and then bound together with resin and placed in bombs. At first these were fired from catapults and later on from cannon. A lit fuse would ensure that the bomb exploded in the air or on landing, burning fiercely and spreading its foul smoke in thick clouds throughout the enemy lines.

In his *Military Encyclopedia* compiled in A.D. 1044 Tsêng Kung-Liang gives the formulae for two poison-gas bombs that would be fired at the enemy by catapult. The key ingredients in his "poisonous smoke bomb" were aconite, wolfsbane and arsenic. The bomb was wrapped up in paper and tied with hemp fiber, then ignited just before it was shot off at the enemy. Tsêng records that the fumes, if inhaled, caused bleeding from the mouth and nose.

Even more horrific was Tsêng's recipe for an excrement bomb:

Human excrement, dried, powdered and finely sifted	15 lb
Wolfsbane	8 oz
Aconite	8 oz
Croton oil	8 oz
Soap-bean pods [to create the black smoke]	8 oz
Arsenious oxide	8 oz
Arsenic sulfide	8 oz
Cantharides beetles	4 oz
Ashes	16 oz
Tung oil	8 oz
	316 oz

This noxious brew was stored in glass bottles until needed, when it was placed in a bomb together with a gunpowder mixture just before firing. Tsêng particularly recommended it for attacking cities and noted that it could penetrate chinks in armor to produce severe irritation and blistering on the affected skin. To avoid the artillery operators themselves falling foul of the deadly poisons, he recommended that they suck black plums and Chinese licorice as a protection.

The medieval Chinese even developed the equivalent of modern tear gas, which they employed with equally deadly effect. Yang Wang Li, in his *Rhapsodic Ode on the Sea-Eel Paddle-Wheel Warships*, describes a naval battle of A.D. 1161:

The rebels of Wanyen Liang came to the North bank of the River in force ... but our fleet was hidden. ... Then all of a sudden a thunderclap bomb was let off. It was made of a paper carton filled with lime and sulfur. Launched from catapults, the bombs came dropping down from the air, and upon meeting the water exploded with a noise like thunder, the sulfur bursting into flames. The carton case rebounded and broke, scattering the lime to form a smoky fog which blinded the eyes of men and horses so that they could see nothing. Our ships then went forward to attack theirs, and their men and horses were all drowned, so that they were utterly defeated.

PERSONAL EFFECTS

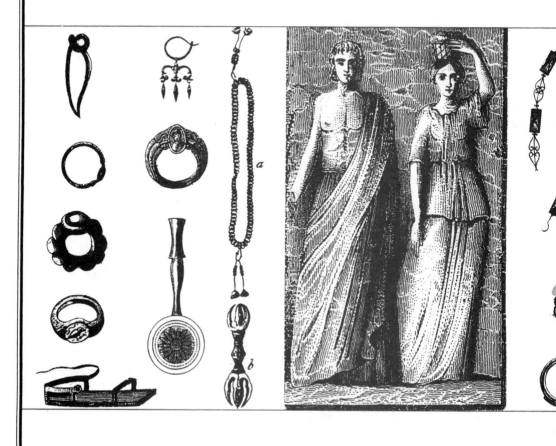

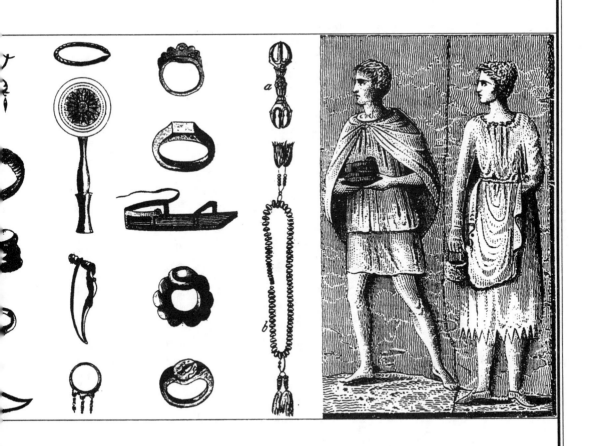

CHAPTER SIX

INTRODUCTION

Vanity comes a close second to necessity for the title "mother of invention." In the twentieth century we are familiar enough with the kind of beauty that depends so much on artificial aids that it can be dismantled into component parts at bedtime. But it is an image that was just as well known in ancient Rome. Nearly two thousand years ago the satirical poet Martial penned the following scathing remarks about an acquaintance of his named Galla; at the time of writing she appeared to be awaiting the delivery of a new wig of imported hair:

> While you stay at home, Galla, and preen yourself in the midst of the Subura [a rather sleazy district of Rome], your hair is at the hairdresser's; you take out your teeth at night as you take off your silk frock, and sleep tucked away in a hundred cosmetics boxes—you don't even have your own face for a bed-fellow. Then you wink at men under an eyebrow which you took out of a drawer that same morning.

Under the Roman Empire, standards of personal hygiene, as well as general pride in one's appearance, reached a height unrivaled in Western culture until our own times. Everything from baths and flushing lavatories (see **House and Home**) and false teeth (see **Medicine**) to the paraphernalia of beautifying oneself, such as wigs and hair nets, mirrors, razors, perfumes and makeup, was available to the Romans in a wide variety of designs. A typical Roman who took pride in his or her appearance would carry around a "pocket set" on a ring attached to the belt, comprising a nail cleaner, tweezers, an ear scoop and a toothpick. A handkerchief, known as a *mucinium*, was also frequently carried for blowing the nose.

Standards rapidly declined in the West after the fall of the Roman Empire during the fifth century A.D. Clothing became simpler and coarser as local materials such as wool and hides replaced imported silks and cotton with the collapse of the Roman economy. The "Dark Ages" that followed were dark indeed with regard to personal cleanliness and hygiene. So, too, were the Middle Ages—"knights in shining armor" paid far more attention to the state of their metallic

A Roman lady keeping her hair tidy with the aid of a hair net. After a fresco at Pompeii.

outer skin than to that of their own. These were times of religious mania and denigration of the body; a perverse concept that suffering was good for the soul meant that in terms of morals the wearing of filthy hair shirts was considered vastly superior to paying any "sinful" attention to one's appearance.

Fashion in clothes seems to have survived the Dark Ages better, and recovered earlier, than personal hygiene. The later Middle Ages and Renaissance period produced some extravagant designs for clothing, and by the sixteenth century wigs, seen as a decadent indulgence in the Dark Ages, once more became respectable in Britain, under the influence of Queen Elizabeth I (1558–1603). At the same time, standards of cleanliness were still at such a low ebb that Elizabeth had to set an example for her subjects by taking a bath once a month. Her successor, King James I, felt no such compunction and restricted himself to the use of a fingerbowl after dinner. The average ancient Greek or Roman would probably have found

A Roman pocket set from London (2nd century A.D.), comprising (left to right) tweezers, nail cleaner, file, ear scoop and toothpick.

it difficult to sit for very long in his company.

The Greeks and Romans, of course, did not have a monopoly on personal cleanliness and grooming. In fact most of the devices and products they used were invented much earlier, by the Egyptians and other Near Eastern peoples. The Romans were actually latecomers to the use of soap, first manufactured around 2000 B.C. by the Babylonians in Iraq. The Ebers Papyrus, dating to about 1525 B.C., gives recipes for lotions used by wealthy Egyptians to prevent "body-odors." The history of shaving takes us well back into the Stone Age of western Europe, possibly as far as 30,000 B.C.

In the East the great civilizations of India, China and Japan had their own traditions of hygiene and fashion stretching back for millennia, which never suffered interruptions as massive as the European Dark Ages. They had a particular interest in oral hygiene, and in the second century B.C. officers of the Chinese court had to pop cloves into their mouths before they could address the emperor. Many things lost in the West after the fall of the Roman Empire, for example parasols, were reintroduced much later from China.

Not all ancient aids to beauty match modern Western tastes, such as the facial scarring or neck stretching favored by some African tribes, or the traditional Chinese binding of female babies' feet to inhibit growth (both practices that continued into the twentieth century). The Maya of ancient Mexico had a great regard for slightly crossed eyes and would hang small beads in front of their children's faces to induce the condition. A widespread custom, practiced in both the Old and the New World, was the artificial deformation of the head. In ancient America the desired shape was a flattened forehead; this was achieved by fastening an infant to a cradleboard soon after birth, with its little head compressed between two boards in such a way as to permanently flatten the skull. The Incas kept the wooden boards in place until the child was three or four years old. In the Near East the custom of head deformation goes back sixty thousand years, as can be seen from some of the Neanderthal skulls found in the Shanidar Cave of northern Iraq. Far less extreme is the very ancient art of tattooing, which went in and out of fashion over the centuries, adopted by widely differing classes in different societies. In the twentieth century it has been popular among sailors and punks, but in ancient Russia, the Balkans and Central America it was a fine art reserved only for the noble-born.

As so much is a matter of shifting taste and fashion, we have highlighted here only those aids to personal appearance that are still widely used today.

MIRRORS

Self-awareness is the hallmark of the human race. We can imagine that even before our remotest ancestors crossed the threshold to become *Homo sapiens*, intelligent—if somewhat apelike—beings would gaze into the still waters of a pool and recognize the reflections there. With nature providing such inspiration it is not surprising that mirrors were invented independently in many different parts of the globe.

In fact almost every ancient civilization had mirrors in some shape or form. The earliest known, those found at the extraordinary Stone Age urban center of Çatal Hüyük, in central Turkey, belong to an almost incredible antiquity (see **Introduction** to **Urban Life**). Ten burials of women, dating to the sixth millennium B.C., contain highly polished mirrors, made of the volcanic glass known as obsidian. Obsidian is extremely hard to work, and the skill required in order for these prehistoric townsfolk to transform it into perfect, scratch-free mirrors astounded the British archaeologists who excavated Çatal Hüyük in the 1960s.

The pre-Columbian inhabitants of the Americas also exploited the reflective properties of obsidian and other minerals in order to make mirrors. The Incas of Peru used a volcanic rock known today as Incas-stone, which, when polished, looks like burnished white steel. The Aztecs of Mexico made mirrors for ceremonial or magical purposes, fashioning them from obsidian or pyrites, nodules of highly reflective crystals. The eyes of Aztec idols were sometimes inlaid with small mirrors of black obsidian to give them an eerie depth.

Metal Mirrors

In the Old World, polished bronze or copper was the most common material for mirrors, beginning with those made by the Egyptians from about 2900 B.C. and by the Indus Valley civilization of India and Pakistan between 2800 and 2500 B.C. In China the manufacture of bronze mirrors goes back to the time of the Shang Dynasty

Typical Egyptian hand mirror, made of bronze, from the New Kingdom period (c. 1550–1070 B.C.). The handle is shaped like a papyrus plant, topped with two small falcons. The bronze disk, now dull, was once highly polished.

(c. 1500–1000 B.C.). Those from the Han Dynasty (202 B.C.–A.D. 220) are the best known, because of their outstanding craftsmanship, including some with gold, silver and turquoise inlays which were carried as belt attachments.

It was also during the Han Dynasty that experimentation with mirrors led the Chinese to discover the principle of the periscope. This, as we can see from a passage in the *Huai Nan Wan Bi Shu*

("The Ten Thousand Infallible Arts of the Prince of Huai Nan"), was appreciated as early as the second century B.C.: "Suspend a large mirror high up, put a basin of water underneath it and you can see the people around you."

Bronze hand mirrors must have been common among the ancient Hebrews as early as the time of their Exodus from Egypt (perhaps around 1400 B.C.), since Moses cast some of the vessels for the Tabernacle from the mirrors that were carried by the women. They presumably picked up the habit in Egypt, where a bronze hand mirror was an essential possession of every lady. At about the same period bronze mirrors were being cast by the miners working at Serabit el-Khadim, in the Sinai. It seems very likely that they used them as a means of casting light into the mine shafts (see **Drilling and Mining** in **Working the Land**).

The art of engraving bronze mirrors was taken to its height in the last four centuries before Christ by the Celts of ancient Britain and France and the Etruscans of central Italy. The Celts favored abstract patterns, usually the spiral shapes so characteristic of their other art. Etruscan hand mirrors, the finest products of their bronze industry, were mostly engraved with pictorial scenes from everyday life, mythology, or a combination of both—such as the goddess Venus at her toilet shown on a fine example of the fourth century B.C.

The Greeks, vain as they were, also loved mirrors, and in the fifth century B.C. invented the "box" mirror. Precursor of the modern compact, it was made of two metallic disks that fitted together, fastened with a hinge; one inner surface was polished to form the mirror, all the other sides being engraved. The Romans used hand mirrors of bronze, copper and silver, decorated in the Etruscan or Greek style. As metal mirrors tarnish easily, a small sponge dipped in powdered pumice stone was often fastened to the casing.

With mirrors, as with everything else, the ancient Romans went to town. The emperors encouraged their manufacture for every conceivable use, from erotic to security purposes. There were wall mirrors made of semiopaque or black glass, as we know from literary references and an actual example screwed to the wall of one of the houses at Pompeii. Sliding mirrors, which moved up and down like windows on sashes, were also made, as well as novelty mirrors—like that described by Seneca, tutor to the Emperor Nero (A.D. 54–68),

The reverse side of a bronze mirror found at Holcombe, England. A fine example of Celtic engraving, it dates from the end of the 1st century B.C.

which had myriad polished facets, repeating the image of the viewer like a crystal. The tyrannical emperor Domitian (A.D. 81–96), justifiably paranoid given the number of enemies he made for himself through his persecution of the senate, lined the gallery in his palace where he took his daily exercise with mirrors of highly polished phenacite (a translucent stone) so that no one could sneak up on him unawares. It did him no good—conspirators trapped him in his bedroom and stabbed him to death.

The Looking Glass

The earliest glass mirrors, like the one from Pompeii and a number found in Egypt, were simply pieces of black or dark glass. But the later Roman Empire produced true glass mirrors: in the same way that modern mirrors have a "silvered" back of reflective metal paint, these were made by applying thin gold, silver or copper leaf to a sheet of glass. As the glass was unpolished, they probably gave rather wobbly images. However, the Romans did not take credit for the invention themselves. According to the encyclopedist Pliny, glass mirrors were first invented in the Lebanese city of Sidon. The tradition is plausible given the unrivaled brilliance of Phoenician craftsmen in the manufacture of glassware (see **Glass Windows** in **House and Home**). Whatever the case, the earliest evidence of glass mirrors as we know them is Roman, of the early third century A.D.

A different method was later developed to produce small pocket mirrors, such as those known from the Gallo-Roman tombs of France dating to the third to fourth centuries A.D. These were made of small, convex pieces of glass (1⅕ to 2 inches in diameter), probably cut from a glass balloon, and filled—after warming to prevent the glass from cracking—with blobs of shiny lead. The result was a mirror that gave a tiny, distorted reflection, rather like that produced by looking into the back of a shiny spoon.

Glass mirrors were still being produced in the last days of the Roman Empire, and as late as the early seventh century A.D. it was noted by Isidore, bishop of Seville in Spain, that "there is no material better adapted for making mirrors." Then silence descends—like so many other skills, the manufacture of true glass mirrors disappeared during the Dark Ages, not only in Europe but in the Middle East as well. The early eleventh-century Arab scientist Ibn-al-Haytham (known to the Western world as Alhazen), in his comprehensive treatise on optics, discussed iron and silver mirrors, but did not even mention glass. The reinvention of glass mirrors had to wait until the thirteenth century.

MAGIC MIRRORS

Techniques of mirror production became extraordinarily sophisticated in the ancient East—so much so that the "magic mirrors" invented in China some fifteen hundred years ago completely baffled Western scientists of the nineteenth and early twentieth centuries.

At first glance these mirrors, which are made of bronze, appear quite normal—the polished side reflects the viewer's face, while the reverse is embossed or engraved with patterns or letters. Yet when held up in bright sunlight they cast an image of the designs on the back of the mirror onto a wall. The whole thing seems impossible—how could light pass through solid bronze?

"Magic mirrors" were even a mystery to prominent scientists in medieval China. In his work *Dream Pool Essays*, published in A.D. 1086, Shen Kua (a physicist, astronomer, engineer and high official) described some examples, obviously considered to be extremely old even in his own day:

> There exist certain "light-penetration mirrors" which have about twenty characters inscribed on them in an ancient style which cannot be interpreted. If such a mirror is exposed to the sunshine, although the characters are all on the back, they pass through and are reflected on the wall of a house, where they can be read most distinctly. . . .
>
> I have three of these inscribed "light-penetration mirrors" in my own family, and I have seen others treasured in other families, which are closely similar and very ancient. . . . But I do not understand why other mirrors, though extremely thin, do not "let light through." The ancients must have had some special art.

"Magic mirrors" first provoked serious interest in the West in 1832. But despite the best efforts of numerous scientists, they defied explanation until 1932, when the British crystallographer Sir William Bragg discovered their secret. Shen Kua had already guessed that their "magic" property relied on a trick, coming close to an explanation of the reflected images when he suggested that there were "minute wrinkles" on the polished surface of the mirror that subtly repeated the design of the reverse: "Although the characters are on the back, the face has faint lines, too faint to be seen with the naked eye."

It remained for Bragg to confirm this with the aid of modern microscopes. To the naked eye the reflective side of the mirror was a smooth convex surface, concealing subtle differences produced during manufacture. The mirrors were cast flat on both sides, except for the embossed or engraved design on the back. The curvature of the mirror side was then produced by careful scraping, the stress of this operation producing minute depressions that matched the bulges on the decorated side. After polishing, a mercury amalgam was applied to augment the tiny differences in surface structure produced by the scraping. The imperfections on the polished side, exactly repeating the designs, were too small to be seen with the naked eye, but, because of the curvature of the mirror, were then greatly magnified in the reflected image.

Joseph Needham, this century's greatest expert on ancient Chinese science, described the subtle techniques used to produce such a brilliant illusion as "the first step on the road to knowledge about the minute structure of metal surfaces."

A Japanese imitation of the Chinese "magic mirror." When held in bright sunshine an image of the raised design on the reverse of the mirror (left) apparently passed through the mirror and was cast on a wall (right).

MAKEUP

The urge to decorate the human figure is probably as old as humanity itself, but the earliest archaeological evidence of the use of cosmetics as we know them comes from the first urban civilizations of the Old World. The women of ancient Sumer, in southern Iraq, painted kohl (made from the mineral antimony or the lead ore galena) around their eyes to make them look larger, still a common practice today, and rouged their cheeks with dyes. In one of the famous tombs at the Sumerian city of Ur, from the end of the third millennium B.C., the excavators found a tiny gold shell-shaped cosmetics case containing a makeup kit.

A young Egyptian lady applying lip gloss with the aid of a mirror.

The ancient Egyptians were even greater believers in cosmetics. As we can see from Egyptian tomb paintings, men as well as women commonly used black kohl and green lapis lazuli or malachite (a copper ore) as eyeshadow. The famous seductress Cleopatra touched up her eyebrows and lashes with mascara and colored her upper eye-

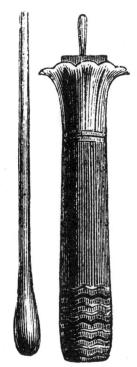

An Egyptian makeup bottle and applicator for kohl (eyeshadow).

lids blue and her lower ones Nile green. Cosmetic ingredients were kept in lumps in little bags of leather or linen and ground on a palette to a fine powder, then applied with a moistened stick of wood, ivory, silver, glass or bronze. Makeup kits containing these contents together with palettes for grinding them into powders are frequent finds in graves of the ancient Egyptian nobility, dating as early as 4000 B.C. Both women and men lightened the skin with yellow ochre, but only women used orange paint to make it darker. A rouge of red ochre and fat was applied to the cheeks and a similar mixture to the lips.

Egyptian women are sometimes pictured with colored nails, an effect possibly achieved using henna, but were not so fascinated by their nails as the ancient Chinese were. They used a coloring agent made from herb juice, which turned their nails a deep red. Until very recently in China it was believed to be extremely dignified to have nails over an inch long—showing that the possessor did no manual work—and rich Chinese fastened special little shields of silver over their elegant claws to prevent them from breaking or chipping.

The people of the Indus Valley civilization (third-millennium B.C. India and Pakistan) were apparently the first to lighten their complexions with a face cream of white lead. Cakes of this cosmetic have been found in urban houses of the period. Similar recipes eventually spread across a vast area from Britain to China. The Greeks imported white lead from the East in the form of tablets, while Roman ladies continued to use it even though doctors had by this time shown it to be highly toxic.

When in Rome

A prime reason for the unfortunate popularity of white lead was its central role in the extraordinary cosmetic preparations of Poppaea, wife of the notorious Emperor Nero (A.D. 54–68), as recorded by writers of the day. She employed no less than one hundred female attendants to maintain her beauty. Every night Poppaea wore a face mask of bean meal, which was washed off in her morning bath of asses' milk. Any beneficial effects were then ruined by the Empress's covering her body with chalk and the poisonous white-lead face cream. She painted her cheeks and lips with red paint, then colored her eyelids, lashes and brows with black antimony, her nails with a mysterious substance called Dragon's blood mixed with fat, and her veins with blue paint. Before this, she would have applied depilatory creams to remove unwanted hair, bean-meal paste and lemon juice to bleach her freckles, powdered pumice stone to whiten her teeth, barley flour and butter to remove pimples and German soap to bleach her hair. Obviously only the rich could afford the time, let alone the money, needed to prepare so elaborately for the day.

Despite having become accustomed to this sort of extravagant makeup on their womenfolk, the senators of Rome were still shocked by the arrival of the boy emperor Elagabalus in Rome in A.D. 218. He had been placed on the throne through the machinations of his grandmother, the sister-in-law of a previous ruler. Elagabalus preferred to dress as a woman, and since he had been brought up in Syria, he did so in the ornate Syrian fashion. He entered the capital in triumph in a highly decorated cart, around which cavorted half-naked eunuchs, freed stable boys and other slaves displaying their affection for him. The new master of the Roman world was himself dressed in long silk robes and a tall hat decorated with streamers; huge earrings covered his cheeks; his eyes were set in concentric rings of blue and gold paint, and his lips were painted blue; his feet were reddened with henna, and his hands and sandals were dripping with jewels. Beside him his mother and grandmother wore symbols of high rank previously worn only by men and had their faces completely masked with white lead, rouge, kohl, blue eyeshadow and lip gloss—a degree of makeup that at the time declared its wearer to be a Roman prostitute. The senators must surely have thought the end of the empire was nigh. Not surprisingly, Elagabalus's reign was a short one.

Such excesses on the part of both men and women were surely in the minds of the early Christian Fathers when they condemned cosmetics. Saint Jerome described them as "poultices of lust," asking, "What can [a woman] expect from Heaven when, in supplication, she lifts up a face that its creator wouldn't recognize?" The makeup industry in the West accordingly went into a long decline, but in the Islamic world it rose to new heights, for around A.D. 840 a famous singer and musician from Baghdad, known as Blackbird, opened the world's first beauty institute in southern Spain. Here he taught hairdressing and the arts of applying cosmetics, removing unwanted hair, using toothpowders and manufacturing deodorants.

Far away from the influence of the Christian Fathers, in Mexico, the fashionable shade for an Aztec woman's complexion around A.D. 1500 was yellow, according to the accounts of the Spanish *conquistadores*. To achieve this, ladies' cheeks were either rubbed with ochre or covered with a face cream based on a waxy yellow substance obtained by cooking and crushing tiny insects. Then as now women were willing to go to any length in the pursuit of beauty.

TATTOOING

It is hard to determine how far back the art of tattooing goes, as human flesh is almost never preserved in the archaeological record. But rare finds of human corpses with the skin intact have enabled us to glimpse early examples of the tattooists' art. The remarkable "Iceman" found melting out of the Similaun glacier, in Italy, in September 1991, having died there some 5,300 years ago, was marked with the oldest tattoos ever seen. They consist of three sets of lines on his back, another group on the right ankle and a cross on the left knee. The powdered charcoal used to create the blue markings was probably applied with small needles.

From Egypt and Sudan there are occasional finds of female mummies with facial tattoos going back some four thousand years. It seems as though these had some erotic significance, since the mummies are thought to be those of concubines, and some centuries later dancers and musicians would sport tattoos of their patron god, Bes, on their thighs. The tattoos found on the mummies were dark blue and were applied, archaeologists believe, by pricking dye into the skin with a device made of fish bones set into a wooden handle.

In 1948 the most remarkable example of ancient tattooing yet found was discovered in the Pazyryk burial mound on the borders of the USSR, China and Mongolia. A man aged about sixty, prob-

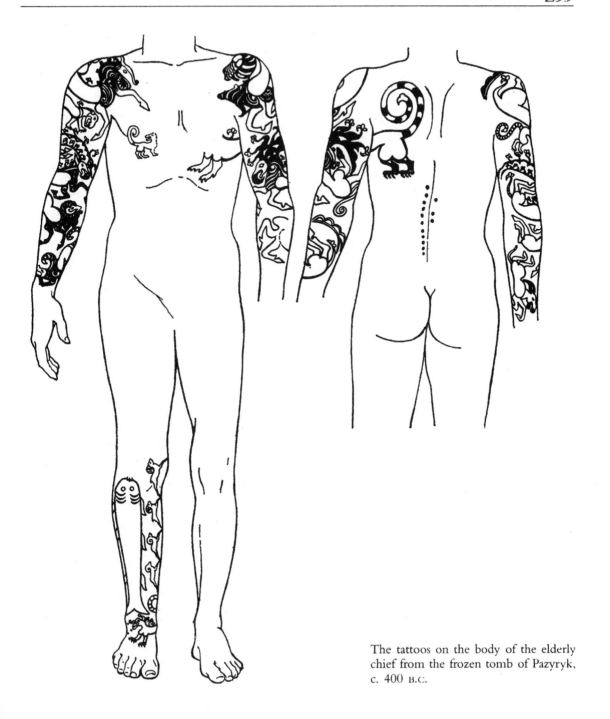

The tattoos on the body of the elderly chief from the frozen tomb of Pazyryk, c. 400 B.C.

ably the chief of a nomadic tribe, had been buried around 400 B.C. in a felt-lined wooden chamber under the mound. After grave robbers had ransacked the tomb, ice flowed in through their exit hole to fill the chamber, thereby miraculously preserving its contents. The surviving skin of the chief was covered with elaborate tattoos, including monsters, a donkey, a mountain ram, deer, birds, a goat, and a fish.

The Pazyryk find dramatically confirms the statements of classical writers that various barbarian peoples to the north and east of Greece regarded tattoos as a symbol of nobility. According to Herodotus, who wrote in the lifetime of the Pazyryk chieftain, the Thracians of the Balkans "consider tattooing a mark of high birth, the lack of it a mark of low birth." Another classical source indicates that the Thracians learned the art from the widely traveled Scythians, a nomadic tribe from Russia, who may well have had cultural ties with the builders of the Pazyryk tombs.

In ancient times the art of tattooing flourished worldwide. It has deep roots in the Far East (and is still highly regarded in Japan), but little is known of its earliest days there, except that tattooing was carried out as a punishment on criminals during the Han Dynasty of China (202 B.C.–A.D. 220). By A.D. 297, however, decorative tattooing had begun in Japan, when Chinese court records note that Japanese men and boys tattooed their bodies. Judging by its representation on terra-cotta figurines, Mayan chiefs in first-millennium B.C. Central America were frequently tattooed. They, too, saw tattooing as the mark of high status.

According to Roman writers, tattooing was rife in Britain. Julius Caesar noted during his expeditions that "All the Britons dye their bodies with woad [an herb], which produces a blue color and gives them a wild appearance in battle," while the third-century A.D. author Herodian described their tattoos as "pictures of all kinds of animals."

Following the conquest of Britain in A.D. 43, Roman legionnaires took to tattooing and spread the custom throughout the empire. The art flourished until the reign of Constantine, the first Christian emperor (A.D. 306–373). Early Christians marked their faces and arms with the sign of the cross, but tattooing later came to be seen as a pagan practice; Constantine banned facial tattoos on the grounds that they disfigured "that fashioned in God's image." In A.D. 787 the Church Council held at Calcuth, in northern England, forbade all tattooing and it then became rare. It did not die out completely, however, even among royalty. After William the Conqueror's momentous victory at the battle of Hastings in 1066, the body of the fallen English king Harold was identified by its tattoos, including the name of his mistress (Edith "Swan-neck") over the heart.

Despite such notable exceptions, the great revival of the art in the West only really came about during the eighteenth century, as a result of encountering masterpieces of tattooing in the Far East and the Pacific, where tattooing had never fallen out of favor.

SOAP

Curiously enough the Romans first became acquainted with soap as a hair dye. According to Pliny's *Natural History* (written in A.D. 79) they imported a substance from Germany called *sapo*, made of goats' fat and beechwood ash pressed into small golf-ball-sized cakes. This was used to dye the hair a sandy red color, more often by men than by women. Only at the end of the second century A.D. did the Romans adopt it as a cleanser. Galen, the most famous doctor of the classical world, said that *sapo* was a better detergent than soda and recommended German soap for washing the body because it was the purest. By this time better-quality soap was being made by boiling potash with fats.

Surprisingly, given the Romans' ignorance of it, soap had been discovered more than 2,000 years earlier in Mesopotamia. Babylonian chemists boiled together oil and alkalis to produce a residue with which people washed themselves. Much later the Phoenicians of the Levant adopted soap; the Scythians of southern Russia washed their hair with soap, according to the Greek historian Herodotus—a habit they may have picked up during their invasion of the Near East in the seventh century B.C.

One shouldn't imagine, however, that ancient peoples who didn't use soap were dirty. Indeed a great number of alternative cleansers

Strigil and oil bottle from Roman London, 1st–2nd century
A.D. Roman bathers began with a cold bath, then entered
warm and hot rooms, finally reaching a sweating chamber.
Dirt sweated out was cleaned off by applying oil and
scraping with a *strigil*.

were employed. The Egyptians and Jews used soda, the Greeks favored bran, sand, ashes and pumice stone, while Roman bathers rubbed themselves down with olive oil.

During the Dark Age after the fall of the Roman Empire, cleanliness was rare in the West but widely practiced in the Muslim world. Hard soap in bars was first produced by the Arabs, and its manufacture became an important industry, especially in Spain, after A.D. 1000. The basic ingredients were olive oil and wood ash; white, odorless, medicinal soaps were very common, while colored and perfumed toilet soaps were exported widely as luxury items. In northern Europe soft soap was made by boiling wood ash with animal fats or fish oils; it smelled terrible and thus was mainly used for washing clothes. But by A.D. 1300 heavily scented soft toilet soap packed in wooden bowls was being produced in London.

Soap was never made in ancient China, but this was not for any lack of desire to keep clean. Instead the Chinese were lucky enough to have a readily available natural alternative rich in saponin, the detergent element of soap. This was the "soap-bean" tree, which produces a mild detergent when the beans are pressed. This extract of soap-beans was made up into balls along with flour, mineral powders and perfumes from the time of the Han Dynasty (202 B.C.–A.D. 220) onward, and was used for washing both the body and the clothes until the twentieth century. When fat soaps were introduced into China, many older people refused to use them because they were much less gentle on clothes than soap-bean extracts, which could keep silk white despite repeated washing.

RAZORS

Prehistoric cave paintings and engravings suggest that man has, for at least the last thirty thousand years, tried to keep neat and tidy by shaving. The earliest razors were probably flint blades, which can give an extremely sharp edge. These would of course have been "disposables," as the shaving edge would soon be blunted. Similar razors made from the volcanic glass obsidian were still being used by the Aztecs of North America in A.D. 1500 and in central Africa in 1900.

Permanent razors were developed along with the invention of metalworking: copper razors became common in both Egypt and India in the third millennium B.C. The ancient Egyptians generally thought that facial hair was a sign of personal neglect, and would shave regularly—though changing fashion occasionally permitted the wearing of neatly trimmed mustaches or small goatee beards. The wealthy would keep a barber on their household staff, but there were also barbers serving poorer Egyptians. The *Satire on Trades*, written about 1700 B.C., describes an itinerant town barber "sacrificing" himself to "chins" and walking "from street to street to seek out those whom he may shave."

Ancient Mesopotamian barbers were held in higher regard, perhaps because they were organized into a guild. Every town had a number of barbers' shops, usually grouped in a single street, which met the needs of the general public. Clients were shaved with a razor and pumice stone and then massaged with oils and perfumes.

Some of the most elaborate razors of ancient times have been found in Scandinavia. These are of bronze and date from 1500 B.C. onward. Those excavated from the Danish Mound Graves of 1300–1200 B.C. were enclosed in leather cases and have finely made horsehead-shaped handles and blades decorated with mythological scenes. The men in these graves, whose bodies were preserved in airtight tree-trunk coffins, were completely clean-shaven. The later Celts, however, were fond of mustaches. According to Julius Caesar (writing in 50 B.C.), "the Britons shave every part of their body except their head and upper lip."

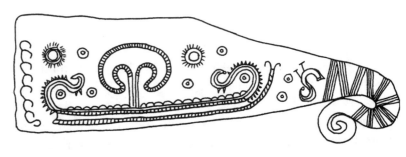

A Scandinavian Bronze Age razor decorated with a scene that may depict a boat with a tree branch as a sail (late 2nd millennium B.C.).

Similar practices are known from fourth century B.C. India, where, according to the writers of the day, the men kept their beards under control with scissors and razors but also shaved off all their chest and pubic hair. Indian women of the time used razors and tweezers to remove unwanted leg hair, as did those of classical Greece, where one alternative was to singe hair off with a lamp—a practice that must have required a strong nerve. Only in the late fourth century B.C. did Greek men start to shave, following the example of Alexander the Great.

The Roman Beard Debate

The Romans were never quite sure whether or not to be clean-shaven. Traditionally, Roman men wore beards, but around 300 B.C. the wealthy Publius Ticinius Maenas brought Greek barbers across from Sicily and introduced a fashion for shaving that held good until the time of the Emperor Hadrian (A.D. 117–138), who revived beards. During this time Roman men would start shaving at the age of twenty-one, the first shave being celebrated by holding a party where gifts were presented to the young man. The hair shaved off would sometimes be placed in a box and kept at home as a treasured memento of his coming of age. Only soldiers and those training to be philosophers could escape the first shave.

Even while shaving was in fashion, older Roman males customarily grew beards. The explanation for this uncharacteristic indecision by the Romans may lie in the quality of their barbers. For shaving customers they employed thin-bladed iron razors, sharpened on whetstones, together with water, but apparently not soap or oil. A common complaint was that they worked far too slowly and that one could spend a lifetime being shaved. The alternative, however, was a speedy but careless barber, perhaps one of those operating on a street corner, who might well inflict several minor injuries. The poet Martial (A.D. 40–104) moaned about one ham-fisted individual:

> He who does not wish to descend to the underworld should avoid the barber Antiochus. . . . These scars on my chin, if you can count them, may look like those on a boxer's face but they were

not caused that way nor by the sharp talons of a fierce wife but by the accursed steel and hand of Antiochus. The he-goat is the only sensible animal; by keeping his beard he lives to escape Antiochus.

Many of these barbers' customers must have resorted to the recipe provided by Pliny in his scientific encyclopedia for a medicated plaster to soothe the pain of razor cuts.

In Europe beards stayed in fashion through medieval times, owing, perhaps, to the horrific experience of Louis VII of France. In 1150 his bishops commanded him to crop his hair and shave off his beard; unfortunately his queen, Eleanor, found the sight so ridiculous that she was unfaithful, and the King divorced her. She then married the Count of Anjou (who later became Henry II of England) bringing as her dowry the rich French provinces of Poitou and Guienne. This was the catalyst for the Hundred Years War, in which the flower of English and French chivalry died.

PERFUME

Nowadays the desire to smell nice for other people supports a vast industry, and so, too, those in the ancient world who could afford it spent fortunes on perfuming themselves. As long ago as 2900 B.C. the Egyptian dead were buried with jars of perfumed oil. The nature of these early perfumes is as yet a mystery, but we do know that a thousand years later the Egyptians ventured far and wide in search of perfumes. Frankincense is a fragrant gum exuded from the cut bark of small trees that grow on the southern coast of Arabia and in East Africa. The pharaohs mounted frequent expeditions to this region, known to them as "the land of Punt," specifically to import frankincense. In the fifteenth century B.C. a fleet sent by Queen Hatshepsut sailed along the ancient Egyptian Suez Canal (see **The First Suez Canal** in **Transportation**) and brought back whole frankincense trees from Punt. They were planted in specially prepared ground, but apparently refused to grow, since similar expeditions continued over the next three hundred years.

In Egypt perfumes and unguents for anointing the body were

made in laboratories within temples. At Edfu, in the Temple of Horus, begun by Ptolemy III in 237 B.C., numerous inscriptions on the walls of the perfume laboratory, a room kept in almost total darkness, clearly show how perfumes and ritual oils were made. The most subtle scents created there took up to six months to mature. One of the most famous Egyptian perfumes was balanos, manufactured in the city of Mendes, in the Nile Delta, and exported from there to Rome. It was made from oil obtained by crushing the kernels found inside fruit of the balanos, or "false balsam," tree, mixed with myrrh and resin.

The Egyptian method of perfuming the body was to shape a solid lump of perfumed fat into a cone and set it on top of the head, or fix it above a wig. During the course of the evening the fat would start to melt and cover wig, clothes and body with a coating of highly scented grease.

A wall painting of c. 1500 B.C from Thebes, in Egypt showing a woman with a cone of unguent on her head.

The distillation of rosewater and other scented perfumes was an Islamic discovery of the ninth century A.D. The perfume industry, which used steam ovens to create its marvels, operated on a very large scale, and its products were exported from Damascus throughout the Muslim world, even reaching China. Imitating her Muslim neighbors, the empress Zoë of Byzantium (1042–1055) turned her bedroom into a home perfume factory by installing braziers; her servants were allotted particular tasks, some distilling the perfume, others mixing it and a third group bottling the scent.

Not all ancient peoples were fond of dousing themselves with perfume. In 361 B.C. Agesilaus, king of Sparta, where perfume was banned, visited Egypt and was entertained at an elaborate banquet. He was so disgusted by the excessive use of perfume by his fellow diners—a practice that he thought decadent and effeminate—that he stormed out. Agesilaus's Egyptian hosts in turn found his behavior uncivilized and uncouth. By contrast Athens had its own perfume market, where scent was sold in specially manufactured vases.

In Roman times perfume was used by all classes, mainly to cover body odors, but also—drunk neat or in wine—to conceal bad breath. Wealthy Romans would perfume various parts of their bodies with different scents, sprinkle perfume on their guests at banquets while they reclined on scented couches and even perfume the walls of their bathrooms.

The ultimate development in keeping the bedroom smelling nice was the "gimbals" perfume burner, invented in China around 100 B.C. Gimbals consist of a series of carefully arranged metal rings, pivoted in such a way that a given object can always be kept level, no matter how it is knocked about. The idea of such a mechanism may have been borrowed from Alexandria, where, a few decades earlier, the engineer Philon constructed a novelty inkwell that could be turned in any direction without spillage (see **Introduction** to **High Tech**). The Chinese, however, credited the invention to one of their own, and this account from the second century A.D. gives an idea of the impression it made on contemporaries:

In Ch'ang-an there was a very clever mechanic named Ting Huan. He made a "Perfume-Burner for Use Among Cushions" otherwise known as the "Bedclothes Censer." . . . He fashioned a contrivance of rings which could revolve in all four directions, so that the body of the burner remained constantly level and could be placed among the bedclothes and cushions. For this he gained much renown.

WIGS

The ancient Egyptians had some curious customs. Their priests believed that hair was unclean and they would remove it from every part of their body. Most ordinary Egyptians also shaved or cropped their heads from childhood on. At the same time, well-to-do men and women would rarely be seen in public without first donning a wig of real or artificial hair. While the reasons behind this ambivalent attitude toward hair are complex, the environment certainly played a large part: In extremely hot climates the less hair one has, the easier it is to keep clean; on the other hand, head coverings are an invaluable protection against the dangerous solar rays that pound down on the inhabitants of the Nile Valley.

Wigs first appear on Egyptian tomb reliefs and paintings toward the end of the Third Dynasty (about 2600 B.C.), but there is little doubt that both hair extensions and wigs proper were experimented with much earlier. Since the pharaohs appointed special officials to oversee the royal wig makers, it has been argued that the wearing of wigs was originally a royal prerogative. Whatever the case, they became the fashion for every well-born Egyptian, male or female. Over the centuries wigs grew in both size and complexity, particularly those worn for festive or official occasions. They reached a peak, almost literally, during the Twenty-first Dynasty. From a tomb of this period comes a fine, if top-heavy example, once thought to have belonged to the royal lady Istemkheb but now thought to have been made for her husband, the High Priest Menkheperre (about 1000 B.C.). Like other ceremonial wigs of the Twenty-first Dynasty it consists of a mass of corkscrew curls with long narrow plaits hanging behind. The exterior of the wig was made of brown human hair, the middle being stuffed with the reddish-brown fiber that surrounds the base of date-palm branches.

A special study of Egyptian hair and wigs is currently being made

Wig of the 21st–Dynasty lady Istemkheb, wife of the High Priest Menkheperre (10th century B.C.). Made of human hair, it is set in ringlet curls with beeswax. It is smaller and less grand than her husband's wig.

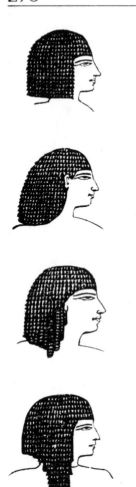

by Egyptologist Joann Fletcher of the Manchester Museum, England, with the help of several laboratories. She has come up with a number of preliminary results, including the observation that the largest, grandest wigs were worn by Egyptian men rather than women. Leaving aside very cheap wigs made of vegetable fiber or grass (popular among the poorer classes during the Roman period), the hair of all the wigs so far examined by Fletcher has proved to be of human origin. Indeed many still contain remains of the human head lice that infested the hair when it was still on the donor! The suggestion frequently made by earlier commentators that the Egyptians made wigs from wool is thus tending to be contradicted. Most wigs analyzed also show traces of beeswax, which was warmed and applied as a pomade before curling.

The Rise and Fall of Wigs

Wig making had spread from Egypt across the Near East and Mediterranean world by the mid-first millennium B.C. Like the pharaohs, the rulers of the ancient Iranian kingdom of Media wore wigs as part of their formal costume. The Greek writer Xenophon described the first meeting between the young Persian Cyrus and his maternal grandfather Astyages, king of Media (585–550 B.C.): "Then he noticed that his grandfather was adorned with pencillings beneath his eyes, with rouge rubbed on his face, and with a wig of false hair—the common Median fashion." Cyrus's small Persian kingdom later absorbed the Medes, and his empire adopted many Median customs—thus we read in Aristotle that wigs were introduced into Anatolia (Turkey) from Persia. The Persian Empire must have introduced wigs to Greece as well, but there they were best known as theatrical props, with a color code to indicate different characters: blond hair for youthful heroes, black hair and beards for villains, and red hair for comic characters.

Wigs were available to Egyptian men in a variety of styles.

Wigs were extremely fashionable among the women of imperial Rome. Messalina, the nymphomaniac wife of the Emperor Claudius (A.D. 41–54), was said to have worn a blond wig as a disguise when she went for her romps in brothels. Faustina, wife of the Emperor

Marcus Aurelius (A.D. 161–180), owned several hundred wigs, which she used for more respectable purposes. The false hair of Roman ladies, like their natural hair, was heavily crimped. Almost any kind or color of hair could be imported from the heads of the many different races on the fringes of the empire. Black hair was shipped in from as far away as India, but the flaxen and red hair of the Germans was the most popular. The Roman poet Ovid joked to a woman friend whose hair was falling out through illness that a new headful was already awaiting her in Germany. The architect Chrysippus noted that fair hair from Germany and Gaul fetched its weight in gold when sold in Rome.

Roman gentlemen also wore wigs, usually to hide baldness. Since men generally combed their hair forward onto their foreheads, a well-made toupee could look very natural—certainly better than the painted-on hair that some balding Romans opted for. Of course some had badly fitting wigs, such as the unfortunate Lentinus, who was mocked in the verses of the satirical Roman poet Martial. And there were always revealing accidents: another Roman poet, Flavius Avianus, described the embarrassment of a bald knight whose wig flew off while riding.

Famous Roman wig wearers include the emperors Otho and Domitian—known to be bald but shown with reasonably full heads of hair on their coins—and the outrageous Caligula, who disguised himself with a variety of wigs during his drunken rampages through the city. Wig wearing could also be turned to political advantage: When the Emperor Caracalla visited the northern frontier in A.D. 214, he wore not only German dress but a blond wig trimmed in the local fashion.

Wigs remained popular throughout the Roman world until the early Christian church tried to stamp them out. The Church perceived them as a clear affront to chastity, especially among women, as they were intended to make the wearer more attractive. The Church Father Clement of Alexandria wrote, around A.D. 200, that no one wearing a wig could receive the laying on of hands from a priest—God's blessing could not penetrate through false hair. His contemporary Tertullian argued that borrowed hair could bring about spiritual contamination:

All wigs are disguises and inventions of the devil . . . if you will not throw away your false hair as hateful to Heaven, let me make it hateful to you by reminding you that it may well have come from the head of a damned person or an unclean person.

In A.D. 692 the Council of Constantinople actually excommunicated a number of Christians for wearing wigs. With growing Church disapproval, wig wearing declined throughout the Middle Ages, and even long hair came to be frowned upon—particularly for men. Wigs were still used in the theater, but they only became popular again in Europe through royal patronage. When the sixteenth-century English Queen Elizabeth hid her graying hair with wigs, they rapidly came back into fashion in Britain. Elizabeth is said to have owned a collection of some eighty wigs of different colors, including auburn and gold. In France King Louis XIII lost his hair from venereal disease in 1624 and began wearing a wig; likewise his son Louis XIV adopted one when his own hair thinned. Courtiers followed fashion, and the way was set for the great wig-wearing era of the late seventeenth and eighteenth centuries.

FALSE BEARDS

Aside from the odd fad for neatly trimmed mustaches and beards, Egyptian men were scrupulously clean-shaven (see Razors). It was usually only their barbarian neighbors, mercenaries or peasants who were depicted in Egyptian art with natural hair sprouting from their chins. Exceptions include a scene painting from the burial chamber of King Ramesses VII, in which an irreverent tomb painter has caricatured the dead monarch by adding several days' growth of stubble to his face.

Yet, with the same paradoxical attitude to beards as they had to hair generally, it was considered a sign of status to *wear* a beard—even during periods when the real thing was deemed unfashionable. Thus false ones were attached to the chin, made of plaited human hair or wool and cut square at the bottom (only gods were generally depicted with beards that curled at the end). Commoners could

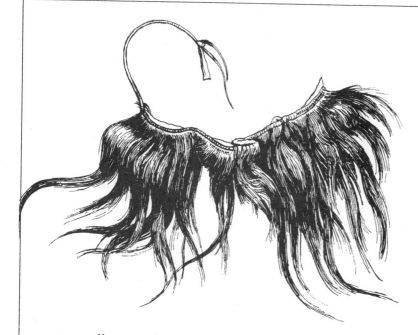

False beard from the frozen tomb of Pazyryk, c. 400 B.C.

wear small ones about two inches in length, while kings sported much bigger beards in the characteristic rectangular shape descending from the chin. We even know of a woman wearing one—Hatshepsut, one of the few queens of Egypt to rule in her own right and famous for her expedition to the land of Punt (see The First Suez Canal in Transportation). She adopted all the titles and imagery of a king and is shown on many reliefs wearing a false beard while presiding over ceremonial occasions.

But false beards were not exclusive to the ancient Egyptians. The old chieftain from central Asia whose frozen corpse is famous for its body decorations (see Tattooing) was clean-shaven; but underneath his head was found his artificial beard, made of human hair and dyed a deep chestnut color.

False beards crop up again in medieval Spain. By the mid-fourteenth century they were so much in fashion that a wealthy gentleman might have possessed a whole range of them in various colors, shapes and sizes to suit different moods and occasions. These "chin wigs," as they later came to be known, were skillfully made and used as disguises by people getting up to all sorts of mischief. In fact the abuse became so widespread that the king of Aragon banned them. At Rouen, in France, false beards were made illegal in 1508, but the edict had to be repeated in 1513. The fact that there were two official efforts to ban them in such a short space of time suggests that they were immensely popular.

CLOTHING AND SHOES

The hunters who colonized the frozen north during the Old Stone Age did not dress in furs slung casually over the shoulder, but in well-tailored garments sewn from animal skins. Archaeological sites dating as far back as 20,000 B.C. have produced eyed needles, the earliest known, made from bone. That their purpose was to make clothes is amply confirmed by other discoveries. From the discolored soil and the pattern of beads and other ornaments found around three male burials at Sunghir, Russia, in 1964, archaeologists were able to deduce that the occupants were buried wearing hats, shirts, trousers and moccasins made of fur or leather. At the contemporary site of Buret, in southern Siberia, a figurine carved from a mammoth tusk was found, showing an individual wearing a fur suit very like traditional Eskimo costume.

Moving to the Late Stone Age, wall paintings at the prehistoric town of Çatal Hüyük, in Turkey, from the late seventh millennium

Twenty-thousand-year-old burial of a man from Sunghir, Russia. The hundreds of ivory beads on the upper part of his body are remnants of the decoration from his cap and shirt.

B.C., show men wearing animal skins, apparently pink leopard skin, with hats of the same material. Fragments of actual flax textiles—notably a girl's skirt—were also found at Çatal Hüyük and were long thought to represent the earliest examples of weaving. However, the recent find of a cache of linen pieces preserved in the dry air of a tiny cave at Nahal Hemer in the Judaean desert has pushed the date of the earliest cloth back to about 6500 B.C.

The oldest surviving cotton, dating to the fifth millennium B.C., comes from the excavations at Mehrgarh on the Indus plain of Pakistan. In the New World, archaeological finds suggest that cotton was being cultivated by the inhabitants of coastal Peru as early as 3500 B.C. At the Sumerian city of Ur, in southern Iraq, textile manufacture had already become a specialized profession by 2000 B.C. Ancient cloth making was a slow procedure, even though the simple loom had been invented by this time. Sumerologist Samuel Kramer has calculated that it took a team of three women eight days to spin and weave a piece of cloth measuring 10½ by 13 feet. Nevertheless these ancient experts could produce splendid results, such as the masterpieces of weaving created by the Egyptians, the most famous linen manufacturers of the ancient world. Microscopic examination of an example from the Memphis area has revealed that each inch of fabric contained 540 threads down and 110 across, all of remarkable smoothness. The cotton fabrics produced by the ancient Peruvians in the last few centuries B.C. were equally fine, with a tighter weave than present-day parachute material (see **Ballooning** in **Transportation**).

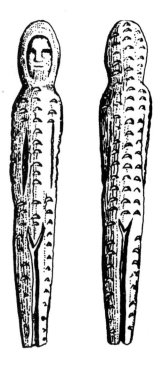

A fur-clad individual carved from mammoth ivory twenty thousand years ago; 4.8 inches high, it was found at Buret, in Siberia.

The Secret of Silk

The most desired material for clothing in ancient times was silk, undoubtedly discovered in ancient China. It is produced by the silkworm, which feeds on white mulberry leaves and then envelops itself in a cocoon made of a single continuous silk filament. The earliest find of a possibly domesticated silkworm cocoon comes from Hsi-yin-ts'un, in southern Shansi Province, dating from before 3000 B.C. Large numbers of silk fragments from the Shang Dynasty (c. 1500–1000 B.C.) show that by then the silkworm was definitely being cultivated—the Chinese silk industry had begun. Mulberry leaves from specially grown bushes were collected for feeding silkworms. The silkworm chrysalis inside the cocoon was killed by steaming, damping with salt or drying in sunlight: the silk could then be unreeled onto a wooden frame and spun. Silk fabrics survive from the Warring States Period (5th–3rd centuries B.C.), and Han Dynasty (202 B.C.–A.D. 220) tombs contain silk robes, skirts, shoes, socks and mittens.

Women gathering mulberry leaves to feed cultivated silkworms—a detail from a Chinese bronze vessel of the Warring States Period (480–221 B.C.).

Silk was being traded to the West by the tenth century B.C., some finding its way as far as Egypt. But it remained a carefully guarded Chinese monopoly until A.D. 552, backed up by the death penalty for attempting to betray the secret, when silkworms were smuggled out in bamboo tubes, along with the secrets of the industry, to the Byzantine emperor Justinian in Constantinople. Muslim craftsmen soon acquired the once-secret knowledge, and the silks they made became famous for their quality, many being exported to the West to be used in churches as coverings for the remains of venerated saints. Indeed even now there are silk covers in some churches that bear the Arabic holy text *La Ilaha illa Allah* ("There is no god but God").

Silk was always reserved for the rich of course, and poorer people had to make do with lower-quality materials. Remains of Chinese clothes made from the fibers of the hemp and ramie plants, both members of the nettle family, can be dated to the third millennium B.C. Hemp-fiber clothes for servants were made in the imperial palace of the Chou Dynasty in the first millennium B.C. Medieval Chinese engineers developed a water-powered multiple spinning frame for ramie and hemp that could handle up to thirty-two yarns simultaneously, to mass-produce fabric from hemp and ramie.

A variety of unusual materials was used for clothing in the ancient and medieval worlds. The inhabitants of Central America made rubber into garments as early as the thirteenth century A.D. if not before. When the Spanish *conquistadores* reached the area, they

Illustration of a water-powered multiple spinning frame for ramie, from the 1530 edition of Wang Cheng's *Book of Agriculture*, first published in 1313. In his day such machines were widespread, and it is assumed that they had been invented at least a century earlier. A waterwheel more than six feet in diameter (extreme right) powered the mechanism via a driving wheel; the end product, collected by the wheel on the far left, was a coarse yarn.

observed the locals also making shoes from rubber by molding it around ceramic foot shapes; how far this practice predates Columbus is unknown. In southern China the thick bark of the paper-mulberry tree, mainly used to make writing paper, was also turned into garments. In the sixth century B.C. Yüan Hsien, a disciple of Confucius, is recorded as having a paper-mulberry hat; actual examples of a paper hat, belt and shoe dating to A.D. 418 have been found at Turfan, in western China. This paper clothing was an excellent insulator; indeed contemporary accounts say it was uncomfortably hot to wear. Remarkably mulberry paper was so strong that it was even used to make armor (see **Human and Animal Armor** in **Military Technology**). Probably the strangest ancient clothing material was asbestos, known from third century A.D. China, where it was called "cloth washable in fire." Occasional eccentrics wore suits fashioned from it, but it was never seen as a practical cloth.

Who Wore the Pants?

Returning to the clothes themselves, detailed evidence for the dominance of fashion, and the existence of fashion "slavery," first comes with the classical civilizations of Greece and Rome. In late republican and early imperial Rome (1st century B.C. to 1st century A.D.) fashionable men dressed with great care for dinner. Some men were obviously as vain as peacocks: the poet Martial wrote of one host who changed his outfit eleven times during dinner, supposedly because of the heat.

Trousers were popular among the Persians of Iran, and in Europe among the Celts and Germans. However, they never caught on with either the Greeks or the Romans, who always thought of them as slightly barbarous, even though the great philosopher Pythagoras was said to have worn pants. As late as A.D. 397 the Roman emperor Honorius legislated against men who dared appear "in the venerable city of Rome" in trousers. However, the ancient Persian word for trousers, *pajamas*, eventually found its way into the English language—during the seventeenth century—as a term denoting pants for lounge and bedroom wear.

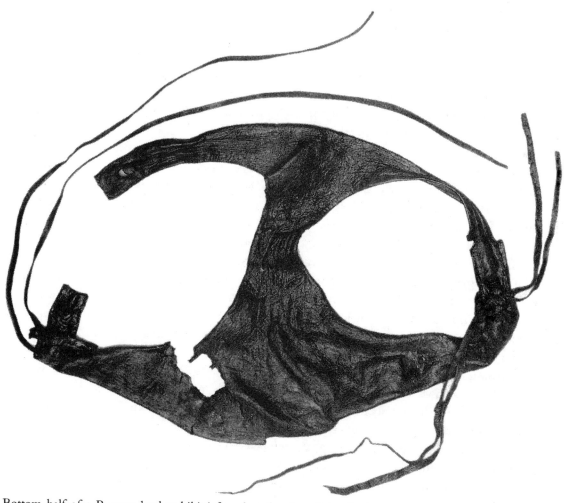

Bottom half of a Roman leather bikini found at the site of the Bank of England, London—1st–2nd century A.D.

One of the most remarkable facts about ancient clothing is that the bikini, which shocked the world when it appeared on French beaches in 1947, was, unbeknownst to its modern creator, a revival of Roman costume. The existence of the Roman bikini is known from a leather bottom fastened by long laces, tied at the hips, found in a first-century A.D. London well and from a mosaic in a Sicilian villa of the early fourth century, which shows agile young women wearing both parts.

SEWING AND KNITTING

Bone needles with eyes were being used as long ago as 20,000 B.C., during the last Ice Age, to sew together skins and furs. The earliest known sewing needles made of iron come from the Celtic hill fort at Manching, in Germany, and date to the third century B.C. No doubt the development of metal needles, much sharper than bone, prompted the invention of the thimble. The tomb of a minor official of the Han Dynasty (202 B.C.–A.D. 220) has been reported by Chinese archaeologists as containing a sewing set complete with a thimble, which would be by far the oldest known example. Despite various claims, there appear to be no examples of Roman thimbles; there are, however, many thimbles from the Byzantine-period levels at Corinth, in Greece, from around A.D. 1000.

The art of knitting, the classic home industry, probably began in Islamic Egypt. The craft grew out of an earlier technique known as nalbinding, or looping, in which a single, eyed, needle threaded with a short length of yarn is used to make a fabric composed of rows of loops. The oldest surviving example of nalbinding—possibly part of a sock—comes from the Roman city of Dura-Europos, in Syria, which was destroyed by the Persians in A.D. 256. The art seems to have flourished in Roman Egypt. Numerous examples of Romano-Egyptian nalbinding dating from the fourth century onward can be found in museums around the world. They are almost invariably socks—some brightly colored and all neatly worked in small loops that closely resemble knitting. The earliest examples of true knitting—a speedier method using two eyeless (but hooked) needles and a continuous piece of yarn—date to the twelfth century A.D. and come from Egypt. Again they are mainly socks, and it seems clear that knitting was developed by Egyptians from the nalbinding tradition.

The oldest surviving European knitwear is a rather more elaborate piece—an intricately decorated cushion from the tomb of the Infante Fernando de la Casa, heir to the Spanish kingdom of Castile, who died in 1275.

Once it had reached Europe, knitting soon became extremely popular, as reflected in the rash of "knitting Madonna" paintings of the late fourteenth century. The best known of these is the altarpiece from Buxtehude, in Germany, painted for the nuns of the Benedictine order, which shows the Virgin Mary knitting clothes for the infant Jesus.

The "Knitting Madonna" painted just before A.D. 1400 by Master Bertram of Minden, for the nuns of Buxtehude, in Germany.

Fancy Footwear

The Romans had strict rules governing the color of shoes people were supposed to wear: for example, only those who had served as magistrates were entitled to wear a particular type of red footwear. The shoes of the wealthy were painted in various colors and often decorated with gold, silver and precious stones. With time, the poorer classes came to imitate these, so that the infamous cross-dressing Emperor Elagabalus (A.D. 204–222) issued a proclamation prohibiting such expensive shoes to any but high-ranking women—and of course himself. Emperor Aurelian (A.D. 270–275) disapproved heartily of painted shoes, which he believed to be effeminate. He therefore forbade men to wear shoes of red, white, yellow or green, presumably the colors favored by dandies of the time.

In thirteenth-century Britain a peculiar fashion for pornographic shoes developed. A popular shoe of the day, the "poulaine," had a turned-up toe that had, over the years, gradually become longer and longer, until some reached twelve inches. These extreme examples had to be stuffed with moss or cork to keep them erect and were attached to the knee with a chain to prevent the wearer from tripping over them. Some men had the tips of their poulaines designed and colored as replicas of erect penises, precise in every detail, and wore them at decadent dinner parties for playing "footsie" under the table. Both the Church and state authorities condemned this exhibitionism, and new laws had to be instituted, restricting commoners to six inches of toe length and nobles to slightly more.

A typical "poulaine" shoe
from medieval England.

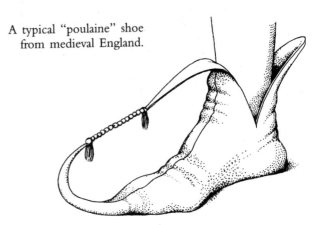

JEWELRY

People must have collected attractive shells or pebbles to make into necklaces since time immemorial, but the Stone Age hunters of Europe and Asia had gone far beyond this by 30,000 B.C. The graves of both men and women from this time often contain large numbers of ornaments, including pendants made of fired clay, necklaces of animal teeth, and elaborately decorated bracelets, necklaces and pendants carved from mammoth ivory.

Even more remarkable jewelry-making skills are evident from the later Stone Age. In the Near East during the sixth millennium B.C. craftsmen were manufacturing beads by drilling through flakes of obsidian, a very hard black volcanic glass. James Mellaart, who excavated the early urban site of Çatal Hüyük in the 1960s, was amazed: "How," he wondered, "did they drill holes through stone beads (including obsidian), holes so small that no fine modern steel needle can penetrate?" Similar beads of polished obsidian form part of a magnificent seven-thousand-year-old necklace found at Arpachiyah, in northern Iraq.

Bracelet carved from a single piece of mammoth ivory, found at Mezin, southern Russia. It dates to 20,000 B.C.

The bead necklace from Arpachiyah in Iraq. Made seven thousand years ago, it consists of obsidian flakes polished on one side, a clay bead, cowrie shells with the backs removed and a stone pendant.

A Golden Age?

Even as the Arpachiyah necklace was being buried, tentative steps were being taken in the development of metallurgy. At first "native copper" (very rich nuggets from a copper-bearing ore vein) was hammered into beads, pins and fishhooks; only later was the ore itself smelted (see **Mining and Drilling** in **Working the Land**). For a long time archaeologists assumed that such early experiments with metalworking took place on a small scale. As a result they were astounded by the massive quantities of gold jewelry discovered in the 1970s at the cemetery at Varna, near the Black Sea coast of Bulgaria—a site that dates to the second half of the fifth millennium B.C. The Varna gold was panned from local rivers, melted down and hammered into a variety of forms, the most exotic being facial or genital (penis covering) ornaments. Four of the richest burials are accompanied by a grand total of 2,200 gold items. One man was buried wearing three necklaces, three massive arm rings on each arm, two earrings, six small hair rings and a number of disks that had once been sewn on his clothes—all of solid gold.

The most famous gold jewelry of the ancient world comes from the mysterious "Death Pits" at the Sumerian city of Ur, in southern Iraq, discovered by Sir Leonard Woolley in the 1920s. Around 2500 B.C. a series of rulers were buried here in awful splendor—surrounded not only by their treasures but also by the bodies of numerous attendants (as many as seventy-four in one burial), who drank poison to accompany their monarch into the afterlife. Many of the men wore headbands—like those worn by Arab men today—consisting of three large beads at the front with a gold chain behind, probably to keep headcloths in position. The women had much more elaborate jewelry, including headdresses of golden flowers and leaves, huge crescent-shaped earrings, chokers around the neck and large dress pins to fasten their robes. The most lavish burial was that of Queen Pu-abi, whose body was covered down to the waist by a cloak of beads made from gold, silver, lapis lazuli, carnelian, agate and chalcedony. She also wore massive earrings, and a superb crown of three golden wreaths strung on triple chains of lapis lazuli and carnelian beads.

The range of techniques used in the creation of these treasures included casting in closed molds, riveting, soldering and the use of gold leaf. Woolley was greatly impressed by the technological mastery demonstrated by the "Death Pits" goldwork at such an early date. As he put it, "The Sumerian craftsman could do very nearly all that can be done by the modern goldsmith and he could do it almost if not quite as well."

Gold continued to be the prime way in the ancient world of displaying one's riches and status—and little has changed. The great kings of the Persian Empire of the sixth to fourth centuries "power-dressed" with lavish amounts of gold jewelry hanging from ears, neck and wrist. The ancient Greek historian Plutarch estimated the value of the outfit that the Persian emperor Artaxerxes II (405–359 B.C.) wore on state occasions as twelve thousand talents (roughly $5.5 million).

Gem Trails

Gold of course was not the only precious material that fascinated the ancients. Indeed in some societies, notably the Chinese and the Aztec, jade was considered to be far more valuable. Jade was first carved in China: in the last few years a series of superb small sculptures of turtles, birds and monsters with huge staring eyes, dating to about 3000 B.C., have been unearthed in Liaoning Province, in northeast China, pushing back the history of jade sculpture by some five hundred years.

The passion for precious stones led to the development of some of the longest and most arduous trade routes of ancient times. The nearest source of jade available to the early craftsmen of Liaoning was at least one thousand miles to the west. But the scope of the ancient traffic in precious stones is best illustrated by the example of lapis lazuli, a beautiful gemstone with a rich blue color. The largest deposits are in southeast Afghanistan; here, as we know from the local site of Mundigak, it was being made into beads as early as 4000 to 3500 B.C. By the mid-third millennium B.C. lapis was the preferred material for inlays in jewelry across the Middle East. To reach southern Iraq, where it was used extensively in the "Death Pits" at Ur, it would have to have been carried 1,400 miles if one followed a straight line across mountain ranges and deserts, and twice as far by any sensible route. It was traded on even farther, being found at Egyptian sites as early as 3000 B.C.

Costume Jewelry

Of course, when the real thing was too difficult to come by, human ingenuity was ready with a solution. The Roman philosopher Seneca (5 B.C.–A.D. 65) reported that there were whole factories in Rome devoted to the production of artificial gemstones. The manufacture of costume jewelry was evidently a highly profitable business. Jewelers of the Roman Empire were particularly adept at imitating emeralds with a specially developed colored glass.

In fact the earliest Egyptian experiments in glassmaking, during the third millennium B.C., seem to have been attempts to imitate rare stones for bead necklaces. After 1500 B.C. "Egyptian blue," a glass paste with a rich blue color imitating lapis lazuli, was widely used for beads, inlaid jewelry and even ornamental additions to walls and ceilings. It may have been invented slightly earlier by the Mycenaeans of ancient Greece, where it was used extensively in palaces as a means of decoration as well as for jewelry. Homer describes the blue glass inlay on the magnificent shield of Agamemnon, king of Mycenae.

The Babylonians and Assyrians of ancient Iraq were highly skilled in the intricate chemistry of colored-glass manufacture. A whole series of Assyrian clay tablets from the seventh century B.C. gives detailed formulas for making glass in a wide range of colors, including "lapis-blue," "sapphire," and "green crystal," as well as recipes for making artificial versions of semiprecious stones, such as carnelian.

The "Lost Art" of Granulation

Other advanced techniques used by ancient jewelers included ways of improving cheaper metals by plating them with gold or silver. Coating an object by simply dipping it into a molten precious metal is not actually a practical method—it is too messy. A much more re-fined technique was commonly in use in the Roman world. First an amalgam of mercury and gold was made into a paint, which was ap-plied to a given object. Heat was then used to drive off the mercury, leaving a perfect gold finish. It is also possible that electroplating was invented in Babylonia during Roman times (see **Electric Batteries** in **High Tech**).

Another remarkable technique used in the ancient world was granulation—a means of manufacturing or decorating jewelry by fixing small spheres of gold onto a surface without the use of con-ventional solder. Credit for its invention seems to go to the Sume-

A fine example of Etruscan granulation—a gold pendant of the 5th century B.C. depicting a horned river god.

rians some 4,500 years ago, as the earliest known example, a small ring made of six gold balls joined together, comes from the "Death Pits" at Ur.

From the Near East the art of granulation spread westward around 2000 B.C. The technique was used in Egypt by about 1900 B.C. and is common in the jewelry of the Mycenaean civilization of Greece from about 1500 B.C. The Phoenicians of Lebanon probably introduced it to the central and western Mediterranean, including Italy, where the Etruscans brought it to a height of perfection in the seventh century B.C. Etruscan craftsmen were able to attach minuscule balls of less than 1/200 of an inch in diameter to almost any surface.

Granulation became less popular during the Roman and Byzantine periods and had died out in European goldwork by about the twelfth century, though remaining in favor in the East.

In recent times Western jewelers attempted to revive the skill, though initially without success. They found it almost impossible to produce the fine granules of gold needed; even if they did, when they tried to solder them onto a surface, the solder either ran and ruined the surface of the piece or bubbled up and shifted the balls. There are various processes that do work, and a range was probably

The intricate process of granulation. Here the tiny gold balls are being pressed into the paste of fish glue, copper hydroxide and water smeared on the surface of the metal.

used in antiquity. One method is to place the gold balls with tweezers, sticking them to the surface with a paste made from fish glue,
copper hydroxide and a little water. Then the object is fired. The
heat burns the glue into carbon, reducing the copper hydroxide to
metallic copper, which alloys with and diffuses into the surrounding
gold, forming a perfect unobtrusive joint. It was not until 1933 that
Western metallurgists rediscovered the secret of this art.

SPECTACLES

One of the most surprising images from the Middle Ages is on a
fresco in a church at Treviso, Italy, painted by Tommaso da Modena
in 1352. The subject, an elderly churchman poring over some manuscripts, is clearly shown wearing a pair of spectacles. The portrait
still remains the earliest irrefutable depiction of eyeglasses. Indeed
glasses seem to have been invented not many years before
Tommaso's painting—there are some literary references going back
to the beginning of the fourteenth century, but they are rare and
point to spectacles being a recent development. Exactly *who* was responsible for their invention, however, has been one of the most
hotly disputed issues in the history of science. Inventors from Italy,
Belgium, Germany, Britain and China have all been put forward as
claimants.

The Italian theory is, however, by far the best. Not that the evidence is any means simple or clear-cut. The story of the invention
of spectacles in medieval Italy is strangely fraught with intrigue, and
modern historians of science have had great difficulty in disentan-
gling its main threads.

The Elusive Inventor

A major problem has been that the whole matter was *deliberately*
confused by scholars in the seventeenth century. Carlo Roberto
Dati, of Florence (1619–1676), wrote the first detailed study of the

subject, entitled *The Invention of Eyeglasses, Is It Ancient or Not?*, in which he gave credit to one Alessandro Spina, a monk and scientist who lived in Pisa until his death in A.D. 1313. Dati admitted that "somebody else was the first to invent eyeglasses," but stated that he was "unwilling to communicate the invention to others." All the same, Spina, according to Dati, had such a brilliant mind that he could re-create "everything that he had heard said or saw done"; he was therefore able to devise his own eyeglasses without the aid of the anonymous inventor.

There the matter might have ended—with the world believing that Spina had put glasses on its nose—until a close investigation of Dati's correspondence and sources was published in 1956 by Edward Rosen, a historian of science at City College, New York. Rosen found that Dati's information had been supplied by a colleague named Francesco Redi, chief physician to the grand duke of Tuscany. The letter still survives in which Redi related to Dati the story of Spina's "invention," quoting for his benefit from the *Ancient Chronicle of the Dominican Monastery of Saint Catherine in Pisa*. The quotation, in Redi's version, says that "whatever he [Spina] saw or heard had been made, he too knew how to make it." Turning to the original *Chronicle*, Rosen found that Redi had tampered with the text, which actually read "whatever had been made, when he saw it with his own eyes, he too knew how to make it." Redi, followed by Dati, had clearly watered down the implication of the original document—that Spina had *seen* the handiwork of another craftsman with his own eyes—and added the idea that he may simply have heard of it.

Rosen showed that other seventeenth-century scholars were in on the conspiracy to credit Spina and downgrade the anonymous inventor of spectacles. He was also able to discover the motive for this bizarre intrigue. The scholars in question were colleagues or admirers of the great Galilei Galileo (1564–1642), part of whose reputation depends on the claim that he invented the telescope. But at the time it was rumored that Galileo had seen an earlier telescope built by the Flemish spectacle maker Johann Lippershey. Galileo himself insisted that he had only heard reports of it, working out the rest for himself "through deep study of the theory of refraction." His friends

(Opposite) Detail from church fresco at Treviso, Italy, painted in 1352 by the monk Tommaso da Modena.

were keen to defend him. In 1678 Redi published a *Letter Concerning the Invention of Eyeglasses* which stated that

> If Friar Alessandro Spina was not the first inventor of eyeglasses, at least it was he who by himself without any instruction rediscovered the method of making them. . . . Exactly the same thing happened, by a certain coincidence of fate, to our most famous Galileo Galilei. Having heard the report that a certain Fleming had invented the long spyglass . . . he worked out a similar one solely with the aid of the theory of refractions, without ever having seen [the original].

And so it was—to save Galileo's face—that Spina was credited with an independent invention of spectacles, while the role of the anonymous craftsman whose work he copied was cunningly underplayed. But who was the mysterious inventor from whom Spina borrowed? Others apart from Spina knew him, and it is even possible to date his invention of glasses to around A.D. 1285, from a passage in a sermon delivered by the Dominican Friar Giordano da Rivalto in 1305:

> It is not twenty years since there was found the art of making eyeglasses which make for good vision, one of the best arts and one of the most necessary that the world has. So short a time is it since there was invented a new art that never existed. I have seen the man who first invented and created it and I have talked to him.

Again, many claims have been made, some of them deliberate hoaxes employing forged evidence, which identify the inventor with known scientists of the thirteenth century, such as the English polymath Roger Bacon (see **Codes and Ciphers** in **Communications**). Rosen, however, has shown that we will probably never discover his name. At best we can see from various clues in the available documents that the inventor was not a monk, like Spina, but a layman and that he most likely resided in Pisa.

In any case his discovery seems to have been rapidly taken up by Venetian craftsmen, manufacturers of the finest glass in the medieval world. Beginning in 1300 there is increasing reference to spectacle lenses in the regulations of the glassmakers' guild, concerned with stamping out the counterfeiting of crystal using colorless glass. The regulations should be a good barometer of the speed with which the novelty of eyeglasses caught on in Venice. If so, we can understand why the elusive inventor whose work Spina copies was so loath to broadcast his discovery—in an era before scientific copyright was enforced, he probably jealously guarded his secret in the hope of making some money before it became too well known.

"A vain book collector wearing spectacles who dusts but does not read his books"—from a publication of 1497.

Spectacles Go East

The intensive research into the origin of glasses—albeit inconclusive—has been enough to scuttle claims that the Chinese were the inventors. For many years the Chinese theory was upheld on the basis of the following passage in the *Tung Thien Chhing Lu* ("Clarifications of Strange Things") by the thirteenth-century writer Chao Hsi-Ku:

> *Ai-tai* resemble large coins, and their color is like mica. When old people are dizzy and their sight tired, so that they cannot read fine print, they put *ai-tai* over their eyes. Then they are once more able to concentrate, and the strokes of the characters appear doubly clear. *Ai-tai* come from Malacca in the western regions.

Since Chao Hsi-Ku's book was written around A.D. 1240—half a century before the earliest Italian references—it was once thought to demonstrate Chinese primacy in the invention of spectacles. During the 1970s, however, it became clear that the passage referring to spectacles was *not* in the earliest copies of the book—it seems to have been added in during the Ming Dynasty (1368–1644). The passage does give a useful clue, however, to the origin of the earliest Chinese spectacles, when it mentions Malacca, a kingdom on the Malaysian peninsula. A Chinese court record from about A.D. 1410 describes how the king of Malacca presented ten pairs of spectacles to the emperor as tribute. At this time Malacca was frequented by Arab and Persian traders, and it seems likely that they brought these early, highly treasured spectacles to Malacca from the West.

The Chinese can claim a first, however, with dark glasses, referred to in the *Records of Leisure Hours* written by one Liu Chhi in the early twelfth century. Made of smoky quartz, they were not worn as a protection against the sun but by judges, who used them to disguise their reactions to evidence as it was read out in court.

SNOW GOGGLES

The skiers of Saint Moritz or Aspen were not the first people to spend so much time on snow that they risked snow blindness. Around two thousand years ago the ancestors of the Eskimos developed wooden goggles to wear while hunting on the ice. The goggles shielded the pupils from sunlight, both direct and that reflected from the snow, while the wide, narrow, eye slits allowed good horizontal vision. Later examples were finely carved from bone and ivory.

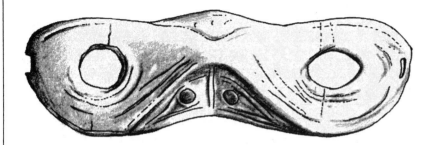

Eskimo snow goggles, carved from ivory between A.D. 100 and 800.

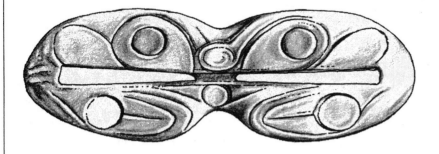

UMBRELLAS

Today most of us will own an umbrella, but in many ancient societies they were the exclusive privilege of noble birth. Our first sighting of an umbrella is on a victory monument of Sargon, king of Akkad (in modern Iraq) dating to around 2400 B.C. Sargon is shown striding ahead of his troops with an attendant carrying a sunshade immediately behind him. From Iraq the idea of the parasol spread west, soon reaching Egypt. Not surprisingly, sunshades became popular all around the Mediterranean and are depicted on the Mycenaean (prehistoric Greek) pottery of Cyprus a thousand years later.

The noble associations of parasols are clear in the Egyptian story of the envoy Wenamun and his unsuccessful mission to Lebanon, written around 1070 B.C., in which a sunshade plays a part. Zakar Baal, prince of Byblos, had a butler standing behind him holding a parasol over his head during his interview with Wenamun. The prince's butler teased Wenamun when the parasol's shadow covered the Egyptian, saying that the shadow of the pharaoh was falling on him. In those days the Egyptian royal family would have been protected by a sunshade when traveling in their chariots.

Parasols were by now used in hot countries halfway around the globe, and Assyrian kings of the early first millennium B.C., as well as the Buddha in India, are frequently depicted on sculptures as being sheltered from the sun's rays by an attendant holding a parasol. By contrast the Roman *umbraculum* (from *umbra*, meaning "shade"), made of light cloth stretched over a wooden framework, was a more democratic affair, and a frequent sight on the streets of Rome. Indeed the poet Ovid thought that women should not go out without "a golden umbrella to ward off the keen sun."

The *umbraculum* was still functionally a parasol, however, offering only protection from the sun. For the ancestor of the modern umbrella, protecting the bearer from both sun and rain, we are indebted to the Chinese. Parasols arrived in China during the Chou Dynasty (c. 1000–221 B.C.). These early examples were of silk and could be used to protect the noble occupants of chariots from sudden rain, but the actual umbrella itself was a product of the Wei Dynasty (A.D.

Egyptian chariot with sun shade—from a tomb painting of the New Kingdom (c. 1550–1070 B.C.) at Thebes.

386–535). It was made of heavy, oiled, mulberry paper and could protect against any weather. The emperor had a red and yellow umbrella while ordinary mortals had to make do with blue ones. In the fourteenth century silk fabric of sufficient strength for umbrella manufacture was developed, but an imperial law of 1368 reserved these fancy new umbrellas for the sole use of the royal family.

The idea of the umbrella spread from China to India, where it was seen by the papal envoy John of Marignolli in A.D. 1340. He reported back to the pope in a letter: "As all the Indians commonly go naked, they are in the habit of carrying a thing like a little tent-roof on a cane handle, which they open out at will as a protection against sun or rain. I brought one to Florence with me."

But John's umbrella didn't catch on. In the end it was only when the humble paper umbrella was brought back to Europe from China by eighteenth-century traders that the umbrella industry revived there after centuries of neglect and false starts.

FOOD, DRINK AND DRUGS

CHAPTER SEVEN

INTRODUCTION

There were, and still are today, some isolated societies that never took up the use of fire in food preparation. But in the broader sweep of things the discovery of fire was the first major breakthrough in culinary history. When this crucial step toward civilization was first taken we will probably never know; we do know that fire had been mastered at least half a million years ago, as the remains of hearths at Choukoutien Cave, in northern China, clearly show. Here "Peking Man," one of our most distant ancestors, left the first possible traces of cooking in the charred bones of numerous species of animals around these hearths—deer, antelope, horse, pig, bison, water buffalo, elephant, rhinoceros and monkey.

Cooking food, initially over open fires, allowed for a greater range of tastes and foodstuffs (some plants are inedible or even poisonous uncooked). The next step forward was in the storage of food and drink. Animal skins, woven baskets and gourds were the first containers, but the invention of pottery marked the greatest advance. For a long time it was assumed that pottery must have been invented in the Near East—"the cradle of civilization"—by the earliest known agricultural societies (see **Introduction** to **Working the Land**). However, in the 1960s surprising evidence came to light in Japan, from a time when the inhabitants were still living as gatherer-hunters. Dating from thirteen thousand years ago the prehistoric clay vessels from Odai-Yamomoto and Ushirono, on the island of Honshu, are the world's oldest pottery. By contrast the earliest Near Eastern pottery, from Iran, dates to around 9000 B.C. Not only is this later, but it is also only sun-dried rather than fired like the earlier Japanese pottery.

Things really got cooking when the preparation of food went beyond the simple roasting of meat, fish and vegetables. The need to preserve food led to the invention of entirely new products and tastes. Drying (by leaving in the sun or smoking over fires) or freezing (if the climate was cold enough) are the simplest ways. Pickling in brine and salting were also developed as ways of saving food for the winter months. These methods of preservation must have been discovered by simple experimentation, while the next major alternative, fermentation, was no doubt the result of accidental contamination by what proved to be highly useful bacteria. These bacteria, by making food go "bad," actually make it last much longer. As part of the agricultural revolution that brought us from the Middle to the New Stone Age, mankind learned to control the process of fermentation, to turn milk into cheese and yogurt, grape juice into wine, and soybeans into soy sauce, all tasty and long-lasting foodstuffs. Allowing grape juice to ferment even longer turns it into vinegar (acetic acid), which, when bottled, can be kept almost indefinitely as a condiment for livening up plain dishes or a medium for preserving other foods by pickling.

Kitchens, bakeries and breweries have always been focal points of technological advance—from techniques for crushing, pressing,

Some of the world's earliest known pottery—a vessel found at the Ishigoya Cave on Honshu, Japan, dating to about 10,000 B.C.

milling and grinding to equipment for boiling, roasting, frying and baking. The first identifiable ovens, dating from 20,000 B.C. were found in Ukraine. These were pits in which hot coals were covered by a layer of ashes; the food was set on top, wrapped up in leaves, and the pit covered with earth. More recognizable ovens made of baked brick were being used by the early civilizations of Sumer (in Iraq) and Egypt about 2500 B.C. These increased in sophistication over time, the Sumerians being the first to develop ovens with cooking ranges on which pots and pans could be placed for boiling and frying. Excavation of the great Sumerian city of Ur produced a number of brick-built ovens and cooking ranges from about 1800 B.C. The excavator, Sir Leonard Woolley, found one so perfectly preserved that he lit the furnace underneath the cooking range and actually prepared food on it.

While the most common fuel for cooking in ancient times was always wood, a surprising piece of information about the use of natural gas in ancient Iran is given in a Greek text of the early second century A.D. (ascribed to Aristotle): "In Media and the district of Psittakos in Persia there are fires burning, a small one in Media, but a big one in Psittakos, with a clear flame. So the Persian King built his kitchen nearby. Both are on level ground, not in high places. They can be seen night and day."

The Persians were clearly taking advantage of natural seepages of gas, presumably piped over short distances to the royal kitchen. Actual drilling for natural gas, mainly methane, was undertaken by the Chinese around A.D. 300; though the gas was mainly used to boil

An Egyptian bakery, as depicted in the tomb of Ramesses III (early 12th century B.C.). On the left, bakers use their feet to knead the dough in a huge bowl. In the center, another prepares fancy-shaped breads. To the right, others fry spiral-shaped doughnuts in a vat of oil, while to the far right a conical oven is being prepared for baking.

brine in salt manufacturing (see **Drilling and Mining** in **Working the Land**), some was also piped in for cooking. An ancient gas cooker has yet to be found, but one never knows. . . .

Almost every corner of the globe has some oil-bearing plant, and in the ancient world corn, sesame, turnip and rape seeds, among others, were crushed to provide nutritious oils for cooking and dressing food in much the same way as they are today. In the Mediterranean olive oil production had already begun by about 3500 B.C.; the earliest evidence of olive cultivation comes from Early Bronze Age Palestine, where farmers developed by selection the modern oil-rich variety of olives from the wild spiny type with smaller fruit. By the time of the Roman Empire (1st to 5th centuries A.D.), olive oil was being consumed in massive amounts throughout the Mediterranean, for use in washing, lighting and the manufacture of perfumes and medicines, as well as in cooking (see **Soap** in **Personal Effects**, and **Introduction** to **House and Home**). Some idea of the immense scale of the ancient olive oil trade comes from a rubbish dump at Rome made from the broken jars used to carry the oil from southern Spain to the capital. The heap is 3,000 feet in circumference and still towers 140 feet above the banks of the Tiber. Archaeologists have determined that it represents as many as forty million original pots carrying a staggering 440 trillion gallons of olive oil.

Honey was the main sweetening agent used by the ancients; wild honey was plentiful, and apiculture as a rural industry had already begun in Egypt by about 2500 B.C. (see **Beekeeping** in **Working the Land**). Fruits were also used as sweeteners, and syrup made from the juice of pressed dates was consumed in large amounts in ancient Iraq.

Perhaps the popularity of honey and dates accounts for the rather surprising absence of sugar in the ancient world as a whole, despite the fact that sugarcane was among the earliest plants to be domesticated in India. It was introduced there from the island of New Guinea during the New Stone Age (5th and 4th millennia B.C.) and has always played a major role in the Indian diet. Its importance is reflected in ancient Hindu myth, which describes the sacred mountain at the center of the world surrounded by seven continents, divided by seven magical seas of produce that reflect the staple needs of humanity: salt, sugar, wine, ghee (clarified butter), milk, yogurt and fresh water.

But sugar was still a novelty to the Romans in A.D. 75, when Pliny's encyclopedia described it as "a kind of honey which collects in reeds, white like gum," noting that "it is only used in medicine." He added that it mostly came from Arabia, though the best quality was from India. The cultivation of sugarcane eventually spread throughout the Near East during the sixth and seventh centuries A.D., and by the eighth century massive quantities were being used by the Arab chefs of Baghdad.

Spices, however, were not so slow in traveling west. The ancient civilizations of the Mediterranean considered them essential to good eating. In a time when only the ultrarich could afford to refrigerate foods by packing them in ice, spices were vital as a way of preserving food or masking the fact that it was not particularly fresh. Most familiar spices are indigenous to the Indian subcontinent and the islands of the Far East, where they must have been used since time immemorial. Still, little is known about when and how they were first cultivated, though chance finds and literary references give some interesting glimpses of the early use of some spices. In 1972 Chinese archaeologists discovered the tomb of a Han Dynasty noblewoman, dating to around 165 B.C. Her corpse was in a remarkable state of preservation, as was the mass of foodstuffs buried with her in bamboo cases, jars and sacks. Ginger, cinnamon and other spices were present, along with grain, meat, fish and fruit. By A.D. 406 ginger was such a favorite spice of the Chinese that the writer Fa-Hsien records in his book of *Travels* that it was grown in pots on board ships so that it could be eaten fresh.

Pepper, which comes from India, makes a surprisingly early appearance in Egypt at the beginning of the twelfth century B.C., when black peppercorns were used as part of the stuffing of Ramesses III's mummy. The Greeks discovered pepper much later, and the earliest classical reference, in the writings of the famous doctor Hippocrates around 400 B.C., suggests that they first knew it as a medicine.

By the heyday of the Roman Empire (1st to 2nd centuries A.D.) the Mediterranean appetite for eastern spices had become a mania. The state-run pepper houses in Rome imported spices through Alexandria, where they were carried from India by middlemen, generally the Arabs. Black and white pepper were imported from the coasts of Malabar and Kerala, in southern India, and cinnamon and its close relative, cassia, from as far away as Vietnam. Cloves and nutmegs had an even more remote origin; they came from the Moluccas, the famous "Spice Islands" of Indonesia over which the great European powers were to fight so fiercely in the sixteenth and seventeenth centuries. To bring these prized spices to Rome required a journey of some six thousand miles. The Roman Senate convened many times to discuss the drain on the economy posed by this massive importation of spices from the Orient.

Turning from the kitchen to the dining table, it seems fairly safe to credit the Chinese with the invention of chopsticks, still the world's second most common eating utensil; literary references in China go back to the third century B.C. at least. The excavation of ancient Egyptian tombs and sites has produced a large number of spoons, and it seems that they were used for eating as well as serving food. But on the whole the Egyptians, like the Greeks and others, used the world's most preferred eating utensil—the fingers. As ancient Greek gourmets liked their food piping hot, this lack of cutlery posed something of a challenge to the greedy. One ingenious gourmet named Philoxenus spent hours hardening his fingers and gargling with near-boiling water so that he could pick up and swallow the hottest tidbit. Having developed these techniques, he was accused of bribing cooks to serve the dishes extra hot when he dined out so that he could gobble them up before the other guests could even taste them. Likewise the Romans used only fingers, spoons and knives. The rich would have their meat cut up for them by a servant called the "scissor."

The table fork is probably an invention of the Eastern Roman Empire based at Byzantium, where forks were in use from the fourth century A.D. onward. From there they made agonizingly slow progress westward into Europe via Greece and Italy. When a medieval Byzantine princess was given in marriage to a doge of Venice, she took along some two-pronged forks; their appearance startled and shocked the Venetians. Despite Italy's status as a model for medieval manners and style, forks failed to catch on in the rest of Europe until the eighteenth century.

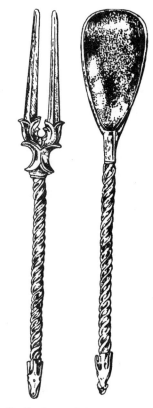

(Left) One of the earliest known eating forks, part of an ornate matching silver set along with the spoon (right), used at Byzantium in the 4th century A.D.

The ancient Athenians invented the original *hors d'oeuvre* table. Other Greeks accused them of meanness, as it meant that the diners had no chance to gorge themselves on each of the treats before they were carried away. The Greeks also began the trend toward gastronomy taken to such extreme lengths by the Romans. By the fourth century B.C. Greek gourmets were already demanding such delicacies as peacock eggs or pork from pigs that had died from overeating.

The inventiveness of Roman cuisine is best expressed in *Satyricon*, one of the few novels to survive from the Roman world. Written by Petronius, who organized entertainments for the Emperor Nero (A.D. 54–68), it describes a feast held in the house of the rich merchant Trimalchio. His guests were offered wild sow with its belly full of live thrushes, a hare dressed with wings to look like Pegasus (the

Roman oven with chimney, in a bakery preserved at Pompeii. At the bottom of the oven is the opening for fuel, while the wide shelf in front of the oven was used for resting the paddles with which the oven was loaded and emptied. The conical shapes in the courtyard are the bases of flour mills.

flying horse of Greek mythology), quinces stuck with thorns to look like sea urchins and roast pork carved into models of fish, songbirds and a goose. Real Roman menus show that the description of Trimalchio's perverse banquet is not unrealistic: The Romans fattened snails on milk until they were too big to retreat into their shells; dormice were kept in pottery jars and fed on nuts until they were plump; and pigeons had their wings clipped or their legs broken while they were fattened on chewed bread. The grossest dishes of all were concocted for the Emperor Vitellius (A.D. 69), who was said to hold at least three banquets a day. One contained a mixture of pike livers, pheasant and peacock brains, flamingo tongues and lamprey roe; the ingredients were specially shipped in from every corner of the Roman Empire.

The excesses of the Roman rich were probably rivaled only by those of the ancient Chinese court, where the emperor had an extraordinary number of retainers occupied in preparing and serving food and drink. In the third century B.C. almost 4,000 servants were employed in the residential quarters of the palace. Of these no less than 2,271, or almost 60 percent, handled food and wine— including 162 master "dieticians" in charge of the daily menus for the emperor and his family, 128 chefs for the family's food and 128 for guests, 62 assistant cooks, 70 meat specialists, 62 game specialists, 342 fish specialists, 24 turtle and shellfish specialists, 335 specialists in grain, vegetables and fruit, 28 meat dryers, 110 wine officers, 340 wine servers, 62 pickle and sauce specialists and 62 salt men.

Of course very few people in history can have enjoyed gourmet food anything like this. Even in the great cities of the past the bulk of the population ate far simpler, largely vegetarian, fare, mainly for economic reasons. But there was also vegetarianism for reasons of principle. As a philosophy it was first developed in India. Here the cow, the mainstay of agriculture, had been elevated to a sacred status at a very early date—hundreds of seals from the urban centers that thrived in the Indus Valley during the third millennium B.C. (see **Introduction** to **Urban Life**) carry majestic portraits of hump-backed cattle. Eventually, because of its value in providing dairy produce, the cow was deemed by Hindus as *aghnya*—"not to be slain." The authors of India's national epic, the *Mahabharata*, composed in the first millennium B.C., pronounced a dire warning to those who ignored the cow's sacred status: "All that kill and eat and permit the slaughter of cows rot in hell for as many years as there are hairs on the body of the cow slain." The *Mahabharata* also contains more general arguments in favor of a meat-free diet: "Those who desire to possess good memory, beauty, long life with perfect health, and physical moral and spiritual strength, should abstain from animal food."

A full vegetarian philosophy was expounded by the Jains and Buddhists. Both religions began in the sixth century B.C. as reactions to the violent and class-ridden nature of early Hindu society, and both had as a keynote a respect for life in all its forms. Jainism, which in its most extreme form today requires its adherents to wear masks to avoid breathing in microorganisms, has always insisted on strict vegetarianism. Buddhism was less strict—one tradition has it that the Buddha himself died from eating some rancid pork. But the basic message was still there, and the vegetarian philosophy spread along with Buddhism to China and Japan. It led to the development of Oriental "monk food," meat substitutes cunningly made of bean curd and gluten for those who missed the taste of flesh. Two thousand years later Western food manufacturers are beginning to catch on to the idea.

An unsolved question of food history is whether ancient Greek vegetarianism was independently invented by the famous mathematician and philosopher Pythagoras (see **Introductions** to **High Tech** and **Music**). During the late sixth century he preached a vegetarian gospel so eloquent that it still resounded in imperial Rome half a millennium later. As he lived only about a generation after the Buddha in India, it is hard to see the similarity in their philosophies as mere coincidence. Pythagoras, like the Buddha, taught that souls reincarnate by transmigration, from one living being to another, and that to kill an animal was to interrupt its long struggle in progressing from a lower to a higher life-form. But even if he was influenced by Indian philosophy, Pythagoras added a new dietary taboo. He believed that the souls of all living creatures began their long spiritual journey as beans. Ironically enough, this staple of the modern vegetarian diet was strictly forbidden by the Pythagoreans.

Apart from vegetarian philosophy, there was little concept of "diet" in the ancient world in the sense that we know it. Some Roman doctors pontificated on the healthiest kinds of bread and the best positions to bake them in the oven, but with little evident rationale. Healthy diet was largely taken care of by common sense, intuition and custom, and despite their profound knowledge in so many other areas, the ancients never really discovered the basics of nutritional science. Inventiveness in the field of food and drink, along with other social pleasures such as drug taking, was mainly directed at developing new ways of satisfying the senses, enjoying some of life's greatest pleasures in as full and as varied a way as possible.

RESTAURANTS AND SNACK BARS

The longest-running restaurant in the world, Ma Yu Ching's Bucket Chicken House, opened for business in A.D. 1153 in the Chinese city of Kaifeng. Despite several changes of dynasty and two revolutions it still serves up cheap and nourishing food to a vast throng of customers.

Ma Yu Ching's establishment is a remarkable survival from the Sung Dynasty (A.D. 960–1279), when restaurants of all kinds blossomed in the country's capitals. They were already flourishing under the previous Tang Dynasty (A.D. 618–906): texts from that era describe the streets of main cities as being full of restaurants "to wait upon wayfarers with food and wine." But with the Sung Dynasty come detailed descriptions, and even paintings, of some of the world's earliest regular eating establishments. The capital of Kaifeng was crowded with them, serving food of every kind and quality. Many Sung Dynasty restaurants boasted of serving "official-style" food, comparable to that prepared by enormous kitchen staffs for the richest families, and some became so good that the Inner Palace itself ordered "takeouts" from them.

Menus in Sung Dynasty restaurants were enormous: one contemporary source refers to a "casual list" of 234 items. Ordering time was mayhem for the staff, as a slightly later source graphically describes:

[Customers] would shout their orders by the hundreds. . . . The waiter then went to get the orders, which he repeated and carried in his head, so that when he got to the kitchen he repeated them. . . . In an instant, the waiter would be back carrying three dishes forked in his left hand, while on his right arm from hand to shoulder he carried about twenty bowls doubled up.

Some restaurants, particularly those in the southern Sung capital of Hangchow, specialized in one particular food—there were restaurants for fish, shellfish, regional cuisines, iced food, vegetarian dishes cooked in the style of Buddhist temples, and even some specializing in dogmeat. Marco Polo, the renowned Italian traveler and gastronome who visited China toward the end of the Sung Dynasty, declared that Hangchow "is the greatest city which may be found in

Restaurants at the capital of Kaifeng, painted by the court artist Chang Tse-tuan shortly before the fall of the Northern Sung Dynasty in A.D. 1126. The liberal use of chairs and tables shows that by the Sung Dynasty the Chinese had made the transition from floor to table eating. The Japanese continued to eat seated on the floor until the 20th century.

the world, where so many pleasures may be found that one fancies himself to be in Paradise." He may of course have also been referring to the prostitutes, both male and female, who thronged the restaurants in numbers so great that, in Polo's words, "I dare not say it."

Fast-Food and Noodle Shops

To all intents and purposes it was the Chinese who invented restaurants as we know them. They also developed some of the earliest "fast food" shops in the world. Under the Sung Dynasty smaller establishments selling snacks blossomed alongside the restaurants.

The favorite fare in these cheaper cookshops was noodles (with or without soup), the archetypal fast food of the Orient. As far as we can tell, noodles were first made in China in the first century A.D. Some three centuries later the Chinese historian Shu Hsi wrote that "noodles and cakes were mainly an invention of the common people," but by the second century A.D. they were so popular that the Han Dynasty emperors were tucking into them. Marco Polo remarked on the noodles that the Chinese were so fond of, giving rise to the suspicion that he introduced the idea into Italy, where it was adapted to create the first spaghetti. Italian pasta chauvinists have of course fought back, arguing on slender grounds that the Etruscans, forerunners of Roman civilization, manufactured macaroni and claiming that the great Roman cook Apicius (see **Cookbooks**) wrote a lasagna recipe, which is untrue. On the other hand, a claim has been proferred for a medieval Italian friar as the inventor of lasagna. It is fair to say that we simply don't know whether the Italians and Chinese invented pasta independently or not. The great noodle controversy remains unresolved and will doubtless rage indefinitely until some firm archaeological evidence is found.

Meanwhile, in medieval Egypt a different kind of fast food was thriving in the bustling streets of Cairo, where there were an estimated ten thousand street stalls at which food was cooked. More enterprising food sellers marched up and down the streets carrying on their heads lighted stoves, with precariously balanced boiling pots or meat roasting on spits. According to Leonardo di Niccolò Frescobaldi, an Italian who visited Cairo about A.D. 1400, the citizens often sat and ate dinner in the street; they would spread out a skin on the ground, set out their bowls and sit down for a meal. The street kitchens served rice and fritters cooked in oil, and meat, especially lamb, chicken and goose, usually in the form of kebabs, as Frescobaldi notes:

> The cooks chop up the meat into little pieces and put it on spits, as we do thrushes, and then they put it into ovens which are open on top; the meat is cooked in an instant. Sometimes they cook a whole sheep, and after it is done, a man carries it on his shoulders, puts a table on his head, and goes through the streets crying "Who wants to eat meat?" Because they have no inns, strangers are obliged to eat wherever they happen to be.

It was the job of an Egyptian government official called the "censor of morals" to ensure that the food sold from these stalls was up to standard and not adulterated.

Classical Snacks

It is unlikely that the etiquette of well-to-do Greeks and Romans would have allowed them to sit down in the street to eat. Indeed most people in ancient and medieval times ate their main meals at home, which is what makes the early Chinese restaurants so exceptional. All the same, there was a busy fast-food trade in the Italian cities of the Roman Empire, catering to the needs of busy middle- and working-class citizens rushing to and from work.

The streets of Roman Pompeii, preserved under volcanic ash since the eruption of Vesuvius in A.D. 79, are lined with "snack bars," which, as we know from literary sources, once served hot food, drinks, bread, cheese, dates, nuts, figs and cakes over the counter at all hours. Wine was also served at these bars, usually mixed with water or sometimes, strangely enough, with seawater. In cold weather a hot mixture of wine and water was served, flavored with spices and often sweetened with honey. The typical Roman snack bar had a counter faced with marble fragments, opening onto the street. Jugs containing wine were set into cool stone to keep their contents fresh. Remains of some two hundred snack and drink bars survive at Pompeii, and in one street near the public baths no less than eight line a single block less than eighty yards long.

From the more legendary end of the classical time line, the apparent remains of a snack bar were uncovered by the excavators of the famous site of Troy, in northwestern Turkey. They come from an archaeological level dated to the thirteenth century B.C., the setting for Homer's epic poems about the Trojan War. Just inside the main southern gateway to the city, the excavators found a small building that opened onto the street through a wide doorway. Inside were three rooms containing a large hearth, a corn mill positioned over a flour-collecting basin, a stone kitchen sink with a drain leading to the street, ovens and storage bins. Remains of charred wheat were found in several places. There were no living quarters attached, and the building's position near the gateway led the archaeologists to suggest that this was a bakery or snack bar, "one of the earliest known." Homer often refers to the great feasts held by the heroes of the Trojan War on both sides, usually the roasting of an ox on spits accompanied by the quaffing of gallons of wine and great social palaver. Perhaps now we should imagine Hector stopping by the gateway to grab some pita bread on his way to do battle with Achilles.

A fast-food shop in Pompeii.

But Troy's claim to the earliest known snack bar is beaten by an example from the great Sumerian city of Ur, in Iraq. In the 1930s British archaeologist Sir Leonard Woolley excavated a number of its side streets, built during the eighteenth century B.C. One private house had been converted by cutting a large window through to the street, opening onto a brick counter about three feet above street level. In the same room was a large bread oven, on top of which was a solid-brick range that held charcoal braziers for grilling meat. The unique arrangement of this kitchen space, open to the street, led Woolley to conclude that it was a cookshop, selling food displayed on the counter to passersby in the street. He also noted, with some astonishment, that the style of the four-thousand-year old oven was "precisely like those used now in Arab houses and cookshops."

COOKBOOKS

The earliest known cookbooks come from Iraq, cradle of the Mesopotamian civilization that produced so many other cultural "firsts." Three recently deciphered tablets in the Yale Babylonian Collection, dating from around 1700 B.C., the time of the great lawgiver King Hammurabi, proved to be recipes. While the handwriting is elegant—the tablets clearly being intended for use in kitchens of the elite, possibly even the royal family—the dishes described sound rather unappetizing to the modern Western palate. They are largely meat stews, such as kid boiled with garlic, onion, fat, soured milk and blood. Perhaps the most acceptable to our tastes is a recipe for a side dish of braised parsnips, although one or two of the ingredients are obscure: "Meat is not needed. Boil water, throw fat in. (Add) onion, dorsal thorn [an unknown plant used for seasoning], coriander, cumin, and kanasu [a legume]. Squeeze leek and garlic and spread (juices) on dish. Add onion and mint."

Culinary literature of a more familiar kind begins with the Greeks and Romans. Athenaeus, a Greek resident of Naucratis in Egypt, is our best guide to the food of the classical world. Writing about A.D.

200, he produced a massive series on the theme of intelligent dinner-table conversations entitled the *Deipnosophistai* ("The Learned Banqueters"). Fifteen volumes, out of a possible thirty, still survive. Imagine a mountain of lifestyle columns from the days of the Roman Empire, all written by the same author, and you have Athenaeus. With food as a foil to conversation, the discourses range over fishing, hunting, farming, natural history, nutrition, aphrodisiacs, medicine, philosophy, literature and straightforward gossip. Some anecdotes and opinions, such as those put in the mouth of the gladiator-physician Galen (see **Introduction** to **Medicine**), were probably gleaned from real dinner guests; other material seems to have been culled from the work of earlier writers. Whatever the case, the result is a fascinating guide to good living—and above all good eating—in the Greco-Roman world.

Athenaeus occasionally refers his readers to existing cookbooks, such as the *Art of Cooking, Gastronomy, Sicilian Cooking, Pickles* and *Vegetables* by various Greek authors. Among these, Archestratus, of the fourth century B.C., was outstanding as an inventor of new dishes; according to Athenaeus, he "diligently traveled all lands and sea in his desire . . . of testing carefully the delights of the belly."

The Great Apicius

Sad to say, none of these ancient Greek recipe books has survived. Some may yet turn up, but in the meantime Greek food, the favorite cuisine of the Romans, has been preserved in the work of Apicius, the author of the world's first extensive cookbook, *On Cookery.* Written sometime in the first century A.D., its authorship is somewhat mysterious. There were actually three famous Roman gluttons called Apicius: one in the time of Julius Caesar (who died in 44 B.C.), another under the reigns of Augustus and Tiberius (27 B.C.–A.D. 37), and a third under the Emperor Trajan (A.D. 98–117). Despite the confusion, the second candidate, renowned for his wealth, decadence and teaching of *haute cuisine*, wins the contest for authorship of *De Re Coquinaria* hands down. Athenaeus gives us a wonderful character sketch of this Apicius—a gourmet whose fanaticism has probably never been matched:

There lived in the reign of Tiberius a man named Apicius, a voluptuary of extraordinary wealth, who gave his name to many kinds of cakes. Apicius lavished countless sums of money on his stomach in Minturnae, a city in Campania.

There he passed his time for the most part eating the very costly prawns of the region, which grow larger than the biggest prawns of Smyrna or the lobsters of Alexandria. He happened to hear that prawns also grew to an enormous size off the coasts of Libya and accordingly set sail from Italy without the delay of even a single day.

After suffering from storms during the voyage across the open sea he drew near land. The Libyan fishermen sailed towards him before he had set out from his ship, for the report of his arrival had already spread, and they brought him their best prawns. Seeing these he inquired as to whether they had any that were larger. But when they replied that none grew larger than those they had offered him, Apicius suddenly remembered the prawns of his own Minturnae and commanded his helmsman to sail back to Italy.

This is not to say that this Apicius was personally responsible for writing the recipe book that bears his name. The Victorian *Mrs. Beeton's Book of Household Management*, which enshrined the standards for modern British cookery from 1861 until the 1960s, was organized by her, but the recipes were prepared by her downstairs staff—the cooks. Likewise we can imagine that Apicius was too busy stuffing his face at dinner parties actually to have written the book himself. This would have been done by his chefs, recording the favorite recipes of their master for his friends, or, more to the point, their chefs.

The manuscripts we have of Apicius's book are in late-fourth-century Latin, and the text seems to have been freely edited over the three centuries since its writing. Some extra recipes were added, including one for spiced beans attributed to the Emperor Commodus (A.D. 180–192). Its vegetarian nature contrasts strangely with what we know of his other dinner-table habits—a psychopathic sadist, Commodus was said to enjoy his meals most when watching prisoners being torn limb from limb in front of him. In addition the surviving edition of Apicius may not be complete. Chapters on pastries and desserts, very popular in the Roman world, seem to be missing. Slightly different versions seem to have circulated in ancient times, as we can see from variations in *The Excerpts of Apicius* prepared by "the illustrious Vindarius" during the fifth century A.D.

The arrangement of *De Re Coquinaria* seems a bit haphazard compared with that of modern cookbooks. Chapter 1, "The Careful Cook," gives recipes for spiced wines, sauces and preserved foods and hints on livening up food that is past its prime. Then follow chapters on dishes of chopped meat (including sausages and dumplings) and vegetables (both cooked and raw). Chapter 4, as the title explains, covers "All Kinds of Dishes," containing curiosities such as nut omelets, creamed calf's brains, lettuce purée and steamed fish custard. Then come chapters on legumes and birds, and one entitled "The Gourmet." This gives some incredibly rich recipes for tripe, pork, truffles, snails, eggs, mushrooms, lungs, cakes and sweet custards. Chapter 8 is entitled "Quadrupeds" (more pork recipes, as well as venison, mutton, veal, lamb, kid, rabbit and hare); 9, "The Sea" (oysters, crabs, lobsters, mussels, sardines, tuna, skate, squid and octopus); and 10, "The Fisherman" (sauces for fish). The rich Romans certainly had a varied diet. Exotic spices, herbs and wine are employed lavishly throughout, and honey is used for sweet dishes in the absence of sugar.

Apicius's cookbook was clearly written by professionals for professionals, as we can see from the general absence of amounts given for the ingredients. This lack of detail has caused great debate. One school of thought has argued that its vagueness masks recipes almost as grim as the ancient Babylonian efforts. Indeed some of the delicacies Apicius recommends, such as roast pig's testicles, may not go down too well today. The other side believed that careful experimentation with Apicius could establish the right quantities, not only for ancient Roman tastes but for use in the modern kitchen. But the matter was not really settled until John Edwards, an English schoolteacher with the required grasp of Latin as well the necessary culinary skills, made a new translation of Apicius and took on the major task of replicating and testing the recipes. The result was published in 1984, and the recipes of Apicius once again grace the shelves of adventurous cooks around the world.

RECIPES FROM A ROMAN COOKBOOK

The best way to appreciate the flavor of Roman food is to try it . . . so here are three examples from the famous gourmet Apicius, written during the first century A.D. and adjusted to the style of modern recipes:

FISH FILLETS WITH TURNIPS IN SAUCE

Sauce

6 medium turnips	1 tablespoon olive oil or butter
2 pounds fish fillets	1 tablespoon flour
1 cup fish stock	¾ cup fish stock
Oil for frying	¼ cup white wine
White wine or cider vinegar	1 teaspoon honey
	½ teaspoon ground cumin
	A few crushed laurel berries or
	¼ teaspoon ground pepper
	Pinch powdered saffron

Peel the turnips and boil or steam until soft. Drain well. Simmer the fish fillets in fish stock until half cooked, drain and set aside. For the sauce, warm the oil in a pan over a low heat; add the flour, stirring well. Gradually add stock, then wine, honey, cumin, laurel and saffron, stirring constantly. Raise the heat and bring to the boil; lower the heat and simmer for 25 minutes, stirring occasionally. Meanwhile mash the turnips and coat the fish fillets with the paste. Deep-fry in oil. When cooked, place the fillets on a serving dish, pour the sauce over them and serve with a dash of vinegar.

The only problematic ingredient in this recipe is the fish stock, or *liquamen*, which is ubiquitous in Apicius's book. *Liquamen* was generally bought by the Romans ready-made from specialist manufacturers and occurs in Apicius's recipes in the same way that ketchup or curry powder might appear in ours. Because of this it is difficult to replicate its flavor. The nearest equivalents today are the fermented fish sauces of the Orient (Thai *nam pla*, Vietnamese *nuoc mam*, Philippino *patis* and Cambodian *tuk trey*). Although these taste more salty than fishy, they can be bought from specialist shops as a *liquamen* substitute. Otherwise, use a light soy sauce, the stock from steamed fish or a combination. Or, if you are feeling adventurous, try the recipe for quick homemade *liquamen* given by the ancient Greek writer Athenaeus: Dissolve enough salt in water until an egg will float in it; add fish scraps and oregano and boil; allow to cool, and strain several times until clear.

ASPARAGUS IN WINE
1   pound asparagus
1½ cups white wine
1   cup vegetable stock
2   teaspoons olive oil
2   teaspoons finely chopped onion
¼   teaspoon ground pepper
1   teaspoon lovage (or celery seeds)
1   teaspoon ground coriander
¼   teaspoon savory
1   raw egg yolk, beaten
Pepper and salt to taste

Finely chop the asparagus shoots and place in a mortar. Pound them into a paste; add the wine and steep for about half an hour. Strain through a colander, reserving the wine. In a pan, mix together ¼ cup of the wine, the stock, olive oil, onion, herbs and spices. Add the asparagus purée, bring to a boil, lower the heat and simmer for 15 minutes. Allow to cool slightly, add the beaten egg yolk and season with pepper and salt.

Apicius added that "You may make a dish of watercress or wild grape [leaves] or green mustard [leaves] or cucumber or cabbage leaves this way."

ROMAN CUSTARD
2 cups milk
¼ cup honey (or sugar)
3 egg yolks, well beaten
¼ teaspoon ground nutmeg or cinnamon

Preheat the oven to 325° F. Mix together the milk and honey and heat in a saucepan. Remove before it boils and stir in the egg yolks. Add the nutmeg, stirring well. Pour into a baking dish and bake for one hour, or until set. Sprinkle with more nutmeg before serving.

REFRIGERATION

The ancient Romans were as fond of putting ice in their drinks as we are today. In the time of the early empire (1st century A.D.), no banquet would have been complete without the provision of lavish quantities of ice or snow for guests to sprinkle into their wine goblets. The passion, almost addiction, that fashionable Romans had for iced drinks and chilled food was frequently denounced by philosophers, who saw it as a clear sign of ever-growing decadence. Seneca, tutor to the young emperor Nero (A.D. 54–68), railed against such fads: "You see skinny youths wrapped in cloaks and mufflers, pale and sickly, not only sipping the snow but actually eating it and tossing bits into their glasses lest they become warm merely through the time taken in drinking!"

Roman doctors agreed with Seneca, believing that the consumption of snow in drinks was dangerously enfeebling and the cause of serious internal disorders. Accordingly Nero decided to chill his drink by plunging the wine bottle into a vessel full of snow—the same principle as a champagne bucket. He also observed, quite correctly, that previously boiled water chills more rapidly than fresh water—the gases in unboiled water retard cooling. But Nero's claim (like so many of his other pretensions) that he was the first to invent a wine cooler is overblown. The Athenians were using pottery vessels specifically made for cooling wine as early as the sixth century B.C.

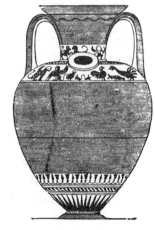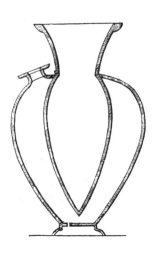

Greek wine cooler of the 6th century B.C. found at Vulci, in central Italy. From the outside it looked like an ordinary jar (left). Inside, however, was a central container for the wine, surrounded by another chamber that would be filled with ice (right).

Icehouses and Snow Shops

For all their love of chilled drinks, food, and even baths, there is no evidence that the Greeks and Romans discovered any method of manufacturing ice from scratch. Where, then, did they obtain the vast quantities of ice and snow that were consumed each day? The simple answer is that it was imported. Donkey trains carried it into the cities, often over considerable distances, from the mountains. The scale of this traffic in snow can be judged from the extravagance of the Emperor Elagabalus (A.D. 218–222), who had a mountain of snow erected in the garden next to his villa during the hot summer months—an early, though rather profligate, attempt at air-conditioning. Centuries later the Mameluke sultans used their efficient postal system to transport snow from the mountains of Lebanon to Cairo, in Egypt (see **Postal Systems** in **Communications**).

Once delivered to the cities, the snow was either sold directly in "snow shops" or stored in underground icehouses until it was needed. In Roman times these were deep pits, filled with snow and covered with straw, an effective insulator. The top layer of snow might melt, but it would freeze again as it seeped down; in fact the pressure formed by the weight of the snow column would turn the lowest level of snow into hard ice. In this way Roman snow salesmen manufactured ice, which they could sell at a much higher price than snow. Poor old Seneca noted that rich Romans would pay through the nose for ice from the bottom of the pits. "Even water has a varying price," he grumbled, seeing it as yet another nail in the coffin of Roman morality. Indeed in Rome ice and snow often cost more than wine. Pliny the Younger, nephew of the famous encyclopedist Pliny, reproached his friend Septimius Clarus for failing to show up for a lavish dinner arranged for him. He specifically complained about the waste of "sweet wine and snow; the snow I'll most certainly bill you for and at a high rate—it was ruined in the serving."

The art of constructing underground icehouses was probably borrowed by the Greco-Roman world from the East, as the first Greek to build them seems to have been Alexander the Great of Macedon (336–323 B.C.). The classical food writer Athenaeus (see **Cookbooks**) notes that during his northern Indian campaign Alexander "dug thirty refrigerating pits which he filled with snow and covered with oak boughs. In this way . . . snow will last a long time."

In the Near East the tradition of building icehouses dates at least as far back as 1700 B.C., when Zimri-Lin, ruler of Mari (a kingdom in northwest Iraq), boasted in one of his inscriptions of having constructed near his capital a *bit shuripim* (icehouse), "which never before has any king built on a bank of the Euphrates." The text also mentions how ice was imported to fill the structure, which was probably an underground pit like the ones built by the Greeks and Romans.

The art of refrigeration also has a long history in the Far East. Under the Tang Dynasty of China (A.D. 618–906), the literary work known as the *Shih King* ("Food Canons") speaks of the rituals surrounding the maintenance of icehouses as a time-honored custom. This involved the annual cleaning of the underground pits in preparation for the annual ice harvest, which was stored until the summer months for use in preserving fruit and vegetables. These practices were evidently long established: The court of the later Chou emperors (4th–3rd centuries B.C.) had an "ice service" with no less than ninety-four staff, whose job was to chill everything from the royal wine to the body of the dead emperor.

Archaeological confirmation comes from the Ch'in Dynasty (221–207 B.C.): The capital built at Hsienyang by the First Emperor, Shih Huang Ti, included a luxury version of the simple ice pit. Made of huge ceramic rings, it was sunk forty-three feet into the ground.

The oldest remains of an icehouse in China—indeed in the world—were discovered in 1976–77 during excavations of the palace at Yongcheng, an ancient capital in Shensi Province. Built in the seventh century B.C., it has a design different from that made for the First Emperor. Ice was stored in a shallow pit nearly square in shape and removed by means of sluice gates made from wooden boards with a layer of rice husks between them for insulation. The gates were reached by a ramp, which went down into the middle of the pit, while a drainage channel built into the ramp drew off water from thawing ice and carried it away to a nearby river. Literary evidence suggests that even earlier icehouses may yet be found in China: there is a possible reference to an icehouse from 1100 B.C.

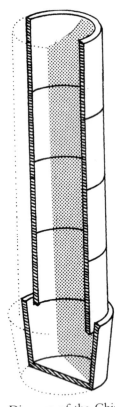

Diagram of the Chinese refrigerator, or ice pit, at Hsienyang built for the First Emperor, Shih Huang Ti (221–207 B.C.). It was made from giant terra-cotta rings about three feet high and five feet eight inches in diameter, and went down forty-three feet below ground level.

Chilling Without Snow

Methods for chilling water (and even making ice) without the aid of imported or stored snow were also developed in the ancient world—naturally enough in two of the hottest countries, Egypt and India. These used simple physical methods: radiation and evaporation. Athenaeus describes how water was being chilled in Egypt in the second century A.D. by placing it on rooftops at night:

> During the day they place the [river] water in the sun, and when the night comes they strain the thick sediment and expose the water to the air in earthen jars set on the highest part of the house, while throughout the entire night two slaves wet down the jars with water. At dawn they take the jars downstairs. . . . They then place the jars in heaps of straw, and thereafter use it without the need of snow or anything else whatever.

In this method the water was "sky cooled," its heat radiating into the atmosphere. All that was needed was a clear night sky, which would allow heat to radiate off and reduce ground temperatures to below freezing point.

Pouring water over the jars throughout the night would encourage heat loss through evaporation, the other technique with which the Egyptians were familiar. Egyptian tomb paintings show slaves fanning large storage jars. The jars were specially made from porous clay, so that a small amount of the liquid inside would seep through and evaporate from the surface. Heat would escape with the evaporating liquid, lowering the temperature of the remainder, just as sweating reduces the temperature of the human body. The same method is used today by peasants throughout the East, and modern experiments show that the technique can cool water to as much as 77°F below the ambient room temperature.

A more effective method of evaporation has been used in India since time immemorial. European travelers of the nineteenth century report how shallow pottery vessels containing water were lined up in rows, covered with straw or sugarcane stalks and exposed to drafts. The evaporation chilled the water and was even known to produce small quantities of ice. Adding a little salt to the water gave it a higher freezing point—a fact noted in an Indian text on the fourth century B.C. An evaporation method similar to the Indian one was being used in Estonia around A.D. 800, as we know from literary evidence.

While these ancient Egyptian and Indian methods may seem rather primitive to us, they are worth remembering. One fine day, when the world reaches its impending energy crisis, they might come in handy for chilling our last remaining stocks of Coca-Cola and champagne.

CHEWING GUM

Chewing gum is today familiar the world over, but for most of its history its use was confined to Central America. When the *conquistadores* invaded the Aztec Empire in 1518, they encountered gum-chewing prostitutes hanging about on street corners looking for business. These women put yellow cream on their faces, colored their teeth red with cochineal, doused themselves liberally with perfume and walked the streets chewing gum.

Chewing gum had been discovered several hundred years earlier by the Maya of southern Mexico. They found that *chicle*, a thick, milky liquid that oozes out of cuts made in the wild sapodilla tree and then hardens into gum, was extremely tasty. The importance of *chicle* to the Maya is clear from their mythology: The culture hero Kukulkan ("the Feathered Serpent"), who conquered the Maya and changed their way of life to such an extent that he became worshiped as a god, was a great chewer of gum.

It is truly ironic that Hernán Cortés, who led the Spanish invasion of Mexico, was mistaken for Quetzalcoatl, the Aztec equivalent of Kukulkan. The god, so it was prophesied, would that year return

Quetzalcoatl, the Aztec version of Kukulcan, the gum-chewing hero of the Maya.

in triumph, but instead Cortés destroyed Aztec society. Among the things that disappeared under the subsequent Spanish rule were the extensive trade routes that brought *chicle* from the forests to the capital. Chewing gum was preserved by the forest dwellers of Mexico until it was discovered in America by Thomas Adams, Jr., around 1870, and independently a few years later by William Wrigley, Jr. A further strange twist in the story is that the growing demand for chewing gum at the beginning of this century was met by the surviving Mayan Indians hunting out more sapodilla trees. In the course of their searches they discovered the ruins of most of the great Mayan cities of the past.

TEA, COFFEE AND CHOCOLATE

Between them these three great drinks spanned a vast part of the ancient world—tea in East Asia, coffee in the Middle East and chocolate in Central America. Of the three the history of tea is by far the best known, with many Chinese works devoted to its study from the Middle Ages onward.

Tea drinking can be traced back to around 50 B.C. in southern China; over the following centuries the custom gradually spread across the whole country. At first tea was used as a medicine—a stimulant (to ward off sleep) or an antidote to the bad effects of overindulgence in rich food or drink—but over the years it came to be appreciated in its own right. The "art of tea" came into being during the T'ang Dynasty (A.D. 618–906). Lu Yü wrote the world's first comprehensive book on tea—the *Tea Classic*, as it became known—in A.D. 780. According to Lu Yü,

> The effect of tea is cooling. As a drink it suits very well people of self-restraint and good conduct. When feeling hot, thirsty, depressed, suffering from headache, fatigue of the four limbs, or pains in the joints, one should drink tea only, four or five times [a day].

The Chinese government started to levy taxes on tea in A.D. 793. At this time the best-quality tea came from the countryside around Shanghai, with the area becoming dominated by tea estates to the point where thirty-five thousand people were employed in picking the leaves or drying them. Over time the importance of tea grew even greater and, during the Sung Dynasty (A.D. 960–1280), senior officials held tea-making contests for which only the highest-quality tea could be used: it cost the princely sum of 1½ ounces of gold per pound.

In Mexico chocolate, first drunk around A.D. 100, was so highly valued by the Maya that it became a form of currency. For once, money really did grow on trees! Unscrupulous merchants attempted to defraud the public by stripping off the husks of cacao beans, carefully filling them with sand and mixing these fakes in with the genuine article. Accordingly, wary customers learned to squeeze each bean before paying up. The cacao beans were roasted, ground and mixed with maize flour and occasionally dried flowers and a little sugar. Small cakes of this paste were shaken up with water in a gourd until a frothy drink was created, which was then drunk in one gulp. Rich people kept the drink in special screw-top pots until it was wanted.

The Aztecs demanded cacao beans as tribute from the coastal regions of their empire. The Aztec rich drank *xocoatl*, a mixture of chocolate, chili peppers, corn flour and water. Their Spanish conquerors thought *xocoatl* was disgusting, and made the sweet chocolate drink we know today by substituting sugar and cinnamon for the bitter chilis.

CACAO GLYPH

A Mayan screw-top jar of the 5th century A.D. found in a nobleman's tomb. The glyph for cacao was identified by David Stuart in 1986. The dried residue inside was analyzed, and shown to be chocolate, by the Hershey Foods laboratory.

Coffee is a latecomer compared with tea or chocolate: it is first mentioned in the tenth century A.D., by the Arabian doctor Rhazes. The earliest coffee was grown in Ethiopia, spreading from there to Yemen in southern Arabia, where the custom of roasting beans began around 1200 A.D. The drink became popular with the whirling dervishes, who drank it to fuel their lengthy dances. In the fifteenth century they introduced it to Mecca. Soon after this, coffeehouses developed there, and grateful Muslim pilgrims spread the habit to every corner of the Islamic world, including Iran, Spain, Egypt and Turkey. Only in 1643 was the first coffeehouse opened in Paris, to be followed by ones in Oxford in 1650 and London in 1652. The first of the famous Viennese coffeehouses was opened in 1683, using sacks of coffee left behind by the Ottoman Turks after their unsuccessful siege of Vienna in that year.

WINE, BEER AND BREWING

Connoisseurs prefer wine that has been laid down for a long time, but even the finest French vintage looks like Beaujolais nouveau compared with the two bottles of wine excavated from a tomb at Xinyang in the Hunan Province of China in 1980. The other finds in the tomb show that the wine was bottled around 1300 B.C. under the Shang Dynasty, making it the oldest known anywhere in the world. Fanatical vintage buffs will be pleased to learn that even older bottles may yet turn up. The latest archaeological research suggests that the art of wine making may be as much as ten thousand years old.

The Near East, together with the Caucasus mountain region south of Russia, is definitely the cradle of wine making. Grapes were being eaten here as long ago as 8000 B.C., as we know from the seeds found at sites in Turkey, Syria, Jordan and Lebanon. These were wild grapes, as shown by the round shape of the seeds. But the more elongated seeds found at sites in Georgia (the Caucasus) dating to around 5000 B.C. indicate that they were from cultivated grapes. They were also found with what appear, by their shape, to be wine-storage vessels. There is little doubt that the Georgians were seriously into the grape by 3000 B.C.—burials were accompanied by vine cuttings wrapped in silver packets.

Current theory is tending to push back the origins of wine making even farther. Archaeological discoveries have shown that grapes were being eaten and, more than likely, pressed for their juice, by 8000 B.C., and there is no reason why the accidental invention of wine could not have happened at an equally early date. Grape skins contain enough yeast and sugar to begin fermentation without any additives—in the right circumstances a container of crushed grapes left for a few days will turn into wine. The pleasant effects of wine would have provided a spur to learning how to monitor and encourage the growth of wild vines. It may even have been one of the motives behind the agricultural revolution of the New Stone Age (see **Introduction** to **Working the Land**) and may have led to the domestication of other crops, such as barley and wheat, which could produce beer. This idea, surprisingly enough, has some scriptural

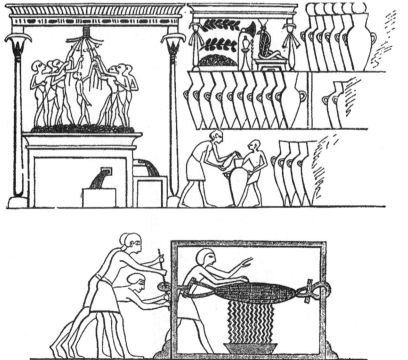

The two methods of ancient Egyptian grape pressing. (Above) Grapes are being trodden in a large container. The juice flows through pipes into buckets and is transferred into jars. (Below) In another technique the grapes were put into a strong bag looped onto a frame, and the juice was wrung out by twisting. This more thorough method may also have been used to give already trodden grapes a second pressing.

support. The book of *Genesis* ascribes a venerable antiquity to viticulture, relating how Noah, shortly after the Flood, planted the first vines and made the first wine—the earliest reference in the Bible to the cultivation of a specific crop.

The Egyptians developed the earliest major wine industry. Around 3000 B.C. wine jars began to be manufactured in large numbers, and hieroglyphic marks on stoppers and seals indicate that royal vineyards had already been established. Later tomb paintings show in great detail every stage of wine production, from the training of vines along trellises to the pressing of the grapes, blending and bottling. The jars were sealed with perforated lids (fermentation locks), which allowed carbon dioxide to escape during secondary fermentation. Still, despite the Egyptians' best efforts, the soil and climate of their country are not really suited to viticulture. Roman wine snobs thought Egyptian wine was foul; the satirical poet Martial said he preferred drinking vinegar.

The Romans favored Greek wine, especially that produced on the islands, such as Rhodes, and it was the Greeks who really perfected the art of wine making. They were, naturally, great wine drinkers, and the storage, serving, mixing and consumption of wine stimulated the finest products of the ancient Greek potter's art. The Greeks also had a popular, but rather silly and extremely messy, game to go with wine drinking, called *kottabos*. According to tradition, this was invented by a light-headed Greek colonist settled in Sicily. Leaning on his elbow during the course of some heavy after-dinner drinking, he bet his friends that he could hit the lamp on the top of its stand with the dregs in the bottom of his cup. This new entertainment for the bored was cashed in on by a crafty merchant, who invented special *kottabos* stands. Precariously balanced on the top of each stand was a bronze disk. The idea of the game was to knock this off with the wine dregs so that it fell onto a lower metal disk, which then rang like a bell.

The phenomenal wine consumption of the Celtic Gauls of France amazed even the hardened drinkers of Greece and Rome. A measure of their capacity can be seen from the famous Vix Crater, discovered in a Gallic princess's tomb of the sixth century B.C. Of Greek manufacture, it was imported into central France from Marseilles. It is a wine-mixing bowl of the finest bronze, made to hold

The party game of *kottabos,* as depicted on an ancient Greek vase. The man on the left hurls the dregs from his flat drinking cup at the target disk on top of the stand. When a player succeeded in knocking it off, it fell onto the plate below, which rings like a bell.

260 gallons of wine, equivalent to 45 amphoras, the standard large earthenware wine containers used by the Greeks. The Gauls became notorious for their addiction to wine. Around 40 B.C. the Greek historian Diodorus of Sicily described how enterprising Italian merchants "transport the wine by boat on the navigable rivers, and by wagon through the plains, and receive in return for it an incredibly high price: for a single amphora of wine they receive a slave—a servant in exchange for a drink."

The Greeks always diluted their wine with water, and it was the foreign habit of drinking it neat that really shocked them. The Greek saying "drinking the Scythian way" meant getting blind drunk and was said of this ancient Russian people because they drank their wine full strength. The Romans followed the Greek custom and carefully mixed their wine with water before drinking. Sometimes chilled water or sieved snow was used (see **Refrigeration**) or, according to taste, hot water. They even invented a prototype of the modern samovar to produce hot water for mixing: a central cylinder was filled with burning charcoal, and this heated the water in the surrounding vessel. These were certainly luxury items, and the notoriously profligate boy emperor Elagabalus (A.D. 218–222) was said to have been the first to have owned silver water heaters. Some far cheaper way of producing warm drinks was undoubtedly used by the "fast-food joints" of Pompeii and other cities (see **Restaurants and Snack Bars**).

Roman wine lovers became connoisseurs, and through time there was increasing demand for vintage wines, which they considered to be at their best after fifteen or so years' storage. However, despite the tsk-tsking of the Romans at the greedy drinking habits of the Gauls, they took full advantage by exporting colossal amounts of wine to France. Archaeologists have calculated that around A.D. 100 the wine-growing estate at Settefinestre, in central Italy, covered some three hundred acres and produced more than 250,000 gallons of wine (equivalent to 4,260 amphoras) per year, most of which was shipped to Gaul. By comparison, the great modern château of Mersault, in Burgundy, farms less than one hundred acres and has storage space in its cellars for some 82,000 gallons of bottled wine.

In addition to exploiting the excessive drinking habits of the Gauls, the Roman rich seem to have done considerable harm to themselves through their own wine consumption. To improve the flavor of wines, they added concentrated wine syrups. These syrups, boiled down in lead containers, were also used extensively in cooking. The toxicity of the syrups resulted in a high degree of gout, lead poisoning and kidney failure among the wealthy.

The Roots of Beer

Beer must have been invented in much the same way as wine—when someone tasted an accidentally fermented mixture, such as a bowl of barley porridge. The invention of bread must also have stimulated the production of beer. The two were made from the same cereals and, as we know from actual finds, a particular kind of bread was popular during the New Stone Age that proved to be an invaluable aid to brewing. The bread was made by allowing the grains to germinate before milling them—germination enables enzymes to convert some of the starch into sugars such as maltose, use-

Men drinking beer through long tubes as shown on a seal from Bahrain, in the Arabian Gulf, of about 2000 B.C. Drinking tubes like this were a Sumerian invention; one end was perforated with tiny holes to filter the beer.

ful in brewing. The same mixture could be used to make beer, by partially baking the bread, then mashing slices of it in water. With the high temperatures of Egypt and Iraq, only a couple of days' fermentation were needed; the bread was then strained out of the mixture, and, voilà, the beer was ready for drinking.

Both bread and beer were already being made and consumed on a large scale by the early urban civilizations of Egypt and Iraq (around 3000 B.C.). The first great beer drinkers were the Sumerians of ancient Iraq. They made wine, from dates as well as grapes, but beer was always their favorite tipple. One Greek legend says that Dionysus, the Greek god of wine, ran from Mesopotamia in disgust because the inhabitants were so addicted to beer. Indeed it was so popular that the Sumerians, founders of Mesopotamian civilization, used up as much as 40 percent of their grain harvest for beer making. Ordinary temple workmen received a ration of 1¾ pints a day, while those higher up were entitled to five times as much, most of which they probably sold to the workmen. Much thicker than modern clarified beers, the beer drunk by the ancient Mesopotamians (and Egyptians) was an important source of nourishment as well as intoxication; rich in vitamin B_{12}, it was an important source of this essential nutrient in a low-meat diet.

Most of the earliest Mesopotamian brewers were women, who made their beer at home and sold it on the premises. The famous law code drawn up by the Babylonian King Hammurabi about 1750 B.C. tried to regulate these informal "bars." There had clearly been complaints, and Hammurabi responded by introducing the world's first legislation to control beer prices. Women who overcharged for beer were to be thrown into the river as a punishment. If they allowed outlaws to drink at their establishments and failed to report them to the authorities, they would be put to death.

Alehouses were clearly hangouts for troublemakers even at this early date. An Egyptian papyrus of 1400 B.C. warned drinkers about the dangers of loose talk in bars: "Do not get drunk in the taverns in which they drink beer, for fear that people repeat words which may have gone out of your mouth, without you being aware of having said them."

The Egyptians made numerous varieties of beer, probably flavored with different herbs. These had colorful names, such as "The Joy Bringer," "The Beautiful" and "Heavenly." In Greco-Roman Egypt doctors recommended tonic beers made with rue, safflower and mandrake.

Sometimes Egyptian drinkers slightly overdid it at parties.

Hangovers from the Ancient World

The quality of ancient Egyptian beer is something we may all be able to judge for ourselves in a year or so. The work is being carried out by a team of archaeologists from Britain (naturally), currently excavating an Egyptian brewery at El-Amarna, the capital built in the mid-fourteenth century by Tutankhamen's father Ikhnaton. The brewery, part of a temple complex built by Ikhnaton's queen, Nefertiti, once manufactured beer and bread for religious festivals. The hope of finding vessels containing enough beer residue for analysis was dashed when it was realized that the ground on which the brewery stood had later been heavily irrigated, wiping clean much of the evidence. Fortunately conditions in other parts of the site—and elsewhere in Egypt—have been more favorable. Delwen Samuel, the team's archaeobotanist, has been able to examine a number of pots containing traces of beer and bread, with the aid of a scanning electron microscope. In some cases the residues are so well preserved that she can study individual starch granules and yeasts, enabling a fairly clear picture to be built up of how ancient Egyptian beer was made. In conjunction with Scottish and Newcastle Breweries, she is now completing an experimental replication of the kind of beer drunk in the age of Tutankhamen.

The Egyptian hieroglyph for "brewer." It is a simplified picture of a man treading the mash for beer making in a giant pot. The mixture was then left to ferment.

The beer of Egypt attracted less scathing comments in the classical world than its wine. Still the Greeks and Romans on the whole clearly disliked beer. Julian, the last pagan emperor of Rome (A.D. 360–363) was persuaded to try some Celtic beer when in France, and wrote it off in a satirical poem, harping on the flatulence brought about by beer drinking. To the wine-drinking classical world, beer was mainly a drink for barbarians. The ancient world, then, must have been full of barbarians, since the manufacture of beer, in some form or other, was almost universal. The Celts of ancient Britain, France, Spain and Germany were skilled brewers, and their traditions continued despite the snobbish tastes of their Roman masters.

To the medieval Germans we owe the invention of today's typical beer, that made by fermenting hops together with the grain. Hops give beer its characteristic bitter taste—ancient Egyptian beer by contrast was quite sweet—and also act as a natural preservative. Tradition has it that hopped beer was being made in Bavaria as early as A.D. 859, but the first certain reference comes from the writings of the early eleventh-century botanist Hildegard of Bingen. Around A.D. 1031 the monastery of Saint Emmeram had special hop gardens near Regensburg and Straubing for its beer production, beginning a long tradition of monastic beer brewing. During the Middle Ages the monasteries took over brewing, turning it into a proper industry. Their methods, involving strict cleanliness and mass production, laid the foundations for all modern brewing techniques and standards.

Another popular drink that we owe to the ancients is sake. Westerners are now quite familiar with Japanese sake, but few realize that it is an example of a type of alcoholic beverage only ever made in China and Japan—a strong beer, made from partly cooked wheat that has gone moldy; the molds convert the starch in the grain into sugar, which is then fermented by adding yeast, then more grain and so on. By 1000 B.C. the Chinese were making beer with an alcohol content of about 15 percent, as compared with 5 percent for normal beer. Sake is merely the modern version of this type of ancient beer.

SPIRITS

It is a great pity that there is no evidence to back up the tradition that whiskey was invented by Saint Patrick, the patron saint of Ireland, around A.D. 450. How far back the manufacture of whiskey (from the Gaelic *uisgebeatha*—"water of life") goes in the Emerald Isle is a mystery, but it is unlikely to have been before the Middle Ages, when distilling as such began in Europe. The first European alcohol seems to have been distilled by the doctors of Salerno, in Sicily, in the twelfth century A.D., for essentially medical purposes. The Sicilians learned the art of distilling from the Arabs, who had been using the method since the ninth century A.D. to create perfumes including rosewater (see Perfume in Personal Effects) and cosmetics such as the eyeliner kohl. In fact the word *alcohol* comes from the Arabic *al-kohl*—rather ironically, as the Arabs never used the method to make intoxicating drinks.

Until very recently it was believed that all spirit distilling grew from its discovery in Sicily. It is now known, however, that primitive distilling, done without the aid of a still, had begun in Central Asia as much as a thousand years earlier. Here, on the northern borders of China, nomadic peoples discovered that a stronger drink could be made from wine when it was left to freeze. A small amount of liquid would not freeze—this was the alcohol. Around A.D. 290 the Chinese writer Chang Hua noted the effects of this crude drink in his *Records of the Investigation of Things*: "The western regions have a wine made from grapes which will keep good for as much as ten years, it is commonly said; and if one drinks of it, one

Mead, a kind of beer made by fermenting honey and water, was popular in the Middle Ages and has now made something of a comeback in certain quarters—a few brands are manufactured commercially in Britain. Some excitement was caused a few years ago when the possibility was raised of resurrecting the mead once drunk in the Scottish Highlands—tradition swathes the drink, made from heather honey, with an almost mystical aura. The recipe was supposed to be such a closely guarded secret that people would defend it with their lives. According to folklore, the last person to know it was a priest of the Picts, one of the ancient tribes of Scotland. The story goes that this priest promised to reveal the recipe to a Scottish king, but only in a place where no one could overhear them. He led

---

will not get over one's drunkenness for days."

The Chinese themselves started to distill brandy by heating wine during the sixth century A.D. As the alcohol part of wine has the lowest boiling point, it can be evaporated out by heating and then condensed by cooling. Hence the Chinese called their brandy "burned wine", just as we do—our word comes from the Dutch *brandewijn*, which has exactly the same meaning. (In fact the art of distilling brandy may have been brought to Europe from China by Dutch sailors.) By the seventh century the Chinese were also distilling whiskey from a fermented mash of cereals.

The traditional underestimation of the Chinese distilling industry is partly due to the fact that little was recorded of it at the time. This was because the brewing and fermenting industries had been nationalized by the Emperor Wang Mang (A.D. 9–23), who formed the "Fermented Beverages Authority," which jealously guarded its techniques. The government monopoly was clearly flouted, as the death penalty was introduced in the fifth century A.D. for private brewers, the world's first known bootleggers. With the invention of distilling, moonshiners created a variety of cover names for their products, such as "the Worthy" or "the Sage." In eleventh-century China, if you had dropped in on your local bootlegger, he might well have offered you a shot of "Wisdom Soup."

Distilling also seems to have been discovered independently in the New World. When the Spanish arrived in Mexico, they found the Aztecs distilling tequila from fermented cactus juice in pottery stills. Charles V of Spain tried to ban the manufacture of tequila in 1529, but, as the world is glad to know, he was unsuccessful.

the king to a clifftop, drew him close with his arm to whisper to him and then jumped off the cliff, taking the king with him.

Would the lost recipe ever be found again? This was the hope of the archaeologists who in 1984 unearthed fragments of a 4,500-year-old pot containing a sticky black substance (evidently the residue from some drink), on the aptly named island of Rhum off western Scotland. Laboratory analysis showed that it had been fermented from a mixture of heather honey, grain (oats and barley) and herbs (royal fern and meadowsweet). A project was then set up with the chemists from William Grant and Sons, a Scottish whiskey distiller, to reproduce the Stone Age drink that left the residue. There were even some hopes of commercially manufacturing the drink, a kind of mead or honey-based ale that must have been related to the legendary brew. But when it came to the grand testing day, with the press and members of the public being served shots from specially made "Neolithic" pots, it was found to have something of a curious aftertaste—"different" was perhaps the best way to describe it. So there may be other explanations why Scottish heather mead died out than the traditional one!

The strangest use for mead, or indeed any alcoholic drink, was invented in Central America by the chiefs of the Classic Mayan Period (A.D. 250–900). They made mead by fermenting honey in a pot along with a toad of the *Bufo marinus* species, which produces intoxicating chemicals in its skin. They then used it as a bizarre enema, administered via a leather or rubber syringe bag fitted with a bone tube during religious ceremonies. Although it sounds like enough to sober up even the hardest drinker, the Maya seemed to love it.

DRUGS

The idea of getting "high" by taking drugs seems to be as old as humanity itself. Recent attempts to explain the remarkable cave paintings and engravings of Stone Age western Europe, going back some thirty thousand years, have argued that they were produced by tribal shamans, or medicine men, during drug-induced trances. Magnificent paintings were still being produced by these methods in Namibia during the nineteenth century A.D. Medicine men of Bushmen tribes went into trances during which they identified themselves with animals in the world around them.

Shamanistic religion survives today in Siberia, where *Amanita muscaria* (fly agaric), a red mushroom with white spots, has long been eaten as part of rituals. Though fatal if taken in large doses, carefully nibbled amounts can produce brilliantly colored hallucinations. The effects are heightened by the drum music accompanying the ceremony. It was also used in ancient times on the other side of the world. A grindstone from Costa Rica of the second century A.D. has carved on it a scene in which a central human figure is flanked by two fly agaric mushrooms.

The Americas are blessed, or cursed, depending on one's point of view, with an ambundance of plants that produce powerful hallucinogenic chemicals. Taking drugs to induce visions became a central part of the ancient American religious life that dominated their worldview and everyday behavior. It is not hard to see in the intricate and fantastic artwork of the Maya, Aztecs, Incas and other pre-Columbian peoples strong elements of the "psychedelic."

Mayan mushroom stone from highland Guatemala, carved between 100 B.C. and A.D. 300.

The Peyote cactus, or mescal, is well known from later cults (and cult novels) in the southwest United States. Its use goes back to at least A.D. 200–300, when it is shown on pottery bowls from Mexico. The hallucinogenic San Pedro cactus contains, when dried, 2 percent mescaline, the highest concentration of this hallucinogen in any plant. This was taken even earlier, as we can see from a stone carving of around 800 B.C. found at the early town of Chavin de Huantar, in the northern Peruvian Andes. The San Pedro cactus is frequently depicted on later textiles from the area, while Moche Culture (200 B.C.–A.D. 600) pots show healers holding slices of it.

The mushroom called *teonanacatl* ("Flesh of the Gods"), which contains the hallucinogen psilocybin, is found on wall paintings from central Mexico of A.D. 300. In the coastal lowlands the Maya made hundreds of "mushroom stones," from 500 B.C. onward. Crowned by umbrella-shaped tops, the stalks are carved with animal or human figures, often shown grinding up mushrooms. The resulting pulp was mixed with water and drunk. Alternatively *teonanacatl* mushrooms were sometimes smeared with honey and served at feasts and banquets, giving the guests fantastic visions. Father Bernardino de Sahagún, an early Spanish chronicler, records such a banquet:

> The first thing that was eaten at this feast was a little black fungus which makes men drunk and gives them visions. . . . They ate it before sunrise . . . with honey; and when they began to grow warm they started to dance. Some sang, some wept, so drunk were they by reason of these funguses; and others did not sing, but sat quiet in the room, thinking. . . . When the drunkenness caused by these funguses had died away, they talked to one another about the visions that they had had.

Evidence for the use of coca, a stimulant rather than a hallucinogenic, goes back much farther. When the leaves of the coca bush are chewed in a wad with lime (i.e., caustic earth), they release cocaine. Partially chewed quids of coca found in the rubbish heaps of ancient Peruvian settlements have been dated to about 2500 B.C., and the faces of carved idols from Colombia with the extended cheeks of coca chewers to around 1500 B.C. Moche Culture (200 B.C.–A.D. 600) pottery from Peru depicts rituals conducted by coca-chewing priests and the equipment involved—a bag to hold the coca leaves and a gourd and stick for applying the lime. Some of the finest goldwork from the pre-Columbian Quimbaya Culture of Colombia consists of containers for the lime chewed with coca leaves. Under the Incas of Peru the use of coca was a privilege of the elite— military officers, scholars, doctors, priests and royal messengers—

A nighttime ritual of coca chewing depicted on a vase of the Moche culture of ancient Peru (200 B.C.–A.D. 600). The three men at the right are dipping into gourds for the lime that they chew along with the coca leaves. The figure on the left is a supernatural being.

who carried their supplies around with them in small bags. The Incas used criminals to cultivate the coca bush, because they found the lowland areas where it grew bad for their health.

Modern forensic medicine has now confirmed the archaeological evidence. Using a test developed for employers in the United States to detect drug abuse in their workers, scientists analyzed locks of hair from 170 Peruvian and Chilean mummies. They found traces of cocaine in about one-third of them, including bodies two thousand years old. Coca leaves must have been chewed by a large proportion of the population in pre-Inca days, to judge from the high number of positive results and the fact that even very young children were cocaine users.

Cannabis

Cannabis (hemp) is now the most widely grown drug in the Americas. Its early use is unclear, although some of the ancient pipes usually thought to be for smoking tobacco (see **Tobacco and Pipes**) may have been used for hemp as well. Accounts of the visions seen by nineteenth-century chiefs such as Sitting Bull and Black Elk suggest cannabis smoking, but we cannot be sure. While native to the Americas in its wild state, cannabis was cultivated there only after Columbus. The most famous of the early cannabis farmers was one George Washington, whose diary entry for May 12–13, 1765, reads, "Sowed hemp at Muddy Hole by swamp."

As a drug cannabis is most at home in Asia, where it has been used from ancient times throughout the continent. Several different ways of ingesting it were developed. The most famous source of cannabis is India, where its medicinal properties were discovered by European doctors, who called it Indian hemp. The traditional Indian method of consuming it was to drink a tea made from powdered cannabis, milk, sugar and spices: it is still made today and called *bhang*. The effects are well known—the taker can be struck by hilarity, sleepiness, amorousness or a raging appetite. The earliest possible references to cannabis intoxication come from the Vedic hymns, religious songs that were probably composed early in the first millennium B.C. The hymns praise the powers of *soma*, a wonderful plant eaten by the gods to make themselves ecstatic and strong. Whether it was cannabis has been hotly debated. The hymns actually say that there were two kinds of *soma*, a terrestrial one (perhaps cannabis) and a version taken by the gods.

The Greek historian Herodotus—whose work is a treasure trove of information on the customs of neighboring peoples during the fifth century B.C.—describes how the Scythians of southern Russia, who had invaded the Near East two hundred years earlier, devised special tents as both vapor baths and cannabis smoking dens (see **Saunas** in **House and Home**). These tents were made of thick felt, with all cracks carefully sealed up. Inside was placed a bowl full of red-hot stones, onto which cannabis seeds were thrown. According to Herodotus, the Scythians would howl with delight as they breathed in the fumes. Sitting in these tents was clearly one of their favorite pastimes. The reference to seeds in Herodotus and other sources is puzzling, since as any cannabis smoker knows, the seeds are by far the least intoxicating part of the plant. But as the flowering heads, the most potent element, also contain the seeds, such confusion is understandable.

Remarkable confirmation of Herodotus's account has been found in the extraordinary frozen tombs of Pazyryk, on the borders of Russia, Mongolia and China, dating to around 400 B.C. (see **Tattooing** and **Wigs** in **Personal Effects**, and **Introduction** to **House and Home**). In one tomb were found two hemp-smoking kits. These consisted of braziers filled with stones and hemp seeds and covered with a cloth, together with a leather flask containing hemp seeds. The method of cannabis smoking described by Herodotus for the Scythians must therefore have been common to all the tribes of Central Asia and Russia.

In ancient China Taoist monks would get "stoned" by inhaling the fumes from hemp added to their incense burners. Large amounts seem to have been used. According to an entry in the Taoist text *Shen Nung Pen Tsao Ching* of the second to first century B.C., "To take much makes people see demons and throw themselves about like maniacs. But if one takes it over a long period of time one can communicate with the spirits, and one's body becomes light."

The Assassins and Hashish

Hashish is a more concentrated product of the cannabis plant, made by pressing the resinous pollen of the flowers. Today, in Afghanistan, for example, the resin is collected by farmers donning leather coats and running through the fields brushing against the plants. The resin is then scraped off the coats and made into cakes. The inventors of hashish are unknown, but likely candidates are the Arabs of the Middle Ages, who were such devotees of the drug that a religious sect was named after it—the notorious "Assassins." They were a fanatical Islamic minority founded by one Hasan-ibn-al-Sabbah, an ambitious Persian nobleman who claimed royal Arabian blood. In A.D. 1090 he gained possession of the mountain fortress of Alamut, which dominated the shortest route from the Caspian Sea to Persia. Hasan's aim was to establish a power base for the Shi'ite branch of Islam within a Sunni-dominated area, using assassination as a political shortcut.

Hasan and his successors as grand master of the Assassins prepared their agents for their missions of terror by persuasive drug-induced visions of Paradise. According to the thirteenth-century Venetian traveler Marco Polo, the fortress at Alamut contained a garden of exquisite beauty, where youths of the sect were brought, dosed with hashish and treated to what must have seemed an endless sexual fantasy come true with the young ladies provided. Having tasted the world to come, the hit men went forth, convinced that they were destined to return to Paradise, whether or not they were killed on their mission. The power of the Assassins was broken by attacks, first from the Mongols on Alamut and later from the Islamic rulers of Syria after the Assassins sought to rebuild their strength there. Eventually taking a peaceful turn, the sect survived. Nowadays it is known as the Ismailis and is led by the Aga Khan, a descendent of the last grand master of Alamut. Notorious as a playboy in his youth, since his assumption of leadership he has become better known for his work at the United Nations with refugees.

Where the famous Middle Eastern water pipe for smoking hashish was invented is uncertain, but the earliest known example, dated to the fifteenth century A.D., comes from the Lalibela Cave near the source of the Blue Nile, in Ethiopia.

The Ancient Opium Trail

Recent archaeological finds have opened a completely new chapter in the history of opium, a narcotic made from the milky juices of poppies. Due to its active ingredient morphine, opium relieves pain, anxiety and sleeplessness. It was once thought that the use of opium in the ancient Near East began only in Roman times. Then, in 1962, Australian archaeologist Robert Merrillees pointed out that a curiously shaped kind of clay jug made in Cyprus between 1600 and 1500 B.C. seems to be modeled on the seedpod of the poppy. These small jugs have a distinctive shape resembling an upside-down poppy head, and some even have parallel stripes painted on the body of the pot, imitating the cuts made by opium farmers in the poppy capsule to allow the white milk to ooze out. The shape and decoration of ancient jars often indicate their contents, and it has been argued that Cyprus was once the main center for opium trafficking in the eastern Mediterranean. History has an unfortunate habit of repeating itself, and the island is once again one of the leading narcotics-trafficking countries in the world.

Cypriot poppy-head jugs have been excavated at numerous sites as far away as Syria, Palestine and Egypt. Evidence for the use of

Scene from a gold signet ring from Mycenae, in Greece, about 1500 B.C. The seated goddess clutches three poppy heads. Other details show that the scene is connected with an opium cult. In the sky the riverlike band represents the Milky Way, which in Greek myth spurted from the breast of the goddess Hera. (Our word *galaxy* comes from Greek *galactos*, "milk.") The goddess on the ring holds her breast, while opium is made from the milky juice of the poppy, giving three linked symbols.

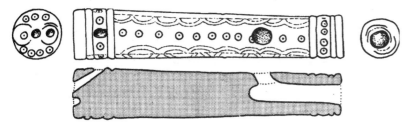

The world's earliest known opium pipe. Dating from about 1200 B.C.,
it was discovered at the ancient city of Kition, on Cyprus. Carved
from ivory, it is 5 ½ inches long. A smoldering piece of opium
would be placed in the mouth of the small hole on the left, and
the smoke produced drawn across the top of the pipe and inhaled
through the larger opening on the right. The two holes are not directly
connected, in order to prevent the smoker from gulping down
the burning opium.

opium in Bronze Age religious cults comes from the Aegean. Figurines made in Minoan Crete during the mid-second millennium
B.C. shows goddesses with crowns made of poppy heads, while a
gold ring found at Mycenae in Greece depicts a goddess handing
poppy heads to the other women.

Cyprus's role in the ancient opium industry has been dramatically
confirmed by another discovery. In 1975 excavations in the ancient
Cypriot town of Kition unearthed a small, delicately carved ivory
pipe, dating to around 1200 B.C. Distinct traces of burning on one
of the pipe's two openings leave little doubt that it was used for
smoking opium. Similar pipes for smoking opium and hashish are
still used today.

The cultivation of poppies may have spread to the eastern Mediterranean from central Europe. Archaeological evidence suggests
that they were first grown by the late Stone Age lake dwellers of
Switzerland around 2500 B.C., since their settlements contain thousands of poppy seeds: however, these were probably made into
poppy-seed cakes or crushed for cooking-oil.

Homer Said "No"

Surprisingly, given their addiction to drinking and almost every
other form of indulgence, the Greek and Roman rich do not appear

to have gone in for drug taking on a large scale. Marijuana was known to them, but as a medicinal rather than recreational drug (see **Introduction** to **Medicine**). Following the Bronze Age Mediterranean custom, however, they did seem to take opium for pleasure. At the Greek town of Aphrodisias, in Turkey, a sculpture shows what appears to be an opium capsule in the hand of Eros, god of love (see **Aphrodisiacs** in **Sex Life**).

· On the whole the Greek attitude to drug taking was negative. The two references to drugs in the poetry of Homer are both unfavorable. In the *Odyssey* Helen of Troy laces the wine of her dinner guests with *nepenthe* ("no grief"): "No one that swallowed this dissolved in wine could shed a single tear that day, even for the death of his mother and father, or if they put his brother or his own son to the sword and he were there to see it done."

Helen had obtained the drug (most likely an opium derivative) in Egypt, where it was used as a medicine. While Homer describes the drug as "useful," he is clearly wary of a substance so powerful that it robs you of your emotions. The same doubt applies to his description of the Lotus-eaters, whose country (probably in North Africa) was visited by Odysseus in his long wanderings after the Trojan War. Some of his crew ate the "lotus" flowers consumed by the locals and drifted into a blissful trance, forgetting all their cares and even the desire to return home to their families. Odysseus had to drag them aboard ship by force and tie them up until the effects of the drug wore off. The plant is unidentified—it was not the Egyptian lotus flower, which is a kind of water lily and has no psychoactive properties.

The taking of hallucinogenic drugs among the Greeks seems to have been largely restricted to the oracle centers, where priestesses gave advice and utterances that they believed to be divinely inspired. Thus Apollo, god of divination, was supposed to have spoken through his priestess at Delphi. While it seems that the oracles had more practical means of obtaining surprising information (see **Pigeon Post** in **Communications**), it is also clear from later classical writings that the priestesses often took powerful hallucinogens to inspire their more mystical sayings. Pliny refers to the use of henbane by the "ancients"; he calls it *apollinaris*, after Apollo. Another prophecy-inducing drug described by Pliny was vervain: "Both kinds [of vervain] are used by the people of Gaul in fortune-telling and in uttering prophecies, but the Magi [of Persia] especially make the maddest statements about the plant: that people who have been rubbed with it obtain their wishes, banish fevers, win friends, and cure all diseases without exception."

At certain oracle centers visitors underwent initiation rites during which they experienced vivid hallucinations of gods or the afterlife (see **Tunneling** in **Working the Land**). Surviving accounts often refer to mysterious drinks offered to the initiates—these were undoubtedly concoctions made from the juices of hallucinogenic plants. A beneficial side effect of the dabbling in drugs at the oracle centers was that the priests and priestesses became intimately acquainted, through lengthy experimentation, with the properties and dose limits of various powerful drugs (see **Introduction** to **Medicine**).

Overall the ancient Greeks, followed by the Romans, stuck to alcohol for their highs. Modern Western society has followed suit. Indeed, alcohol is the only drug sanctioned by the Judeo-Christian religions and even plays a part in Christian ritual (although only a token amount is drunk during Communion). Each civilization seems to have had its own preferred "poison," which, when its use was limited and controlled by social custom, was not seen as a problem.

As far as we can tell, ancient peoples generally took their drugs in contexts that made their consumption much safer. It is highly unlikely that Aztec teenagers taking *teonanacatl* mushrooms containing psilocybin (a close natural equivalent to LSD) did themselves serious harm—they would have taken the drug in the midst of their family, enjoying a shared experience of heightened perception and glimpses into a spiritual realm.

There is much that modern society can learn from the history of drugs. For example the ancients generally took their drugs in a "whole," unrefined form. It would be oversimplistic to blame all the bad effects of modern drug taking on the overrefinement of drugs, but consuming them in their more natural states is certainly far less dangerous. Take too many mushrooms and the body protects itself by making you throw up, while old-fashioned opium smokers would fall asleep before they could overdose. Today's refined products—heroin, for example, is two stages removed from the natural form of opium—are a very different matter.

TOBACCO AND PIPES

Christopher Columbus must have been very confused in the weeks following his arrival in the New World on October 12, 1492. Not only had he failed to reach the Orient as he had expected, but locals kept offering him heaps of shriveled-up leaves. He had no idea what to do with these gifts until a few weeks later. In his *Historia de las Indias*, Bartolomé de las Casas recorded how two of Columbus's crew, Luis de Torres and Rodrigo de Jerez, became the first Europeans to witness tobacco smoking:

> These two Christians met many people on the road, men and women, and the men always with a firebrand in their hands, and certain herbs to take their smokes, which are some dry herbs put in a certain leaf, dry also, after the fashion of a musket tube made of paper. . . . These are lit at one end, and at the other they chew or suck and take in with their breath that smoke which dulls their flesh and as it were intoxicates and so they say that they do not feel weariness. Those muskets, or whatever we call them, they call *tobacos*.

An ancient Mayan relaxes while puffing on a cigar-shaped pipe. A plug of grass inserted before filling the pipe with tobacco acted as a filter and stopped ash falling into the mouth.

According to the Maya of Mexico themselves, tobacco was a cure-all, good against asthma, nasal congestion, headaches, indigestion, toothache, boils, snakebite and even problems experienced during childbirth. To the north the Aztecs mixed tobacco with pulverized charcoal, then added flowers and other sweet-smelling substances. While the poor wrapped this blend in leaves to make cigars, as did the Indians encountered by Columbus's men, the rich smoked it in tube-shaped pipes made from silver, tortoiseshell, or painted and gilded wood. The last Aztec emperor, Montezuma (1502–1520), indulged in a relaxing smoke every night before going to sleep.

Long before the Spaniards arrived in America, tobacco had been used across the continent, with pipes being manufactured in both North and South America. The earliest in the Americas seem to be those found on Marajó Island, at the mouth of the Amazon, and at Poverty Point, in Louisiana, both sites dating to the early first millennium B.C.

The most elaborate ancient pipes were those of the Hopewell culture (100 B.C.–A.D. 700) of the eastern United States. Its craftsmen carved their pipes out of rare stones imported from sources spread across North America. They created a new form—the platform pipe—in which the bowl was centered on a rectangular tablet, with the smoke being drawn through a hole at one end of the tablet. The bowl was generally carved in the shape of an animal or human figure: the human head, the female nude, panthers, bears, beavers, alligators, birds of prey, wildfowl and frogs are among the various subjects portrayed. These are true masterpieces of miniature stone sculpture, showing a skill and command of the medium not achieved in Europe until hundreds of years later.

A fine carved stone pipe, four inches long, from the Hopewell culture of the eastern United States. Such pipes were used two thousand years ago by the Hopewell Indians. The bowl of the pipe is in the back of the toad.

Tobacco in Ancient Egypt

Archaeology constantly produces surprises. Even the reexamination of old discoveries can provide new evidence shattering long-held beliefs. Everybody knows the story of Walter Raleigh's introduction of tobacco smoking to Europe in the late sixteenth century. Indeed until recently it was still generally taken as fact that tobacco, native to the Americas, was unknown to the ancient Old World—until, that is, the mummy of Ramesses II took a trip to Paris.

The mummy of this great warrior pharaoh of the thirteenth century B.C. was found in 1881. It was originally taken to Cairo Museum, where, after a preliminary examination by anatomists, it rested until the French government arranged for the Pharaoh to come on a seven-month "state visit" to Paris in 1979. There Ramesses' body was subjected to an intense investigation by a team of doctors, microbiologists, botanists and other experts. Opening up his stomach, they found that his intestines and other internal organs had been carefully scooped out and replaced by a stuffing of vegetable matter, including plantain, stinging nettles, flax, black pepper seeds, chamomile, wheat and, to the great amazement of the botanists, chopped tobacco leaves.

The French botanists who identified the tobacco leaves in Ramesses' stomach suggested that they were included, along with other herbs, to kill insects and help stop putrefaction. What other herbal uses, if any, the Egyptians may have had for tobacco are as yet unknown. It seems unlikely that they smoked it—given the hundreds of Egyptian sculptures and paintings depicting everyday life, one would have thought that at least one smoker would have been spotted. Still smoking was known in the Mediterranean world long before Columbus. Pipes, dating from Ramesses' time, have been found in Cyprus, but they seem to have been used by opium rather than nicotine addicts (see **Drugs**).

Pharaoh Ramesses II of
Egypt (13th century B.C.).
The stuffing inside his
mummy was found to
contain tobacco leaves, the
first evidence that it was
grown in the Old World
before Columbus.

URBAN LIFE

CHAPTER EIGHT

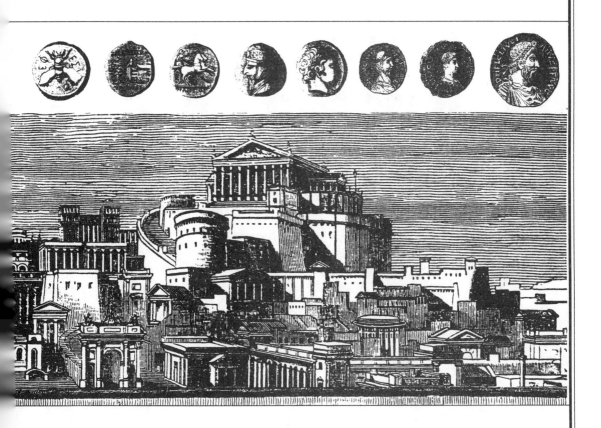

INTRODUCTION

E ven before the retreat of the last Ice Age, around 8000 B.C., the first towns were growing in the Near East. Beginning as small villages where seminomadic hunters would stay on a seasonal basis, they developed into true urban centers with the rise of farming. Jericho, in southern Palestine, can claim to be the oldest, its beginnings going back some ten thousand years. By about 7000 B.C. it was a town with some two thousand inhabitants and impressive fortifications (see **Introduction** to **Military Technology**).

The next two millennia saw a rash of towns springing up throughout the Near East and eastern Mediterranean. Çatal Hüyük in Turkey is the most extraordinary; it can rightfully be described as one of the greatest wonders of the prehistoric world. Founded around 6700 B.C., it covered some thirty acres (four times the size of the Jericho of its day), with carefully built mud-brick buildings arranged in blocks around narrow streets. The inhabitants were buried in plastered wall seats inside their houses. Two special buildings appear to be shrines, containing huge mural paintings of people and animals in a vivid naturalistic style—scenes show processions, hunting, dancing and even a town being threatened by a volcanic eruption. For its size, sophisticated layout and remarkable wall paintings, Çatal Hüyük is unmatched anywhere in the world at this early date.

By the third millennium B.C. we can begin to talk in terms of cities, the huge urban complexes of the great river-valley cultures of Egypt, Mesopotamia and the Indus Valley. The earliest of all seems to have been Uruk, in Mesopotamia (modern Iraq), which had a population of fifty thousand by 3000 B.C. At the hub of these cities' political and social life lay the palaces and temples—great civic buildings equivalent to modern town halls and cathedrals. The temples were often the most important structure, both for economic as well as the obvious religious reasons. The temples owned vast tracts of land and received a constant stream of offerings to the gods from the citizens.

The cities of ancient Mesopotamia were dominated by *ziggurats*, temples built on artificial mountains of brick, whose crumbling remains still dot the landscape of Iraq. These proto-skyscrapers were meant to facilitate the visits of the gods to earth, and likewise the communion of people with higher powers. The first true ziggurat, with three levels surmounted by a temple to the moon god, was built at Ur around 2000 B.C. It is a masterpiece of construction— thousands of mud bricks were massed together and cased in with baked bricks. Weeper holes at regular intervals allowed excess water to drain off during the rainy season and prevent the whole structure from busting. It was built to last, and the bulk of the first two stages still survive today.

The greatest of all the ziggurats was the temple of the god Marduk at Babylon, described in the eyewitness account of the Greek historian Herodotus around 450 B.C. Crowned by a temple decorated with gold, it had eight levels connected by a spiral staircase, with seats provided for climbers to rest during their ascent. Some two hundred feet high, it seems to have given rise to the biblical legend of the Tower of Babel, the ambition of whose builders was to "reach unto heaven" (Genesis 11:4). Indeed everything at Babylon seemed to have been built on a grand scale. Herodotus had clearly never seen anything like it before: "Babylon lies in a wide plain, a vast city in the form of a square with sides nearly fourteen

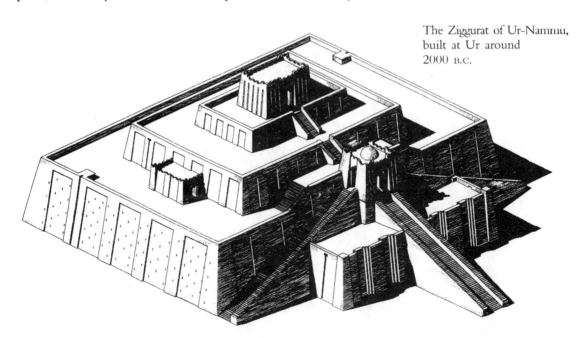

The Ziggurat of Ur-Nammu, built at Ur around 2000 B.C.

miles long and a circuit of some fifty-six miles. In addition to its enormous size it surpasses in splendour any city of the known world." The city had one hundred gates, with doors cast from bronze, while the main city wall was so broad that there was room for two rows of single-roomed buildings on the inner and outer sides—in between them was a road wide enough for a four-horsed chariot to turn around.

The figures for the teeming populations of such ancient metropolises are much greater than those of medieval Europe, and even comparable to modern Western cities. During its heyday Babylon must have housed at least half a million souls. In the time of Christ some 900,000 people lived in Rome; in 1526 after invasions and plagues only 55,000. Likewise the bustling city of Alexandria, in Egypt, had some 600,000 inhabitants around 50 B.C. but a mere 35,000 by A.D. 1300.

While many ancient cities simply sprawled as they expanded, others were carefully structured to absorb their large populations. Latter-day town planners are usually taught that the father of their profession was the Greek architect Hippodamus of Miletus. In the fifth century B.C. he invented the right-angled planned city with a pattern of streets crossing each other at ninety-degree angles and oriented to the points of the compass. New colonies sent out by Greek cities were laid using Hippodamus's ideas, unlike the haphazard and unplanned development of the founding cities.

Archaeology has shown, however, that Hippodamus was merely the reinventor of this idea. On the other side of the world a new capital of the Qin kingdom of China was founded at Yongchang in 677 B.C. The city was based around a network of two sets of absolutely straight parallel streets, each some two miles long and fifty feet wide, one running east-west and the other north-south. More than a thousand years earlier (in the 19th century B.C.) the Egyptian town of El Lahun had a classic Hippodamian layout. Even older examples have been discovered in India. Between 2800 and 2500 B.C. the Indus Valley civilization flourished in India and Pakistan: one of its most striking features is the remarkable uniformity in the layout of its towns. The earliest known example of a town laid out with a right-angled grid comes from Rahman Dehri, on the western Indus plain. This massive settlement, with a population of twelve thousand, was divided in two by a road running northwest-southeast; at right angles to this were the houses, laid out in blocks on either side of straight, narrow lanes. This model of town planning has been dated to around 3500 B.C., three millennia before Hippodamus's inspiration.

Social planning was also needed, in order to control the large population concentrations in these urban centers. The earliest known police force was that developed in Egypt, where by the Fourth Dynasty (c. 2500 B.C.) every province had a government of-

A very early example of town planning from Lothal, a city of the Indus valley civilization of India, c. 2500 B.C. Buildings were aligned on the cardinal points and set out beside streets laid out at right angles to each other.

Egyptian policemen detain a suspect for questioning—a scene from the 12th dynasty tombs at Beni Hasan (18th century B.C.).

ficial with the title of "judge commandant of the police." Later on we know that the Madjoi, a Nubian tribe, made up the bulk of the Egyptian police force. Around 1320 B.C. Pharaoh Horemheb created a "River Security Unit," with powers to board and search the boats of suspected smugglers. Other specialized units were developed to guard the riches in the tombs of the pharaohs, notably those in the Valley of the Kings, at Thebes, where security was organized by a "chief commandant of the police of the City of the Dead."

Police forces gradually developed through out the ancient world. By 367 B.C. one had been started in Rome; under the Han Dynasty of China (221 B.C.–A.D. 220) the policeman was a well enough known figure to be used as a pointer on a water clock. A highly elaborate policing system was in use in India by the third century A.D., according to written records. To keep law and order the governor of each city had under his command a force of police, secret agents and watchmen, also used as a fire brigade. The secret police had to keep a register of the income and outgoings of all citizens and to note the description of any strangers.

Another innovation contributing to public safety was lighting the town at night. Street lighting seems to have been introduced in some cities in the East during the Late Roman Empire. The Church Fathers Libanius (A.D. 314–393) and Saint Jerome (A.D. 345–420) report that the Syrian capital of Antioch was lit up at night by oil lamps hung on ropes over the streets, while Saint Basil the Great saw something similar at Caesarea, in central Turkey, in A.D. 371. An even more advanced system of street lighting may have existed in medieval China, where literary references imply that in Szechuan Province, the center of China's natural-gas industry, bamboo pipes were used to bring gas from the wellheads to light the town.

The existence of street lighting a thousand years ago is not that surprising given that so many of the familiar features of modern urban life were known in the ancient world, from sewers and plumbing to banks and money.

SEWERS

Almost as soon as people started living in towns some nine thousand years ago, it must have become obvious that, unless something was done, they were going to end up knee-deep in their own filth. Early attempts to deal with this unwanted material presumably just involved carting it away and dumping it on the fields, but this was a problem that, as cities became ever more crowded, could not always be dealt with so simply.

Nevertheless the honor of building the world's earliest sewage system was not claimed by one of the great urban centers of the ancient world: instead it belongs to the Neolithic villages of the Orkney Islands off the north coast of Scotland. At the beginning of the third millennium B.C. the inhabitants of sites such as Skara Brae had built drains fourteen to twenty-four inches high, lined with stone slabs, running from toilets in separate small rooms within their houses. It seems as though the drains ran away under the settlement to the nearby cliff, where they discharged their contents into the sea.

Shortly later on the other side of the world, in the Indus Valley civilization of India and Pakistan, the city architects had to deal with a problem of sewage disposal on a far larger scale. Their solution was to build networks of brickwork effluent drains following the line of the main streets. These generally ran past the houses down one side of the streets. Typical drains were seven to ten inches broad, cut some two feet below ground level, with U-shaped bottoms lined with brickwork, and were covered with stone slabs, wooden boards or loose bricks that could easily be removed for cleaning. Where two drains met there were cesspools with steps leading down into them to allow for a periodic cleaning out.

Palaces in the ancient world were always equipped with elaborate networks of drains for sewage disposal. Around 2000 B.C. Knossos, on Crete, had drains running from each part of the palace into a main stone-lined sewer, into which a series of channels led rainwater, so that the violent rainstorms that periodically hit the island would blast the sewers clean. In Mesopotamia (modern Iraq), palaces of the first millennium B.C. had drains that were often lined with

A typical street drain from the remarkable civilization of Mohenjo-Daro (Pakistan) c. 2500 B.C.

blocks of asphalt. The main sewers were so large that their ends had to be covered by a grating to stop burglars, presumably blessed with a strong stomach, from sneaking in.

Rather less sophisticated (and less healthy) methods were typical of early classical Greece and Rome. At the height of Athens' glory, in the fourth century B.C., pipes ran from the houses into cesspools in the street, which were periodically emptied by private contractors. In Rome the enormous *Cloaca Maxima*, built by the Etruscan kings of Rome, which cut through the middle of the city, was an open drain for six centuries until the Emperor Augustus enclosed it in 33 B.C. (This work was carried out because of the smell rather than on public-health grounds.) The tunnel created was large enough to drive a horse and cart through.

The *Cloaca Maxima* at Rome, where it enters into the river Tiber.

The refurbished *Cloaca Maxima* was just one of many magnificent brickwork sewers built by the Romans in all the major cities of the empire. Even with the fall of the Roman Empire the tradition of sewer building was kept alive in monasteries and nunneries. It was normal for the monks' or nuns' communal lavatory to be sited above a stream diverted into a brick-lined drain. These streams would be covered over for some considerable distance away from the site. It is now clear that many of the "secret passages" that appear to link monasteries to nunneries and have thus fed our worst suspicions about illicit activities in medieval times were in fact nothing more than sewers.

PIPES AND PLUMBING

A proper water supply is one of the things Western civilization takes for granted today, but for most of human history people were forced to rely on the nearest spring or stream. With the growth of towns after 7000 B.C. it became clear that these sources were no longer sufficient, so pipes and plumbing came into being.

The first cities to be equipped with waterpipes belonged to the Indus Valley civilization of India and Pakistan. By 2700 B.C. they used earthenware pipes of standard sizes with broad flanged ends, making it easier to join sections together with asphalt to stop leaks. Within a few hundred years plumbing was found widely in both the Near East and Europe. The world's earliest metal pipes come from the Fifth Dynasty mortuary temple of Pharaoh Sahurê at Abusir, in Egypt (around 2450 B.C.): about a quarter of a mile (1,300 feet) of tubing made from folded copper sheets has been recovered from the site.

One of the most impressive of the early water-supply systems was that constructed for the palace of Knossos on Crete around 2000 B.C. Water was carried across bridges (aqueducts) from a spring in the mountains seven miles away; it was then brought into the palace through terra-cotta pipes two to three feet long. The clever tapering design of the pipes increased the water pressure, preventing sediment buildup. Handles on the outside were tied together with ropes to keep the pipes from coming apart, and the joints between pipes were sealed to prevent leakage.

Later peoples, especially the Romans, built great aqueducts, many of which survive today, to bring clean water into their cities. Less well known are the ingenious systems of pipes sometimes used as an alternative to these massive engineering works. The most remarkable of these was that of the Greek city of Pergamum, in Turkey, con-

The 4,500-year-old pyramid temple of Sahurê at Abusir, site of the world's first metal piping.

Pottery drainpipes from the palace of Knossos, in Crete, installed in the 2nd millennium B.C.

structed in the second century B.C. Water was carried from a source 1,200 feet above sea level, from which it was taken down across two valleys and then piped up again to reach the citadel, the summit of which is only 130 feet below the height of the spring.

Archaeologists have traced the route of the two-mile-long pipeline by identifying the stone "sleeves" in which the pipes actually carrying the water were set. Lead was the most likely material for the pipes, made by folding a rectangular sheet of lead into a cylinder and soldering it. (Such pipes were usually made in lengths of just under ten feet and in ten standard diameters, and joined together by soldering on a collar or sometimes simply by pushing one into the neck of the next.) The Pergamum system must have been extremely well maintained in order to keep up the natural water pressure necessary to carry the supply up to the citadel.

The best known of all ancient water supplies is that of Rome. Because of the city's enormous population (around one million in A.D. 100) great efforts were required to meet its demands for water. Accordingly the water commissioner of the city was one of the most

A bronze water valve from the palace of the Roman emperor Tiberius (A.D. 14–37) on the island of Capri.

important imperial appointments. The ex-governor of Britain, Julius Frontinus, given the job in A.D. 97, found to his horror that up to three hundred gallons of water a day per head of population was apparently being consumed, something like six times the amount used in a modern city. Although much of this was supplied to the emperor, official buildings and fountains, about half went into private homes.

Part of this extravagance could be put down to wastage, as few houses had stopcocks or taps, so the water ran continuously. Even worse, Frontinus found that the quantity supplied was nowhere near matched by the income from water taxes. These were levied on private householders who had applied successfully to the emperor for permission to connect up to the public supply. A bronze nozzle stamped with an official mark was fitted to the mains and a lead, earthenware or wooden pipe from the nozzle took water to the home of the owner, who paid taxes according to the bore of the pipe.

Starting at the source, Frontinus found that landowners through whose lands the aqueducts carried the city's water had diverted part of the flow to irrigate their fields. He soon put a stop to this, but was unable to solve the problem of corruption in his own workforce. The seven hundred slaves responsible for running the 250 miles of conduits inside the city had allowed illegal, untaxed, pipes to be connected up underground. Despite the best efforts of honest men like Frontinus, it is doubtful if such practices were ever stamped out.

Control over urban water supplies continued to be a matter of great concern in the Roman world. Official regulations recorded for Constantinople in A.D. 382 specified the three diameters of pipe that could be connected up to the public system for private use. The largest buildings were allowed two-inch pipes to fill their more substantial baths; medium-sized homes could attach one-and-a-half-inch piping if they had baths; while small houses were only permitted pipes of a half-inch diameter.

After the fall of the Roman Empire the idea of plumbing survived in private buildings, especially the great monasteries, but attempts to supply whole cities were not resumed in the West until 1190, when a network of lead pipes running into Paris was laid down.

APARTMENT BUILDINGS

The crowded apartment buildings of today's cities are not such a modern phenomenon as might be imagined. By the fourth century B.C. Rome, for example, had already reached such a point of over-crowding that building upward had become necessary. The writings of the first-century B.C. architect Vitruvius clearly set out the thinking behind building blocks of apartments:

> In view of the present importance of Rome, and the unlimited number of citizens, it is necessary to provide dwellings without number. Therefore . . . necessity has driven the Romans to build high. By the use of stone piers, crowning courses of burnt brick and concrete walls, high buildings are raised with several storeys, producing highly convenient apartments. And so with walls raised high through various storeys the people of Rome have excellent dwellings without any hindrance.

Other writers were not so convinced of their merit and instead emphasize the dangers of fire and collapse. These early apartment blocks were often flimsy constructions consisting of timber frames covered by sun-dried mud-brick walls with wooden staircases. The highest-quality structures took advantage of the remarkably strong waterproof cement made by the Romans after the third century B.C. from *pozzolana*, a volcanic ash found near Naples, mixed with lime and pebbles or potsherds.

It seems as though the apartments in Rome, its port of Ostia and other towns and cities of the empire were generally built above rows of shops in the town centers. They were entered directly from the street using steep staircases, up which tenants had to trudge with buckets of water filled from the communal tap at ground level. Only in the best apartment blocks did landlords provide plumbing with the pressure necessary to reach the upper stories. Toilets connected to the sewage system were, however, standard on the upper floors— their contents would be flushed down with buckets of water.

How high were these buildings? No source gives a direct answer

A reconstruction drawing of apartment buildings at Ostia (the port of ancient Rome) with a fast-food shop on the ground floor.

for Rome, but one clue is provided by a strange event recorded by the historian Livy: he writes that in 218 B.C., the year of Hannibal's march on Rome, an ox fell from the third story of an apartment building on the Forum Boarium. Outside Rome there is further vital information from the town of Herculaneum, completely buried by the eruption of Vesuvius in A.D. 79. An apartment building with a frontage of 265 feet and a surviving height of 40 feet has emerged from the volcanic ash; reinforcing at the base of the structure shows that it was probably originally much higher. The final piece of evidence comes in a Roman letter found at the Egyptian city of Hermopolis, which gives detailed directions to a seven-story building of apartments.

Attempts were eventually made to curb the throwing up of ramshackle tenements, first by the Emperor Nero after the disastrous Roman fire of A.D. 64. He restricted new apartment blocks to seventy feet in height, which would still have allowed enough room for five or six stories and perhaps some four hundred occupants. In fact it was really only in the second century A.D., after the Emperor Trajan had further reduced the allowed height of apartment blocks to sixty feet, that the kind of "excellent dwellings" envisaged by Vitruvius were built. By this time nearly all blocks of apartments were concrete-built, and sometimes marble-clad; such buildings came to be the normal form of housing for all social ranks.

The decline of the West after the end of the Roman Empire meant that population pressure was removed, and only in medieval times did blocks of flats start to return. By the thirteenth century they were once more being constructed, and in the port of Acre, in Lebanon, people still live in the five-story apartment buildings built at that time by the Crusaders.

During the period of urban decline in Europe, however, apartment buildings were being constructed in North America. Pueblo Bonito, dating to around A.D. 1000, is the most impressive of the nine massive **D**-shaped buildings of Chaco Canyon, in New Mexico.

Nineteenth-century reconstruction of the town of Pueblo Bonito, built by the Indians of the Anasazi culture around A.D. 1000.

Covering an area of 550 by 295 feet, it held at least 800 rooms and probably contained a population of more than 1,200. The steplike design, rising to a height of four stories at the rear wall, allowed residents to use the roofs of lower rooms as open balconies. Pueblo Bonito remained the largest apartment building in North America until overtaken by those of nineteenth-century New York.

FIRE ENGINES

The urban centers of the Roman Empire were some of the worst firetraps in world history, with thousands of wooden buildings, including wooden-framed apartment buildings, crammed together.

The first major technological development in fire fighting came in Alexandria, home of many ancient inventors (see **Introduction to High Tech**), where, in the third century B.C., Ctesibius made a water pump for putting out fires. An improved design was produced in Alexandria by Heron in the first century B.C. His pump had pistons worked by a rocker arm pivoting on a post in the center. The outlet pipe was fitted with a device enabling the nozzle to be tilted up or down and swiveled around in any direction. This was not just an idea on paper: the remains of actual Roman fire engines based on this design have been found at Bolsena, in Italy, and at Silchester, in England.

However, this invention did not by itself guarantee an end to destructive fires. In Rome, fire fighting, using fire engines following Heron's plan, was mostly left to a body of slaves called the *Familia Publica*. Not surprisingly they were unwilling to expose themselves to danger and were notoriously slow to respond. This gap in services allowed some extremely unsavory characters to enter the firefighting business, such as Crassus and his private team of slaves. He would rush to the scene of fires and try to buy up burning houses; if the owners rejected this forced sale, then Crassus's men held back; as the houses burned to the ground, Crassus made steadily lower bids, and only when the owners gave in did Crassus finally order his firemen into action.

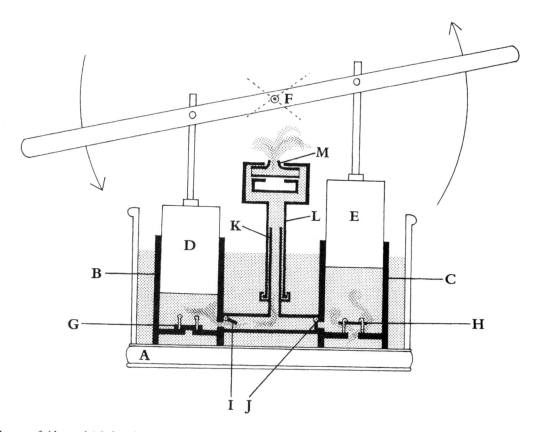

Heron of Alexandria's hand-powered firefighting device. The apparatus was housed in a tank (A), larger than the one shown in this drawing, and carried on the back of a wagon to wherever the emergency was. Attached to the bottom of the tank are two cylinders (B) and (C), containing pistons (D) and (E), which are moved up and down by the same pivoted handle (F). When piston (E) is raised, the change in water pressure causes a valve (H) at the bottom of the cylinder to open, and waters flows in from the tank. At the same time piston (D) thrusts down, closing valve (G), opening the flap valve (I) and forcing the water into the central pipe (K/L). The increased pressure keeps the flap valve (J) to the other cylinder (C) closed, so the water can only escape through the central pipe and out through the top. The jet of water can be directed up or down by swiveling the mouthpiece (M). The central pipe is made of two parts. The upper half (L) is attached to the lower (K) by a special joint that allows it to be turned and the jet of water aimed through 360 degrees.

After a quarter of Rome burned down in A.D. 6 the Emperor Augustus created the *Vigiles*, a corps of fire fighters made up of seven thousand freed slaves, who won the right of citizenship after six years' of service. A century later free-born men were entering the service for the prestige it brought—by then it was ranked equivalent to an elite military unit. The *Vigiles* had the power to break into a house if they thought a fire had erupted there, while their commander was entitled to have a householder flogged if he considered that a fire had been caused by the owner's negligence.

The *Vigiles* were a successful firefighting force for four hundred years: in all that time the only serious disaster was the great fire of A.D. 64, but here they may have been ordered to stand aside by Emperor Nero so that he could clear the ground for his scheme of building a new Rome, grand enough for his megalomaniac visions (see **Zoos** in **Sport and Leisure**). He blamed the early Christians for starting the fire, carrying out a wave of cruel persecution.

Surprisingly, the example of the *Vigiles* was not widely followed elsewhere in the Roman Empire. This seems to have been primarily due to political reasons, as the letters exchanged at the beginning of the second century A.D. between Pliny the Younger, son of the encyclopedist, and the emperor Trajan demonstrate. Pliny, the governor of the province of Bithynia (northwestern Turkey), was keen to set up a local fire brigade after public apathy and a lack of equipment had led to a minor fire getting out of control. Trajan, however, clearly thought that the danger of losing political control over the firemen was the greater:

> It has occurred to you, Plinius, to follow the example of other places and consider the possibility of setting up a fire brigade. I remember that your whole province . . . has been troubled by societies of that kind. Whatever name you give them, and with however good a reason, men who are banded together will become in a very short while a political club. Your duty then is to be content with getting things ready to assist in controlling fires, and if the situation is serious enough to get the public to lend a hand.

With the fall of the Roman Empire large cities disappeared in the West and with them the need for fire engines and firemen. When crowded cities arose once more in the Middle Ages, they were for a long time firetraps. It was not until the end of the fifteenth century that the fire engine was reinvented, following much the same design as that of Ctesibius 1,700 years earlier.

BANKS

Banks as such were almost certainly invented in ancient Iraq. The first lending banks seem to have been started as early as the second millennium B.C. by the sacred prostitutes of Babylonian temples. As their earnings accumulated into massive stores of wealth, they were in a position to lend money for business ventures conducted by temple staff and their families. Likewise in early Greece the great temples had a virtual monopoly on banking, starting off by holding valuables on deposit for the wealthy and then moving into moneylending.

Commercial banking of the more usual kind had begun in Babylonia by the seventh century B.C. and blossomed enormously at the end of the sixth. Loan arrangements, written on clay tablets, were straightforward. If the social standing of the borrower was good, he simply returned the loan (an amount of silver) with interest, usually charged at 20 to 30 percent a year. Where the standing of the borrower was more dubious, the bank would charge no interest but would take control of some property (e.g., a house, plot of land, or a slave) as security. This provided income for the bank while the loan was outstanding and was kept if the money was not repaid. That there were many defaulters we know—it was the custom to smash the loan tablet when a debt was paid off, so the dozens that survive must represent those unfortunates whose property and goods were seized.

With such favorable terms banking dynasties, such as Egibi and Sons of Babylon, made colossal fortunes, soon outstripping the temples, and perhaps even the state, in wealth. During the fifth century B.C. the Murashu family of the city of Nippur was charging interest at a rate of 40 to 70 percent to landowners trying to meet the high taxes levied by the Persian Empire, then ruling Babylonia. Not surprisingly the Murashu bank came to own or hold in mortgage most of the land and canals around the city, as well as herds of cattle and fisheries. Notorious as "loan sharks," they also engineered massive real estate swindles to bankrupt the tenants of state-owned land with the collusion of corrupt officials. The Murashu business seems to have been closed down by the Persian government in 417 B.C.

It is fairly certain that the sixth century also saw the first entry of the Jews into high finance. The members of the Egibi family business often bore good Babylonian (pagan) names, but seem to have been of Jewish origin. *Egibi* appears to be a Babylonian equivalent of *Jacob*, and it is likely that the prosperous banking family were descendants of the middle-class deported from Judah by Babylonian conqueror Nebuchadnezzar in 587 B.C., who had somehow "made good" in the land of their captivity.

The most sophisticated banking system in the ancient world was that developed a few centuries later by the Greek Ptolemaic kings of Egypt (see **Introductions** to **High Tech** and **Communications**). A network of royal banks, employing thousands of clerks, spread across the kingdom—all organized from a head office in the capital of Alexandria. There were provincial banks in each district capital and even local branches in small towns. This state bank system served three main functions: to help in collecting state taxes and revenues; to take money on deposit from individuals and repay it on demand; and to act as a *bureau de change*. Service charges were probably the bank's prime source of profit, rather than interest on loans. In Roman Egypt (after 30 B.C.) the fact that supervising a bank was a compulsory public service suggests that there were slim pickings for bankers.

Banking was less well developed in the rest of the Roman Empire, where it was a private system largely in the hands of Greek businessmen, based, in the eastern Mediterranean, at the major temples. Banks disappeared in Europe after the collapse of the empire in the fifth century A.D.

By the ninth century the Arab world had developed a sophisticated banking network. Sources mention banks with head offices in Baghdad and branches in other cities. They carried on business mainly through the use of the check (*sakk* in Arabic), guaranteed by bonds and transferred by letter of credit, so that it was possible to draw a check in Baghdad and cash it in Morocco. Most bankers in the Islamic world were Christians or Jews, because of the Koran's ruling against the levying of interest on loans.

The beginnings of European banking can be traced back to A.D. 808, when Jewish merchants in northern Italy banded together to create a primitive form of bank for depositing cash. Modern banks first developed in Europe in late twelfth-century Venice, when the great maritime state levied a special tax on its citizens to pay for a war it was waging simultaneously with both East and West. This heap (*banck* in German) of cash was looked after by a special committee that eventually transformed itself into a bank.

COINS AND PAPER MONEY

While the history of money officially begins with the invention of coinage in the late seventh century B.C., it is hard to imagine how the complex urban societies of the ancient world could have managed for so long without a convenient means of exchange. Simple barter, such as trading a number of goats for their equivalent value in grain, would have been far too cumbersome. In fact money had a long prehistory.

One of the earliest forms of money was the cowrie shell. Many bronze drinking vessels from Shang Dynasty China (c. 1500–1000 B.C.) carry inscriptions recording gifts of strings of cowrie shells. A string of cowries, possibly five or ten shells, was a basic monetary unit in Shang times. The tomb of Queen Fu Hao at Anyang, the Shang capital, contained some seven thousand cowries—she was certainly an extremely wealthy woman. Cowries were precious essentially because of the great distance of Anyang from the Pacific coast, where they were collected. According to tradition, the importance of cowries continued even after the introduction of coinage— around 600 B.C. the prime minister of the kingdom of Tsu, in modern Honan, is said to have issued metal coins in the shape of cowrie shells.

In Egypt and Mesopotamia during the second millennium B.C. a more recognizable form of currency was used—rings of bronze that could easily be carried around when threaded together onto a larger ring. Their value, if there were any doubt, could be measured simply by weighing them. Weights of metal were standardized in ancient Babylonia and frequently bore an official government stamp authenticating them. Such stamped weights represent the first step toward coinage proper. In the reign of Sennacherib of Assyria (705–681

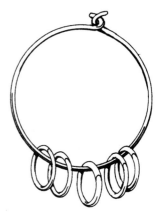

Egyptian money rings from Tanis, the "small change" threaded on a larger ring that would be worn around the neck.

B.C.) there is a reference to half-shekel pieces of silver known as Ishtar heads—these were presumably stamped with a portrait of the goddess Ishtar, making them a kind of proto-coinage.

According to the textbooks, the first true coins appeared in the kingdom of Lydia, in western Turkey, about 650 B.C. They were small bean-shaped slugs of electrum (a natural alloy of gold and silver found in the rivers of Lydia) stamped with the lion insignia of King Gyges on one side and punched with marks certifying the weight and purity of the metal on the other. From there the novelty rapidly spread west to the Greek city-states and east to Persia.

However, other, far-distant countries have equally good claims to have produced the earliest coinage. Indian coins were first produced in the seventh century B.C. by the Magadha Empire; these were either small, bent bars or flat pieces of silver, and like those of Lydia these coins were punchmarked. The symbols punched into the coins were those of the government or those of traders and bankers.

Very different from any other coinage were the cast coins of China. The knife-shaped coins of the Pacific coast state of Ch'i may date as far back as the ninth century B.C.; the Chou Dynasty later copied the idea, producing spade-shaped money. The earliest securely datable evidence for Chinese coinage is the court record of a monetary reform carried out by King Jingwang of Chou in 524 B.C. That Chinese coins developed completely separately from those of the West is proved by the fact that, apart from the gold coinage of the state of Ch'u, minted between 500 and 200 B.C., all Chinese coins up to modern times were cast rather than stamped.

It comes as no surprise that counterfeiting began almost as soon as the first coins appeared. The world's oldest counterfeit money is a copy of a sixth-century B.C. silver coin from the Greek island of Aegina. The fake has a copper core plated with silver. The technical accomplishment represented by this coin is amazing, considering that coinage had only just been introduced into Greece.

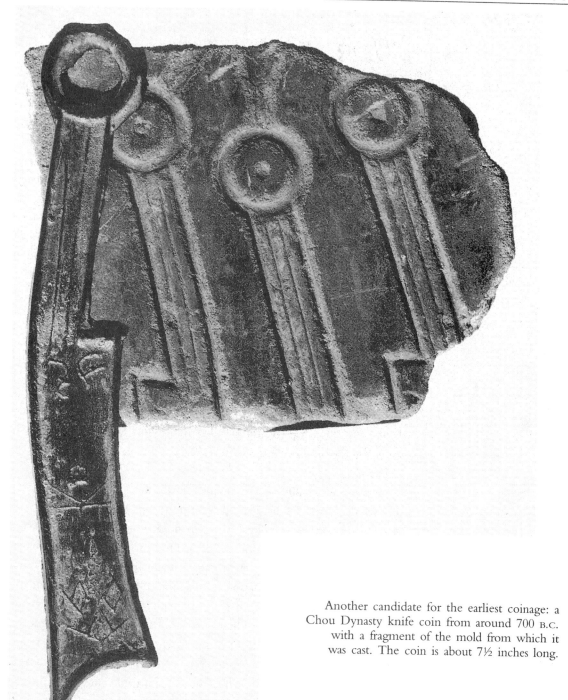

Another candidate for the earliest coinage: a Chou Dynasty knife coin from around 700 B.C. with a fragment of the mold from which it was cast. The coin is about 7½ inches long.

"Flying Money"

One development that is unquestionably Chinese in origin is the use of paper money. A first attempt to introduce it took place during the reign of Emperor Han Wu-ti (140–87 B.C.), after constant military expeditions against the Huns in Mongolia had exhausted the empire. The private minting of money had debased the coinage so much that there were continual dramatic swings in its face value. Wu-ti called in most of the coinage by issuing treasury notes each worth 400,000 copper coins. The notes, which were made from the hide of a white stag—an extremely rare creature—were a foot square and bore a special pattern. This extraordinary experiment in central banking was doomed to failure, however, as the supply of white stags was very limited.

The Chinese made a fresh start with paper money around A.D. 800, at which time it was called "flying money" because of its tendency to blow away. This early paper money was not fully exchangeable but was a certificate given to merchants by private banks in exchange for cash; issued in the capital, it could then be redeemed for cash by the merchants when they returned to the provinces. The idea was to stop highwaymen relieving merchants of their cash on the journey home. In 812 the government took over this activity.

Around A.D. 1000 Chinese banks started to issue fully convertible banknotes licensed by the government, but in 1023 these were withdrawn and only official notes printed by the government were allowed. This money had a notice printed on it to the effect that it was good only up to a certain date, usually three years later, which must have made its circulation pretty rapid when its time was nearly up. Little thought was given to backing the currency issue with adequate gold reserves, and inflation soared during the twelfth century. This made counterfeiting an attractive proposition, and in 1183 a printer, who had produced 2,600 fake notes in six months before being caught, was sentenced to death. The idea of paper money slowly spread west to less well developed economies: the Mongols printed Chinese-style notes in Iran in 1292, and the earliest European paper money was printed in Sweden in 1601.

A print from the earliest
surviving block for printing
paper money, dating to
the 11th century A.D.

WORKING THE LAND

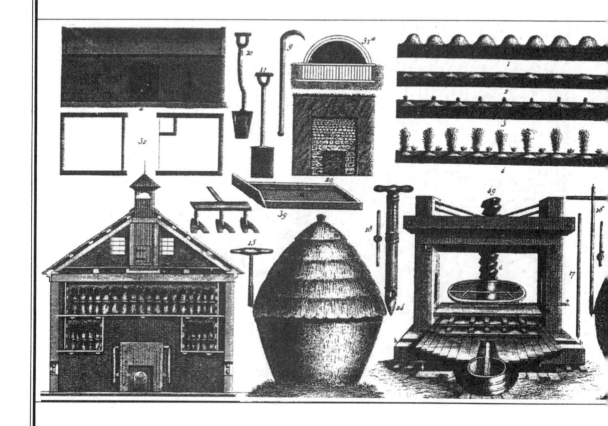

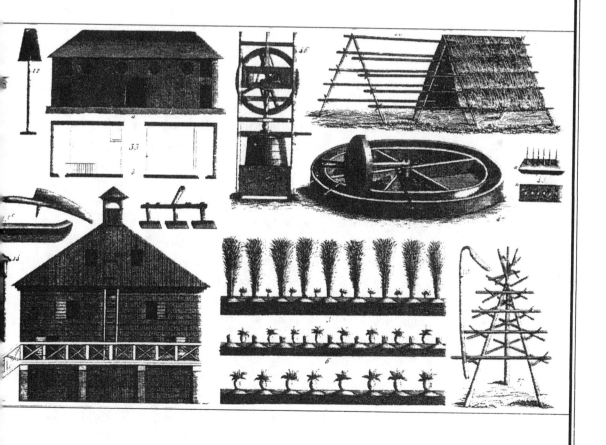

CHAPTER NINE

INTRODUCTION

Originally the whole world was "the countryside," until the human race began to explore it. Once we did, things started to change, and the story of mankind's attempts to exploit, tame and modify the wilds with the aid of technology is almost the history of humanity.

Gathering and hunting, of course, began as soon as the first humans reached out for some food. The earliest surviving tools, largely flint axes and scrapers, were all used for hunting, chopping meat and cleaning skins, though we should not take for granted the traditional picture of the evolution of humanity solely in terms of "big-game hunting." The archaeological record is biased in this respect. As the collecting of berries, fruit, grubs and shellfish requires few tools (the most important being leather bags and woven baskets), stone weapons and large animal bones are much more likely to be preserved.

More sophisticated hunting weapons were developed during the latter part of the Old Stone Age. Today the boomerang is associated solely with the Australian Aborigines, who, archaeological evidence has shown, have used it as a weapon for some ten thousand years. But in ancient times it was employed on virtually every continent. The oldest boomerang in the world, dating to twenty-one thousand years ago, was found in Obłazowa Rock Cave in southern Poland. Made of wood, it has a span of some twenty-eight inches. The bow and arrow seems to have been invented at much the same time. The site of Parpalló, in eastern Spain, has produced the earliest flint arrowheads from eighteen thousand years ago. The oldest known fishing hooks and "fish gorges" (straight splinters of bone that were baited and attached to fishing lines) date back fourteen thousand years, and come from Stone Age sites in Europe and South Africa.

In much later times hunting also became an organized sport for the idle rich, and the great kings of the Near East frequently went on safari. In northern Syria stocks of elephants were kept in game parks and were hunted by the pharaohs of Egypt and the Assyrian kings of northern Iraq. Tiglath-pileser I of Assyria (1115–1077 B.C.) was a noted huntsman. His inscriptions boast that he killed a total of 120 lions on foot, and 800 from his chariot. On one expedition he slew four wild bulls and ten "mighty bull elephants"; on another occasion he took a boat ride on the Mediterranean and killed a narwhal. The Roman nobility, as is well known, were not so sporting: they imported tens of thousands of exotic animals from every corner of the empire to watch them being slaughtered in the grotesque spectacles of the Roman arena.

The countryside as we know it—fields, pastures, canals, ponds and orchards dotted between woodlands and the wilds—only began to take shape with the invention of agriculture during the New Stone Age. This phenomemon is often called the Neolithic Revolution, although it is now known to have been a far more gradual development than was once thought.

Toward the end of the last Ice Age, around 10,000 B.C., communities in the Near East started to harvest cereal grains with flint sickles and grind them with stone mortars and pestles. By 9000 B.C. intensive collecting of cereals had spread to villages in northern Syria, which apparently lay outside the natural habitat of the plants involved. This would have required grains to be deliberately sown by humans. The final stage in domesticating cereal plants must have been taken by 8000 B.C.—the grains found in deposits at Jericho, in Palestine (see **Introduction** to **Military Technology**), and at other settlements in Iran and Syria of this date are much larger than those produced by wild plants.

Through similar processes a host of plants and animals were domesticated, some in other areas of the Near East and others much farther away:

Peas and lentils in southern Iran, northern Syria and Jericho, in Israel, around 8000 B.C.
Beans and squashes in various parts of South America by 7500 B.C.
Taro (an edible root) in New Guinea around 7000 B.C.
Sheep and goats in northern Syria about 7000 B.C.
Rice in China by 7000 B.C.
Cattle and pigs in southern Turkey around 6500 B.C.
Potatoes about 6000 B.C. in the Chilca Valley, near Lima, Peru

These dates are only the current ones—new finds tend steadily to lengthen the timescale for the domestication of various plants and

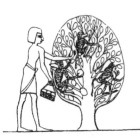

Ancient Egyptian farmers seem to have trained monkeys to help them in fruit picking, as seen in this tomb painting from around 1900 B.C.

animals. It has already been claimed, for example, that the horse was tamed thousands of years before the dates given here for other animals (see **Introduction** to **Transportation**).

The domestication of plants and animals went hand in hand. Stored-up grain from the autumn harvests kept herds alive during the winter, while the dung of grass-eating animals was, and still is, the best fertilizer known to mankind. Plowing made the whole business more effective. Early evidence of ards, the simplest plows, comes from the picture writing of Uruk, in Iraq, dating from before 3000 B.C. At almost the same time, in many areas of Europe (as far away as Britain), plows were being pulled by oxen, leaving furrows in the ground that are now being painstakingly revealed by archaeologists. This rapid transfer of technology, which seems to have been accompanied by the first milking of cows, the invention of the cart (see **Introduction** to **Transportation**) and the earliest production of wool from sheep, has been termed the Secondary Products Revolution. Animals became an increasingly important part of farm life; to judge from Egyptian tomb paintings, even monkeys were trained to be useful, being sent by farmers up trees to collect fruit.

The botanical knowledge of the ancients came through practical experience. George Sarton, a leading historian of science, has explained how the ancient Babylonians discovered the sexuality of the date palm by trial and error. Male date palms produce no fruit, and we can imagine an ancient date farmer, not wanting to waste precious water supplies, uprooting the male plants and discarding them. He would soon have learned his mistake, as the next year he would have had no crop at all. The two sexes needed each other, and the next stage was the discovery that fruit production could be increased by climbing up the male palms, collecting the flowers and attaching them to those of the female plants.

Sarton argues that the Babylonians must have discovered the sexuality of the date palm some four thousand years ago, when they were already cultivating it heavily. The modern theory of the sexuality of plants was only expounded at the end of the seventeenth century. As Sarton remarked, "This is the outstanding example of application preceding theory; in this case the application was already completed by 2000 B.C., if not much earlier, the theory formulated only in A.D. 1694!"

The other essential for agriculture is of course water, and creating artificial means to supply it where lacking was among the first engineering feats of the ancients. The world's earliest irrigation works were probably those dug in the fourth millennium B.C. by the early farmers of Geokysur, in southern Russia. They created a network of canals, each up to a mile and a half long, about four feet deep and up to ten feet wide.

By about 2800 B.C. there was an Egyptian "department of irrigation," which may have been called in to oversee the building of the world's earliest dam. This is the Sadd al-Kafara ("Dam of the Pagans" in Arabic), near Helwan, twenty miles south of Cairo, dating from around 2500 B.C. Its remains have survived until today. The dam was 348 feet long, rising to a maximum height of 37 feet above the valley floor. It was immensely solid, made of two walls of imported masonry each 78 feet thick at the base; between the two

Assyrian sculpture of the 9th century B.C.—a winged genie carries out the ritual fertilization of a sacred palm tree by inserting the male flower into the female.

walls was a gap of 118 feet filled with sixty thousand tons of gravel from the riverbed. The reservoir formed by the dam was to supply drinking water for the workers at nearby stone quarries.

Many early canals were dug to irrigate the fields of Iraq as farmers tried to spread out from the narrow river valleys where Mesopotamian civilization began. Parts of the first recorded canal can still be seen—as the Al-Gharrif waterway, which runs from the river Tigris; it was originally cut by a governor of the Sumerian city of Lagash before 2500 B.C., as we know from a surviving account of his project. In 690 B.C. King Sennacherib of Assyria constructed a masonry dam on the Atrush River. The water was carried in a thirty-six-mile-long winding canal to his capital at Nineveh. Where the canal had to cross the river Jerwan, Assyrian engineers built an aqueduct three hundred yards long and fifteen yards wide, supported on five arches. The canal had an asphalt base, covered by a concrete bed and topped off with a limestone pavement; the sides were lined with more than two million limestone blocks.

In China a major canal designed by the hydraulic engineer Zheng Gou was built in Shensi Province in 221 B.C. Ninety-three miles long, it directed the water from a river across a massive plain, irrigating some 350,000 acres. The system was designed not just for irrigation but also for fertilization; the canal diverted from the river at a point where the maximum silt can be drawn off, allowing the alkaline soils of this vast area to become fertile.

The Nabataeans, who lived in southern Jordan and the Negev Desert of Israel from the first century B.C. to the first century A.D., developed very elaborate irrigation systems. They carefully husbanded the rainfall from occasional showers, blocking off seasonal streams to divert water to the fields. Around the ancient city of Ovdat, in the Negev, there are seventeen thousand dams in an area of fifty square miles.

The Romans built vast dams and reservoirs. The *Piscina Mirabilis* at Bacoli, in Italy, covered an area of almost 2,400 square yards. Lined with waterproof cement, it would be divided by gates into separate sections for cleaning. Near Emperor Nero's villa at Subiaco there is an unusual example of a dam built for recreational purposes. This had walls 45 feet thick, behind which was a lake 1½ miles square. The dam was so well made that it didn't collapse until A.D. 1305, after 1,250 years of service.

The ancient Sri Lankans were undoubtedly the greatest builders of irrigation works of all time. According to tradition, King Panduwaasa constructed the "Giant's Tank" at Anuradhapura in 494 B.C.: it has an area of 223 square miles, equal to that of Lake Geneva; its retaining dyke is 15 miles long and 300 feet wide at the floor. The huge tank at Kalawewa was built in A.D. 459 by King Dhaatusena. It is 40 miles in circumference and held back by a forty-foot-high dam made of granite blocks tightly fitted together. The Kalawewa Tank supplied water along a canal to the capital at Anuradhapura 60 miles away. King Parakrama Bahu the Great (A.D. 1153–1186) built or restored 165 dams, 3,910 canals, 163 major reservoirs and 2,376 minor tanks. His program of irrigation works culminated with a reservoir called the Sea of Parakrama. Even in its present reduced state it still covers 5,940 acres and irrigates 18,200 acres; the retaining dyke is 8½ miles long and 40 feet high.

Another way of channeling water supplies was to build tunnels, and this, too, the ancients did on a grand scale. The massive network of underground canals called *qanats*—built by the Persians to bring water from the mountains to the desert—and the extraordinary tun-

A Nabataean dam of the Roman period, 1st to 3rd centuries A.D., at Wadi el-Jilat, in Jordan.

nel drilled through a mountain by the Greeks of Samos to supply their city with spring water are two of the greatest civil-engineering feats of world history. The almost uncanny skills developed by the ancients in drilling through solid rock were also used to produce some of their most awesome religious monuments, including tombs and temples chiseled into mountainsides.

Little, it seemed, deterred ancient engineers in their quest to exploit the resources of nature. Mining has an almost incredible antiquity, going back at least some twenty thousand years, the date of the earliest flint mines in Australia. After the Neolithic and Urban Revolutions of the ninth and fourth millennia B.C., city dwellers mounted expeditions into the wilds, often hundreds of miles from home, to prospect for and mine valuable mineral resources. Flint, metals, precious stones, salt and coal were all mined. Even drilling for natural gas had begun 1,500 years ago.

Ancient peoples' handling of the environment was on the whole far more balanced than our own. There were of course bad aspects. The deforestation and desiccation of the landscape is not a modern phenomenon. While the problem in North Africa is largely due to global climatic changes, it was considerably aggravated by the intensive farming of the Romans. Their domesticated goats also played a part. When let loose in large numbers, goats will eat any green thing in sight and can rapidly destroy whole forests by stripping the buds from every tree and bush struggling to reach the light. The huge consumption of timber by the smelting works around the silver mines of Laurion, near Athens, played a large part in the disappearance of the forests of Attica.

On the other hand all ancient farming was by definition organic. The use of phosphates and other chemicals that produce quick results but long-term disaster is a modern curse. The idea of conserving the world's natural resources is not a twentieth-century invention. Many ancient civilizations learned their lesson the hard way.

For example, the oyster lagoons of the Kuangtung (Canton) coast, in southern China, fostered a thriving pearl-diving industry from the late second century B.C. onward. But by the second century A.D. greedy divers had overworked the lagoons so much that the oyster population dwindled to nothing; the industry collapsed and the region nose-dived into poverty. Then Mêng Ching, appointed governor around A.D. 150, suspended diving for a while and banned "former evil practices." The oysters returned, and with them prosperity, to Kuangtung. As Joseph Needham, leading historian of Chinese science, remarked, Mêng Ching "stands out as a successful exponent of nature-protection and fisheries conservancy." In fact he was so respected that later fishermen worshiped him as the patron deity of their industry.

Further, the ancients, especially the Chinese, were way ahead of us in the use of biological pest controls in farming. The technology used in ancient times was also far more "appropriate," yet it could be suprisingly advanced. The husbandry of every kind of animal, from the more usual cows and sheep to the highly specialized skills needed for bees, fish and oysters, was developed to a fine art. Reaping machines were invented in France under the Roman Empire. They used no gasoline, and the waste products of the "tractors," which were donkeys, provided beneficial mulch for the fields. Windmills were widely used in the Middle Ages, and watermills, as archaeologists have recently shown, were very common features of the landscape under the Roman Empire—they provided safe, renewable energy and, including the setting-up costs, a very cheap way of milling flour for an empire.

There are valuable lessons to be learned from the experience of the ancients: by avoiding their mistakes and emulating their successful management of the environment, we could better handle what is left of the natural world.

THE REAPING MACHINE

Food crises caused by ever-rising demand and shortages of supply are not just a modern worry: they have occurred in the past wherever large urban populations have arisen rapidly. One ancient solution to this problem was the reaping machine, developed by farmers in the Roman province of Gaul during the first century A.D. The reaper was needed to cope with the large demand for grain in an area with unpredictable weather during the short harvest season and a local shortage of agricultural labor.

The *vallus*, as the reaping machine was known to the Romans, was described by the encyclopedist Pliny in A.D. 77: "On the vast estates in the provinces of Gaul very large frames fitted with teeth at the edge and carried on two wheels are driven through the corn by a donkey pushing from behind; the ears torn off fall into the frame."

A reconstruction drawing of the *vallus*, invented in Roman Gaul, in use.

Relief sculptures on local Gallic tombstones give further information on how the reaping machine worked. Pliny's "teeth" were rows of very sharp knife blades set close together on the leading edge of the cart at a height slightly below that of the heads of grain. The grain fell into a boxlike hopper behind the "teeth." Texts other than Pliny suggest a larger version of the reaping machine, the *carpentum*, which was pushed by an ox.

Yet despite the obvious benefits of a successful mechanical reaper, it failed to be adopted widely in the Roman Empire. This may seem particularly strange given that running an agricultural estate was the only career, outside serving in the army, that the upper classes of Roman times deemed worthwhile. Whole manuals on efficient farm management were written for the gentry by Greek, Carthaginian and Roman landlords.

In fact there was nothing about the reaper itself that led to its neglect; rather this was a result of one key factor in Roman life: its slave economy. While the Romans certainly did not lack inventive ability, they did fear change. With their estates farmed by vast numbers of slaves, the danger of social upheaval was always present and might have become acute if slaves started to be displaced by machines. Accordingly the Gallic reaper fell out of use, providing a classic example of an ancient invention that should have had a far greater impact than it actually did.

Similar social factors seem to have been at work in medieval northern China, where a "push-scythe" was described in 1313, the only mention of a mechanical reaper in Chinese history before the introduction of Western technology. It looked like a primitive lawnmower, but with a single, fixed blade. The device never caught on because of the attitude taken by most Chinese bureaucrats toward laborsaving devices in agriculture. With an enormous mass of peasantry available, they saw no need to save on labor, and in any case successful mechanization would only have put the peasantry out of a job.

The mechanical reaper was not used again anywhere until the early nineteenth century, when the ancient Gallic version was to play a crucial part in its reappearance. A description of the Gallic reaper was preserved in the work of Palladius, a fifth-century Roman writer on agriculture. A translation and reconstruction drawing was published in J. C. Loudon's *Encyclopedia of Agriculture* in 1825, and caught the eye of John Ridley, a young Londoner who later emigrated to Australia. There, inspired by an acute labor shortage and the example of the Gallic reaper, he devised his own reaping machine in the summer of 1843.

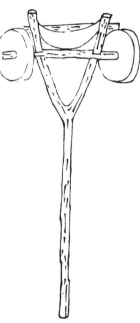

鐮 推

A Chinese drawing of the push-scythe. Cut ears of grain would have to be picked up off the ground by the operator.

WATERMILLS AND WINDMILLS

Like so many other inventions, the idea of harnessing the energy of water seems to have come to several different peoples at the same time. Consequently no one can really be sure where it was discovered.

The traditional claimants are the Romans, since the architect Vitruvius refers to watermills around 20 B.C. The earliest direct archaeological evidence for a Roman watermill comes from Venafro, a village near Pompeii. Here the eruption of Vesuvius in A.D. 79 buried a vertical waterwheel, leaving behind an impression in the lava.

On the other hand a slightly earlier watermill occurs in an account written in the late first century B.C. by the Greek geographer Strabo, describing the riches of King Mithridates of Pontus, on the Black Sea coast of Turkey. He and his kingdom had been captured by the Roman general Pompey in 65 B.C.: "It was at Cabeira that the palace was built, and also the watermill and here were the zoological gardens, and, nearby, the hunting grounds, and the mines." The palace at Cabeira was built by Mithridates' predecessor in 120 B.C., so its watermill may well belong to the early first century B.C.

China can also put in a claim, although the evidence is less straightforward: historical records state that a highly complex water-powered device for blowing bellows in a metalworking machine was invented there in A.D. 31 by Tu Shih, an administrator. It is highly unlikely that the Chinese leaped straight to this level of development, so there must have been technically simpler watermills quite a bit earlier. Records exist of the use of water for powering pounding machines in A.D. 20, but again this is nowhere near long enough for the technological advances implied by Tu Shih's bellows to have occurred.

Finally, from Denmark, dams, reservoirs and the races dug to channel water to what were apparently two horizontal watermills have been traced by archaeologists at Bølle, in Jutland. The dating of these remains is uncertain—they are probably of the first century A.D. or later—but the fact that they belong to horizontal rather than vertical watermills may point to an independent invention.

Wherever they came from originally, watermills went on to play a major part in the ancient economy, especially for grinding corn. Because they are rarely mentioned by Roman writers, their importance has generally been underestimated by historians. A steady stream of archaeological finds now shows clearly that in fact they were a normal feature of villas, forts and towns throughout the empire as far north as Hadrian's Wall. The idea that their full impact came only in medieval times is completely wrong.

The largest complex of Roman watermills was at Barbégal, near Arles, in southern France, dating to around A.D. 300. Water from the river Durance was brought down a steep hill on an aqueduct that split into two millraces, each descending the hill in eight steps with a waterwheel set in both branches on each step. The output of this impressive piece of engineering, which may have been state controlled, has been estimated at twenty-seven tons of grain daily, enough to feed 12,500 people.

An invention certainly developed in Rome is the shipmill, created by the great Byzantine general Belisarius in A.D. 536. He was engaged in a campaign to reconquer Italy for the Eastern Roman Empire from the Goths. Belisarius and his army were trapped in Rome by the Goths, who, by cutting the water supply to the city's

The Roman flour mill at Barbégal. Its sixteen overshot waterwheels produced over thirty horsepower for turning the grinding stones. Water, carried to the top of the mill by an aqueduct, was chaneled through races over the waterwheels; the flow could be regulated by using an outer channel as an overflow. The inset shows how the power of the water was transmitted to the grindstone within the mill.

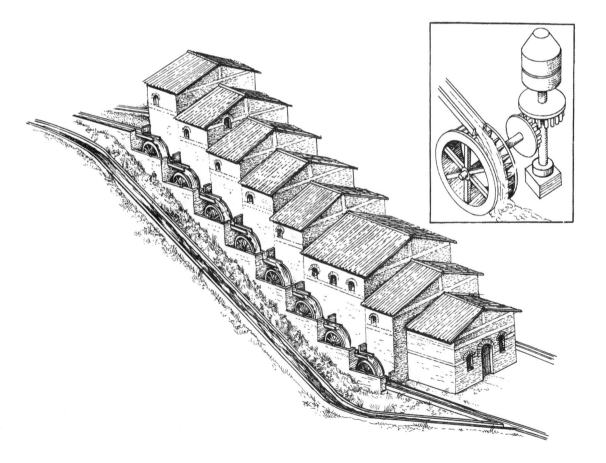

cornmills, hoped to starve out the inhabitants. Belisarius had the in-spired notion of mounting the mills on barges and floating them on the Tiber, which was far too big a river for the Goths to divert. They tried to wreck his invention by floating trees and dead Ro-mans down the river to catch in the paddles, but the wily general installed a large chain upstream to catch anything thrown into the river, and the city held out.

Watermills were extremely important in the Islamic world, where vast amounts of grain were produced. At Mosul massive shipmills, built from iron and teakwood, were moored to the banks of the riv-ers Tigris and Euphrates during the tenth century A.D.; each mill had two pairs of stones working around the clock and could sustain an output of ten tons a day to feed the 1.5 million mouths of Bagh-dad.

Harnessing the Wind

The great Alexandrian engineer Heron (see **Introduction** to **High Tech**) designed a wind-powered organ, driven by a wind vane, in the first century A.D. There is no evidence, however, that it was ever built outside of Heron's workshop. And strangely enough, although they were fully aware of the principle through the work of Heron, the Romans never exploited the windmill; still, most of their empire was blessed with rivers that could be used to provide water power when needed.

After a long period of neglect the windmill was eventually devel-oped in the Islamic world, stimulated by the need for cornmills in areas without steady flows of running water. Heron's original would have been known to Islamic engineers, who actually preserved more knowledge of Greco-Roman technology than their European coun-terparts. But the inventor of the Islamic windmill, inspired though he may have been by reading Heron's work, produced a completely different design. The shaft holding the blades was upright rather than horizontal and was housed in a vertical mud-brick tunnel with a door and flues to catch the wind. The result was a structure that looked something like a modern revolving door.

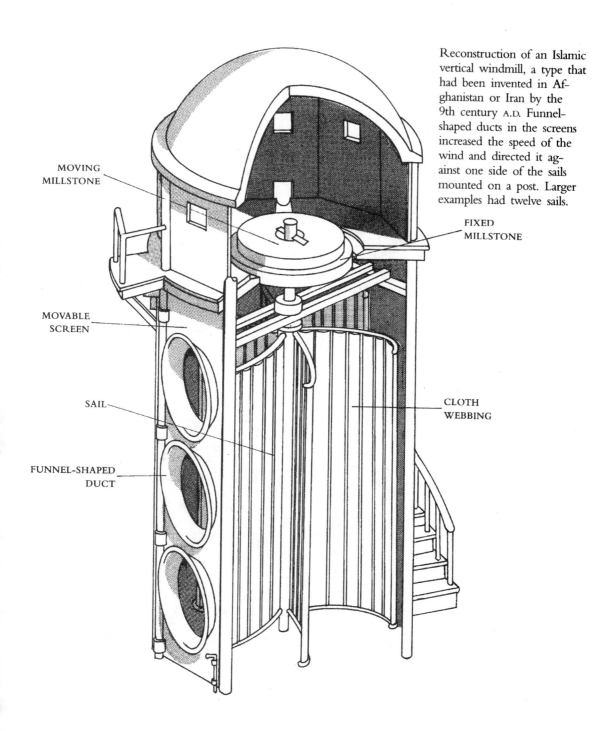

Reconstruction of an Islamic vertical windmill, a type that had been invented in Afghanistan or Iran by the 9th century A.D. Funnel-shaped ducts in the screens increased the speed of the wind and directed it against one side of the sails mounted on a post. Larger examples had twelve sails.

MOVING MILLSTONE

FIXED MILLSTONE

MOVABLE SCREEN

SAIL

CLOTH WEBBING

FUNNEL-SHAPED DUCT

According to tradition, the inventor of the Islamic windmill lived in Iran during the time it was conquered by the Arabs in the mid-seventh century A.D. The second caliph, Omar (A.D. 634–644), levied heavy taxes on the windmills that the Iranian inventor had constructed—supposedly he was so incensed that he murdered the caliph. How much truth lies behind this colorful story is hard to know, but archaeology has shown that mud-brick structures for housing windmills were being built in Afghanistan and Iran by at least the ninth century. From here the design spread throughout the Arab world, and beyond, to India and China.

The idea of the windmill had reached England by A.D. 1137, where it underwent a significant change. Effectively a new kind of windmill was invented (rather like Heron's) in which the shaft was horizontal rather than vertical. But there was also horizontal rotation: In the post-mill, the most popular design, the whole mill structure could be turned to catch the wind. This was so successful that windmills soon became a familiar sight in the countryside of medieval Europe.

One of the earliest pictures of a Western windmill, painted in England about A.D. 1250. The miller is turning the mill on its post to face into the wind.

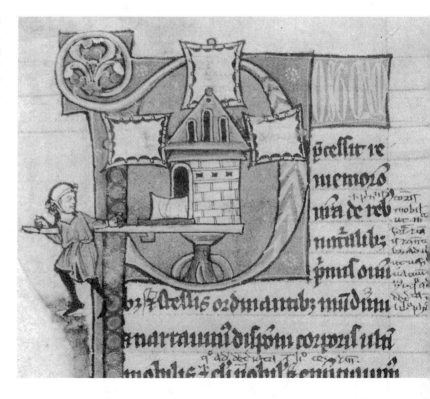

PESTICIDES

Ever since people started to grow crops, a state of war has existed between the human race and the insects that feed off the fruits of our labors. Today there are real fears that the use of artificial pesticides has gone too far and has become a major threat to health. Researchers are therefore now reexamining the vast variety of natural defenses against crop-devouring insects, developed by farmers over thousands of years. The organic pesticides used in ancient times were far weaker than modern ones, killing only the insects at which they were aimed, while even more specific protection was offered by utilizing "biological weapons," creatures who are the natural predators of pests.

The earliest records of pesticides come, not surprisingly, from China, the wealthiest agricultural society of the ancient world. In the third century B.C. special officers were appointed to take charge of pest control. In later years numerous schemes were produced to take advantage of the natural properties of various animals and plants.

Whole books survive from the Han Dynasty (202 B.C.–A.D. 220) onward, describing the precautions that should be taken to safeguard crops at every stage of the agricultural cycle, from sowing to storing the harvest. Pesticides were sprayed on the ground before planting to kill off insects and weeds and on the seeds themselves to keep birds from eating them. During their growing period, crops were again sprayed to keep animals away. When the grain harvest was gathered in, it was sprayed with pesticides once more before being stored. Lamps were hung on peach trees to protect them from insect infestation: attracted by the light, the insects were burned up. Poisonous oils were injected into trees to kill off boring insects. In the home as well the war against pests continued. Leaves of the rue plant were placed in books as protection against bookworms, and mosquitos were kept at bay by burning herbal mixtures.

In the Roman Empire many similar schemes were developed in the same period, some of which were recorded by the encyclopedist Pliny in A.D. 77. He reports the practice of steeping seeds in noxious liquids before sowing to protect them against insects, burning garlic to drive off birds and caterpillars, and scattering poisonous plants, leaves and shoots in granaries to ward off rats and insects. The crushed cypress leaves mixed with seeds, for example, would have given off hydrocyanic acid, an effective insecticide.

Biological Pest Control

The Chinese were, however, far ahead of the Western world in natural pest control. In the countryside frogs were always a forbidden food because they ate insects. Praying mantises were released in gardens among the chrysanthemums to drive away leaf-eating insects.

The most remarkable and economically important of the ancient Chinese biological weapons was the yellow citrus killer-ant. Its use is described in Hsi Han's *Records of the Plants and Trees of the Southern Regions*, written in A.D. 340:

> The Mandarin Orange is a kind of orange with an exceptionally sweet and delicious taste. . . . The people of Chiao-Chih sell in their markets [carnivorous] ants in bags of rush matting. The nests are like silk. The bags are all attached to twigs and leaves, which, with the ants inside the nests, are for sale. The ants are reddish-yellow in color, bigger than ordinary ants. These ants do not eat the oranges, but attack and kill the insects which do. In the south, if the mandarin orange trees do not have this kind of ant, the fruits will be damaged by many harmful insects, and not a single fruit will be perfect.

Defending orange trees against pests became a small business in southern China. In the twelfth century the ants were trapped by filling a pig or sheep bladder with fat and hanging it up next to an ants' nest. Once the ants had moved house to the bladder, it was taken off to market to be sold to fruit growers. To help the ants spread through an entire orange grove, the owners would build miniature bamboo bridges connecting the trees. With the current revival of interest in natural pesticides, perhaps the yellow citrus killer-ant may one day spread around the world—with a little help from humanity.

BEEKEEPING

From its first discovery honey has been one of humanity's favorite foods. Spanish rock paintings of c. 6,500 B.C., predating the arrival of farming in western Europe, show honey being robbed from wild bees' nests. Actual beekeeping, however, took several thousand years more to develop.

The earliest certain evidence of beekeeping—as opposed to the collecting of wild honey—comes from ancient Egypt. The Fifth Dynasty (2550–2400 B.C.) sun temple of Neuserre, at Abu Ghorab, contains a stone relief showing all the stages of honey manufacture, from taking honeycomb out of the hive to straining honey into storage jars. Egyptian hives were made of hollow cones of dried mud, while later Greek and Roman ones were fashioned from terra-cotta. The Egyptians offered enormous quantities of honey to their gods: one list from the reign of Ramesses III (12th century B.C.) amounted to fifteen tons in 31,092 jars, probably a year's output from five thousand hives. Honey was also a very common ingredient in Egyptian medicines, presumably to make them easier to swallow.

The Hittites of ancient Turkey were great beekeepers, although at times rather uscrupulous ones, as suggested by the fact that around

A honey-collecting scene from a rock painting at La Arana, in Spain, dating to around 6000 B.C. Honey from a wild bees' nest at the top of a tree is scooped out into a bag.

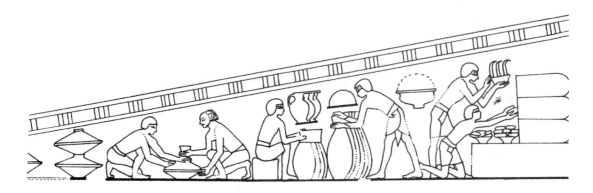

A stone carving from the 5th Dynasty (c. 2500 B.C.) sun temple of Neuserre, at Abu Ghorab, in Egypt. It depicts a set of hives with a head beekeeper kneeling in front of one of them, collecting honeycomb in a dish; some of his assistants drop pieces of honeycomb into a large pot, while others strain honey from a basin into a jar.

1500 B.C. they had to introduce a system of fines for stealing other people's swarms. In later times beekeeping spread south from Turkey to Iraq, encouraged by government policy. In the eighth century B.C. Shamash-resh-usur, Assyrian governor of Suhu on the river Euphrates, proudly set up an inscription to record his role in this movement:

> Bees which gather honey, which no one among my predecessors had seen or brought down to Suhu, I brought down from the mountains [of eastern Turkey] and established [in my town]. They collect honey and wax. I understand how to extract the honey and wax, and the gardeners also understand it. Let any person who comes along later ask the old men of the country whether it is true that Shamash-resh-usur the Governor of Suhu introduced bees.

Honey lost none of its attraction in later times and was widely used for sweetening foods (see **Introduction** to **Food, Drink and Drugs**). Beekeeping was established in the Americas by the Maya, who had the great advantage of dealing with stingless bees. Only the Spartans of ancient Greece rejected honey, presumably because they thought of it as decadent; they especially loathed honey baked into cakes, which they scorned as "no food for free men." At the other end of the scale (where most of us would doubtless place ourselves), the Romans were so fond of honey that they said to someone whom they wished well, "May honey drip on you."

FISH AND OYSTER FARMS

Gluttons for seafood of all kinds (see **Cookbooks** in **Food, Drink and Drugs**), the Romans perfected the art of fish farming to a degree unrivaled until modern times. The idea itself goes back much farther, however. Fishing was a favorite pastime of the Egyptians, and while keen anglers would fish in the Nile itself, lazier folk, if they could afford it, had the river and its produce brought into their back gardens by means of small canals. Tomb paintings going back to the Eighteenth Dynasty (1550–1300 B.C.) show Egyptian noblemen fishing in their gardens to relax.

This Egyptian practice was really fish channeling rather than farming; the first fish farms proper seem to have been invented by the Greek colonizers of Sicily. Here, in the fifth century B.C., Gelon (ruler of Syracuse) had a colossal tank, nearly 4,200 feet around and 30 feet deep, dug at Agrigentum. According to the historian Diodorus, "Into it the waters from rivers and springs were conducted and it became a fish-pond, which supplied fish in great abundance to be used for food and to please the palate; and since swarms of swans also in the greatest numbers settled down upon it, the pool came to be a delight to look upon."

Scene from an ancient Egyptian tomb painting: a nobleman "goes fishing" the easy way—from a canal dug between the Nile and his estate.

Roman Fish Farms

At about the same time early Roman farmers were breeding fish in their ponds and experimenting with keeping marine fish in freshwater lakes. Some species can be transferred in this way and successfully reared, but gourmets still claimed to be able to tell the difference. One of their favorite jokes concerned a notorious food snob, who, on tasting a sea bass raised in a river, spat it out and exclaimed, "Damn me, I thought I was eating fish!"

The problem of farming sea fish was solved around 95 B.C. by one Licinius Murena, hailed as the real inventor of the fish farm. According to Pliny, he spent a small fortune digging a channel from the sea, cutting straight through a mountain near Naples, to supply seawater to the fish tanks on his estate. For this feat Licinius earned the nickname "Xerxes in Roman Dress," a reference to the Persian emperor famous for his megalomaniac feats of canal building (see **Introduction** to **Transportation**). Others, with land by the sea, simply built their tanks directly on the coast.

Both fresh and sea water tanks for fish (called *piscinari*) became the rage, and any self-respecting villa owner had a *piscina* to show off to visitors. Like their Egyptian predecessors, sedentary Roman gentlemen would "go fishing" at home. The Roman poet Martial lampooned one Apollinaris, who "could always furnish his table no matter how stormy the sea, by letting down a line from his couch to catch turbot or bass." Many Roman nobles took a keen personal interest in the welfare of their fish, often treating them as pets rather than food. Some were actually too fond of their fish to eat them and instead bought fish for the table from the market. Antonia Augusta, daughter of Cleopatra's lover, Mark Antony, and mother of the emperor Claudius (A.D. 41–54), kept in her *piscina* a favorite lamprey, which she decorated with golden earrings. One nobleman, Quintus Hortensius, was said to be more concerned about the health of his fish than that of his slaves, while another, Vedius Pollio, took this trend to its logical extreme and fattened his fish on the flesh of delinquent slaves.

The excesses of the Roman nobility scandalized the experts on practical farming. Varro, who wrote a handbook on agriculture in the late first century B.C., thought that the mania for *piscinari*, by that time often more like glorified aquaria than fish farms, had gone beyond a joke, and he condemned the practice altogether. Columella, whose book *On Farming Matters* appeared around A.D. 60, agreed that the snobbery involved was repugnant, but stressed that fish farming was still a practical and honest way for poorer landowners to earn extra income. His book gave detailed instructions on the kinds of fish suited to different coastal terrain and on how to build and maintain tanks. The best results, he stressed, came when their construction allowed for fresh, cool water to be brought in by the tide each day—the water must never become warm or stagnant. Bronze grids on the channel entrances would let the sea in but keep the "water flock" from straying. Columella devoted a great amount of space to correct feeding and—modern fish farmers take note—emphasized the importance, both for a good product and for the well-being of the fish, of making their environment as natural as possible:

> If space allows, it will not be amiss to place in various parts of the pond rocks from the sea shore, especially those which are covered with sea weed and, as far as the wit of man can contrive, to represent the appearance of the sea, so that, though they are prisoners, the fish may feel their captivity as little as possible.

Some extra tips on fish farming were provided by Pliny. The mullet had a reputation among the Romans of being gullible (as well as particularly tasty), and Pliny described how, in the mating season, a male mullet with a fishing line tied to its gills was released into the sea. It was then reeled in back to the farm, bringing with it a number of interested female mullet; the opposite method was used during the laying season, when a female fish would be sent out to lure in male mullet. Pliny's decription and other accounts of fish breeding dispense with the modern myth that the Roman *piscinari* were not fish farms but merely tanks for storage.

The literary descriptions are now nicely complemented by archaeological evidence. Underwater archaeologists have explored and mapped a number of *piscinari*. By Pliny's time Roman fish farming had spread throughout the Mediterranean; the mullet-catching method he described was used as far afield as southern France and Lebanon. Indeed the best examples of working fish farms have been

A carefully designed fish farm in the Roman style near the port built by King Herod on the coast of Palestine. The connections between the main tank, smaller tanks and the sea were all protected with sluice gates, to control the movement of fish and the flow of water. At the south eastern corner is a round hole bored into the rock at sea level—the lapping of waves at the hole would have had the effect of a syphon, sucking new seawater into the system through the other channels.

found outside Italy, including a number on the coasts of eastern Crete. Cut into the rock, they have from two to ten compartments. The archaeologists who explored them were reminded of Varro's words: "Who, however, goes in for fish tanks and does not have a row of them? For just as painters have large boxes with compartments for keeping their pigments of different colours, so they have tanks with compartments for keeping the varieties of fish separate."

Varro, in his sarcasm, omitted to add that the reasons for this were entirely practical. After interviewing modern Cretan fishermen, marine archaeologist Costis Davaras decided that the purpose of such compartments was to keep apart kinds of fish that would eat or harm each other. Columella stressed the need, for example, to separate off moray eels, since they occasionally indulged in frenzied attacks on other fish.

The most extensive fish-farming complex from the Roman world

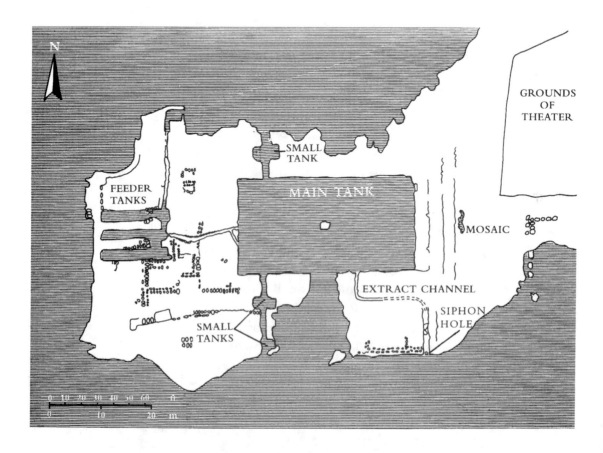

is that found at Lapithos, in northern Cyprus. Here a series of six tanks separated into three groups was dug out of the coastal rock. The main tank is 88 feet long and 45 feet wide, with three channels connecting it to the sea; the flow of water into the tank was regulated by sluice gates, made from stone slabs slotted into grooves that could be raised and lowered. Once thought to be the largest from the ancient world, the Lapithos tank was surpassed by a discovery made in 1964. Exploring the coast around the magnificent ancient port of Caesarea, in Israel, underwater archaeologist Alexander Flinder discovered an enormous *piscina*. Like the port itself, it was almost certainly built between 22 and 9 B.C. by King Herod the Great, then the most powerful ally of Rome in the East. The main tank at Caesarea is 115 feet long and 58 feet wide, with several feeder tanks and a complex system of channels connecting it to the sea. Overlooking it is a terrace with remains of a mosaic floor and decorated columns. Archaeologists still hope to find nearby the remains of a seaside villa built by Herod. It is curious to imagine Herod, the notorious tyrant of the New Testament, taking a break from his main business of exterminating political rivals by dabbling in fish farming at his holiday villa!

Oyster Farms

Fish were not the only aquatic food cultivated in ancient times. About 50 B.C. one Fulvius Lippinus began farming water snails. He specialized in producing giant edible snails, using selective breeding and special diets, including wine-enriched food. According to Varro, Fulvius produced snails whose shells could hold eighty quarts, though this is hardly believable.

Oysters and mussels were also much in demand in the ancient Mediterranean world, not only as food but for the pearls they contained. The Roman writer Suetonius noted that reports of the pearls produced by the mussels of British rivers helped convince Julius Caesar to mount his invasion of Britain in 55 B.C. On his return to Rome he dedicated a breastplate decorated with British pearls in the temple of the goddess Venus. Roman ladies vied with one another over the number of large pearls they could cram onto their hair, neck and hands. Cleopatra, queen of Egypt, was reputed to have owned the two largest pearls ever seen, one of which (worth about one million dollars in today's terms) she swallowed to win a bet with Mark Antony that she could squander more money than he on a single banquet.

Not surprisingly, then, the Romans began to farm oysters in much the same way that they did fish. The first to develop oyster farms, around 95 B.C., was the ingenious businessman Sergius Orata, whose invention of heated ponds led to the development of the hypocaust central heating system (see **Central Heating** in **House and Home**). According to Pliny, his motive was not gluttony but pure avarice: Orata made an emormous profit from the enterprise, having carefully selected the tastiest varieties of oyster for his breeding stock.

It was left to the ancient Chinese, who rated the value of pearls above that of gold and second only to jade, to develop the knack of producing cultured pearls. Natural pearls are formed by oysters around an irritant that they have swallowed, but one may have to open dozens of oysters before finding one with a developed pearl. The natural process can be speeded up by dropping small pieces of foreign matter into an open oyster, which then coats it with nacre. Just when the Chinese began oyster farming is uncertain, but by the Middle Ages they were cultivating them by placing small false pearls into the oysters, as shown by court records from A.D. 1086:

Hsieh Kung-Yen, an Executive Official of the Ministry of Rites, found out a way of cultivating pearls. The way this is done now is to make "false pearls." The smoothest, roundest and most lustrous of these are then selected and inoculated into fairly large oysters kept in clean sea water, as soon as they open their valve. The clean sea water is repeatedly renewed, and at night the oysters take up the best influences of the moon. Then after two years real pearls are fully formed.

Whether Hsieh Kung-Yen really invented the art is doubtful, according to Joseph Needham, the leading expert on ancient Chinese science. References in other texts suggest that the Chinese understood the natural processes involved hundreds of years earlier. In modern times the Chinese have practiced the art of cultivating shaped pearls by dropping small carvings or molded wire into oysters. They may well have discovered this technique as early as A.D. 489, when one text records how the emperor was presented with "a white pearl shaped naturally like the image of a meditating Buddha."

DRILLING AND MINING

Drilling for oil is a business we associate so strongly with modern times that it is surprising to learn that the industry in fact dates back eight hundred years. Natural seepages of crude petroleum from underground oilfields had been exploited much earlier for oil lamps (see **Introduction** to **House and Home**), but these became insufficient to meet demand. In the twelfth century A.D. oil wells had been drilled both in Europe, at Pozzuoli, near Naples; and in the Islamic world at Baku, on the western coast of the Caspian Sea. Around 1250 the Venetian merchant Marco Polo visited the Baku wells, reporting that "a hundred shiploads might be taken from it at one time." Baku oil had clearly been tried out for a variety of uses, some more successful than others, as Polo noted: "This oil is not good to use with food, but it is good to burn and is also used to anoint camels that have the mange. People come from vast distances to fetch it, for in all the countries round it they have no other oil."

In China industrial drilling goes back even farther, although not for petroleum. The brine wells of Sichuan Province were first commissioned by Ling Bing, a local official of the Warring States Period (480–221 B.C.). These were probably wide, fairly shallow, wells, but by A.D. 300 drilling operations for both brine and natural gas (mainly methane) were being carried out side by side, down to depths of 650 feet. The gas was ignited and used to boil the brine, evaporating the water and leaving the desired product, salt.

The depth of wells increased over the years, reaching 850 feet under the Tang Dynasty (A.D. 618–906), but the method of drilling them remained unaltered. At the top of the borehole a shaft was dug with spades until hard rock was reached. The shaft was then filled

The ancient Chinese technique of drilling for brine—as depicted on a molded brick of the Late Han Period (A.D. 25–220). On the left is the derrick, with a winding gear at the top to raise and lower the drill bit. At the bottom right other figures are evaporating salt from the brine.

with carefully prepared stones, pierced with holes in the middle, stacked one on top of the other up to ground level and perfectly centered so that a long hole, eight to fourteen inches in diameter, extended down through them to the rock. Then the drilling began, using a drill with a cast-iron bit suspended from a derrick by bamboo cables. The bit was lifted by a man jumping onto a lever and came crashing down when he leaped off again. Through this laborious process one to three feet a day could be drilled. When the brine was reached, it was collected in a bamboo tube with a valve at the bottom and hoisted to the surface. Around the boreholes huge evaporation pans heated by natural gas were set up to produce the salt. The salt industry became a state monopoly in 199 B.C., when government officials were posted at the wells to supervise drilling.

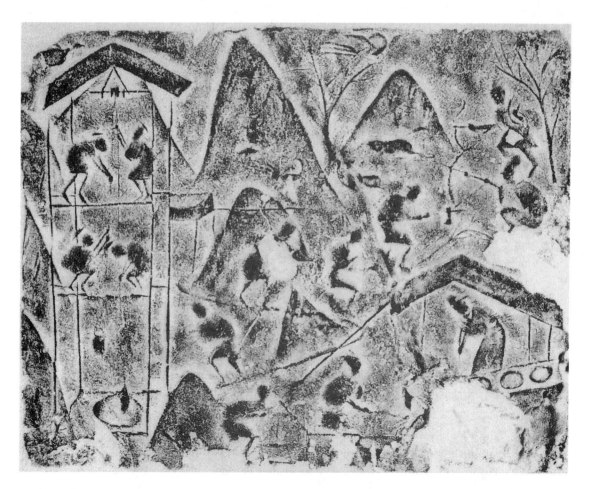

Flint, Metal and Coal Mining

Mining for solid raw materials, as opposed to drilling for liquids and gases, has a far longer history, stretching well back into the Stone Age. Ancient Australians exploited the limestone cave at Koonalda as a flint source some twenty thousand years ago. The cave was one thousand feet long and could be reached only after a difficult two-hundred-foot descent from the plain above the entrance. Artificial lighting must have been used, since the cave lies well beyond the reach of sunlight.

Flint was a vital raw material for tools for thousands of years in the Old World, and when the demand for flint axes (used to clear woodland by early farmers) exceeded the supply available from surface workings, mines were developed. Around 4000 B.C. a whole series of flint mines was opened up in western Europe. The miners dug shafts thirty to forty feet deep to reach seams of high-quality flint in chalk or limestone, which they followed by digging galleries with red-deer antler picks, removing the debris with shovels made from the shoulderblades of cattle. Hundreds of shafts were excavated at some sites, and many thousands of flint axes produced.

Strangely, mines may have been dug in search of metal ores before they were used to find flint. Copper mining had started at Rudna Glava, in Serbia, by 4500 B.C.; the miners followed ore veins down some sixty feet, digging out the shafts with antler picks; in the following centuries they excavated thousands of tons of ore as the Balkans became one of the major sources of copper for the ancient world.

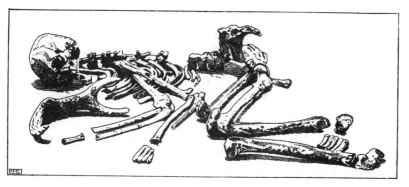

Skeleton of a Stone Age miner discovered in the flint mine at Obourg, in Belgium. He was killed around 3500 B.C., when the roof of the mine collapsed. Next to his hand is the antler pick he was using to extract flint from the seams.

The Egyptians were skilled miners and highly proficient at organizing mining expeditions into the mountainous desert regions on either side of the Nile Valley. As early as 2600 B.C. the pharaohs were sending mining gangs beyond Egypt to the Sinai Peninsula to work its rich deposits of turquoise. It used to be thought that the Egyptians also initiated copper mining in this area, but it has now been shown that the mines of the Negev, in northern Sinai, had already been worked by the local inhabitants for two hundred years or so before the Egyptians came. Large shafts, up to twenty feet in diameter, were bored down through the sandstone and connected by a system of underground galleries. This "large-scale, sophisticated enterprise," in the words of its modern excavator Beno Rothenburg, "is the earliest shaft-and-gallery copper mining system known so far." Later the Negev copper mines were taken over by the Egyptians, who exploited them on and off for the next two thousand years. During the mid-tenth century B.C. King Solomon of Israel seems to have taken control of the copper mines (perhaps with the permission of his Egyptian allies), which thus provided one of the sources of his fabled wealth.

For gold the Egyptians prospected mainly in the eastern desert of Egypt and in Nubia to the south. While we know little from archaeology about their gold-mining practices, an indication of the scale is given by the stupendous amounts of gold dedicated by the pharaohs in their temples. Around 1450 B.C. Thutmose III made an offering in the Temple of Amen, at Thebes, of 13½ tons, equivalent to some $150 million at today's market price. The Nubian gold mines were still being worked in the time of the Ptolemies (3rd–1st centuries B.C.), whose last and most famous ruler was Queen Cleopatra. The ancient Greek author Diodorus left a detailed account of the Nubian mines. Unlike the copper and turquoise mines to the north, which were manned by professionals, the gold mines were worked by convicts and political dissidents sent there as a punishment. Chained together, they were worked to death to provide the insatiable Egyptian court with gold. The Egyptian passion for gold also inspired the world's earliest known mining map (see **Mapmaking** in **Transportation**).

Coal mines existed in China as far back as the Western Han Dynasty (202 B.C.–A.D. 9) and were well developed by A.D. 1000. The ancient mine at Hebi, in Henan, has a main vertical shaft 150 feet deep, with a series of galleries running off at intervals to reach the seam. The mined coal was placed in baskets, pulled to the top by ropes. On the other side of the world, in Britain, the Romans extracted coal from A.D. 100 onward, although no actual mine sites have been found so far. The coal was used as fuel for hypocaust systems (see **Central Heating** in **House and Home**) and in smithies and workshops. Large amounts were supplied to the army, and tons of coal have been found in forts on Hadrian's Wall, abandoned around A.D. 400 when the Roman legions left Britain. At Housesteads fort the guardhouse had even been converted into a coalshed around A.D. 300.

Refreshments are delivered by rope to busy Greek miners—scene from a clay plaque of the 6th century B.C. found at Corinth.

The importance of mines as a source of wealth in ancient economies is amply demonstrated by the case of Athens. The silver and lead mines at nearby Laurion had been worked from 2000 B.C., but in 483 B.C. a rich new vein of silver was opened up, which provided the money to build the fleet that three years later destroyed the massive Persian armada at the great Battle of Salamis. The mines were state owned but leased out to private contractors, who used a mixture of free and slave labor. Galleries were driven up to 330 feet in from the side of the hill, following the seams. These are low and narrow, just large enough for a man to crouch in. Laurion is in a very dry area, so an elaborate system of water storage was developed to provide the water needed for washing the ore (a necessary part of the silver-extraction process) and for the workforce to drink. This involved digging large water cisterns into the rock and lining them with cement, to conserve the winter's rainfall for the summer months.

Roman mines were normally run by the state, with a workforce composed largely of slaves and convicts; the Spanish copper mine at Córdoba had a shaft 688 feet deep, while at El Centenillo Roman miners cut galleries 3,500 feet long and 650 feet deep, chasing seams of silver-bearing lead. At sites such as Las Herrides, in Spain, metals were produced on a vast scale—here 270,000 tons of slag from silver-ore extraction were left behind after 350 years of mining. The Romans took the techniques of mining to a height unrivaled until the later Middle Ages. According to George Sarton, a specialist in ancient science, "they developed new ways of flushing, pitting, driving galleries, sinking shafts, lighting and ventilating, draining, propping, hauling, and surveying," as well as better iron tools. At Spanish mines remains have been found of water screws, the device invented by Archimedes for raising water (see **Gardens and Gardening** in **Sport and Leisure**) and waterwheels fifteen feet in diameter. Both were driven by manpower, and they enabled mines to be constantly drained—and hence greater depths to be reached.

On the other hand the working conditions of the miners were generally appalling, though some efforts were made to improve their lot. The enlightened emperor Hadrian (A.D. 117–138) ordered that baths be established at all Roman mines so that the workers could at least wash themselves.

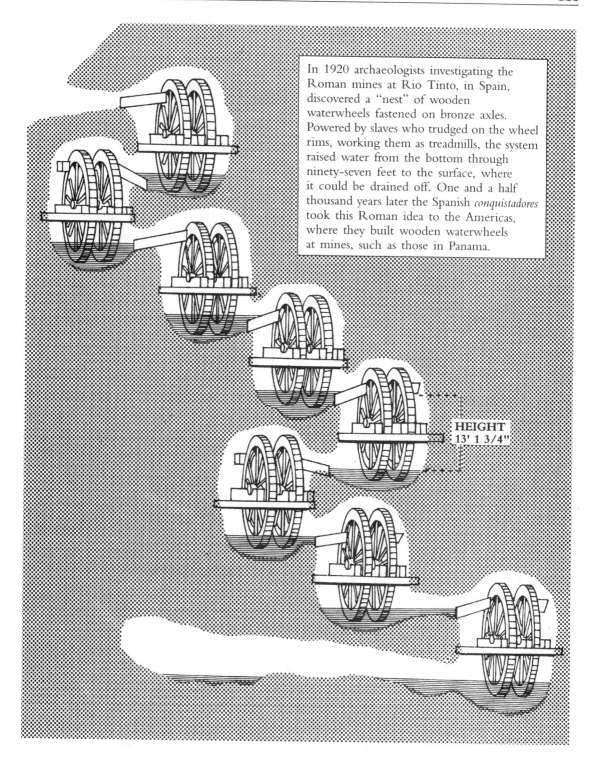

In 1920 archaeologists investigating the Roman mines at Rio Tinto, in Spain, discovered a "nest" of wooden waterwheels fastened on bronze axles. Powered by slaves who trudged on the wheel rims, working them as treadmills, the system raised water from the bottom through ninety-seven feet to the surface, where it could be drained off. One and a half thousand years later the Spanish *conquistadores* took this Roman idea to the Americas, where they built wooden waterwheels at mines, such as those in Panama.

HEIGHT
13' 1 3/4"

Hot Rocks With a Dash of Vinegar

The possibility that a chemical means of boring through rock was available in ancient times is suggested by an intriguing story given by the Roman historian Livy. Describing Hannibal's famous crossing of the Alps in 218 B.C. (see **The "Claw" of Archimedes** in **Military Technology**), Livy states that the descent from the mountains was too steep without a path:

> It was necessary to cut through the rock, a problem they solved by the ingenious application of heat and moisture; large trees were felled and lopped, and a huge pile of timber erected; this . . . was set on fire, and when the rock was sufficiently heated, the men's rations of vinegar were flung upon it, to render it friable. They then got to work on the heated rock and opened a sort of zigzag track to minimize the steepness of the descent, and were able . . . to get the pack animals, and even the elephants, down it.

Part of the story is easily understandable. "Hot mining" is a standard and age-old mining technique. If sufficient timber is available, rock faces can be heated to the point where they begin to crumble and fracture, making the chiselers' job much easier. The technique is mentioned in Diodorus's account of the Egyptian-controlled gold mines in Nubia. But what role did the vinegar play in the Hannibal story?

The passage has completely baffled historians and has been the subject of a long-standing controversy. The word used by Livy, *acetum*, definitely means "wine vinegar." It is true that the acetic acid in vinegar can eat away some rocks, such as limestone, and it has been suggested that if the rocks encountered by Hannibal's army were limestone, vinegar could have been used to dissolve them. In practical terms, however, for every ton of limestone dissolved in this way, Hannibal would have had to be carrying twenty-four tons of vinegar! Besides which, the Alps are not made of limestone. Others have argued that Livy's text is corrupt and that the word was originally *aceta*, which has been interpreted as meaning an "ice pick" or "blowpipe." Herbert Hoover, thirty-first president of the United States and an experienced mining engineer, ventured to suggest that the original Latin did not read *infuso aceto* ("pouring on of vinegar") but *infosso acuto* ("furious digging"). He did admit, however, that his version, while it made sense from a miner's point of view, made rotten Latin. Finally, the Swedish engineer Gösta Sandström argued that the smell of acids released from the wood when it was burned ("wood vinegar") wafted over to the soldiers, who, ignorant of mining techniques, assumed that actual vinegar was being used; they passed the story on, and Livy was therefore reporting a "200-year-old veterans' tale."

None of these theories takes into account other classical references to the use of vinegar in mining. The Roman encyclopedist Pliny also refers to this use of vinegar: "Tunnels are bored into mountains. . . . When quartz occurs one seeks to blast it with fire and vinegar . . . the resulting steam and smoke often fill the tunnels." Similarly the Roman architect Vitruvius wrote, "Even rocks of lava, which neither iron nor fire alone can dissolve, split into pieces and dissolve when heated with fire and then sprinkled with vinegar."

It is silly to suggest that scientists such as Pliny and Vitruvius were merely following a corrupt text or a misunderstanding on Livy's part. They were clearly discussing a familiar mining practice. Heating up rocks rich in quartz and then rapidly cooling them with water will certainly help them fracture along natural planes, and according to John Healy, author of the standard work on Roman mining, adding vinegar to the water might well speed up the disintegration caused by the heat. Until practical experimental work is done—the one factor missing among the welter of discussion and speculation—we will still be unsure if Healy has hit on the right solution at last.

Mirrors as Mining Tools?

Another mystery concerns how ancient mines were lit. Torches and oil lamps were used of course, but in confined spaces they would have made the atmosphere suffocating by consuming the oxygen and filling it with smoke. Many Egyptian tombs cut deep into the sides of mountains (see **Tunneling**) have white-painted walls that show not the slightest trace of soot left from lamps. How, then, were they illuminated for the artists who decorated them?

A clue comes from a trick used by modern guides working in Egyptian tombs. They place sheets of tin foil on large pieces of cardboard and stand at strategic points in the tunnel to reflect sunlight in for the tourists. In the strong sunshine of Egypt the illumination provided by this simple method is extremely effective.

The invention of a similar technique in ancient times is quite conceivable, and evidence for the use of mirrors in mining has now come to light from the Egyptian turquoise mines carved into the mountain of Serabit el-Khadim, in the Sinai. In 1977 archaeologists cleared out one of the mines, worked around 1500 B.C., and discovered a wide range of molds for casting metal tools, left behind by the miners. These included—as might be expected—shapes for axes, adzes, chisels and knives, but also several molds clearly designed for casting the simple type of Egyptian bronze mirror (see **Mirrors** in **Personal Effects**). As the excavator remarked, "The discovery of mirror molds as an everyday utensil . . . seems unusual in an assemblage that consists mainly of tools." He suggested that the mirrors were offerings made by the miners for dedication at the nearby temple of Hathor, their patron goddess, but this seems rather a forced explanation, and it has been argued that they were really made to reflect light into the mine shafts. The alternative theory—that the Egyptian miners of the Sinai used, or even invented, the technique still employed today by tourist guides—is far more attractive.

TUNNELING

The skills used in their mining activities were also employed by the ancients in impressive tunneling work. Civil engineers such as those of the Roman Empire were little daunted by obstacles such as mountains if they wanted to build a road or a watercourse. Obstructions were simply hacked and chiseled away by sheer sweat and strength of numbers.

The Egyptians were past masters at tunneling. The great tombs of the Egyptian pharaohs and their queens, built in the Valley of the Kings during the New Kingdom (1550–1075 B.C.), were excavated from solid rock. Almost as soon as a new pharaoh ascended the throne, he would commission the lengthy task of building his tomb. Gangs of slaves worked for years to complete these magnificent edifices, using copper saws and chisels, and drills that ground down the rock with a paste made of emery powder and water. The tomb of King Merneptah (son of the great Ramesses II), built about 1220 B.C., was begun with a tunnel, 350 feet long, after which a long shaft was sunk, opening into another stretch of tunnel 300 feet long, which led to the burial chamber. The tomb of his grandfather, Seti I, has an entrance tunnel 700 feet long.

About a thousand years later the building of rock-cut temples became a craze in India. The trend of building extravagant temples and monasteries by tunneling into the sides of mountains was begun by the Buddhists and later continued by Brahmin priests of the Hindu faith. They have often misleadingly been described as "cave tem-

Façade of Buddhist temple at Bhaja (Maharashtra State, western India), chiseled out of solid rock during the 2nd century B.C. Such masterpieces of Indian architecture have often been wrongly described as "caves," when they were actually man-made tunnels.

ples," masking the fact that they are completely man-made. At Ellora, in western India, six miles of underground tunnels were chiseled out of solid rock between A.D. 200 and 600. Their façades decorated with intricate sculptures, these Indian temples remain the world's most beautiful and elaborate tunnels. Surprisingly the oldest Buddhist examples, dating from the second century B.C., are also some of the most sophisticated, something that particularly struck Peter Brown, an authority on ancient Indian architecture: "Strangely enough there are no evidences of experimental undertakings or trial cuttings, no gradual growth, no progressive stages . . . the art emereges in a fully matured state. In fact the earliest examples are the most perfectly aligned, planned and wrought of all these excavated halls, with every line mathematically straight and every angle true."

Underground Water Supplies

Extensive tunnels were also built for irrigation purposes, particularly in the Middle East, where many countries have chronically low rainfall. Much of Iran, for example, receives as little as six to ten inches of rain a year. In such a hot, dry climate water conduits above ground, such as the great aqueducts built by the Romans, weren't very effective. So a network of underground water tunnels called *qanats* was developed, bringing water from natural underground reservoirs in the highlands to the arid areas where it was needed for agriculture. Individual *qanats* could be up to twenty miles in length. As they were dug, shafts were let to the surface at intervals of between 65 and 500 feet, for ventilation and to remove the debris from tunneling. The full extent of this system of underground aqueducts is some 170,000 miles. But when was it made?

The greatest expansion of the *qanat* network of Iran belongs to the medieval Islamic period—indeed it is still a mainstay of modern Iranian agriculture. Nevertheless its origins are much older. *Qanats* are known from ancient Armenia before 750 B.C., and the Assyrians of Iraq copied them during the reign of King Sennacherib, in the late eighth century B.C. The Iranian *qanat*s were certainly operating by the time of the Persian Empire (6th–4th centuries B.C.), and the Persians introduced the technique into Egypt after they conquered it in 525 B.C. In 1968 H. E. Wulff, reporting on his intensive study of the *qanat* network of Iran, marveled at the scale of a system that "rivaled the great aqueducts of the Roman Empire," as well as its longevity—begun at least twenty-six hundred years ago, it is still in use today.

Drilling Through Mountains

The need for water supplies was also the motivation for the famous tunnel built on the island of Samos in the late sixth century B.C. by Eupalinus, the first civil engineer we know by name. The historian Herodotus, writing around 450 B.C., rated Eupalinus's tunnel as one of the "greatest building and engineering feats in the Greek world." It was more than half a mile long, short perhaps by *qanat* standards, but in this case it was driven clean through the base of a hill nine-hundred feet high. The remains of the tunnel were discovered in 1882 and closely match Herodotus's description. The main tunnel is about six feet high and six feet wide. At the bottom ran a channel some two feet wide and thirty feet deep, laid with pipes that brought fresh spring water into the city of Samos.

Examination of the tunnel showed that it was built—like the modern Channel Tunnel now connecting England and France—by digging simultaneously from both ends. The junction in the middle is not quite perfect, yet somehow the excavators managed to meet. The great historian of ancient science Dr. George Sarton was clearly puzzled by this achievement: "How did . . . Eupalinus solve the mathematical problems involved? We can only guess; did they have instruments to measure azimuths and differences of levels?"

Presumably they must have had such instruments, though we only have descriptions of them from much later writings. The works of the great Alexandrian engineer Heron, of the first century A.D. (see **Introduction** to **High Tech**, and **Fire Engines** in **Urban Life**),

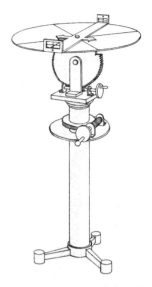

A reconstruction of the *diopter*, a sophisticated surveying instrument, as described by Heron of Alexandria in the 1st century A.D. The device pivoted on a circular plate and was moved around horizontally by a worm screw driven by a small handle. The top plate was mounted on a toothed semicircular plate and was adjusted in the vertical plane by another worm drive. Angles could be read off the top disk, scored with two lines crossing at right angles. The top part of the device was removable and could be replaced by a water level. Like a modern theodolite, it was used in conjunction with a sighting staff with a sliding marker. By using trigonometry, heights could then be calculated by lining up the sights, the marker on the staff and a distant point.

give the mathematical formulae for accurately "piercing a mountain with a straight tunnel" when the entrance and exit had already been selected, and a description of the instrument required for the job. Called the *diopter*, it was very similar to the modern theodolite used by surveyors, with gear systems enabling the sights to be turned in both the vertical and the horizontal planes. Whether earlier versions of such a sophisticated instrument were available in the time of Eupalinus, six hundred years before Heron, is an open question.

The tunnel built by Eupalinus on the island of Samos, Greece, in the 6th century B.C. Cut for over half a mile through limestone, it is remarkable for its straightness. It was a service tunnel for the maintenance of a deep water channel.

A Tunnel Straight to Hell

One of the most technically ambitious tunneling projects of the ancient world was not made for any mundane purpose but for an arcane religious one. It is connected with a legend described in Rome's national epic, the *Aeneid*, written by her greatest poet, Virgil, shortly before his death in A.D. 19. The hero Aeneas, a refugee from the Trojan War, wandered the seas till he arrived on the coast of central Italy, where, in a cavern near the Crater of Avernus, he consulted a weird prophetess called the Sibyl. From there she led him to " a deep rugged cave, stupendous and yawning wide," and conjured open the gates of Hades. Aeneas entered and followed a long road downward until he came to the Styx, the river of the Underworld. He witnessed the most awesome sights, including the dreaded boatman Charon ferrying the souls of the dead over the Styx to their future abode. He saw the shades of his old comrades-in-arms from the Trojan War, and, in a long interview with the spirit of his father, learned of the glorious destiny planned for his descendants, who were to found Rome and conquer the world.

Virgil's account of the Underworld, like so many other legends, proved to have some foundation in fact. In 1870 the German adventurer Heinrich Schliemann enthralled the world when he discovered the city of Troy, the focal point of Homer's *Iliad*. But hardly known at all is the equally extraordinary discovery made in 1962 by two retired naval officers, one English, the other American, Robert Paget and Keith Jones. Convinced that the description of the Underworld in the *Aeneid* was based on an actual location, they spent two years clambering in and out of more than a hundred holes and craters in the volcanic area of Avernus before they came across an opening in the ground under an ancient temple in the Gulf of Baiae (facing the island of Capri). Despite warnings from an archaeologist that entering it would be fatal (because of poisonous gases), they climbed down on ropes and found it to be safe. They had actually found the entrance to the Hell described in the *Aeneid*, which turned out to be a mind-bogglingly ingenious system of man-made tunnels.

The first section of tunnel was clear, and Paget and Jones followed it down for 200 yards, until they were standing 140 feet below the surface. Here they found the "River Styx," an underground stream fed by two boiling volcanic springs, and that, by some unknown contrivance, maintains a constant level. Though one main branch was blocked with soil, Paget and Jones managed to trace the major twists and turns of the tunnel complex, including an upper chamber, which formed the Inner Sanctuary. The total length of the passages they explored is 290 yards, all cut through solid rock.

From ancient accounts, historian Robert Temple has reconstructed the way in which the complex was run by priests as an Oracle of the Dead, where people were given "guided tours" of the Underworld. Dosed with hallucinogenic drugs (which were used in many ancient cults—see **Drugs** in **Food, Drink and Drugs**), they would be guided by chanting figures in black robes through the complex tunnel system in the flickering light cast by smoky lamps. The effect was elaborately stage-managed—vistors were no doubt convinced that they really had been to Hell and back. The priests must have made a good living from customers wanting a glimpse of the afterlife or a chance to talk to their dead relatives. So much for the use of the tunnels. What bothered Temple, however, were the unanswered questions regarding their construction.

The Oracle of the Dead was apparently suppressed by Marcus Agrippa, the right-hand man of the Emperor Augustus (27 B.C.–A.D. 14). He cut down its sacred grove for ship timber and blocked one of the tunnels with soil. Temple calculated that thirty thousand man-journeys were needed for this task, stressing that only a few people could work at a time in the narrow tunnels—all this for the comparatively simple job of filling in one part of the complex with soil. Temple wisely shied away from the task of estimating the time needed to excavate the entire system from solid volcanic rock. It must certainly have taken years by any normal means. Moreover, the tunnels were perfectly cut, with, as Temple notes, "no traces of exploratory or random passages, or of false starts."

But the greatest mystery is how the project was planned. The first stretch (408 feet) of the entrance tunnel is oriented toward the sunrise on Midsummer's Day (the summer solstice). As well as taking this factor into account, whoever built this model of the Underworld must also have known that by drilling through the rock they would hit an underground stream, complete with boiling water, to play the part of the river of Hell. Yet, as Paget and Jones showed by introducing fluorescent tracer chemicals into the water, the "River Styx" is not connected to any of the springs visible outside the cavern. In his classic work on architecture, the Roman author Vitruvius included a chapter on the subject "How to Find Water," by clues from soil, vegetation and the lay of the land. However, such methods (along with water divining, or dowsing, which Vitruvius doesn't mention) are hardly exact. Everything seems to have been fully conceived in advance, but how did the builders manage to combine a solar orientation with the feat of drilling down to reach an invisible underground stream?

A final mystery, as Temple points out, is that we have no idea who built the Oracle of the Dead at Baiae. If, as he suggests, the same location was the model for the description of Hell used in Homer's *Odyssey*, composed in the eighth century B.C., then it would predate that. Or—as seems more likely—did Homer's account inspire the builders? The writings of the classical Greeks and Romans seem to have been exhaustively studied by academics—but it is clear that many fundamental questions still remain unanswered.

The mysterious tunnel complex at Baiae, Italy, used in Roman times as an Oracle of the Dead, where visitors took simulated trips to the Underworld. All the passages are twenty-one inches wide and six feet high. Along their walls are five hundred niches for oil lamps.

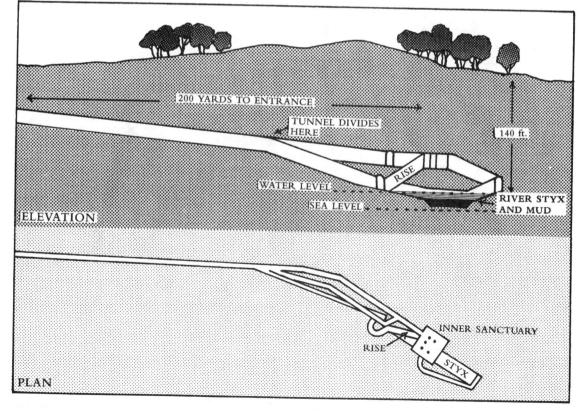

HOUSE AND HOME

CHAPTER TEN

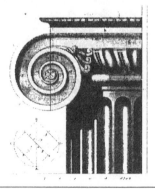

INTRODUCTION

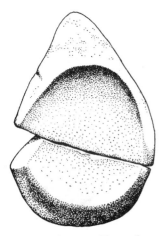

Palaeolithic stone lamp from La Mouthe, Dordogne (France). Stone lamps like these, filled with animal fat and using wicks made of vegetable fiber, were used by the creators of the famous cave paintings of prehistoric France.

As soon as our most distant ancestors began to stay in the same place for any length of time, they started to create windbreaks and shelters from the weather. Some kind of indoor fire was a basic requirement, even in the Early Stone Age, and perhaps as early as 1.5 million years ago, fire was tamed in East Africa. Some archaeologists see this as a crucial evolutionary step: huddling around the hearth for warmth would have increased the bond among the group, essential to the cooperation needed to defeat larger, stronger, animal competitors for food. Controlling fire eventually led to the invention of cooking, another leap forward for the human species (See **Introduction** to **Food, Drink and Drugs**).

But for all their undoubted ingenuity, Stone Age people never solved the smoke problem caused by an open fire. Even the builders of the remarkable mammoth-bone houses of eastern Europe and Ukraine simply left a hole in the roof through which the smoke could escape. A solution had to wait until the invention of the chimney. The Greek Theophrastus (a pupil of Aristotle) mentions a chimney in the early fourth century B.C., but this is almost the only record of one in the ancient world. Chimneys only really became common in the twelfth century A.D., especially in cold, damp Europe. By this time the great halls where the rich held court had been raised to the second story of buildings—so the hearth was moved from the center of the room to a position against an outside wall and provided with a hood to collect the smoke; this was then channeled up through the walls to a chimney. Of course underfloor heating, as used by the Chinese and the Romans, avoided these problems. The Romans also developed glass windows as a way of letting in light while keeping in heat.

In hot climates the opposite problem—overheating—was a cause for concern. Around A.D. 180 the famous Chinese inventor Ting Huan made a man-powered mechanical fan to air-condition the Han Dynasty palace. In the eighth century A.D. the Tang emperors of China would sit in the "Cool Hall" on hot days, where a large water-powered fan kept the air moving.

The second great domestic invention was lighting. The hearth would provide some light, but rarely enough to work by. The Late Stone Age artists of France and Spain came up against the lighting problem when, for reasons that are still mysterious, they began decorating the deepest recesses of caves. About twenty thousand years ago they started to hollow out stone lamps, which they filled with animal fat and lit. The remains of these blackened lamps can be found in large numbers of the floors of the caves.

Forms of lighting changed only gradually over the millennia. Ordinary homes in ancient Egypt some 3,500 years ago were lit by simple pottery bowls containing a wick floating in oil; richer people and royalty used more elaborate lamps, many made from semitranslucent calcite to create magnificent effects.

From the fifth century B.C. crude oil deposits were exploited in Iran and India, but Pliny, writing around A.D. 75, is the first actually to describe a petroleum-rich liquid being used as fuel in lamps. Centuries later, in medieval China, gasoline was mopped up with straw, placed in pots and used for burning in lamps. In the Muslim world the oil fields of Baku, in Azerbaijan, were developed on a commercial scale by A.D. 850, with much of the product going to domestic lighting (see **Drilling and Mining** in **Working the Land**).

Most of the world, however, continued to use fat-burning lamps or candles for lighting the home. In the mid-first millennium B.C. the Etruscans of central Italy were great manufacturers of elaborate candelabra, some over four feet high, standing on three legs and often topped with human figures. Smaller candelabra stood on tables and generally had a decorative figure supporting the central column at the base. The Etruscans even produced novelty pieces with expanding tops. The Chinese were manufacturing less smoky and smelly candles from beeswax by the fourth century A.D.

What must have been almost everlasting asbestos wicks, for use with candles made of animal fat (apparently including whale blubber), are recorded in a rather exaggerated Chinese account from 300 B.C.:

In the second year of King Chao of Yen, the sea-people brought oil in ships, having used very large kettles for extracting it, and presented it to him. Sitting in the "Cloud-Piercing Pavil-

A 6th-century B.C. candelabra with an expanding top made by the Etruscan craftsmen of central Italy. The candle height could be adjusted to throw light wherever it was needed.

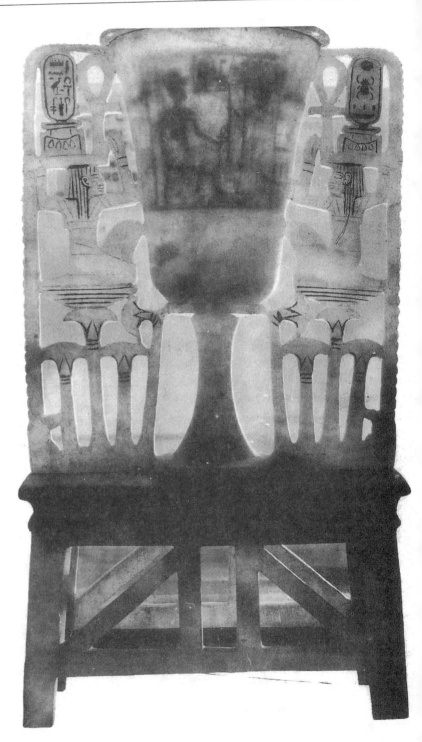

A beautiful lamp found in the tomb of the boy pharaoh Tutankhamen (late 14th century B.C.), carved out of semitranslucent calcite crystal.

on" he enjoyed the brilliant light of the lamps in which the dragon blubber was burned. The light was so brilliant that it could be seen over thirty miles away; and its smoke was colored red and purple. The country people, seeing it, said "What a prosperous light," and worshiped it from afar. It was burned with wicks of asbestos.

The great problem with burning any material, whether for heating or lighting, is how to get it started. In the Stone Age, people used sparks from flint or friction methods. The simplest of these was to rub two sticks together, but more sophisticated methods were developed. One was to twirl between the hands a pointed wooden stick (or "drill") pressed down onto another piece of wood. Much greater speeds could be achieved by winding a bow-string once around the drill and making a sawing motion with the bow. In Bronze Age Europe, around 1500 B.C., a fire-lighting "kit" was developed, consisting of a specially shaped "strike-a-light" flint tool, iron pyrites for striking and dry moss to catch fire.

The Chinese eventually came up with two more technologically advanced solutions: piped natural gas and matchsticks.

Of these the use of natural gas must be the most surprising method of combining heating and lighting (with the added advantage of always being lit) known from the ancient world. In Szechwan Province natural gas was being piped out to areas around the sources as early as the year A.D. 347, according to the geographer Ch'ang Ch'ü in his *Records of the Country South of Mount Kua*:

At the place where the river from Pu-p'u joins the Huoching River, there are fire wells [producing natural gas]; at night the glow is reflected all over the sky. The inhabitants wanted to have fire and used to ignite the outlets with brands from household hearths; every so often there would be a noise like the rumbling of thunder and the flames would shoot out so brilliantly as to light up the country for miles around. Moreover, they use bamboo tubes [pipelines] to "contain the light," conserving it so that it can be made to travel from one place to another, as much as a day's journey away from the well, without its being extinguished. When it has burned no ash is left, and it blazes brilliantly.

The gas, which came from a depth of 2,000 feet, had to be mixed with air before it would burn safely, producing permanent flames 1½ feet high which were used for both heating and lighting the home. In the tenth century A.D. the Chinese had anticipated the modern portable butane gas cylinder by pouring petroleum products (recovered as a by-product of drilling for gas) into bamboo tubes.

According to one Chinese tradition, matches were invented in A.D. 577 by the ladies of the northern Ch'i court of China while they were under siege by nomads from Central Asia. In A.D. 950 the scientist T'ao Ku gave a less dramatic account of the origin of matches in his *Records of the Unworldly and the Strange:*

> If there occurs an emergency at night it may take some time to make a light and light a lamp. But an ingenious man devised the system of impregnating little sticks of pinewood with sulfur and storing them ready for use. At the slightest touch of fire they burst into flame. One gets a little flame like an ear of corn. This marvelous thing was formerly called a "light-bringing slave," but afterwards when it became an article of commerce its name was changed to "fire inch-stick."

The basics of the ancient house were warmth and lighting; once these had been developed, the invention of fancy furniture inevitably followed. Some of the oldest furniture in the world comes from Stone Age Scotland, around 3000 B.C. The beds made by the inhabitants of the Orkney Islands, off the north coast of Scotland, may not look all that impressive, but they were probably very comfortable. They were built on the ground against the houses' thick stone walls, with the ends and side away from the wall formed by thin stone slabs. The interiors were filled with soft mosses. The inhabitants of these prehistoric villages also made dressers or sideboards from thin stone slabs, placed so that they were the first thing a visitor saw on stepping through the door into the house. One of these Orcadian sites, Skara Brae, has also produced the world's earliest known lavatories and sewage-disposal system, a tribute to how seriously the inhabitants took their domestic comforts.

The Egyptian bed of this time was a low wooden framework resting on four animal-shaped legs, with the space inside filled by soft flax string closely plaited and tied to the sides and end of the bed to give it some spring. This was covered by folded linen sheets forming a mattress. Later Egyptian beds often had woven rush bases and a footboard at one end. Instead of pillows the ancient Egyptians used headrests, consisting of a curved neckpiece set on a pillar that stood on an oblong base; these were mostly made of wood, though some were of stone. Some headrests have been found still wrapped in several layers of cloth to give them a soft surface. The Egyptians also made armchairs and wooden or woven reed chests for storing the bed linen. Roman beds were higher off the ground than ours, and steps or a stool were needed to climb into them.

To the Egyptians furniture was a clear indication of status. Wealthier citizens reclined on slant-back chairs or inlaid stools with plush cushions, and dined from individual low tables. The Egyptians also invented the folding stool. Two found in the boy pharaoh Tutankhamen's tomb (late fourteenth century B.C.) were made of ebony, with leather covers and gold decoration. There are a large

A five-thousand-year-old stone house at Skara Brae, in the Orkney Islands. As one looks down from the top of the surviving walls, the entrance is below, the beds to the left and right, the dresser at the back and a fireplace in the middle.

Child in a high chair (or potty?), from an Athenian vase painting of the late 5th century B.C.

number of folding stools from this date in Egypt, and a picture of one from Crete. Strangely, almost identical stools are found in Danish graves of the same date. Is this a case of far-flung Egyptian influence or one of independent invention?

Greek dining couches of the fifth century B.C. were a standard size of five feet two inches long by two feet six inches wide, and stood on short legs. The Greeks didn't have cupboards, but they did make a kind of dresser or sideboard, which they placed opposite the door, to display valuables such as gold and silver cups and statuettes. In displaying their treasures like this they were the heirs to the long tradition started by the Orkney Islanders. The Romans copied both dining couches and sideboards from the Greeks, but usually displayed their prized possessions on tables.

While the floors of poorer homes were usually bare or straw covered, those of the wealthy were often richly carpeted. Ancient sources record that Princess Artystone, daughter of the Persian emperor Cyrus the Great, owned a carpet factory around 500 B.C. A century later the Greek soldier Xenophon wrote that the city of Sard, in Iran, prided itself on the quality of its carpets, which "yielded" underfoot. These were bought by the court, where only the emperor was allowed to walk on them. On the stone reliefs of the Apadana staircase at the Persian capital of Persepolis, carved in the fifth century B.C., we can see tribute in the form of fabrics and textiles being brought to the king.

A superb Persian carpet of this date has been preserved in one of the frozen tombs at Pazyryk in Central Asia, famous for its remains of a tattooed chief with a false beard (see **Tattooing** and **Wigs** in **Personal Effects**.) It is a woolen pile carpet of extremely fine texture, with 520 knots per square inch (as compared with 80 knots for coarse woolen carpets and 800 knots for the finest silk examples of today) and is decorated with figures of riders, deer and griffins. Some fragments of an even earlier Persian knotted-pile carpet, dating to around 650 B.C., were found in the frozen tomb of Bash-Adar, about one hundred miles from Pazyryk. While these examples were imported from farther west, recent investigations in the Sumbar Valley, of Turkmenia, suggest that the art of carpet making may have come from Central Asia in the first place. Archaeologists have uncovered several women's graves, dating to around 1500 B.C., which contain

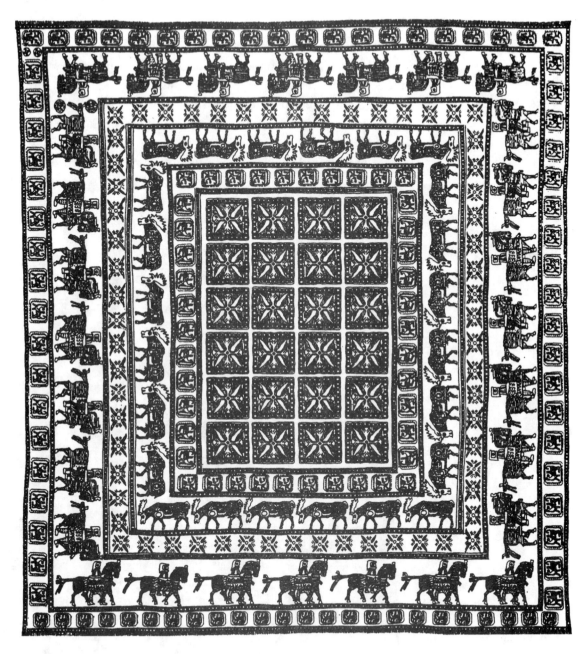

Reconstruction of the Pazyryk carpet, woven in Persia around 500 B.C. Nomads ride or lead their horses in the outer band, while a procession of elk walk around the inner band.

unusually shaped bronze knives of a form identical to the traditional tool used by carpet makers for trimming down the pile.

Perhaps the most extravagant carpet of the ancient world was that made for Chosroes I (A.D. 531–579), king of Persia. It was lavishly adorned with precious gems, depicting every sweet-scented blossom, to console him during the winter months when no flowers bloomed. When the Arabs sacked the palace of Ctesiphon in A.D. 637, they dismantled Chosroes's carpet to sell the gems incorporated in it.

The walls of Greek and Roman houses were often painted, sometimes with scenes providing us with remarkable insights into ancient

A Viking ironing board? Whalebone plaques, like this one from a burial at Grytóy, in northern Norway, are frequently found in women's graves along with glass smoothers.

life. The modern habit of decorating walls with illustrated paper seems to derive from the Chinese practice of hanging up large printed paper sheets on the walls depicting animals, flowers and landscapes. Although traditional Chinese writers condemned such decoration as vulgar, the custom impressed Christian missionaries, who brought the idea back to Europe at the end of the fifteenth century. The earliest wallpaper in Europe was made in 1481. Louis XI of France (1461–1483) commissioned one Jean Bourdichon to paint fifty large rolls of wallpaper with angels on a blue background. The paper was hung in the royal château at Plessis-les-Tours.

Domestic life consists of more than just decoration and furniture, and even the roughest of ancient cultures welcomed that homey touch. Worrying about finding a nicely ironed shirt before going off raiding is the last thing we would expect of the blood-soaked, destructive Vikings of legend, but many Viking ship burials actually contain objects believed by archaeologists to have been for ironing clothes. These are flattened balls of glass, often burned on the base, which would have been irons, and whalebone plaques decorated with animal heads that seem to have been ironing boards. However, ironing equipment is found only in women's graves, so something of the macho image of Viking legend does survive. According to Scandinavian folk history, the boards were carved by young men during long winter evenings, to be given to the girl they loved as a proposal of marriage.

For the most basic domestic matters, necessity was certainly the mother of invention, but the desires for comfort and novelty steadily took over as the driving force for change in the home.

MAMMOTH-BONE HOUSES

Not all, if indeed any, "cavemen" actually lived in caves. Because they traveled widely, the people of the Old Stone Age generally lived in temporary structures, such as tents. Few traces of such shelters have been found by archaeologists, but some of the occasional more permanent dwellings do survive in a surprisingly complete state.

Front (left) and back (far right) views of one of the mammoth-bone houses discovered at Mezhirich in Ukraine, as reassembled by archaeologists. When it was in use, fifteen thousand years ago, it would have had a waterproof covering of mammoth hides.

The most impressive dwellings of the Old Stone Age, which the Flintstones would have been proud to call home, are the mammoth-bone houses of the Ukraine plain. The best-known settlement is that at Mezhirich, near Kiev, found by a farmer in 1965 when he was digging out a new cellar six feet below his home. He discovered a fifteen-thousand-year-old village of at least five sturdy houses, measuring from thirteen to twenty-one feet across and constructed entirely from the remains of mammoths.

The largest limb bones were used to form a massive foundation wall, often arranged in decorative herringbone patterns; smaller bones were set above this in intricate arrangements to make up the upper wall. The roof of the house was made from a framework of tusks, over which mammoth hides would have been stretched. The floor was hollowed out, and at the center was a hearth. The biggest of the houses at Mezhirich used up the bones of nearly a hundred mammoths.

These sturdy structures protected the roving mammoth hunters from the elements during the winter months, being reused for many generations. They took considerable communal effort to construct— archaeologists think it would have taken ten men five or six days to build one, let alone the time involved in collecting the bones, a vastly greater effort than erecting a tent of the time would have required. This suggests that the houses were important tribal bases where the leaders lived together with their followers, in groups of anything up to fifty people.

Some other Ukrainian houses from the same period were even larger: at Kostienki one building was 115 feet long and 50 feet wide, with eleven hearths along its length. Some idea of the wide-ranging contacts of the tribal leaders living in these mammoth-bone houses is given by the finds from the buildings, which include seashells brought from the Mediterranean, four hundred to five hundred miles away, and amber from the Baltic, one hundred miles distant.

CATS AND DOGS

No living creature of course has ever been "invented" by mankind. But dogs and cats as we know them today would not exist without the human race. Both were tempted from the wild by the benefits of human company into a partnership that has not only lasted thousands of years but has also resulted in major psysiological changes: Wolves became dogs, and wild felines became cats. Since they must have made the first moves, dogs and cats in a sense invented themselves. We in turn invented the concept of the pet: unlike other animals, we have always kept dogs and cats, as much for the simple reason that we like them as for the help they give us.

The title of "man's best friend" is well deserved by the dog, the result of our earliest definite success at domestication. There is evidence from as early as 11,000 B.C., some two thousand years before farming began (see **Introduction** to **Working the Land**), that wolves, attracted by the campfires and food scraps of hunters, were becoming tamed and beginning the transformation into dogs. The scene of these first contacts is Palestine, where a number of Stone

Greyhound- and dachshund-type dogs as depicted in Egyptian tombs at Beni Hasan around 1900 B.C.

Age sites have produced the skeletal remains of very early dogs. The most interesting find came in 1977 at the site of Ein Mallaha: An old person, probably a woman, was found in a grave buried with a puppy three to five months old, which must have been a pet. At this and other settlements nearby the regular presence of dogs is also shown by the condition of small gazelle bones that are corroded as a result of having passed through the digestive system of dogs.

By cornering, bringing down, or even retrieving prey, early dogs were invaluable to hunters, and it is likely that the process of wolf taming took place independently in several parts of the world, from the Americas to Australia. Dog remains from about 9000 B.C., nearly as old as the Palestinian examples, were discovered near the Stone Age settlement at Seamer Carr, in Yorkshire, England. In North America the hunters known from the extensive Koster site, in Illinois, were giving favorite dogs a respectful burial from 5500 B.C. onward.

The earliest evidence for the development of different breeds of dogs comes from Egypt, where a painted pottery dish of around 3500 B.C. shows a man holding on leashes a pack of dogs resembling greyhounds. By the second millennium B.C. paintings show this greyhound type (known today as the "Pharaoh Hound") as well as a breed resembling the dachshund. Around the same date a mastiff type of dog was being bred in the Near East for hunting large game, such as wild boar and antelopes. The greyhound and mastiff types of hunting dog appear to have continued to the present day almost unchanged.

A mastiff straining at the leash—from a hunting scene that decorated the walls of the palace of the Assyrian king Assurbanipal at Nineveh (northern Iraq) in the mid-7th century B.C.

The greatest dog breeders of the ancient world were, without doubt, the Romans. They developed specialized fighting and hunting dogs, sheepdogs, guard dogs and lapdogs. These varied greatly in height, build and skull shape. Various provinces of the Roman Empire produced different breeds of hunting dog, some renowned for their strength and others for their speed in the chase. Hunting dogs underwent intensive training, starting when they were puppies. Sheepdogs were bred in Greece and imported from there to Italy; Greek watchdogs were also popular in Roman households. The lapdogs of the Romans were the world's first completely nonworking dogs, kept entirely because they were cute. The Chinese were later developers of this idea, breeding, for example, the Pekingese around A.D. 700.

Cats Chase Mice Inside

The early history of feline "domestication" is less clear, especially since it is generally acknowledged that cats, unlike dogs, exploit their "owners" far more than the other way around. In fact they flagrantly defy almost every definition of the word *domesticated* as it applies to other animals.

Cats also changed less than dogs when they began their partnership with human beings. Some have even become useful companions in their natural state, such as the tame cheetahs used in gazelle hunting in the Middle East and India over the last thousand years. Some Egyptian pharaohs, including Ramesses II (of the 13th century B.C.), were depicted with tame lions running alongside their war chariots. While some of these depictions may be symbolic, another relief of Ramesses II includes the royal lion with its keeper in the fairly mundane surroundings of the military camp along with the horses and oxen. Taming lions was hardly beyond the resourceful ancient Egyptians, and the sight of a lion prowling around Pharaoh's chariot would have had a valuable psychological effect on loyal and rebellious subjects alike.

A fierce Roman guard dog shown on a mosaic buried at Pompeii in A.D. 79 by the eruption of Vesuvius. In case the picture wasn't warning enough, the artist has added the words *cave canem*—"beware of the dog."

The more ordinary, but no less interesting, domestic cat *(felis catus)*, is a descendant of the wild cat *(felis sylvestris)* that still roams the remoter regions of Europe, North Africa and the Near East.

This wild cat really differs very little from the "tame" kind—it is slightly larger and tougher, and the two "species" can still interbreed, as can wolves and dogs such as huskies.

The striped coat of the wild cat shows that the ancestor of all our present breeds of domestic cat was probably a tabby, although claims have been made that some varieties, such as the Persian and Siamese, were descended from other wild species, including the sand cat, which inhabits deserts from the Sahara to northern India. Whatever the case, and with or without our help, cats have diversified into a number of distinctive breeds. The range is, however, far less extreme than that for dogs, who, through intense selective breeding, have now been stretched so far apart on the genetic elastic that holds a species together that a tiny female chihuahua could never bear the pups of a German shepherd. Cats have somehow defied this kind of intensive genetic manipulation—perhaps in part because, given half a chance, they will always sneak out the back door to select their own mates.

Still cats have some use for us, and the first thing that seems to have attracted cats to human beings was our development of crop cultivation around 8000 B.C. Mice were attracted into storehouses by the grain they held, and cats followed the mice. Barns, with their supplies of rodents and comfortable straw bedding, are still a paradise for cats. It must have been shortly after the agricultural revolution of the New Stone Age that cats took their next step and invented the tame housekeeper.

Cat remains have been found at the prehistoric town of Jericho, in Palestine, dating from around 7000 B.C. (see **Introduction to Urban Life**), but the best candidate for the first domestic cat comes from Khirokitia, in Cyprus, around 6000 B.C. This must have been a domestic cat, since its teeth are different from those of Mediterranean wild cats—and, since there were never any wild cats on Cyprus, it (or its ancestors) must have been imported or at least jumped ship, not something a wild cat would do.

The best evidence for early domesticated cats comes of course from Egypt, where tomb paintings and sketches from about 1550 B.C. onward depict cats in a variety of scenes—joining in the hunt for wildfowl in the marshes, chewing up mice and playing under the chairs of their "owners." How much the Egyptians cared for their cats is well illustrated by a papyrus giving a formula for removing poison from cats stung by scorpions.

But the sheer weight of the ancient Egyptian evidence can be misleading. For one thing there is nothing to support the assumption that the Egyptians were the first to domesticate the cat—the Cypriot example precedes the earliest Egyptian evidence by more than four thousand years. Second, the ancient Egyptian respect for the cat has been blown out of all proportion. A trivial but persistent mistake of general-knowledge quizzes is to ask the question: Which animal did the Egyptians worship? The expected answer—the cat—is not wrong so much as the question is. The Egyptians worshiped, or rather held as sacred, a number of animals. Cats, whose mummified bodies have been discovered by the thousand, were obviously highly regarded. But so were ibises, vultures, crocodiles, hawks, cobras, dung beetles and a multitude of other creatures. The myth of Egyptian cat "worship" seems to date back to the days of the early Roman Empire. In 59 B.C. the Greek writer Diodorus witnessed the mobbing (and, it seems, lynching) of a Roman diplomat who had accidentally killed a cat while visiting Egypt. According to Diodorus, the same would have happened if an ibis had been the victim, but you won't get any points for giving this answer in a quiz.

None of this gainsays the fact that the Egyptians were remarkable cat lovers. Indeed the Greek historian Herodotus tells how in the

This Egyptian tomb painting of around 1300 B.C. shows a cat sitting under its mistress's chair, happily gnawing on a bone.

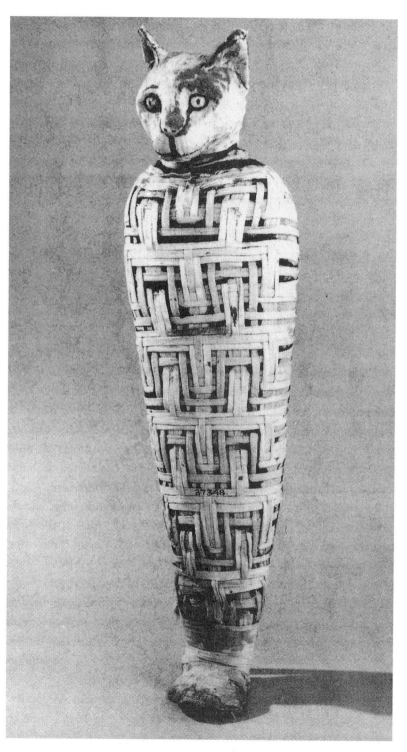

One of many thousands of cat mummies from ancient Egypt, this fine example of the embalmer's art dates to Roman times, after 30 B.C.

fifth century B.C., if a cat died in a house, everyone living there shaved off their eyebrows as a sign of mourning. (If a pet dog died, they would shave their entire body!) Maybe not all of our modern cats descended from Egyptian ancestors, but the custom of keeping them as housepets certainly spread from Egypt to Europe via the Greeks and Romans, and thence around the world. It has even been argued that the word *puss* comes from Pasht, the name of the Egyptian cat-headed goddess, but it doesn't seem that this name was in common use in Egypt itself. The usual Egyptian word for cat was *miw*. Its pronunciation needs no explanation.

LAVATORIES

The third millennium B.C. could be called the Age of Cleanliness, for during that time both toilets and sewers were invented in several widely separated parts of the world. What are probably the world's earliest lavatories, dating to around 2800 B.C., have been found at the picturesque Late Stone Age village of Skara Brae, on the Orkney Islands, off the north coast of Scotland. Recesses in the stone walls of the houses appear to be toilets, since they are connected to drains running away from the houses (see **Sewers** in **Urban Life**).

At roughly the same date, on the other side of the globe, lavatories were also being built into the outer walls of houses at the large city of Mohenjo-Daro, in the Indus Valley of Pakistan. These first Western-style lavatories were neatly built out of bricks, with wooden seats on top. The toilets had their own vertical chutes through which the sewage fell into the street drain outside or into a cesspit. Sir Mortimer Wheeler, director general of archaeology in India from 1944 to 1948, was so struck by this invention that he was prompted to write, "The high quality of the sanitary arrangements could well be envied in many parts of the world today."

More communal facilities were developed in ancient Mesopotamia. A palace at Eshnunna, on the Diyala River, in Iraq, dating to around 2300 B.C., had six toilets with raised seats of baked brick set

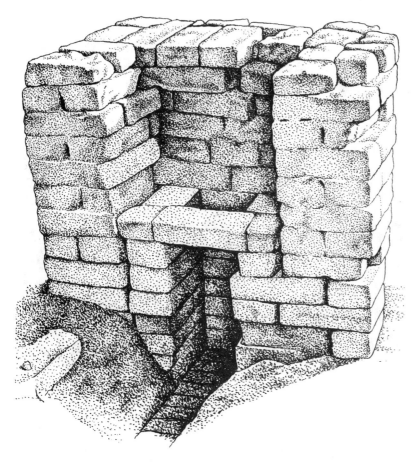

The Western-style sitting toilet may well have been invented by the Indus Valley civilization more than four thousand years ago. Neatly built brick lavatories like this were found in many of the houses at the ancient metropolis of Mohenjo-Daro (Pakistan), dating to around 2500 B.C.

in a row. These were connected to bitumen-lined drains leading into a brick sewer three feet deep. Each of the lavatories had a large water vessel placed next to it, containing a pottery dipper that was used to flush the toilet after use.

In Egypt the earliest surviving toilets are found in the bathrooms of tombs, put there for use in the afterlife. Egyptian lavatories were less sophisticated than those elsewhere: They were rarely connected to sewage systems and had to be periodically cleaned out by hand. At El-Amarna, capital of the heretic King Akhenaton (mid-14th century B.C.), one toilet has a rather chilly-looking limestone seat with a keyhole-shaped hole, which presumably had a large removable jar underneath. In other cases brick supports held a wooden board with hollow spaces on either side of the seat that held sand for throwing down the toilet after use. One Egyptian "first" seems to be

Limestone seat from a lavatory at El-Amarna, Egypt, carved in the mid-14th century B.C.

the portable toilet—a wooden stool with a large slot in the middle, for use with a pottery vessel below it, was found in the tomb of Kha, the senior official of the workmen's community at Thebes. It dates to the mid-fourteenth century B.C.

Some of the most hygienic early toilets were constructed by the Minoan civilization of Crete, discovered by the British archaeologist Sir Arthur Evans at the turn of this century. The best example comes from the palace at Knossos, dating to the mid-second millennium B.C. Off one of the main apartments Evans found a small toilet containing a drain connected to the main sewer running below the palace. The walls were lined with easily cleaned gypsum slabs, and from slots in the wall above the drain Evans deduced that there was once a wooden seat. The drain hole is offset from the center of the seat, and Evans noted that there would have been enough space on one side of the seat to hold a large water jug for flushing after use. Below the drain opening is a curious projection from the rear wall, which Evans surmised "may have been used for the attachment of a balance flap to shut off the escape of sewer gas." Evans was suitably impressed: "As an anticipation of scientific methods of sanitation, the system of which we have here the record has been attained by few nations even at the present day."

The Romans must have gotten the idea of toilets from the Greeks, but they developed them into grand communal facilities. Public baths normally incorporated lavatories, since they were not only popular places but also had large amounts of free water available (see **Baths**). Public lavatories independent of baths were also found in the busier sections of towns, usually near the Forum and at major crossroads, where they could make a profit by charging a small entrance fee. By A.D. 315 there were 144 public lavatories in Rome.

Although Roman municipal lavatories normally provided separate facilities for men and women, they were far more public than their modern equivalents, since they weren't divided into cubicles. They were thought of as convenient meeting places, and the Romans felt no embarrassment at chatting to friends there or even swapping dinner invitations. Ten to twenty people could be seated in comfort around three sides of a room. The waste fell into a drain below the seats, to be washed away by running water. Sometimes the toilets

Public lavatory at Ostia, the port of ancient Rome. This fine example of Roman sanitation was built during the refurbishment of the public baths in the 4th century A.D.

had elaborately carved marble seats (cold to sit on but easy to keep clean), along with washbasins and decorative statues. One men's toilet at Ostia, the port of Rome, even contained an altar dedicated to Fortuna, Goddess of Luck.

Hand-flushed lavatories, usually with one or two seats, have been found in grander Roman houses, sometimes on the second floor. Amazingly toilets set above tubs or cesspits were often built into kitchens, as in some houses at Pompeii. This was the result of ignorance rather than lack of money: the idea was to collect kitchen and human refuse together in one place before it was shoveled out and removed. However, those living in apartments (see **Apartment Buildings** in **Urban Life**) depended on portable chamberpots or the public facilities.

TOILET PAPER

Roman public lavatories provided small sponges on the end of sticks for cleaning, which could be rinsed before use in a water channel running in front of the seats. Outside the towns some Romans used moss instead, as shown by excavations of latrines at the fort of Bearsden, in Scotland. Moss was still popular in medieval times (as found by archaeologists at the port of Bergen, in Norway), along with hay and straw, which were kept in heaps next to the lavatories in the castles and monasteries of western Europe. Moss and hay must have been fairly scratchy, but more hygienic than sponges, given that they couldn't be reused.

The imperial court of China had different standards of course, and although no old lavatories have been excavated there, we do know that the Chinese made the world's first toilet paper. The earliest mention dates to A.D. 589, when Yen Chih-T'ui, a court official, wrote, "Paper on which there are quotations or commentaries from the *Five Classics* or the names of sages, I dare not use for toilet purposes." Presumably less-exalted literary works were used instead. Eventually the court decided it needed special toilet paper, and by A.D. 1391 the Bureau of Imperial Supplies was churning out a massive 720,000 sheets of toilet paper every year, each measuring two feet by three feet, for the imperial court. They also made another 15,000 special sheets, three inches square, "thick but soft, and perfumed" for the exclusive use of the imperial family through the year.

SAUNAS

Given the popular image of the Vikings as a race preoccupied with looting, pillaging and raping, it can be quite hard to imagine how they enjoyed themselves during their time off. But one pleasure the hardy warriors from Scandinavia certainly indulged in was the sauna. Modern saunas are a Finnish invention and descend directly from the Viking steam baths of the Middle Ages.

Remains of Viking saunas have been excavated at a number of sites in northern Europe, scattered between Scotland and Poland, and dating back to the tenth century A.D. onward. Viking saunas have also been excavated at their western colonies in Iceland, Greenland and even Canada. Here, in Newfoundland, archaeologists working on the settlement founded by the Vikings at L'Anse aux Meadows around A.D. 1000, identified a bath house containing a large number of brittle burned stones as a sauna, the earliest known from North America.

The Vikings' love of steam baths had a highly practical basis, given the cold northern climate in which they lived. But did they

Medieval Germans enjoying a sauna. Like modern Scandinavian bathers, they use bundles of twigs to tone up the skin. From a 13th-century manuscript.

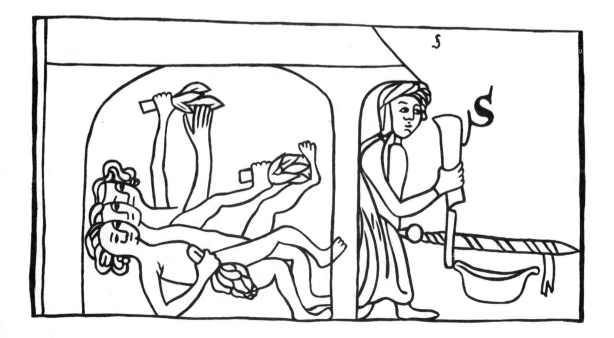

invent the sauna? It seems not. Nor did the Romans, although the hot rooms in their public baths were rather like saunas in that the bather sat in a heated room and sweated his pores clean (see **Baths**). Still, the mechanics were rather different from saunas, in which the heat is increased by throwing water on heated coals or stones to give off steam. (You can't always see the water vapor because it is super-heated by the coals.) The Roman hot room *(caldarium)* was heated from underneath by great furnaces (see **Central Heating**) and sometimes had running hot water.

It seems that the ancestors of the Viking sauna were actually pre-Roman and belong to a far older tradition of sweat or steam bathing, once widespread in ancient Europe. From classical sources it can be detected in countries as far apart as Portugal and Russia. The Greek geographer Strabo (1st century B.C.) tells how the inhabitants of the Duoro Valley, in Portugal, used red-hot stones to make steam baths. In the fifth century B.C. Herodotus described how the Scythian tribes of Russia never took ordinary baths in water but only vapor baths in tents. Though omitting to mention the addition of water to the hot stones, he gives a marvelous description of how the Scythians got high on hemp (cannabis) in these sauna tents:

> And now for the vapor-bath: on a framework of three sticks meeting at the top they stretch pieces of woolen cloth, taking care to get the joins as perfect as they can, and inside this little tent they put a dish with red-hot stone in it. Then they take hemp seed, creep into the tent, and throw the seed onto the hot stones. At once it begins to smoke, giving off a vapor unsurpassed by any vapor-bath one could find in Greece. The Scythians enjoy it so much that they howl with pleasure.

The Prehistoric Saunas of Britain

But the earliest evidence for the use of saunas may come, surprisingly, from Britain, where recent archaeological research suggests that they were in use more than two thousand years before the Vikings. Archaeologists have long been puzzled by the meaning of the numerous "burnt mounds"—artificial hillocks of heat-cracked stones and charcoal—dotted throughout the British countryside. Little is known about these mounds other than their date—radiocarbon testing of the charcoal shows that most were formed in the later second millennium B.C., during the Bronze Age of Britain, when Stonehenge was still in use as a ritual monument.

Burnt mounds are always found near a water source, and, as excavation has shown, are sometimes accompanied by a rectangular trough or basin made of stone, clay or wood. These basins were obviously designed to hold water, but their purpose is less obvious. The sites overall have been frequently interpreted as feasting places, with the heaps of burned stones thought to be "pot boilers"—stones cooked in a fire until they were red-hot, then thrown into the basins to bring the water to a boil. Using pot boilers was a common enough method of cooking in societies that did not have metal pans that could be placed on the fire.

But were all the burnt mounds of the British Isles simply heaps of used pot boilers from cooking sites? Lawrence Barfield and Mike Hodder, archaeologists leading a Birmingham University project on the burnt mounds of the West Midlands, have arrived at a very different conclusion. They dug one mound near Birmingham, but despite "total sieving of tons of excavated material" they did not find a single piece of animal bone—surprising if this were indeed a cooking site. Vegetables may rot away, but bones and shells often remain. Had they been eaten away by acids in the soil? No—a pH analysis of the soil showed that it was neutral. And why was there no trace of the other clutter left from cooking, such as broken bowls?

The answer seems to be that the "burnt mounds" were not necessarily cooking sites. In 1987 Barfield and Hodder concluded that the mound they excavated near Birmingham was actually the debris left from a sauna. This would account for the mass of burned stones, the location near a river and the absence of any food or settlement debris. The sweathouse itself could have been a flimsy structure, like the tents used by the Scythians of classical times. Barfield and Hodder have gone on to identify all of the burnt mounds as sauna remains, which may be going too far. Some mounds in Ireland do have animal remains, while Irish folklore tells how the heroes of bygone days boiled water in pits with red-hot stones for both cooking and bathing. It is quite possible that some mounds are the remains of feasting sites *and* sweathouses.

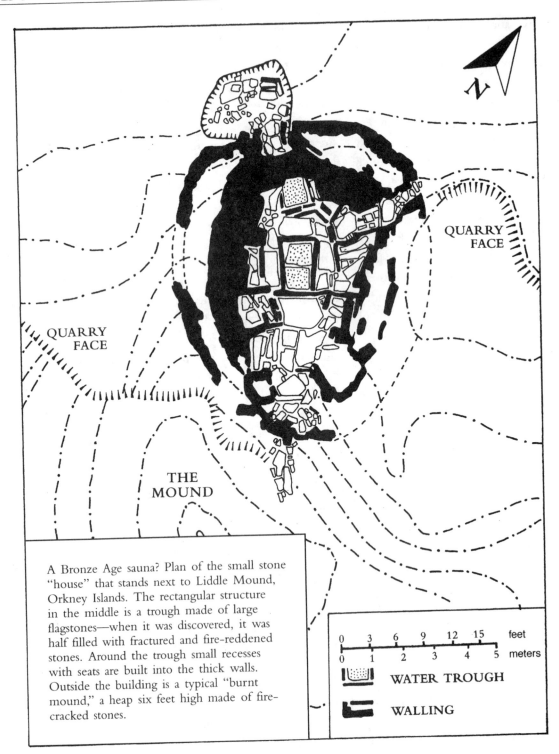

QUARRY FACE

QUARRY FACE

QUARRY FACE

THE MOUND

A Bronze Age sauna? Plan of the small stone "house" that stands next to Liddle Mound, Orkney Islands. The rectangular structure in the middle is a trough made of large flagstones—when it was discovered, it was half filled with fractured and fire-reddened stones. Around the trough small recesses with seats are built into the thick walls. Outside the building is a typical "burnt mound," a heap six feet high made of fire-cracked stones.

0	3	6	9	12	15		feet
0	1	2	3	4	5		meters

WATER TROUGH

WALLING

The most convincing prehistoric sauna is Liddle Mound, on one of the Orkney Islands, north of Scotland. Next to a classic mound of burned stones is a Bronze Age building carefully built from rough stones that is certainly not big enough to be a house. Inside are cubicles with stone benches, but they are too small to lie down in. In the center of the room is a large rectangular stone basin, which must have been made to hold water.

The overall impression is that there is far too little elbow room inside the building at Liddle for it to have been used as a kitchen—there is just enough space for a few people to creep in and sit in the cubicles around the basin. As the excavator pointed out, someone throwing hot stones into the basin to cook would have received a steam bath automatically. It seems highly unlikely that anyone would try to cook in such a stifling atmosphere, and this seems to be confirmed by the finds from the mounds of burned stones. It contained no organic remains apart from nettles, crowberries, heather and grasses. The use of medicinal herbs in conjunction with saunas is well attested from ancient and modern times—for example, they can be thrown on the stones and their vapors inhaled, as the Scythians did with cannabis. Nettles (eaten or rubbed on the skin) are particularly useful in alleviating rheumatism and arthritis, and we know from the skeletons of the Orkney Islanders that such complaints, particularly osteoarthritis, were common there in prehistoric times. Only a little imagination is needed to see Liddle Mound as a local center where the Bronze Age farmers of this chilly island would go to ease their aching bones with both sweating and herbal treatment.

Mayan Sweathouses

Far from the snows of Russia and northern Europe, where Scythians, Vikings and others huddled together for a thermal treat, the inhabitants of Mesoamerica had independently invented the sauna. When the *conquistadores* arrived, they found that steam-bathing was an extremely popular pastime among the Aztecs of Mexico and the Maya of the Yucatán Peninsula. The Aztecs called their steam baths *temazcalli* and frequently used them as therapy for fevers, cramp, rheumatic pains, postnatal conditions and the effects of poisonous bites and stings. As in the Old World, invigorating or medicinal herbs were often used in conjunction with the steam treatment.

Using early travelers' accounts and the surviving *temazcalli* in Mexico, archaeologists have been able to identify the plan of ancient Mesoamerican saunas. The excavated examples generally have enclosing buildings, with benches for the bathers and drains to carry off water. The sauna itself was a small, stone-built room, its most distinctive feature being a firebox either sunk beneath the floor or built into one of the outside walls. The method of heating was much the same as in the Old World: the fire was lit, and water thrown directly onto the coals or onto stone slabs forming an interface between the firebox and the room.

Sweathouses following this plan were built by the Toltecs of Mex-

Reconstructed exterior of a Mayan sauna building from Piedras Negras, built during the Late Classic Maya Period (A.D. 600–900).

ico, whose empire (10th to 12th centuries A.D.) preceded the Aztecs. The sway of the Toltecs extended to the Yucatán Peninsula in the south of Mexico, where Toltec culture fused with that of the native Maya. It was here, it seems, that the Toltecs learned the art of sauna building. Dozens of Mayan sweathouses have been discovered in Yucatán. At one site alone, Piedras Negras, there are eight, dating to the Late Classic Maya Period (A.D. 600–900).

The accounts of early travelers were summarized in 1780 by the Spanish Jesuit historian Francisco Javier Clavijero, who prepared a detailed record of Mexican steam bathing, a tradition that has scarcely changed in more than a thousand years:

> The *temazcalli* is so common that in every place inhabited by the Indians there are many of them. . . . When any person goes to bathe, he first lays a mat within the *temazcalli*, a pitcher of water, and a bunch of herbs, or leaves of maize. He then causes a fire to be made in the furnace, which is kept burning, until the stones which join the *temazcalli* and furnace are quite hot.
>
> The person who is to take the bath enters commonly naked, and generally accompanied for the sake of convenience, or on account of infirmity, by one of his domestics. As soon as he enters, he shuts the entrance close, but leaves the air-hole at the top a little time open, to let out any smoke which may have been introduced through the chinks of the stone; when it is all out he likewise stops up the air-hole. He then throws water upon the hot stones, from which immediately arises a thick steam to the top of the *temazcalli*. While the sick person lies upon a mat, the domestic drives the vapors downwards, and gently beats the sick person, particularly on the ailing part, with the bunch of herbs, which are dipped for a little while in the water of the pitcher. . . .
>
> The sick person immediately falls into a soft and copious sweat, which is increased or diminished at pleasure, as the case requires.

BATHS

The ancient Greeks and Romans regarded bathing as one of the hallmarks of civilization. The description of the baths of Atlantis takes a prominent place in the account of the fabulous lost city written by Plato in the fourth century B.C.:

> The springs they made use of, one kind being cold, another of warm water, were of abundant volume, and each kind was wonderfully well adapted for use because of the natural taste and excellence of its waters. . . . They set reservoirs round about, some under the open sky, the others under cover to supply hot baths in the winter; they put separate baths for the kings and for the private citizens, besides others for women, and others again for horses and all the other beasts of burden, fitting each out in an appropriate manner.

The "Great Bath" at Mohenjo-Daro in Pakistan, built around 2500 B.C.

Even horses, it seems, had special baths in Atlantis! Plato's yarn is largely a work of fiction, an ancient Greek vision of the wonders of the prehistoric past. One of its main aims was to stress to the Greeks that their own culture was a relatively young one and that other, equally great, civilizations had preceded it. Many have seen in the Atlantis story echoes of the great civilizations of the Copper and Bronze ages—if so, it was not too far off the mark, at least as far as human bathing went.

One of the most striking discoveries made at Mohenjo-Daro, in the Indus Valley of Pakistan, was its "Great Bath." Built around 2500 B.C., Mohenjo-Daro was one of the world's earliest great urban settlements, dominated by a citadel on a huge artificial mound. On the summit was the "Great Bath," thirty-nine feet long, twenty-three feet wide and nearly ten feet deep. The pool was lined with bricks neatly laid in two layers with an inch-thick damp course of bitumen in between. At each end large flights of steps led down into the water, and in one corner was an outlet leading to a huge drain. The Bath was surrounded by porticos leading to rooms, some containing smaller baths. The purpose of the Great bath is uncertain—was it for royalty or priests, for practical or ceremonial use? Perhaps it was simply a great public bath, a forerunner of the great meeting places built by the Romans more than two thousand years later.

Whatever the case, washing was clearly an important part of daily life at Mohenjo-Daro. Almost every one of the hundreds of private houses excavated had its own bathroom. Usually located on the ground floor (though a few were on the second), these small rooms had finely made brick pavements, sometimes with a surrounding curb. In these the citizens would douse themselves with buckets of water, which flowed away through a hole in the floor down chutes or pottery pipes in the walls into the municipal drains. These little brick structures of the Indus Valley civilization were the first purpose-built bathrooms in the world.

What makes these early bathrooms so outstanding is the fact that the earliest civilizations, including that of the Indus Valley, were generally situated by rivers, where people could easily bathe without building special structures. Even the Egyptians, fastidious about washing, rarely seem to have had special bathrooms. Still, we have relatively few ordinary houses to judge from—temples, palaces and tombs have proved too great a lure for excavators in Egypt. The city of Akhetaten (El-Amarna), the short-lived new capital founded by the heretic Pharaoh Akhenaton in the mid-fourteenth century B.C., is something of an exception. Here a number of private houses have been fully excavated, and some—not always those of high officials— had bathrooms next to the main bedrooms. They were identified by the limestone slab set in the floor at one corner of the room, with other slabs placed vertically against the mud-brick walls to protect them from splashing. The floor slabs have a shallow depression with a spout at one side to conduct the water into a large pot set into the floor or to a drain leading out of the house.

A more familiar style of bathroom began to appear in Mesopotamia in the second millennium B.C. At the palace built about 1700 B.C. at Mari, on the Euphrates (northern Iraq), excavators found numerous rooms containing terra-cotta bathtubs; many were accompanied by simple squatting toilets (a hole in the floor leading to a sewer), and some were even heated by small fireplaces with flues to carry off the smoke from the fire.

The bathtub, a simple enough idea, was probably invented independently in several areas. The Minoans of ancient Crete were certainly among the first to make them. When their civilization was rediscovered at the turn of the century by the great British archaeologist Sir Arthur Evans, his Edwardian public was particularly struck by finds showing how conscious the Minoan royalty were of cleanliness. In the great palace of Knossos, begun about 1700 B.C., Evans found both bathrooms and toilets (see **Lavatories**). The bathrooms had sunken floors reached by steps, and some still contained their elegant terra-cotta bathtubs. Strangely enough, for a palace with an elaborate drainage system, the bathrooms had no plumbing, but since the tubs had handles, it seems that they were simply carried out to be emptied.

The ancient Cretans seem to have been so fond of bathing that they were often buried in clay coffins that, except for the addition of a lid, are indistinguishable from bathtubs—they even have plug holes. From Crete the bathtub passed to the other great Aegean civilization, that of the Mycenaeans on mainland Greece. The palaces of Mycenae, Tiryns and Pylos, built during the fourteenth to thirteenth centuries B.C., all have royal bathroom with tubs. There were some improvements here on the Minoan plan; the bathroom at Tiryns has a floor made of a single enormous slab of stone, inclined so that overflow from the tub would run into a gutter and off through a drain.

Oddly enough, the royal bathtubs of these times crop up in later Greek traditions about the Heroic Age (their term for the Bronze Age). For example, King Minos of Crete (whose name was used by Evans to coin the term *Minoan*) was supposed to have been murdered

The "Queen's Bathroom" at the Minoan palace of Knossos, Crete, as it would have appeared about 1400 B.C. Minoan bathtubs, like the one found in this room, were generally too small to lie down in and were used like old-fashioned hip baths. Reconstruction by the excavator, Sir Arthur Evans.

The hot room (right) and cold room (far right) from the smaller baths at Pompeii, 1st century A.D.

while taking a bath at the court of the king of Sicily. His enemy, the inventor Daedalus, ran a concealed pipe through the roof above the bath and scalded Minos to death by pouring boiling water on him.

The Greeks, then, were heirs to a long tradition of civilized bathing. During classical times they used a variety of systems including pools, bathtubs, footbaths and even the first proper showers. But it was left to the Romans to take bathing to the ultimate heights of luxury and sophistication, through their unrivaled brilliance in plumbing (see **Pipes and Plumbing** in **Urban Life**), and their highly efficient heating systems, which enabled baths to be warmed twenty-four hours a day (see **Central Heating**). The great baths were sumptuously decorated using lavish amounts of marble and were arguably the Romans' finest architectural achievements. The state allocated huge budgets for the building and maintenance of public baths, which functioned as the main community centers of every town. These huge complexes included rooms for conversation, poetry and philosophy, ringed by rows of shops. As a result city dwellers rarely had their own baths at home, though the rich built bathrooms into their villas in the countryside.

The Roman baths were all variations on the same essential plan, a division into rooms of different temperatures, with separate ones of each kind for men and women. On entering, bathers went first to the *apodyterium*, or undressing room, then into the first hot room (the *caldarium* or *sudatorium*). These did not always contain pools—rather, they were superheated rooms where the bathers sat and sweated. They moved from there to the *tepidarium*, or warm room,

SHOWERS

The simple bathrooms of ancient India, Egypt and Mesopotamia were all rudimentary showers. They did not contain baths, and people washed by pouring water over themselves or having it administered by a servant standing behind a low wall.

But the first real showers, with plumbed-in water, were invented by the ancient Greeks. After exerting themselves in the stadium, ancient Greek athletes would freshen up in the kind of shower depicted on an Athenian vase of the fourth century B.C. Two shower rooms are shown, occupied by four rather muscular young ladies. Piped-in water sprays down on the bathers through showerheads shaped like the faces of boars and lions. Near the top is a rack or pole over which the girls have draped their garments and towels. The whole scene is amazingly contemporary: apart from the animal showerheads, it would not be out of place in a modern gym.

The remains of a whole complex of shower-baths were excavated in a gymnasium at Pergamum, a rich Greek metropolis in western Turkey. The last phase of its construction dates to the early second century B.C., at which time seven bathing units were in use. Water ran from an overhead mains system onto the bathers and then flowed from one bath to another, running from the last one into a drain. Thus the system, like a modern shower, provided a footbath as well.

Shower scene from an Athenian vase of the 4th century B.C.

to cool off before taking a plunge into the *frigidarium*, which contained baths of cold water. The rounds could be done to suit one's taste, and the skin was cleaned by sweating, body rubs or oil and scraping as well as by bathing (see **Soap** in **Personal Effects**).

After the fall of Rome its great tradition of public baths was maintained in the East by the Byzantine Empire and then by the Arab and Turkish conquerors of the Roman provinces in Egypt, Palestine, Syria and Anatolia. They took over some of the old baths lock, stock and barrel, renovating and adapting them. In this sense modern Turkish baths actually preserve a continuous tradition going back to Roman times.

In Europe, however, baths became steadily less popular in the centuries that followed the collapse of the Roman Empire (see **Introduction** to **Personal Effects**). The Vikings, rather surprisingly, seem to have been one of the cleanest peoples during this dark age of personal hygiene. They built special houses for steam bathing throughout the northern European lands that they colonized (see **Saunas**) and even managed the feat of building an open-air bathing pool in, of all places, Iceland!

This was not through foolhardiness. At Reykholt you can still see today the pool owned (and possibly built) by the great Icelandic poet and historian Snorri Sturluson (A.D. 1178–1241). Made of blocks of stone neatly fitted together, it is fed by an underground conduit with hot water from a natural hot spring about a hundred yards away. A stone slab slotted into the conduit controls the water supply. A contemporary saga describes Snorri and his comrades discussing politics as they sat in the pool. At the time the country was racked by civil strife, and Snorri, in a rash attempt to become the earl of Iceland, eventually got himself into a very different kind of hot water. He was murdered in his house in A.D. 1241, but he left his open-air hot bath as a testimony to Viking ingenuity.

CENTRAL HEATING

A simple method of central heating has long been in use in northern China. A raised floor space for sleeping is built over the area where the fire or oven is situated. This primitive central heating has been known since Han Dynasty times (202 B.C.–A.D. 220), as shown by the pottery models of houses buried in the tombs of landowners from this period.

Similar forms of underfloor heating in the West may have inspired the breakthrough achieved by Caius Sergius Orata, a Roman businessman who lived about 100 B.C. Orata was a fish and oyster farmer, and to keep his stock healthy, he invented a way of keeping warm the large tanks *(piscina)* in which they were kept (see **Fish and Oyster Farms** in **Working the Land**). This involved raising the floor of the tanks on stacks of tiles like short pillars. A nearby furnace produced hot air, which passed underneath the tanks and kept the water at a steady temperature.

Within two generations Roman engineers had applied Orata's bright idea, known as the *hypocaust*, to the heating of public baths (see **Baths**). The natural extension was to use the idea to heat whole buildings as well, using a variety of systems based on the hypocaust. Orata's invention became one of the most popular elements of Roman civilization and was enthusiastically adopted by the native aristocracies of wet, cold northern Europe. Two main types of hypocaust were available, one using pillars and the other pipes.

The pillar hypocaust is the best known, and the one preferred by the Romans themselves. The floor of the room rested on sturdy ceramic-tile pillars, up to three feet high. The shallow basement created was connected to a furnace, which, when lit, sent a flow of warm air under the floor, gently heating it. In addition hot air could be directed up through hollow flue tiles built into the wall to provide all-around warmth.

The piping method was much cheaper, since it didn't involve building a basement. Hot air was channeled from the furnace into the center of the room via a metal pipe set into the floor, then out to the corners and up through the walls with more piping. Although less expensive, this method was also less satisfactory, since it provided an uneven distribution of heat within the room and made it far more difficult to regulate the temperature.

The most impressive central heating systems were those of the great public baths of the Roman Empire. The heat generated there was often so intense that the bathers had to wear thick-soled sandals to avoid burning their feet. In the third century A.D. the baths at Constantinople used gasoline to provide underfloor heating and hot water. Subterranean vaults contained vast numbers of glass or earthenware lamps filled with gasoline, which enabled the required air and water temperature to be reached more quickly than the usual wood-burning method. Massive amounts of gasoline must have been consumed, since just one of these baths, the "Kaminia," was used by more than two thousand people a day.

HOT AIR

Diagram of a Roman pillar hypocaust. Hot air from a furnace outside the house circulates around the tile pillars on which the floor is built, up through flues in the wall and out through openings on the roof.

GLASS WINDOWS

Glass making originated in the Near East around 2000 B.C., with the manufacture of beads, seals and ornamental inlays. Techniques improved, and by 1500 B.C. extremely fine glass vessels were being produced in Egypt. The Phoenicians of Lebanon, blessed with the silica-rich sands needed for glass making, became the greatest glass manufacturers and exporters of the ancient world.

The first century B.C., when the Phoenicians were drawn into the orbit of the Roman world, saw an explosion in the development of glass technology. One great technical breakthrough was the invention of glassblowing, probably achieved in Palestine—the earliest blown glass in the world was found at Jerusalem in a deposit dated to 50 B.C. by the pottery found in it. Glassblowing led to the production of massive numbers of cheap glass vessels in a variety of colors, replacing metalwork in middle-class homes.

Glassblowing was also rapidly adapted—probably by Roman technicians—into ways of producing window glass. In the main method glass would be blown into a large bulb, which by swinging and manipulation was made into a sausage shape. By opening and stretching the ends the sausage was turned into a cylinder, which was split lengthwise with a red-hot iron. The split cylinder was then reheated on the flat bed of an oven and drawn flat with pincers. The result— not quite as flat as modern window glass—was then mounted into frames.

Some of the best evidence for early glass windows came from the Roman time capsule of Pompeii, buried under volcanic ash by the eruption of Vesuvius in A.D. 79. The oldest windowpane at Pompeii comes from a building in the Forum that dates to about 60 B.C.— the pane itself is somewhat later. It is disk-shaped, half an inch thick and fixed in a bronze frame that turned on two pivots attached halfway up, so that the window could be opened. The largest windowpanes at Pompeii were in the public baths—forty by twenty-eight inches and half an inch thick, mounted in bronze frames. They were frosted on one side, probably by rubbing with sand. Much more of-

Blown-glass windowpane of Roman date; twenty-one by twelve inches, it was trimmed from a larger sheet.

ten, however, the glass was made in small pieces to be fitted into richly ornamented wooden frames.

The Pompeiian examples aside, window glass was relatively rare in the warm Mediterranean in Roman times, becoming more common as one moved north. Enormous glass factories were established at Trier and Cologne, in Germany, during the second century A.D. In Britain glass windows were common after A.D. 50, when England and Wales were incorporated into the empire, and are found at many native settlements and even beyond the frontiers of the empire in southern Scotland. This should not lead us to imagine that the poor could ever afford glass windows; rather it means that the old tribal chiefs had become rich men under the empire. In A.D. 270 a merchant of Alexandria was able to show off his wealth by having the windows in his house fitted with panes of glass set in pitch. Glass windows were an expensive luxury that their owners protected with curtains, blinds or wooden shutters.

Window glass continued to be made in the East even after the fall of the Roman Empire. Most windows here were made by blowing glass into spheres and then snipping them while still hot into the form of shallow bowls with broad rims. The resulting "crown-glass windows" are still made in Palestine today. In the West the use of window glass was restricted to churches for many centuries. Stained-glass windows were also developed, their original function being to display Bible stories to the illiterate. After details of the scenes were painted on the glass, it was fired to fix the colors. In Britain glass making was slow to revive; the church historian Bede wrote that in A.D. 675 the abbot of the new monastery at Monkwearmouth had to send for glassmakers from France to make its windows and sacred vessels. Only around 1200 did the industry really pick up again in Europe, when royal palaces started to be glazed.

A bluish-green windowpane of the "crown-glass" type common in the eastern Roman Empire (side view). Found at Jerash, in Jordan, it was made around A.D 400. It is about 5 ½ inches in diameter.

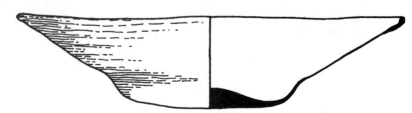

Reconstruction of a small stained-glass panel found in the Riverside Guest House of Saint Bede's monastery at Jarrow, dating to the 8th century A.D.

One invention that may have occurred during the Roman Empire to be promptly lost again is unbreakable glass. A number of Roman writers record that an inventor went to the Emperor Tiberius (A.D. 14–32) with this remarkable discovery, hoping for a reward. According to one version of the story, Tiberius inquired whether the secret was confined to the two of them, and when the inventor assured him of this, he promptly ordered the man to be executed. The emperor took this drastic step because he believed that the invention would put glassmakers out of business.

KEYS AND LOCKS

The *Odyssey*, the great epic poem written by Homer in the eighth century B.C., has barely any references to the technology of the time apart from the weaponry used by its heroes. One exception is a curious passage describing how Penelope, wife of King Odysseus, unlocked the strongroom of their palace to find his bow. Using "a cleverly made copper key with an ivory handle," she opened the door: "She quickly undid the strap attached to the door-handle, thrust the key through the hole and, with a well-aimed thrust, shot back the bolt. The key did its work. With a groan like a bull roaring in the meadow, the doors flew open before her."

The key described by Homer was clearly very different from the kind we use today. But exactly how did it work? By combining clues from archaeological evidence and domestic scenes painted on Greek pottery, we can get a pretty clear picture. There was no lock as such—just a fairly ordinary sliding wooden bolt on the inside of the door. To shut it from the outside, the bolt was yanked into place by a rope that passed through a hole in the door; this was then tied into a complex knot on the doorknob. To open the door, the knot was undone, and the bolt slid back by pushing an L-shaped "key" through a second, larger hole until it caught on a slot cut in the bolt.

In the evolution of security devices the "Homeric lock" stands between the simple bolt and proper key-and-lock mechanisms. To-

day we know that the history of the key began long before Homer's
time. But when the evidence was first uncovered, the archaeologists
involved were completely baffled by what they found.

In 1887, during his excavations at Kahun, an Egyptian town of the
nineteenth century B.C., British archaeologist Sir Flinders Petrie discovered a cylindrical piece of wood with each end pierced by a hole. This
puzzling find was set to one side as a minor mystery awaiting a solution. But similar wooden objects continued to turn up at other sites in
Egypt. A fine example came from the tomb of Kha, chief architect to
the pharaoh in the mid-fourteenth century B.C.; his wife, Meryt, was
buried with him, and in a basket containing part of her trousseau were
a dressing gown, braids, a comb, and one of the mysterious wooden
objects. At the beginning of this century bronze versions began to be
discovered in Sicilian tombs of the tenth century B.C., again primarily
in the graves of women. These associations inspired suggestions that the
objects were hair curlers or spools for winding thread.

Such theories, while plausible, were never generally accepted.
The true purpose of these peculiar objects was discovered in 1907,
when similar wooden pegs were seen being used as keys in Ethiopia.
This ancient style of key, probably invented in Egypt or Nubia
around 2000 B.C., is elegant in its cunning simplicity—so cunning in
fact that without the vital clue provided by its survival in Ethiopia
for four thousand years, the purpose of the mysterious objects from
Egypt and Sicily would certainly never have been guessed (see page
470). After many years of debate it is now generally agreed that they
were keys and that their frequent discovery in female graves reflects
the role of women in ancient society as mistresses of the home. Like
Penelope in the *Odyssey*, the women of Egypt and Sicily were in
charge of the household keys.

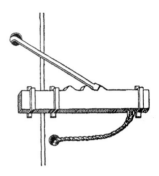

The "Homeric lock," the
earliest kind used by the
ancient Greeks (view
from inside of door). The
bolt was drawn across by
pulling a rope from outside.
It was slid back by
poking a special key
through another hole above
the lock, which caught on
a slot or protuberance on
the bolt.

The Tumbler

Yet however ingenious, these early key-and-bolt arrangements were
still relatively easy to pick—with enough thought and effort an instrument of the right shape could be forged to slide back the bolt.
They were eventually overtaken in the Near East by the next major
development in security systems—the invention of simple tumblers.

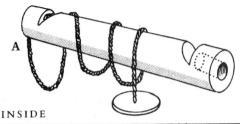

A

INSIDE

(Left) Wooden key of a type once used in Egypt and Sicily. Deceptively simple, it might have a modern expert on security systems scratching his head. The key works by the manipulation of two strings, one (A) attached to the key. The

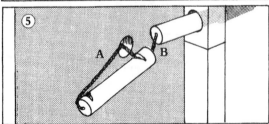

B

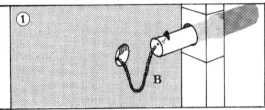

B

other (B) is fixed to the end of the bolt. *To close the door* (which is inward opening), string B is threaded through the keyhole and pulled from the outside, drawing the bolt across (1). The string is left hanging outside (2). *To open the door*, string B is threaded through one end of the

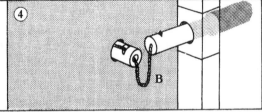

B

A

key (3), and the key popped through the hole (4). The disk attached to the free end of string A prevents it from falling all the way through. String B is pulled from the outside, drawing the end of the key up against the bolt (5). While

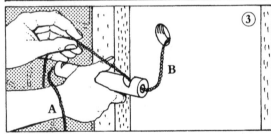

B

A

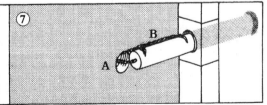

A   B

string B is held fast, A is yanked (6). This draws the key hard against the bolt (7), pushing it back into its mounting. The door can now be pushed open, and string B is removed from the end of the key in order to enter.

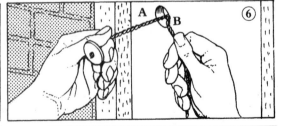

B

A

An open-bottomed box containing wooden pins was fixed onto the inside of the door above a bar; the pins fell into holes in the bar when it was drawn across, holding it fast. To unlock it, one wiggled into the holes the matching teeth of a wooden key (shaped rather like a giant toothbrush), which pushed the pins up out of the way. The bolt could then be drawn back with the key. To make them more difficult to pick, such locks were often placed on bolts on the inside and were reached by passing one's hand through a hole in the door.

Locks with pin tumblers seem to have been invented in Iraq, where the earliest known example was found in the ruins of the palace of Khorsabad, built by the Assyrian king Sargon II (721–705 B.C.). This cunning device spread quickly throughout the Near East. The large wooden key used in such locks is referred to by the Hebrew prophet Isaiah (Isa. 22:22), a contemporary of King Sargon: "And the key of the house of David will I lay upon his shoulder." By the sixth century B.C. the tumbler lock had reached Egypt. From there the "'Egyptian lock," as it came to be known, spread via Greece to Europe. It was a great improvement on earlier systems and has been described as the only major improvement in European domestic architecture during classical times. The modern Yale lock is essentially a more sophisticated version of the Egyptian lock and

(Opposite) The mysterious Egyptian Key.

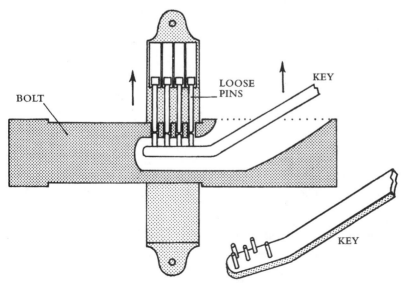

The simple tumbler or "Egyptian" lock, invented in the early 1st millennium B.C., and the distant ancestor of modern Yale locks. The bolt is held fast by wooden pins, which drop from a box into slots; the key lifts the pins out of the way and is then used to draw the bolt.

works on the same basic principle. The tumbler lock also seems to have been developed independently by the Chinese, since examples have been found in tombs of the earlier first millennium B.C.

It was the Romans, or possibly Greek locksmiths under the Roman Empire, who developed the first modern-looking keys and locks. They introduced the widespread use of metal, a great improvement on wood, casting locks from iron and keys from bronze. And while the old-fashioned tumbler lock was still common, they also invented the first rotary locks. In these the key is rotated in the lock, which contains projections known as wards to ensure that only the right key will fit. If gaps of exactly the same shape as these wards are cut into the key, it passes safely over them and pulls back the bolt—the same principle is used in the mortice locks that are common today. Curiously enough the Romans don't seem to have invented a lock that could be opened from both sides of a door, so they generally only locked doors after leaving the room.

Roman key for rotary lock.

Padlocks

The modern padlock was invented in 1831, but almost two thousand years earlier the Romans were making keys small enough to be worn on finger rings. Perhaps this gave them the idea of tiny portable locks. Roman padlocks consisted of two parts that pressed together. Although rotary-type padlocks were made, they were never as popular in the Roman world as the barb-spring padlock, with its eminently simple design. The padlock case has a rectangular hole at one end into which the bolt is pushed; the bolt has a central spine, riveted to the tip of which are one or more springs that splay out to give the appearance of barbs. As the bolt is pushed into its case, the springs are compressed, but once inside, they spring out again, making it impossible to pull the bolt out. It can only be opened by inserting an L-shaped key through a slit at the opposite end of the case; the key has a square hole in it that slips over the bolt and compresses the springs, enabling the bolt to be withdrawn.

A Roman iron barb-spring padlock was found at the palace of Fishbourne, in Sussex, built by a local king friendly to the Romans shortly after their conquest of Britain in A.D. 43. Such padlocks were designed to work with chains to secure a door or gate, as are modern examples. The keys were of iron or bronze. The most elaborate keys combined a bronze handle with an iron shank. This meant that the handle could be mass-produced by casting in a mold and then assembled with an individually made shank.

The Chinese seem to have (independently?) invented padlocks at much the same time as the Romans. The *History of the Later Han Dynasty* (A.D. 25–220) refers to padlocks attached to a chain for securing prisoners. Some Western and Chinese padlocks became highly elaborate, but more everyday padlocks were in universal use among merchants and traders of the ancient world. Even the Viking corner shops of medieval Europe were protected by padlocks after closing hours.

COMMUNICATIONS

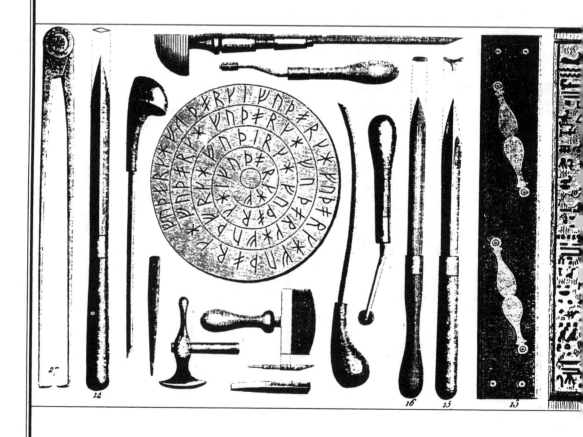

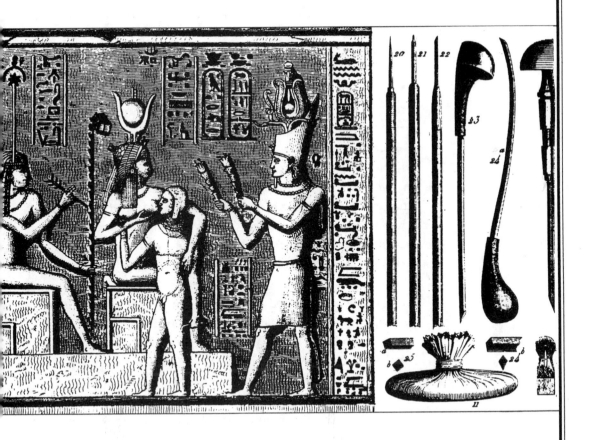

CHAPTER ELEVEN

INTRODUCTION

One of the first acts of an isolated human being, whether a prisoner in extended solitary confinement or a shipwrecked survivor on a desert island, is to start a calendar by scratching off each day on a stone or tree. To keep track of time is a fundamental human need. Outside of language and pictorial art, calendars mark the first steps of the human race to record and communicate information. Our Stone Age ancestors may have started making calendars as much as thirty-five-thousand years ago.

The rich cave paintings of western Europe created over the next twenty-five-thousand years incorporate a number of enigmatic signs. Their meaning eludes us today, but to the painters they were clearly a symbolic means of communication. Other tantalizing symbols were employed by the Azilian hunters of France and Spain around the end of the last Ice Age, about 8000 B.C. Using red ocher, they decorated hundreds of pebbles with dots, bars and wavy lines. Their purpose is unknown, but they may have been tokens, counting aids or even gaming pieces.

In the Near East, farming societies contemporary with the Azilians were developing more sophisticated tokens. Two sites in Iran dating to about 8500 B.C. have produced a series of clay spheres, semispheres, cones and disks, incised or punched with dots and lines. It is thought that they were used for simple accounting of animals and crops. Such tokens became common throughout the agricultural belt, from southwestern Turkey to Pakistan, and the range of different types increased over time. By the fourth millennium B.C. the urban civilization of Sumer, in southern Iraq, had developed a complex accounting system, involving 250 different types of tokens. They would be sealed in groups into clay envelopes and used as the equivalent of modern invoices to accompany trade goods.

These curious containers may have played a vital role in the next quantum leap in communications—the invention of writing. The first step was to record the nature and quantity of the tokens inside by impressing small symbols on the outside of the envelope—this would soon have made the tokens themselves redundant. A clay tablet with impressed symbols did the job more easily. At some point it was realized that this method could be used for far more than accounting. Pictorial symbols (pictograms) can show almost anything, from objects to actions. Thus writing was invented, the first clear examples coming from the Sumerian city of Kish around 3500 B.C. A similar process seems to have brought about the invention of Egyptian hieroglyphic writing around 3000 B.C., though its development was more sudden.

Pictograms rapidly evolved into more stylized forms of writing, such as the cuneiform script of ancient Iraq. This could be speedily written by pressing a wooden stylus into soft clay tablets, which were then sun-dried or baked. Over time the script became increasingly phonetic, expressing sounds rather than depicting objects. As soon as cuneiform was able to communicate complex grammatical forms, we find it being used to compose works of literature. Sumerian tablets from around 2300 B.C. include the work of the world's earliest known poet—Enheduanna, the daughter of King Sargon of Akkad. Installed by her father as high priestess of the moon god at the city of Ur, she wrote several hymns in honor of the great temples and gods of the land, signed by her as priestess of the moon god; there is even a picture of Enheduanna from Ur.

Clay tablets are a very durable way of storing information—a carefully baked tablet is as hard as pottery and will last almost indefinitely. Indeed the clay libraries of the Assyrian and Babylonian civilizations of ancient Iraq will endure the effects of a nuclear holocaust better than our own records, which are largely dependent on fragile media such as paper and magnetic storage. All the same, clay tablets are rather cumbersome as a means of communication. Clay letters were posted in ancient times (in clay envelopes, naturally) but there remained a need for a less bulky and more portable medium.

Papyrus, first made by the Egyptians around 3000 B.C., fitted the bill. It was found in ready supply in Egypt, as papyrus reeds once covered large areas of the marshland formed by the Nile Delta. Strips of the membranelike pith of papyrus reeds were beaten flat, laid across each other at right angles and put into a press. The resulting sheets were then polished with stone rubbers. By the early first millennium B.C. at the latest, Egypt was exporting rolls of papyrus to other states. The invention of a similar kind of primitive paper seems to have happened independently, though much later, in Meso-

A selection of the accounting tokens used at Uruk in ancient Iraq (5th–4th millennia B.C.) which gave rise to the earliest pictographic writing. From top to bottom are the tokens symbolizing beer, bread, wool, rope and a rug.

PICTOGRAM MEANING	c. 3500 B.C.	c. 2500 B.C.	c. 1800 B.C.	c. 900 B.C.	c. 700 B.C.	WORD AND MEANING
STAR						AN "Heaven" / DINGIR "God"
BOWL						NINDA "Food"
THE EARTH						KI "Land"
HEAD						SAG "Head"
MOUNTAIN						KUR "Mountain"
MOUTH & FOOD						KU "To eat"
STREAM						A "Water, in"
MOUTH & WATER						NAG "To drink"
LEG						DU "To go" / GUB "To stand"
BIRD						MUSHEN "Bird"

The development of the cuneiform script. The first column shows the pictograms invented by the Sumerians of southern Iraq around 3500 B.C. Rotated through ninety degrees, they became progressively more stylized over the millennia, the last column reflecting the signs used by the Assyrians and Babylonians about 700 B.C. Cuneiform ("wedge-shaped") takes its name from the shapes that make up the signs, created by impressing a square-ended stylus into soft clay.

america. Examples of paper begin to appear at Teotihuacán in Mexico from the fifth century A.D. Ancient Mexican paper was most commonly made from fig bark, by soaking and beating the fibers into thin sheets, which were coated with chalky varnish and stone-polished. This was manufactured on a large scale, and by the time of the Aztec Empire of the fifteenth century A.D., its bureaucracy went through an estimated 480,000 sheets a year.

More advanced techniques are needed to produce the kind of paper we use today, which is made from the thin sheets of sediment left when vegetable pulp settles out from a watery solution into a flat mold. The water is drained off and the deposit is removed and dried. The technique was first invented by the Chinese, but exactly when is uncertain. The alleged writing paper from a tomb near Xian, dating to the first century B.C., is now known to be a fake, so the earliest paper is probably that found in 1942 under the ruins of a northern Chinese watchtower, destroyed during a rebellion in A.D. 110. From the same date we have literary references to the manufacture of hats, clothes and even armor from mulberry-bark paper (see **Clothing and Shoes** in **Personal Effects** and **Human and Animal Armor** in **Military Technology**). For a long time the Chinese closely guarded the secret of paper manufacture and tried to eliminate other Oriental centers of production to ensure a monopoly. But the technique eventually spread to India in the seventh century A.D., and then to the Arabs through prisoners captured in a battle at Samarkand in A.D. 751. They, too, kept it a secret, and Europeans did not learn how to make paper until the twelfth century.

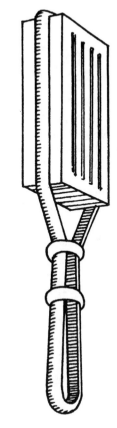

Stone paper beater from Mexico, 5th century A.D. Such beaters were used to hammer fibrous masses of plant material, usually fig bark, into thin sheets.

Of course techniques and materials for recording information are only of use when there is an educated class. In many ancient societies this was an extremely small percentage of the population, most of whom were tied to the land. Being literate provided the only access to careers beyond farming, laboring and basic soldiering, and to be a "scribe" meant that you had made it in the ancient world. It was a highly specialized job. In ancient Egypt a scribe's education began as early as age five. The complexity of many prealphabetic scripts meant that they took years to learn, and even kings would brag about their mastery of the scribal art.

The earliest known formal schools in the world appear to be those that the Sumerian king Shulgi claimed to have established at

Seated Egyptian scribe, writing on a papyrus with reed pens. On the table in front of him is a tray containing various colored inks and a water pot.

Nippur and Ur, in Iraq, shortly before 2000 B.C. The remains of an actual Sumerian school, dating from about 1700 B.C., were discovered sixty years ago at Ur by the great archaeologist Sir Leonard Woolley. In part of a priest's house run as a boys' school, Woolley found some two thousand clay tablets, bearing students' exercises, mathematical tables, religious texts, inscriptions copied from local monuments and vocabulary lists. Discipline at such schools was harsh. In Iraq "a man in charge of the whip" was on the school staff to keep the pupils at their books. Still, only sons of the well-off had to endure the hardships of scribal training. The first civilization to introduce compulsory education appears to have been the Aztec. According to the Spanish conquerors of Mexico, all Aztec boys, whether of low or of noble birth, had to attend school. Oddly enough, girls were only encouraged to go to school if they were the daughters of tradesmen or peasants.

What we know today as higher education was started by the Greeks. The famous Academy of Athens was founded by Plato in 387 B.C. in the hope of reforming public life. His aim was to train citizens who, after rejoining society, would become a new, educated ruling class. Ethics and philosophy were therefore high on the curriculum. Under Plato's successors the Academy's scope broadened out into an encyclopedic approach to learning, cataloguing information from all branches of knowledge. The example of Athens was followed by the foundation of similarly open institutions in Beirut and Antioch, and the famous "Museum" at Alexandria. Female students and professors were as welcome as men. By the third century B.C. the Alexandrian Museum had become the scientific center of the Mediterranean world (see **Introduction** to **High Tech**). Although eclipsed by Alexandria, the Athenian Academy lasted a further eight hundred years, only being closed in A.D. 529 by order of Emperor Justinian, who saw it as a hotbed of paganism. By this time the Eastern Roman Empire was formally Christian, and the University of Constantinople had already been in existence for a hundred years as an exclusively male and Christian center of learning, in opposition to the ancient pagan institutions.

Such bitter religious conflicts seem to have been absent from ancient Chinese higher education, which was dominated by the ethical system of the great philosopher Confucius. The Imperial University at Lo-yang was founded in 124 B.C. by Emperor Wu-ti, with fifty students, the enrollment increasing to three thousand by 10 B.C. and ten thousand by A.D. 30. Several professorships were created, the holders of these posts being known as "scholars of great knowledge." In A.D. 276 a National Academy was also set up. The two bodies developed different roles, the university educating the descendants of the imperial clan, while the academy took "talented common people."

Little is known about the origins of higher education in India, but the Buddhist University at Nalanda is well known from the accounts of travelers who visited it. In the early seventh century A.D. the Chinese scholar Hsüan Tsang wrote that the complex consisted of four hundred buildings and that there were ten thousand students. Admission to Nalanda University was by a rigorous oral examination, which 80 percent of candidates failed. The university's reputation was so high that many people pretended to have studied there; eventually the problem of bogus graduates became so acute that the university had to issue a degree certificate in the form of a clay seal.

Home of the earliest writing and schools, Iraq also produced the first libraries. From Sumerian times local temples collected large archives of religious texts, which were dutifully preserved and copied over the centuries. While there were collections made by earlier kings, the first great national library we know of was built by the Assyrian king Assurbanipal (668–627 B.C.) within the grounds of his palace. The Assyrians were heirs to three thousand years of Mesopotamian civilization, and it was the task of Assurbanipal's librarians to gather together all the surviving knowledge of their ancestors. They were instructed by the king to search out new acquisitions diligently: "Hunt for the valuable tablets that are in the archives and which do not exist in Assyria and send them to me. I have written to the officials . . . and no one shall withhold a tablet from you, and when you see any tablet . . . about which I have not written to you, but which you think may be profitable for my palace, seek it out, pick it up, and send it to me." Much of the material gathered was in the dead Sumerian tongue. To help the scribes translate these texts into Assyrian, hundreds of word lists and dictionaries were prepared. The Assyrians' efforts mean that Sumerian literature is accessible to scholars today. The scientific study of languages began in Mesopotamia more than two and a half thousand years ago.

Assurbanipal, king of
Assyria (668–627 B.C.),
founder of the
world's first great
library, at Nineveh in
Iraq.

The libraries of Mesopotamia also took the first steps toward information science. Texts running over more than one tablet were
carefully numbered in sequence, the last tablet often containing a
brief summary of the contents as well as the copyist's name. Tablets
were stored together on shelves in bundles or baskets carrying clay
labels, while catalogs were prepared giving the contents and location
of the various baskets.

The Greek rulers of the Near East during the Hellenistic period
became even more fanatical book collectors than the Assyrians.
Ptolemy I of Egypt (323–283 B.C.) initiated the most famous library
of antiquity at his capital of Alexandria by collecting some 200,000
scrolls. Its organization was completed by his son Ptolemy II, an ardent patron of the sciences, while the post of chief librarian was held
by a succession of scholars gleaned from throughout the Hellenistic
world. The main Alexandrian Library was built for the use of the
scholars of the Museum, but an adjunct library to house the overflow of manuscripts was founded shortly afterward for the use of
other readers. As one snobbish scholar at the Museum complained,
it "gave the whole city the opportunity to philosophize."

Ptolemy III (247–222 B.C.) was a notorious bibliomaniac. He ordered anyone arriving in the port of Alexandria to hand over any manuscripts he or she was carrying. If these works were not already in the Library, they were seized, and the owner was recompensed with a copy on cheap papyrus. Ptolemy was not averse to dirty tricks to acquire original texts. He persuaded the Athenian authorities to lend him the official copies of the plays written by their greatest dramatists, Aeschylus, Sophocles and Euripides, giving them an enormous deposit of gold as security. Once they were in his hands, he decided to keep them. The furious Athenians were fobbed off with the gold and some worthless copies of their priceless literary treasures.

Eumenes II (197–159 B.C), ruler of the Greek kingdom of Pergamum, in western Turkey, established a rival library to the one in Alexandria. To nip this threat in the bud, the Ptolemies banned the export of papyrus, a move that Pergamun countered by developing a parchment industry. (Parchment was made from specially prepared animal skins by a lengthy process of soaking, scraping, stretching and smoothing. The product was cut into sheets to form scrolls.) This incident gave rise to the notion that the Pergamites invented parchment; indeed the word *parchment* comes from the name *Pergamum*. But while the Pergamites may have refined the technique, archaeological finds show that the invention of parchment is much older; examples made from camel skin in the seventh century B.C. were found near Hebron, in Palestine, in 1969.

The lust for information kept the royal libraries of Alexandria and Pergamum at odds until both cities were conquered by the Romans. Where their books disappeared to is something of a mystery. The Greek writer Plutarch repeats a story that Mark Antony seized the 200,000 volumes of the Pergamum library in 41 B.C. and presented them as a gift to Queen Cleopatra, last ruler of the Ptolemaic dynasty. But this satisfying resolution, in which the two rival libraries were eventually combined, is spoiled by Plutarch's comment that he does not believe the tale. In any event by this time the Alexandrian Library seems to have amassed half a million scrolls, and it continued to thrive under Roman rule. The story that Julius Caesar destroyed the Library in 48 B.C. is a myth that came about through the misunderstanding of a reference to a large quantity of papyri stored in the docks being burned during his siege of the city.

The real enemies of the Library were the later Christians, who systematically ransacked the pagan centers of learning during the fourth to fifth centuries A.D. According to tradition, the Library was burned by the Arabs after their conquest of Egypt in A.D. 640, but by that time there would have been precious little surviving of this once great home of ancient science. In fact if it were not for the Arabs, who preserved much of Greek learning in their own universities, passing it on to the Western world during the later Middle Ages, we would be ignorant of many of the scientific writings of the Greco-Roman world.

We should not imagine, however, that ancient communication skills were limited to reading and writing, linguistics, information studies or the transmission of literary and scientific classics. When the ancients used codes and ciphers, the skill actually lay in obscuring the message so that only those possessing the key could read it. With more straightforward information, especially military communications, the key element was the speed with which the message could be passed. Postal services, the pigeon post and even telegraphs were invented to take the news ever faster from A to B. The achievements of ancient societies in communications probably bring them closer to us than any other technical field: not only are the systems they invented remarkably similar to those of our recent past but their obsession with preserving records of their own times provides us with an invaluable source of information about life in the ancient world.

CALENDARS

In April 1963 science writer Alexander Marshack was completing a book on the rise of human civilization when he came across an article about a small scratched bone found at Ishango, a Stone Age site at the headwaters of the Nile in Central Africa. The bone dated from about 6500 B.C., three thousand years before the flowering of Egyptian civilization and the earliest hieroglyphic writing. For some reason the object seized his imagination. Following a hunch, Marshack immersed himself in a study of the scratches on the bone, and within the space of a day he felt that he had intuitively "cracked the code" hidden in its markings—it was a calendar, with the marks representing numbers of days for the successive lunar phrases from the first crescent of a new moon, waxing through to a full moon and then waning to a crescent and invisibility again.

Not sure whether to believe it himself, Marshack went on to examine dozens of similar examples from Stone Age sites, particularly from the famous painted caves of western Europe. The same patterns seemed to recur, and despite initial skepticism, his work began to be taken seriously by archaeologists. Marshack also produced a persuasive case from the evidence of contemporary "primitive" cultures, such as the Yakuts of Siberia and the isolated islanders of Nicobar, off Malaysia, whose "calendar sticks" bear a close resemblance to his prehistoric examples. Some of these are compelling. The markings on an inscribed eagle bone from Le Placard, in France, dating to between 13,000 and 11,000 B.C., are clearly careful and deliberate engravings, not accidental scratches. Marshack's lunar theory makes good sense of the patterning in the tiny strokes on the bone. He later tracked down a second eagle bone, thought to have been lost, from the same cave; the remarkable similarity of the markings on the two bones convinced Marshack that his theory was correct.

Few archaeologists would accept all of Marshack's claims. Some of the markings he has discovered, particularly earlier examples, may in fact be accidental scratches; and not all of the objects on which there is deliberate notation are necessarily calendars. Some may involve different kinds of communication altogether (like the "letter sticks" used by Amerindians). All the same, Marshack's work has been responsible for a quiet revolution in our understanding of the prehistoric mind. The possibility that Stone Age lunar calendars existed, dating as far back as 30,000 B.C., is no longer considered outrageous by the archaeological establishment.

Marshack is certainly right in thinking that the earliest calendars were lunar. The moon is of obvious importance to societies that survive by hunting and fishing and may hunt certain creatures at night. Unlike the sun, the moon does not just automatically appear as a

A Stone Age calendar? One side of an engraved eagle bone fragment from Le Placard (Charente), France—the careful notching may be a detailed record of lunar phases.

glowing orb once a day but follows a more mysterious pattern, a fact that must have impressed our *very* earliest ancestors some 500,000 years ago. The spawning of important food sources such as fish, tur-

Sculpture of a fertility goddess from the rock shelter of Laussel (Dordogne, France), carved from a limestone slab between 27,000 and 20,000 B.C. She seems to be a moon goddess—the crescent-shaped horn she holds has thirteen clear lines, the same as the number of lunar months in a year. The number thirteen is still associated with lunar magic and is either unlucky or lucky, according to one's point of view.

tles and other marine animals is tied in with lunar movements, since these control the tides. And it could hardly have escaped prehistoric women's notice that the female menstrual cycle is roughly equivalent to a lunar month of 29½ days.

Measuring the Year

On the other hand the sun is hardly an inconspicuous object, and it, too, has its pattern, defining the annual succession of seasons. While it is easier to count the days of the moon's phases than those of the sun (the former are smaller, and the changing phenomena are clearly visible), it is also simple enough to work out that something over 350 days elapses between the returns of seasonal events, such as spring tides or the long days of summer. It is also not too difficult to determine with some precision the length of the year in days.

Discovering the exact length of the year requires patience and ingenuity, but only very simple equipment. All that is needed is a flat horizontal surface, which can be prepared on the ground in a number of ways—by water leveling, or by hand and eye (checked by rolling pebbles across the surface). The only other equipment necessary is a straight stick planted vertically in the ground. To check that it is upright, a plumb bob (a weight on the end of a string) can be used.

How to calculate the length of the year with a stick and some pebbles: Level the ground, plant the stick vertically. Beginning one morning in late summer, mark the tip of the shadow cast by the stick with a pebble or similar marker. Add more markers during the day until the evening. Repeat the procedure every day. The curve formed by the markers will gradually move farther away from the stick until a day comes when it begins to move back again. This turning point (B) marks the winter solstice (shortest day of the year). Continue making curves with the pebbles until another turning point (A) is reached, marking the summer solstice (longest day of the year). Count the number of days it takes for the shadow curve to move from A to B and back again—it should be 365.

With this set up, a hypothetical prehistoric scientist could then have experimentally determined the length of the year. Each day as the sun rose, the stick would cast a shadow, the end of which will gradually move toward the stick as noon approached and then moved away on the other side in the afternoon. The curve (a parabola) traced by the top of the stick could be traced on the ground. As the year progressed, these curves would move farther away in winter (when the sun casts the longest shadows) and nearer in the summer (when the sun is more directly overhead). With a series of curves like this (marked out on the ground with stones), it would now be easy for our patient prehistoric scientist to observe by counting that the whole cycle (from shortest to longest shadows) took 365 days. He or she would also have discovered at the same time precisely when in the year the longest and shortest days (the solstices) fell and when the turning points between these (the equinoxes) occurred.

None of this scenario is far-fetched. It is perfectly natural given ancient man's fascination with the behavior of the sun. Ancient people would have been inspired to investigate the matter by the simple observation that trees and tent poles cast longer shadows in winter than they do in summer. In later times it was through studying the behavior of shadows that the Egyptians invented the first timekeeping devices (see **Clocks** in **High Tech**) and that the Greek geographer Eratosthenes accurately determined the circumference of the earth (see **Mapmaking** in **Transportation**).

And so, at a very early date, mankind would have had at its disposal accurate knowledge of both the moon's phases and the length of the year—enough to devise a workable calendar, one might think. However, that's where the *real* difficulties began. The intrinsic problem with all calendars is that our superficially harmonious solar system is actually a complete jumble. One revolution of the earth on its axis makes a day, but 365 of these revolutions are not equal to the time it takes for the earth to orbit around the sun (i.e., one year), which amounts to 365.242199 days. Likewise a lunar month is not a round figure—it is actually 29.53059 days. And while there are roughly twelve lunar months in a year, they add up to only 354.36706 days—eleven days short of the solar year.

Try taking all these factors into account in one system and you will have quite a headache. Devising a calendar that works has thus been one of mankind's longest quests.

The earliest calendar for which we have written evidence is the one invented by the Sumerian civilization of southern Iraq. By around 3000 B.C. the Sumerians had devised a relatively simple calendar of two seasons (winter and summer) divided into twelve months of twenty-nine or thirty days each. It was regulated by actual observations of the moon, with each new month beginning on the evening of the disappearing crescent moon. To resolve the difference between the seasonal and lunar years, the Sumerians simply added what is known as an intercalary month whenever it was needed to keep the system in step with reality.

This custom of intercalation, first attested among the Sumerians in the twenty-first century B.C., remained (and still remains in a different form) the standard method for regulating the calendar. By this date they had also introduced a notional year of 360 days, based on rounding off the lunar month to 30 days and multiplying by twelve. This fit their sexagesimal mathematical system (based on the number 60, in contrast to the more widespread decimal system, based on 10, in use today). Although 5 days short of the seasonal year, a 360-day year divided neatly by 6 into 60 and became the basis for all of Sumerian calendrical and astronomical philosophy. Following the Sumerians, we still divide the sky, and in fact every circular object, into 360 mathematical degrees.

About the same time as the early Sumerian calendrical experiments, the sun and moon were being carefully observed from the huge circles of rough-hewn stones erected by the megalith builders of western Europe (see **Ships and Liners** in **Transportation**). These sightings were doubtless tied in to an agricultural calendar. But speculations that sophisticated programs of astronomical observations were used to regulate a remarkably accurate calendar are wide of the mark. They arise from modern astronomers imagining themselves as ancient scientists trying to solve problems they would be interested in if transported back into the past. The real function of astronomical alignments at stone circles was probably to impress the public on ceremonial occasions. The prime example of this is at

Stonehenge, where, around 2000 B.C., massive stone blocks (weighing up to fifty tons apiece) were arranged in a circle and inner horseshoe that had its axis aligned on the midsummer sunrise to create a truly memorable spectacle. For viewers standing with their back to the "Altar Stone" and looking through the open end of the horseshoe, the rising sun would be framed by a double "window" formed by three massive stones in the circle and two upright blocks at the beginning of the processional avenue.

In ancient China the surviving "oracle bones," used by the rulers of the Shang Dynasty to foretell the future during the fourteenth to thirteenth centuries B.C., show that the Chinese had a lunar calendar similar to that of the Sumerians. To twelve lunar months of twenty-nine or thirty days a thirteenth month was added every two or three years to keep things in step with the solar year. Later on, the importance of having a reliable calendar grew after it became tied to astrology, and one of the emperor's prime responsibilities was to ensure that the calendar was running accurately. Accordingly, around a hundred calendrical reforms were initiated by the imperial court from the first unification of the empire, in 221 B.C., to the end of the Ming Dynasty in A.D. 1644, at a rate of about one reform every twenty years.

World Cycles

Meanwhile in the New World the advanced Zapotec civilization of Mexico had been working independently on calendrical problems during the early first millennium B.C. At the city of Monte Alban, in highland Mexico, the Zapotec calendar is described in great detail in a series of inscriptions running across the walls around the chief ceremonial courtyard. Perhaps the Zapotecs despaired of linking lunar movements with solar, for they adopted an entirely different system from that of the Old World. Instead of basing their calendar on a lunar year of roughly 354 days, they adopted for their religious festivals a sacred calendar of 260 days, the origins of which are quite obscure. However it was arrived at, it allowed the Zapotecs, and later the Maya and Aztecs, to develop the calendar into a bizarre nu-

merological game of ever-increasing complexity. After fifty-two repeats of the normal 365-day year, seventy-three of the 260-day year
would have elapsed, at which point the two calendars would start
again on the same day. This "Calendar Round" of fifty-two years
(18,980 days) became an integral feature of ancient Mexican culture.
When the Spanish conquered Mexico in the sixteenth century, they
reported that the end of a fifty-two-year cycle was greeted with
great despondency; it was feared that the sun might rise no more,
and the arrival of the first dawn of a new Calendar Round was the
cause for great celebration.

Even more complicated number juggling was indulged in by the
Maya, who, a thousand years after the Zapotecs, worked with a
360-day year called a *tun*, divided into 18 months of 20 days; the 5
days left over from the 365-day year were counted as "days of ill
omen." They also had a lunar calendar and the same 260-day period
as the Zapotecs, which they called a *tzolkin*: days within it were referred to by intermeshing a cycle of 20 named days with the number
sequence 1 through 13. Each day had its own omens and associations, so the cyclic calendar worked as a sort of permanent fortune-
telling machine, guiding the destiny of the Maya nation. They later
combined the *tzolkin* with a lunar calendar, in a cycle of 405 lunar
months or 46 *tzolkins* (11,960 days). The same calendar system was
still being used by the Aztecs when the Spanish *conquistadores* arrived
in Mexico, in 1519.

The Mesoamericans in general were great believers in calendrical
cycles, and the Maya established a "Long Count" as part of this
scheme. The basis of the "Long Count" was the *tun* of 360 days,
with twenty *tuns* making a *katun* (7,200 days), twenty *katuns* making
a *baktun* (144,000 days), and thirteen *baktuns* forming a "Great Cycle" (1,872,000 days, or about 5,130 years), at the end of which the
Maya believed that they and all things would cease to be. According
to the standard interpretation, the "Long Count," and with it the
world, will end on December 24, A.D. 2011.

A calendrical cycle has also been detected in Egypt, though in this
case it is supposed to have come about by accident, not by plan. The
sun god was always the most important deity in the Egyptian pan-
theon, so naturally a solar year of 365 days was respected. But as the
solar year is 365¼ days, the Egyptian calender would have fallen out

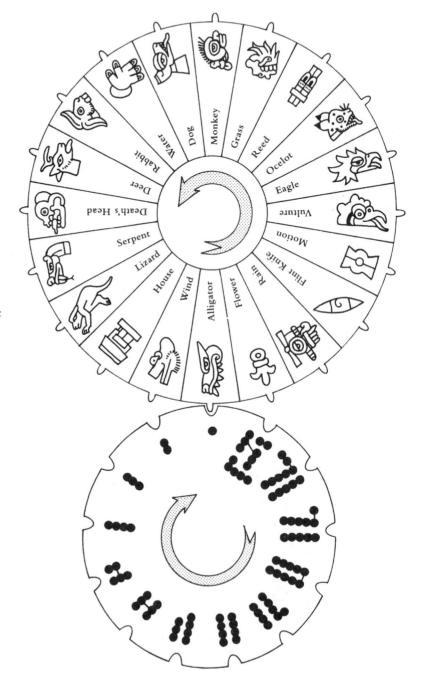

Diagram showing the Aztec version of the ancient Mesoamerican 260–day year. A series of twenty named days (right-hand wheel) interweave with the numbers one through thirteen (left-hand wheel). Thus the first "week" of the year began with the day "One Alligator" and lasted for thirteen days, after which the number wheel had revolved completely and the day wheel moved on by twelves places, making the second "week" start with "One Reed." Thirteen days later the third week began with "One Death's Head" and so on, until by the end of the year each day sign had started a week.

of step with reality by one day every 4 years. After 730 years the situation would have become acute, since the summer and winter months would have completely reversed with respect to the seasons.

Such a slipping calendar can indeed be detected in Egypt between the third century B.C. and the second century A.D., but there are good reasons to believe that before this the Egyptians, like other ancient Near Eastern peoples (including the Sumerians, Babylonians and Jews), made regular calendar reforms to keep things in step. This is exactly what we find during the well-documented Ptolemaic period in Egypt (323–31 B.C.), when Egypt was ruled by a dynasty of Macedonian Greeks descended from Alexander the Great's general Ptolemy.

Alexander himself tried to impose the Macedonian calendar on the Egyptians: it kept lunar and solar periods "in sync" by the occasional addition of a month. But the Egyptians would have none of it. Under Ptolemy III (247–222 B.C.) the link between the Macedonian calendar and the lunar months was formally severed, and the Macedonian system was made to conform to the Egyptian. Under the same ruler a convocation of priests drew up a decree (in 238 B.C.) that prescribed the addition of an extra day every four years into the Egyptian calendar to soak up the awkward quarters of a day in the solar year of roughly 365.25 days. This simple idea, the basis for the calendar we use today, didn't catch on, and it took the military might of Rome to pound this solution into the Egyptians and then impose it on the world.

Enter Julius Caesar

The Romans were interested in the Egyptian dilemma because they, too, were having acute problems. In an attempt made in 153 B.C. to correct their calendar, the Romans had shifted the beginning of the year from March 1st to January 1st, with the result that the numerical names for the months were rendered completely nonsensical. The seventh to tenth months became the ninth to twelfth and we are still stuck with the results—the names for September, October, November and December are derived from the Latin numerals and reflect their positions in the pre–153 B.C. Roman calendar. Worse still, the Romans had struggled for centuries with a lunar year of 355 days, with an extra "month" of 22 or 23 days inserted into February every couple of years or so. This vital adjustment was the responsibility of the College of Pontiffs, whose decisions were all too often influenced by noncalendrical factors, such as tax collectors wanting to extend the year to rake in more money, or important politicians with undesirable appointments such as governorships in distant poor provinces wanting the year cut as short as possible.

Julius Caesar was determined to end these abuses and settle the problems of the Roman calendar once and for all. Rome's rapid expansion from a state centered in Italy to the dominant power throughout the Mediterranean had only aggravated the situation: Each of the subject peoples conquered had its own system of calendrical reckoning, and the only solution was to devise, and impose by law, a new universal system. So on his visit to Egypt in 48 B.C., in between the ups and downs of his love affair with Queen Cleopatra, Caesar sat down for long hours of discussion with the Egyptian savants. Of particular help was the Alexandrian astronomer Sosigenes, whose advice was to drop the lunar calendar altogether and start again from scratch using the Egyptian solar year of 365 days. But this time, as Caesar and Sosigenes agreed, an extra day would be added to February every fourth year. Then the calendar year would not slip with respect to the solar year. The invention, borrowed from the failed Egyptian reform of 238 B.C., is still followed today when we add a 29th day to February every leap year.

Julius Caesar introduced his new calendar to the Roman world on January 1, 45 B.C. To get it started, he had to decree that the previous year (46 B.C.) would last 445 days in order to restore the relationship between the civil calendar and the agricultural year. Yet despite the clarity of the Julian solution, the Roman pontiffs in charge of the calendar misconstrued it and busied themselves adding an extra day to February far too often. By the reign of Emperor Augustus, Caesar's nephew, things were sliding into chaos again, and a new edict had to be issued in 8 B.C. banning February 29th for a few years in order for the calendar to correct itself. Eventually the Julian calendar began to operate properly thoughout Europe and the Mediterranean world. The efforts of Julius Caesar and Augustus were well rewarded. The Roman months of Quinctilis (July) and Sextilis (August) were renamed in their honor.

Difficulties persisted, however, because the Julian calendar year of 365.25 days was still not precise enough. The actual seasonal year is slightly shorter (365.242199 days). This error of eleven minutes and fourteen seconds is not enough to create any major problem within a lifetime, but sufficient to throw the calendar significantly out of step when several centuries had rolled past. By the sixteenth century A.D. the discrepancy amounted to some 10 days, causing widespread concern. The pope had to take over where Julius Caesar left off, imposing another calendar reform by papal edict. In 1582 Pope Gregory XIII decreed that the extra day of a leap year should not be added into the last year of a century unless it was exactly divisible by 400. Thus 1600 was to be a leap year but not 1700. The formula

A member of Pope Gregory XIII's commission on calendar reform demonstrates the mismatch between the Julian calendar (outside the zodiac) and the seasonal year (inside the zodiac). The decision of Gregory's commission, in 1582, resolved a problem that had accumulated from tiny inaccuracies in the Julian calendar and laid the foundations for our own.

THE DAYS OF THE WEEK

Much as some scientists like to scoff at astrology, every Friday evening when they leave their labs and say "See you Monday," they are unwittingly following an ancient astrological system. Not only our names for the days of the week but also their number and order go back to the astrological beliefs of the Babylonians. By about 700 B.C. they had devised a seven-day week, linked to the major planetary gods, which we still follow today.

The modern Western world owes its calendar to the Romans, who, after struggling for some time with an eight-day week, eventually opted for the Babylonian system widely used in the Near East from the third century B.C. onward. The Romans simply exchanged the names of the Babylonian planetary gods for their Roman equivalents. Thus the day of Nabu, the Babylonian god of scribes, became the day of Mercury, the Roman god of communication. The Latin names of the

PLANET	ANCIENT DAY-GODS			MODERN WEEKDAY NAMES		
	Babylonian	*Roman*	*Saxon*	*English*	*French*	*Italian*
SUN	Shamash	Sol	Sun	Sunday	Dimanche	Domenica
MOON	Sin	Luna	Moon	Monday	Lundi	Lunedi
MARS	Nergal	Mars	Tiw	Tuesday	Mardi	Martedi
MERCURY	Nabu	Mercurius	Woden	Wednesday	Mercredi	Mercoledi
JUPITER	Marduk	Jove	Thor	Thursday	Jeudi	Giovedi
VENUS	Ishtar	Venus	Frigg	Friday	Vendredi	Venerdi
SATURN	Ninurta	Saturnus	Saturn	Saturday	Samedi	Sabato

The modern European names for the days of the week are derived, via the Latin and Anglo-Saxon languages, from the ancient Babylonian planet gods. Shamash was the Babylonian god of the Sun, Sin the Moon god, Nergal the god of Mars and so on. The only exceptions are Italian *Sabato* (Saturday), derived from the Jewish word *Sabbath,* and the Italian and French names for Sunday, which mean "the Lord's day" and are derived from Latin *Dominus* ("Lord"). Even the last has pagan associations: during the Late Roman Empire Jesus Christ was identified with the Roman god Sol. Christ merely took the sun god's place in the sequence.

days are still followed closely by the French and Italians, for example the Roman day of Mercury simply became *Mercoledi* in Italian. In the English language the translation went one stage farther, when the pagan Anglo-Saxon ancestors of the English borrowed the day system from the Romans and adapted it to their own gods. In this northern European system (also used by the Vikings), Jove or Jupiter (the thunder god) was known as Thor. Hence Babylonian Marduk's day became the Roman day of Jove, French *Jeudi* and our Thursday (Thor's day), and so forth.

But why do the days follow this particular order? Beyond the sun-moon pairing at the beginning, the sequence seems to be utterly random. And while the choice of seven days reflects the number of the planetary gods, the sequence does not re-

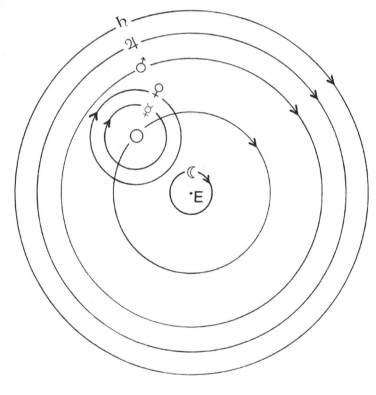

| ○ SUN | ☾ MOON | E EARTH | ☿ MERCURY |
| ♀ VENUS | ♂ MARS | ♃ JUPITER | ♄ SATURN |

The solar system, as conceived by the Greek philosopher Heracleides in the 4th century B.C. His scheme, which follows the traditional order of the planets established by the Babylonians, reveals the profundity of ancient astronomical knowledge. Saturn, Jupiter and Mars are given in their correct order of distance from Earth, Saturn being the outermost planet known to the ancients. Venus and Mercury are given after the Sun, to distinguish them from the outer planets. Again this is correct: the orbits of these two "inner" planets lie between Earth and the Sun. The Moon, naturally, is the nearest "planet" to Earth and is given last. The only mistake as such in the sequence is the position of the Sun, which, according to the ancient system followed here, was thought to orbit Earth like the other planets.

flect their traditional order, based on the ancient understanding of the solar system's arrangement: Saturn—Jupiter—Mars—Sun—Venus—Mercury—Moon.

How does one explain the discrepancy? The answer comes from another great invention of ancient astrology, which we still follow—the division of the day into twenty-four equal units of time, or hours. The planet gods, in their traditional order, took turns presiding over the hours. For example, Saturn controlled the first hour of Saturday, to be followed by the other six gods until the seventh hour. The cycle began again with Saturn on the eighth, fifteenth, and twenty-second hours. The twenty-third and twenty-fourth would be dedicated to Jupiter and Mars, and the first hour of the day after would fall to the Sun god, the next in the sequence. The Sun god therefore presided over that day.

A simple device for calculating the day names from the planetary hour gods was invented. By arranging the gods on the points of a seven-sided figure one can read off the order of the day gods by following the diagonals. When, and by whom, this handy geometrical trick was first invented is unknown, but an example of such a figure is provided by a graffito found at the Roman city of Pompeii.

Thus the names, number and order of our days of the week are dictated by the logic of ancient astrology. As if this were not enough of an affront to modern scientific tastes, when the Babylonians invented the seven-day week, they anticipated the findings of twentieth-century biologists. It has recently been discovered that the human body is governed by a seven-day biorhythm, which is detectable from small variations in blood pressure and heartbeat as well as response to infection and even organ transplants. The same biorhythm affects other life-forms, even simple organisms such as bacteria.

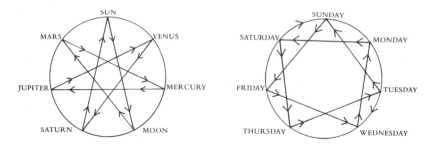

(Left) A simple device for calculating the order of the days of the week from the traditional order of the planets. The planets are arranged in their ancient astronomical sequence around a seven-sided figure (heptagram). Starting with the day of the Sun, following the arrows along the diagonals will bring you to the Moon (Monday), Mars (Tuesday) and so on. (Right) Diagram for the reverse procedure, to derive the order of the planets from the weekdays.

is somewhat crude, but good enough to keep things working within an accuracy of 1 day in 3,300 years.

But Gregory's reform, like Caesar's, had to reset the clock before it could work. He decreed that ten days would have to be skipped, making the day after the fourth of October the fifteenth. Understandably many people, specifically Protestants, were not too keen on the idea, and Britain, together with her American colonies, only followed suit in 1752, by which time eleven days had to be caught up. Thus George Washington, born on February 11, 1732, later celebrated his birthday on the twenty-second of the month.

The next time you fill in your diary, spare a thought for the scientists and sages who have been wrestling with the problem of providing us with a workable calendar for the last thirty thousand years.

THE ALPHABET

Today we take the alphabet so much for granted that we forget what an extraordinary invention it is. In the Western world it has enabled mass literacy on a scale unknown in any previous or contemporary society. The value of the alphabet lies in its elegant simplicity—its ability to express a wide range of vocal sounds with as few as twenty-six letters. Compare this to the forty-nine thousand or so characters, each expressing a different word, that have to be learned by a literate Chinese person. As classical scholar A. C. Moorhouse quipped, "We say 'As easy as A.B.C.' No one ever said 'As easy as Chinese ideograms, or Egyptian hieroglyphics.' "

With the benefit of hindsight it perhaps seems strange that the ancients took so long to develop an alphabet, struggling for thousands of years with impossibly cumbersome scripts such as hieroglyphics. But the alphabet seems obvious to us only because it is so familiar. It is one of those chance inventions that is beneficial *because* it is simple—but that doesn't mean it was easy to invent.

The origins of the alphabet are in fact shrouded in mystery, and only the later stages of its history are reasonably clear. The Cyrillic

The alphabet was invented by the Phoenicians of Lebanon and borrowed by the Greeks and Etruscans during the 8th century B.C. The Latin script of the Romans was developed from the Etruscan and Greek, and our own alphabet is a modernized version of the Latin. The letters compared are shown in the Greek order.

PHOENICIAN	CLASSICAL GREEK		ETRUSCAN	LATIN
K ⴹ	Alpha	A	A	A
9 9	Beta	B		B
⌐ 1	Gamma	Γ	>	C G
◁ ◁	Delta	Δ		D
⋺ ⋺	Epsilon	E	Ǝ	E
Y	Digamma	F	⅂	F
ⵣ I	Zeta	Z	‡	Z
ⷚ H	Eta	H	⊟	H
⊗	Theta	Θ	⊙	Th
ⵥ	Iota	I	I	I J
⵾ ⵾	Kappa	K	ⵦ	K
⎰ ⎰	Lambda	Λ	⅃	L
ⵯ ⵯ	Mu	M	ⵯ	M
⵲ ⵲	Nu	N	ⵤ	N
ⵣ ⵣ ⵣ	Xi	Ξ	⋈	X
○ ○	Omicron	O		O
⎦ ⎦ ⎦	Pi	Π	⅂	P
Φ Φ P	Qoppa	Ϙ	Q	Q
ⵍ	Rho	P	ⵍ	R
W	Sigma	Σ	ⵞ	S
X	Tau	T	⵮	T
	Upsilon	Υ	V	U V
	Phi	Φ		Ph
	Chi	X		Ch
	Psi	Ψ		Ps
	Omega	Ω		Ō

alphabet used today in Russia and some parts of eastern Europe is believed to have been invented in the ninth century A.D. by saints Cyril and Methodius, Byzantine missionaries to Czechoslovakia. It is based on the Greek alphabet, with the addition of some extra letters. The modern Western alphabet (shared by English, French, Spanish, German and Italian, among other languages) is identical to the Latin script used under the Roman Empire; the only difference is the medieval addition of *J*, *U* and *W*. (The Romans made do with *I* and *V* to represent these sounds.) This far back things are certain, as is the common origin of the Greek and Latin alphabets.

During the eighth to seventh centuries B.C. different alphabets mushroomed around the Mediterranean world—Greek in the Aegean, Etruscan (parent of the Latin script) in central Italy and an Iberian version in Spain and Portugal. Without doubt they were all borrowed from the alphabet used by the Phoenicians, the ferociously mercantile sea traders who lived on the coast of what is now Lebanon and Syria. It is here that we can take the history of the alphabet farther back by several centuries.

The Phoenicians were a Semitic people closely related to the Hebrews and Arabs—the Phoenician, Jewish and Arabic scripts are likewise closely related on the alphabetic family tree, but when exactly the three groups began to separate is still not known. Archaeological evidence from sites in Palestine and Syria shows without doubt that the parent of all these scripts, known as the Western Semitic alphabet, was current by at least the fourteenth century B.C. The alphabet was almost certainly invented by the Semitic peoples of the Levant—but where, when, how and why?

Simplified Hieroglyphics?

In 1906 Sir William Flinders Petrie, the great pioneer of Near Eastern archaeology, discovered examples of a previously unknown script in the Sinai Desert, between Egypt and Palestine. Dating from about 1500 B.C., the letter shapes of these enigmatic inscriptions are more pictorial than the usual alphabetic forms and certainly close to the hieroglyphic. It seemed for a while that the mystery of the alphabet's origins had been solved. The Proto-Sinaitic texts, as they came to be known, were hailed as the "missing link" between Egyptian hieroglyphics and the alphabet. Some scholars even believed that these were the writings of the Israelites during their wanderings on their way from Egypt to the Promised Land.

So the theory developed that the alphabet arose when Semitic peoples simplified hieroglyphic signs borrowed from the Egyptians.

It agreed with the tradition preserved by classical writers that writing was invented in Egypt and taken from there by the Phoenicians to Greece. The story has some sense in it, as the Semites may have first become acquainted with the art of writing by rubbing shoulders with the Egyptians. But attractive as these speculations were, more recent finds show that the Proto-Sinaitic inscriptions were not the earliest attempts at alphabetic writing, examples of which have now been found in Palestine dating back as early as 1700 B.C. Rather than being a "missing link" in the development of the alphabet from hieroglyphics, the Sinai script is actually an Egyptianized version of an already existing alphabet. It seems to have been developed by Semites involved in the Egyptian copper-mining industry in Sinai (see **Drilling and Mining** in **Working the Land**), who modified their alphabet under the influence of Egyptian writing.

Under closer examination the Egyptian theory of alphabetic origins falls apart completely. The Egyptians did have alphabetic characters within their hieroglyphic writing system to represent simple sounds rather than whole words, but none of them look remotely like the early Semitic characters. For example the nearest thing that the Egyptians had to a sign for the Hebrew letter *aleph* was a drawing of a vulture; their equivalent of *beth* was drawn as a foot and so on. It places quite a strain on the imagination to transform these into the Semitic forms. The ancestor of the Hebrew *aleph* means "ox" and was clearly drawn as a simplified ox's head, while *beth*, meaning "house," was drawn as an open square shape, like a simple dwelling.

An Accidental Invention?

The Sinai inscriptions proved to be a false trail, and the Egyptian theory of the alphabet has now been generally abandoned. We are left with the conclusion that sometime in the early second millennium B.C. the ancestors of the Phoenicians invented the alphabet from scratch. But how on earth did they manage to hit on such a brilliant discovery?

As Professor Cyrus Gordon, a senior expert on ancient Semitic languages, has pointed out, the ancient Phoenicians seem to have preempted modern linguistic science by nearly four thousand years.

Today we are aware of what linguists call the phonemic principle, which states that any language in the world can be broken down into a limited number of distinctive sounds, usually ranging from about twenty-five to thirty-five. (English is unusual in having as many as forty-four phonemes.) This means that, with one symbol for each phoneme, a system of some thirty basic signs is adequate to express most languages. The original Phoenician alphabet of some twenty-nine characters fits the bill perfectly. We then have to assume that the early Phoenicians had a highly sophisticated knowledge of linguistics—which is of course not impossible.

But if the alphabet is a deliberate invention, crafted by some ancient genius, why does it have such a strange order? If you were to invent a phonetic system, it would be more logical to place the vowels and similar-sounding consonants together in groups. The Arabic version of the alphabet does just this, grouping together the semi-vowels (glides) *w* and *y*, the three letters for the different *s* sounds, and so on. But this was a later rearrangement of the old Semitic order. The original alphabet as it stands looks like a complete jumble.

Gordon has suggested an intriguing alternative for the origin of the alphabet—that it was an accidental invention drawn from a system of notation that originally had nothing to do with expressing sounds. As he points out, the numerical values of the ancient Semitic alphabet were of equal importance to their sound values. Thus Hebrew *aleph*, *beth* and *gimel* also mean 1, 2, 3 and so on. When the Arabs rearranged the alphabet, they kept the old numerical values intact, even though their sequence was destroyed; thus their alphabet reads, numerically, 1, 2, 400 and so on. Likewise the Greek language still preserves as numbers three letters (*digamma*, *qoppa* and *sampi*), which have long ceased to have any phonetic value. These facts underscore the alphabet's value as an accounting tool as much as a linguistic one.

The numerical significance of the letters is merely a step toward the core of Gordon's theory, which is that the signs of the alphabet were originally created to designate the days of a lunar month. He argues that by chosing a different-sounding word (for an animal, object, etc.) for each day of the month, the ancients arrived at a system of thirty words, first used for calendrical notation and mathematics. By using different-sounding words they accidentally invented a phonetic system at the same time—it was later appreciated that words could be formed by using the initial sounds of the calendrical words, and the alphabet was born. The twenty-nine or thirty letters of the original Phoenician alphabet known from the fourteenth century B.C. do indeed match the number of days in a lunar month (see **Calendars**). But what about the order and the names of the letters?

Alphabet and Zodiac

A theory similar to Gordon's, proposed by orientalist Hugh Moran, argues that the names for the letters of the alphabet were drawn from an ancient lunar zodiac. His primary evidence comes from an ancient Chinese zodiac with twenty-eight constellations marking divisions of the horizon. The full moon appears in the sky at different positions throughout the year, moving against the background of constellations, and the ancient lunar zodiac appears to have been drawn up as a way of fixing the calendar for agricultural purposes.

It may seem bizarre that ancient lunar symbols from China, a culture that never developed an alphabet, can throw light on the origins of the Phoenician script. All the same, Moran seems to be on the right track. The first two signs on his Chinese calendar are symbols for an ox and for a woman. The first is made of six stars in the shape of an ox head and is identical to the ancient Semitic *aleph*. The second is formed of four stars and resembles the Semitic *beth*, which means "house"; or *bath*, "daughter." Though they only occasionally fall in sequence like the first two characters, numerous other parallels are found by Moran in later parts of the alphabet. He concludes that this zodiac, also widely used in Burma, India and other parts of the Orient, was actually invented by the Sumerians of Iraq. It does not seem far-fetched that the Phoenicians knew of a related lunar zodiac, from which they drew their alphabet.

In many ways a zodiacal origin for the alphabet makes perfect sense. The Phoenicians were exceptional navigators as well as traders—one can easily imagine them developing a calendar from their practical knowledge of the constellations. A calendar is obviously a useful tool for traders, as is a system of numerical notation. As in Gordon's theory, it would only have been at a later stage that

天の河全図

紫微垣

The traditional Oriental lunar zodiac of twenty-eight signs, redrawn from a modern Chinese-Japanese source. According to one theory, a similar zodiac inspired the first alphabet, that developed in the Near East during the early second millennium B.C. Many of the constellations resemble alphabetic characters. The Milky Way, or "River of Heaven," runs across the upper part of the sky plan; at its left-hand end (arrowed) is the Chinese constellation *Niu*, "The Ox," which closely resembles the first letter, *aleph* (also meaning "ox"), of the early Phoenician alphabet.

the twenty-nine or thirty symbols of the system were also found to be useful as a way of expressing language. Though Moran takes his case too far, trying to find a precise correspondence between the alphabet and the Chinese lunar signs, he may well have hit upon the key to the origin of the alphabet.

Moran's case has been taken up by David Kelley, professor of archaeology at the University of Calgary, in Alberta, and a specialist in the Mayan civilization of Mesoamerica. Kelley has analyzed many of the zodiac-based calendars from the eastern regions of the Old World and found a broad similarity with those of the New World, particularly that used by the Maya. Do these go back to Stone Age times and the common ancestor of the Mongoloid and Amerindian races, or, as Kelley suggests, to much later contacts involving maritime traders, such as the Phoenicians? For Kelley the most convincing evidence of Moran's zodiac-alphabet theory is an intriguing link that exists between the Mayan calendar and the Semitic alphabet.

The Mayan day list includes the sequence HAND (hieroglyph)—
lamat—WATER (hieroglyph). In the Phoenician-Hebrew alphabet we
find the sequence *kaf,* meaning "hand," followed by the letter *lamed,*
and then *mem,* meaning "water." The order of the hand and water
hieroglyphs might have been suggested by constellations that were
interpreted in the same way in different parts of the globe. But the
occurrence of the *lamat/lamed* sound in between, unless it is due to
sheer coincidence, is harder to explain.

Proof of any meaningful contact between the Old and New
Worlds before the Viking settlement of the tenth century A.D. has
long been sought. Does the invention of the alphabet, of all things,
provide some tangible evidence of such pre-Columbian contacts?
Only time, and much further research, will tell.

CODES AND CIPHERS

The idea of hiding secret meanings and messages in a text is almost
as old as the art of writing itself. Numerous documents from the ancient Near East show how its educated elite prided themselves on
their skill with riddles, puns, puzzles and cryptograms. Professor Cyrus Gordon, a leading expert on ancient Semitic languages, summed
up the delight that those in the know took in such wordplay: "For
the sages it was not enough to handle plaintext. Plain language is for
plain people. The elite must also master riddles and cryptograms,
and grasp the deeper meaning concealed beneath the surface meaning."

As early as the Bronze Age, Near Eastern scribes were being
trained in the art of deciphering scrambled or coded texts. A clay
tablet from Ugarit, in Syria, dating to around 1200 B.C., gives a simple student's exercise, the object of which was to find the answer to
a four-letter anagram. This of course was beginners' stuff; much
more complex systems were handled by the ancient scribes. The
most popular was the use of the acrostic, in which a hidden message
can be read by putting together the first letters of each line or verse
of a text. A fine example is given on a clay tablet from Iraq of the

mid–second millennium B.C., known today as the *Babylonian Theodicy*. The text reads well as a poem of twenty-seven verses (each one of eleven lines) giving wise counsel in religious matters. But it was also cunningly written so that the initial syllables of each verse formed an acrostic revealing the name and credentials of the author: "I Shagil-kinam-ubbib, exorcist, am a worshiper of God and King!" Other tablets use not only acrostics but also telestics, in which the last syllables also spell out a message.

It was against this Near Eastern background of clever text manipulation that the ancient Hebrews developed one of the first systematic ciphers, a method known as *temurah* ("exchange"). The twenty-two letters of their alphabet were divided into two parts, one being placed above the other: letters were then exchanged with those falling above or below them. Numerous combinations could be made, depending on where the alphabet was split and in which direction the letters were laid out, each system being referred to by the first two pairs of letters that resulted. The simplest method was to fold the alphabet in the middle so that the first two letters, *A* and *B*, fell over the last pair, *T* and *SH*—giving the method the name of Atbash.

K	Y	T	Ch	Z	W	H	D	G	B	A
כ	י	ט	ח	ז	ו	ה	ד	ג	ב	א
ל	מ	נ	ס	פ	ע	צ	ק	ר	ש	ת
L	M	N	S	P	'	Ts	Q	R	Sh	T

The ancient Hebrew Atbash code, used for literary purposes. The alphabet is divided in two, the first (top) half running from right to left; the second half is placed beneath it running from left to right. Words are encoded by simply reading off the adjacent letters, for example, A = T, B = Sh and so on.

The use of *temurah* ciphers can be detected in the Bible, in at least one instance concealing a political message. A prophecy of Jeremiah (25:26), made in the early sixth century when Judah was under Babylonian military occupation, includes a curse of doom on the rulers of all nations of the world, ending with "the king of Sheshach." At face value this name is gibberish. When read using the Atbash cipher, however, the letters *SH SH CH* spell out *B B L*, or Babel, a familiar term for the oppressive Babylonian regime. The study of *temurah* became an integral part of the cabala, the mystical branch of

Judaism that crystallized during the Middle Ages; cabalists used the ciphers either to implant secret meanings in their own writings or to attempt to wring new interpretations from seemingly innocent biblical writings.

Military Applications

The great Roman general Julius Caesar (99–44 B.C.) devised a substitution cipher similar to one form of the Jewish *temurah*, in which the fourth letter of the alphabet stood for the first, the fifth for the second and so on.

A more technical encoding device was being used by the rulers of Sparta as early as 400 B.C., for sending secret messages to military commanders on campaign. This was the *scytale*, a wooden staff; the Greek historian Plutarch describes how it worked:

> When the Spartans send out an admiral or general they make two round pieces of wood exactly alike in length and thickness, so that each corresponds to the other in its dimensions. They keep one themselves, while they give the other to their commander. . . . Whenever they wish to send some secret and important message they make a long and narrow scroll of parchment and wind it round their *scytale*, leaving no vacant space but covering its surface all round with the parchment. After doing this they write what they wish on the parchment while it is wrapped round the *scytale*. When they have written their message they take the parchment off, and send it to the commander. He cannot make sense of the message—since the letters appear to be random—unless he winds the parchment strip around his own *scytale*. When its spiral course is restored perfectly he reads round the staff and so deciphers the complete message.

A Spartan military encoding device from 400 B.C. The message is concealed among letters written on a narrow scroll or parchment while wrapped around a wooden staff or *scytale*. The text appears to be complete nonsense, but when it is wrapped around a *scytale* of the same thickness as that which it was written on, the message appears.

The military uses of ciphers remained dominant until medieval times, although some interest was also taken in them for their own sake. The first treatise on cryptography was produced by Aeneas Tacticus in the fourth century B.C. Hundreds of years later codes still fascinated the leading thinkers of the day: The great English philosopher Roger Bacon (A.D. 1214–1294) cracked several simple cipher systems that were in contemporary use in Italy.

More secure codes than the Spartan one, which could be broken by a little trial and error once the method had been discovered, were developed by the medieval Chinese. The *Essentials from Military Classics*, compiled in the eleventh century A.D., recommended assigning different messages to the first forty characters of a particular poem known to both parties. When the first character in the poem was written at a prescribed point in an otherwise innocent message, the recipient would know, for example, to send more supplies. Such codes were essentially unbreakable but were limited in their scope.

The great invention in the history of codes, which expanded their range enormously, was made by Leon Battista Alberti (A.D. 1404–1472), an illegitimate member of a wealthy Florentine family. He gained renown as a painter, architect, musician and athlete and as a writer on law, art, philosophy and science. In addition to all this, he was the father of Western cryptography. Alberti's crucial invention was the mechanical cipher disk, which worked as follows: Within a large outer disk was a movable inner disk, both with alphabets running around them. In order to read the code, the receiver had to know how the disks should be positioned relative to each other. This information was given at the beginning of the message.

In our day of computerized code breaking one might imagine that all the ancient ciphers would have been cracked long ago. However, at least one medieval masterpiece of the art remains unde-

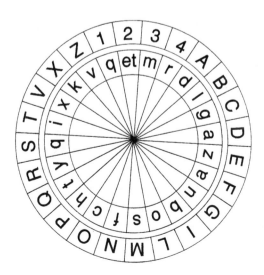

Alberti's mechanical cipher disk of the 15th century A.D. To encipher a message, you aligned a selected key letter on the inner ring with a letter or number on the outer ring. The text was then converted from the letters on the inner ring to the corresponding letters and numbers on the outer ring. To decode the message, the opposite procedure was followed, with the key pair having been given at the start of the text.

SHORTHAND

It is difficult to say when the very earliest shorthand was developed. But according to his ancient biographer, the soldier and writer Xenophon (around 430–354 B.C.) invented shorthand when drafting the memoirs of his mentor, the great philosopher Socrates. Greek shorthand later became very popular among the civil service of the Roman Empire, until Marcus Tullius Tiro, a freed slave employed by the politician Cicero, invented the first Latin shorthand system in 63 B.C. Cicero, a lawyer, politician and orator, was noted for his long-winded speeches, so the motivation behind Tiro's invention is clear. His Latin shorthand continued to be used in Europe for more than a thousand years.

Shorthand had also been invented in China by the tenth century A.D., as we can see from an anecdote told by the great Arab physician al-Razi (about A.D. 850–925). Al-Razi received as a houseguest a Chinese scholar who had learned fluent Arabic during his stay in Baghdad. A month or so before he had to return to China, he asked al-Razi if he could dictate to him the entire sixteen volumes of Galen, a classical medical treatise from the ancient Greek world preserved by the Arabs (see Introduction to Medicine):

> I told him that he had not sufficient time to copy more than a small part of it, but he said: "I beg you to give me all your time until I go and to dictate to me as rapidly as possible. You will see that I shall write faster than you can dictate." So together with one of my students, we read Galen to him as fast as we could, but he wrote still faster. We did not believe that was he getting it [right] until we made a collation and found it exact throughout. I asked him how this could be, and he said, "We have in our country a way of writing which we call shorthand, and this is what you see. When we wish to write very fast, we use this style, and then afterwards transcribe it into the ordinary characters at will." And he added that an intelligent man who learns quickly cannot learn this style in under twenty years.

Al-Razi's guest had been using "grass writing" (*tshao shu*), or Chinese shorthand. Surprisingly this complex ancient Oriental system directly influenced the development of modern European shorthand. During the Middle Ages Latin shorthand fell out of use, having become associated, through its use of symbols, with witchcraft. The first modern attempt at shorthand was made in 1587 by Timothy Bright, who invented a system of English shorthand using ideograms arranged in vertical columns, inspired by Chinese "grass writing."

The mysterious medieval Voynich manuscript, which has foiled all attempts to decipher it since its discovery in 1912.

ciphered. In 1912 William Voynich, an American dealer in rare books, found a peculiar manuscript in the library of a Jesuit school in Italy. The text is readable, being in the familiar Latin alphabet, but makes absolutely no sense, since it is written in an unknown code. Inscribed on fine parchment, it measures six by nine inches, with

204 pages surviving, and contains a number of illustrations, mostly of plants.

The manuscript was first discussed in 1666, at which time it was thought to have been written by Roger Bacon to hide his heretical thoughts from Church authorities. Voynich arranged for copies to be sent around the world to the leading code breakers of his day, but all their attempts at decipherment failed. In 1962 a panel of experts decided, on the basis of the kind of parchment and ink used and the style of the handwriting, that the manuscript was probably written about A.D. 1500. Various solutions have been proposed, including the theory that it was written in an artificial language or that Bacon used it to cover up his invention of the telescope and subsequent observation of unsettling facts about the heavens. None of these has been accepted, and the Voynich manuscript, which today sits in the rare-book vaults of Yale University, remains a mystery.

BOOKS AND PRINTING

In 1455 Johannes Gutenberg printed his famous Latin Bible at Mainz, in Germany. This was the first book printed in the West and was long thought to be the earliest in the world. Along with gunpowder, printing was hailed as one of the great inventions of medieval Europe; but like gunpowder, it was really a Chinese invention, only reinvented in Europe hundreds of years later.

The book, as such, is far older than the printed word. In the West, by the end of the first century B.C., the old idea of tying wooden waxed tablets together with thongs to make what the Romans called a *codex* was combined with the new invention of folding parchment. The folded sheets were stitched together, and the paper was placed between thin wooden boards to form covers, bound up to make a *volumen*. Initially these primitive books were used as notebooks, perhaps first by Julius Caesar (102–44 B.C.), or as ledgers, but by the third century A.D. they had been adopted by Christians for their religious works.

The oldest books (in the form we know today) so far discovered were recently found at Dakhla Oasis, in western Egypt. The dry conditions of the desert can preserve organic materials almost indefinitely, and in 1989 archaeologists lifted from the sand two books with wooden leaves bored for binding them together with cord, dating to around A.D. 375. One is a schoolboy's copy of a well-known volume of political essays by the fourth-century B.C. philosopher Isocrates of Athens, the other a farmer's account book used over a four-year period, giving a fascinating glimpse into everyday life.

The ancestors of Chinese books were bound blocks of bamboo or wooden tablets. They probably go back as far as the sixth century B.C., although the earliest surviving examples, inventories of mortu-

Bookcase and bookworm, from a Byzantine religious manuscript of the 8th century A.D. The figure depicted is the biblical writer Ezra "the Scribe" (5th century B.C.).

ary goods in a tomb at Sui-hsien in Hupei Province, date to 433 B.C. The bamboo strips, or "slips," were made by splitting bamboo rods; each slip was then formed into a rectangular shape by polishing and was finally bound together with silk strings or ribbons. Most Han Dynasty (221 B.C.–A.D. 220) texts were written on rolls of wooden strips. Before use, the sap was removed from the wood by drying and the wood was then cut to standard sizes. Each slip carried a single column of writing on one side only. Long tablets 19 inches high were used for the classics, and shorter strips 9¼ inches high for lesser works and copies of civil service regulations.

A thousand years later books began to appear on the other side of the world, in America. Mexican painted books were made from a strip of paper, deerskin or cloth, up to thirteen yards long and six to seven inches high, folded in concertina style like a modern map. The two ends of the strip were glued to thin wooden sheets, which formed the book covers. Both sides of the strip were written on: the strip was read from left to right, and each page from top to bottom.

A Mexican screen-fold book of the 15th century A.D., recording royal genealogies of the Mixtec people.

THE EARLIEST MOVABLE TYPE

Archaeology occasionally produces unique finds that seem to defy explanation. Among the best known of these is the mysterious clay disk found in 1903 by the French excavators of the Minoan palace at Phaistos, in Crete. Found with pottery dating to about 1700 B.C., the circular disk carries a hieroglyphic text that was impressed into the soft clay on both sides before it was baked. There are forty-five different signs on the disk, with lines dividing up words, while the text runs in a spiral out from the center.

The Phaistos Disk.

The most remarkable feature of the disk is the fact that the various signs are identical each time they occur. Stamps, carved from wood or cast in metal, must have been used to print the symbols into the clay—the world's earliest example of movable type. As for its contents, since nothing else in the same script has ever been found, they have proved impossible to decipher. The Phaistos Disk remains a mystery, despite nearly a century of speculation.

From Printing to Publishing

Six hundred years before Gutenberg the Chinese took the crucial step in the development of the book by applying an earlier invention—printing. Its origins lie in India. A Chinese monk who visited India around A.D. 670 reported that "the priests and laymen in India . . . impress the Buddha's image on silk or paper." As a result of such contacts the idea of printing as a way of communicating the written word came to be developed by the Chinese.

Woodblock printing on silk and paper started in China during the late seventh century, but the earliest printed texts to survive come not from China but from neighboring countries. In A.D. 764 the Empress Shotoku of Japan ordered one million miniature three-story wooden pagodas to be made, which were distributed equally to the ten leading temples in the kingdom. A sheet of paper bearing a printed charm was placed in each of these tiny pagodas. These were thought to be the oldest existing printed matter in the world until the discovery in 1966 of similar printed charms in the Sokkat'ap pagoda of the Pulguska Temple in Korea, which were made between the years 704 and 751. They were produced in order to try to gain the Buddha's favor for the newly created kingdom.

But the first complete printed book, dating to A.D. 868, does come from China. Discovered in the Tunhuang caves of western China in 1909 by the British explorer Sir Aurel Stein, it contains the complete text of the Buddhist classic the *Diamond Sutra*, translated from Sanskrit into Chinese. It was made from individual sheets stuck onto a backing to form a scroll 17½ feet long and 10½ inches wide. According to the printer's inscription, it was "reverently made for universal free distribution by Wang Jie on behalf of his parents on the fifteenth of the fourth moon of the ninth year of Xian Long."

During the ninth century woodblock printing became quite common in China, but the printing of books only really took off after A.D. 1000. Movable type made from earthenware was produced by Pi Sheng in A.D. 1041, but it was unsuitable for large-scale printing due to the fragility of the type. Metal cast type was invented in Korea around A.D. 1300. Such movable type never really caught on with the Chinese because of the thousands of characters in their language, so they stuck with woodblock printing. The best wood was that from pear trees, as it has an even and smooth texture, is not too hard and can be carved in any direction with few problems of going against the grain. Chinese printing did not involve a press as it did in the West. The woodblock was inked, paper was placed on it and the back of the paper was gently brushed. The medieval Chinese brought woodblock printing to new heights, even experimenting with two-color printing (black and red) as early as A.D. 1340.

Chinese printing grew quickly thanks to some extraordinary pioneers. Of these the greatest was Prime Minister Feng Tao, who decided in A.D. 931 to print the eleven classics of the Confucian doctrine. He pressed on despite dramatic political upheavals involving five changes of dynasty and ten reigns in twenty-two years. Taking up 130 volumes, and sold by the Chinese National Academy, they were the world's first official printed publication. By A.D. 1000 paged books in the modern style had replaced scrolls—a good 450 years ahead of Gutenberg.

Frontispiece of the world's oldest surviving printed book, the *Diamond Sutra,* published in China in A.D. 868. The Buddha is shown surrounded by disciples, divinities, monks and officials in Chinese dress.

ENCYCLOPEDIAS

The first attempt to produce a work containing information on all fields of human knowledge was made around 350 B.C. by the Athenian Speusippus (407–339 B.C.). Having succeeded the great philosopher Plato, his uncle, as director of the Academy at Athens, Speusippus attempted this daunting task to help with his teaching. No doubt he drew on the great series of scientific works written by Aristotle (384–322 B.C.), another member of the Academy, which make up a kind of encyclopedia themselves.

Speusippus's pioneering work does not survive, unlike parts of the earliest Chinese attempt, the *Erh Ya*. This was more of a collaborative effort, built up over the fourth to second centuries B.C. The core of the work existed as early as the sixth century B.C., but sections were still being added as late as the first century B.C.

The greatest of the early Chinese encyclopedias was *Master Lü's Spring and Autumn Annals*, completed in 239 B.C. According to the biography of its sponsor, Lü Pu-Wei,

> At that time in the State of Wei there was the Lord of Hsin-Ling; in Ch'u there was the Lord of Ch'un-Shen; in Chao there was the Lord of Phing-Yuan; and in Ch'i there was the Lord of Mêng-Ch'ang. All of them had been members of the lesser gentry, and they delighted in having [around them] visiting [scholars] by means of whom they could compete with each other [in argument]. Lü Pu-Wei was ashamed that Ch'in, with all its power, was still not equal [to these states in scholarship], so he summoned scholars to come, and entertained them lavishly, until he had three thousand visitors whom he supported. At this time among [the entourage of] the feudal lords there were many disputing scholars who wrote books which were spread throughout the world. Lü Pu-Wei now had all his guests record what they had learned, and he collected their discussions to form eight "Observations," six "Discussions" and twelve "Records" [totaling] more than 200,000 words. He maintained that all matters pertaining to Heaven, earth and the myriad things [in the universe] were contained [in this work]. He entitled it "Master Lü's Spring and Autumn Annals."

Lü was so confident that this encyclopedia contained the sum to-tal of human knowledge that he offered a reward to anyone who could fault his masterpiece:

> It was displayed at the gate at the market-place at Hsien-Yang, where 1,000 catties of gold [about 1,250 pounds weight] were suspended above it, and notification was issued to the travelling scholars and visitors of the feudal lords that any one among them who could add or subtract a word would win this treasure.

The most important and influential of the ancient encyclopedias was, by contrast, penned by a single man, the Roman administrator and soldier Pliny the Elder (A.D. 23–79). Pliny's *Natural History*,

The Roman encyclopedist Pliny, from an illuminated Italian manuscript of his *Natural History* produced in Rome about A.D. 1460.

completed in A.D. 77, comprised 2,500 chapters in thirty-seven volumes. According to him, it contained twenty-thousand noteworthy "facts" culled from two thousand works by more than one hundred authors. The *Natural History* covered astronomy, climate, minerals and metals, trees, flowers and plants, land animals, birds and fishes, human biology, the history of the arts and sciences, agriculture, medicine, and physical and historical geography. Pliny's life of public service came to an unfortunate end in A.D. 79. As commander of the Roman fleet at Misenum, in the Bay of Naples, he went ashore at Pompeii to investigate the explosion of Vesuvius, only to be suffocated by the fumes of the volcanic eruption that destroyed the town.

Later encyclopedias grew in size, with the Chinese statesman Li Fang compiling a collection of abstracts from 1,690 works in fifty-five sections at the end of the tenth century A.D. The largest encyclopedia ever seen was, however, that commissioned by Emperor Chu Ti around 1410. A staff of two thousand compiled this great work, which was written out by hand in 11,095 volumes totaling 917,480 pages—far more imposing than the heftiest reference work of today. Only three copies of this ultimate encyclopedia were ever made, none of which survives.

POSTAL SYSTEMS

Could you post a letter in 2000 B.C.? If you were rich enough and lived in Iraq or Egypt, the answer would be yes. By that date simple postal systems had already been set up by the extraordinarily well-organized civilizations of the Near East.

The Assyrians of northern Iraq had one of the most efficient services. In the nineteenth century B.C. they had a thriving commercial empire, reaching into central Turkey, whose lifeblood was the postal system operated between the homeland and its trading bases abroad. Excavation of the Assyrian merchants' colony at Kultepe, in Turkey, has uncovered a mass of correspondence, accounts and legal documents (more than sixteen thousand texts in all) left by the traders who lived there.

An ancient letter (left) and envelope(right) from Iraq. Both are made of clay and are inscribed in the Babylonian cuneiform script of about 1700 B.C. The envelope bears the impression of a seal depicting a religious scene.

The letters were written in the cuneiform Assyrian script on small clay tablets, about three inches square and were enclosed, naturally, in clay envelopes bearing the name and address of the recipient. The correspondence, which spans six generations, is a mine of information on day-to-day life in this ancient trading center. Letters were not all about business—some concerned family gossip and other personal matters. One Assyrian merchant was harangued by his womenfolk at home for being preoccupied with business and neglecting his religious duties. References in the letters such as "by the earliest messenger" and "by return of post" show that the service was fast and reliable. It was also secure: Money was sent to and fro using the couriers who ran the post, as this letter demonstrates: "I have received your instructions and the day the import of your tablet was made known to me, I provided your agents with three minas of silver for the purchase of lead. Now, if you are still my brother, let me have my money by courier."

The other great civilizations of the Old World were not far behind the Assyrians. The pharaohs of Egypt had started a royal courier service by about 2000 B.C., carried first by river and then overland after the establishment of relay stations around 1900 B.C. The system was taken abroad as the Egyptian Empire expanded, and by the fifteenth century Egyptian couriers were running back and forth between the pharaoh and his vassal rulers in Syria and Palestine and his fellow kings in Mesopotamia and Anatolia. The excavation of El-Amarna in Egypt during the 1890s uncovered the remains of the royal archive from the mid-fourteenth century B.C., including letters addressed to Tutankhamen. Like the Assyrian letters from Kultepe, they were written on clay in the cuneiform script. The language used, surprisingly, was not Egyptian but Babylonian, then the *lingua franca* of the Near East.

In China a royal postal service using mounted couriers was established early in the Chou Dynasty (1027–221 B.C.). It was proverbial for its speed. The great sage Confucius (551–479 B.C.) wrote that "the influence of righteousness travels faster than royal orders by stages and courier." In the following centuries references to the imperial postal service become more frequent, and many new features were added, such as the use of feathers attached to letters to indicate "Urgent!"

The First "Pony Express"

But the first really extensive postal system was set up by the Persians of ancient Iran some two and a half thousand years ago. They had great need of it—by 539 B.C. their enormous empire already stretched from the Aegean Sea to northern India, and a rapid means of sending dispatches between the emperor and his widely scattered governors was essential. The founder of the empire, Cyrus the Great (550–530 B.C.), built its major artery of communication—a "Royal Road" running from Sardis, near the Aegean coast of Turkey, to Susa, the capital of Persia. A highly efficient postal system operated along the Royal Road, using a system of horse riders working in relay that was curiously similar to the great Pony Express. Postal stations holding fresh supplies of horses and riders were set up along the Road a day's ride apart, each run by a superintendent with a staff of grooms. The entire 1,600-mile route from Sardis to Susa could be covered by the relay riders in nine days.

The Greek historian Herodotus (5th century B.C.) marveled at the speed of the Persian express riders: "Nothing in the world can travel as fast as these Persian couriers . . . nothing puts them off accomplishing at top speed the distance which they have to go, not snow, rain, heat, nor darkness. The first rider delivers his dispatch to the second, and the second passes it on to the third; and so on from hand to hand along the whole line, like the light in the Greek torch-race."

The Persians were masters of organization, and the basic structure of the postal system they established survived in various guises for almost two thousand years. After the conquest of the Persian Empire by Alexander the Great, in 330 B.C., Iran saw numerous changes of dynasty until the time of the Sassanids (A.D. 224–636), a dynasty of native rulers who claimed descent from the original Persian emperors. During this time the empire no longer included Turkey but still stretched from Syria to India, and the old royal postal system continued its work despite all the changes in government.

Only a few years after the death of the Prophet Muhammad (A.D. 632), the Sassanid dynasty had fallen to his followers, and Iran entered the Islamic world. Over the following centuries the Near East was ruled by religious leaders, the caliphs of Baghdad, who reformed and reorganized the old Persian postal system. Nine hundred and thirty postal stations were set up around their empire at average intervals of 7½ miles within Iraq and Iran and fifteen miles in the provinces. Mules and camels were used as well as horses, while the function of the local postmasters changed radically—they were now government agents whose responsibility was to provide intelligence on local affairs and reports on other local officials.

This curious mixture of postal and intelligence network was abolished by the Turks when they invaded in A.D. 1063, only to be revived again under the Mameluke Turkish rulers of Egypt and Syria in the thirteenth century. Under Sultan Baybars it was reorganized and improved by the introduction of pigeons (see **Pigeon Post**) and a system of warning beacons stretching from Iraq to Cairo (see **Telegraphy**). The delivery of ice for cooling drinks (see **Refrigeration** in **Food, Drink and Drugs**) became a regular part of the service in Mameluke times; the Egyptian treasury paid for camel trains to transport snow from the mountains of Syria all the way to Cairo. The medieval Arabic postal system was finally destroyed by the Mongol invader Timur in A.D. 1400.

In the meantime the Mongol rulers of eastern Asia had established a fast courier service throughout their empire, partly inspired

by the Persian and Arab models. When the Venetian merchant
Marco Polo visited China in the reign of the Mongol emperor
Kublai Khan (A.D. 1260–1294), he was amazed by the organization
of the royal postal system. Based on a massive network of relay riders
and post stations, it was remarkably similar to the old Persian "pony
express." Polo reported that there were ten thousand stations spread
throughout the Khan's empire and that no less than 300,000 horses
were used. In emergencies postal riders could requisition more
horses wherever they happened to be, simply by displaying their
badge of office. They were reported to cover up to 250 miles a day,
with a speed that Polo could only describe as "marvelous." The
Chinese system continued to thrive over the following centuries,
and in A.D. 1402 the Ming emperor Yung Lo opened the imperial
service to the use of private individuals.

Marco Polo's amazement at the imperial Chinese organization is
understandable. Though the Venice of his time had a rudimentary
postal system, Europe had yet to recover from the collapse of the
magnificent service that the Romans had once operated throughout
their empire. The Roman post, known as the *cursus publicus*, once
ranked in efficiency with the old Persian system. It, too, was almost
exclusively for the use of the state and was first organized by the
Emperor Augustus (27 B.C.–A.D. 14). A vast network of couriers,
staging posts and ships, it was the means by which the emperor and
his officials kept in touch with local governors and administrators. It
also helped to weld the empire together into one community by acting as a news service—reports of the latest public affairs in Rome
would be carried to the far corners of the empire and eagerly read
there.

But valued as it was, the *cursus publicus* could not survive the disintegration of the empire during the fifth century A.D. In the East
the remnants of the system continued in use under the rulers of the
Byzantine Empire or were incorporated into the Islamic postal services. In the West it slowly declined. The Frankish kings who took
over the Roman province of Gaul (France) tried to resurrect the
postal service of the Western Empire by using the old Roman post-
houses, but even the great Frankish emperor Charlemagne failed to
make it a permanent institution. Despite his best efforts, the system

collapsed again with his death in A.D. 814. In postal organization, as in so many other things, Europe fell far behind the East, where the ancient art of organizing complex communications systems was maintained until modern times.

PIGEON POST

Many ancient Romans were, like Andy Capp, keen pigeon fanciers. Pliny, writing in the first century A.D., complained how the passion for pigeons among his fellow citizens had become a mania: "Pigeon fancying is carried to insane lengths by some people: they build towers on their roofs for these birds, and tell stories of the high-breeding and pedigrees of particular birds."

But the Romans did not appreciate pigeons only for their looks or racing abilities. Although Pliny was rather sniffy about the subject, he himself admitted that tame pigeons could be of great practical value in "important affairs." The noble Brutus, besieged by Mark Antony in the city of Modena (northern Italy) during 44 to 43 B.C., still managed to communicate with his allies by tying his dispatches to the feet of pigeons. What use to Mark Antony, Pliny wryly noted, "were his rampart and watchful besieging force, and even the barriers of nets that he stretched in the river, when the message went by air?"

Pliny also mentioned the more unusual use of swallows as carrier birds. One Roman gentleman who was particularly fond of chariot racing would catch swallows from a nest at his country home and take them to the races in Rome. To give his friends advance results, he would paint the birds with the color of the winning team (see **Introduction** to **Sport and Leisure**) and release them to fly back to their nest. As Pliny noted, swallows were excellent carriers, as their speed meant they were rarely caught by predators. The Greeks were using pigeons for similar purposes, from sending love letters to reporting the winners at the Olympic Games, from at least the fifth century B.C.

Airmail Service

But neither the Greeks nor the Romans invented the art of carrier-pigeon training, the credit for which belongs to the Near East. The earliest mention of domesticated pigeons comes from the civilization of Sumer, in southern Iraq, from around 2000 B.C. Most likely it was the Sumerians who discovered that a pigeon or dove will unnerringly return to its nest, however far and for however long it is separated from its home. The first actual records of their use as carrier birds come from Egypt. By the twelfth century B.C. pigeons were being used by the Egyptians to deliver military communications. And it was in the Near East that the art of pigeon rearing and training was developed to a peak of perfection by the Arabs during the Middle Ages.

This scene, from the temple at Medinet Habu, built by Ramesses III (early 12th century B.C.), shows the symbolic use of pigeons in the great religious festival held annually to reinforce the kingship. Four pigeons are released to announce to the gods of the four quarters of the world the revitalization of the pharoah's power.

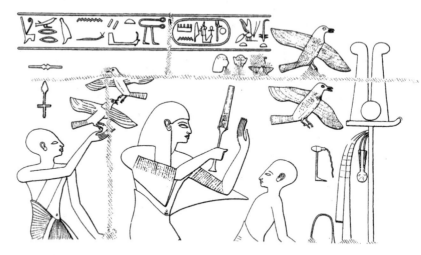

The caliphs who ruled the Moslem Empire after the death of Muhammad in A.D. 632 developed the pigeon post into a regular airmail system in the service of the state. Postmasters in the Arab Empire were also the ears and eyes of the government (see **Postal Systems**), and with the local postal centers stocked with well-trained pigeons there was little chance of the caliphs failing to be warned of potential troublemakers in the provinces. The state airmail was occasionally employed for more lighthearted purposes. Aziz, the caliph of North Africa between A.D. 975 and 996, one day had a craving for the tasty cherries grown at Baalbek, in Lebanon. His vizier arranged for six hundred pigeons to be dispatched from Baalbek, each with a small silk bag containing a cherry attached to its leg. The cherries were safely delivered to Cairo, the first recorded example of parcel post by airmail in history.

The Arab pigeon-post system was adopted by the Turkish conquerors of the Near East. Sultan Baybars, ruler of Egypt and Syria (A.D. 1266–1277), established a well-organized pigeon post throughout his domains. Royal pigeons had a distinguishing mark, and nobody but the Sultan was allowed to touch them. Training pigeons for postal work became an industry in itself, and a pair of well-trained birds could fetch as much as a thousand gold pieces. The royal pigeon post was also invaluable as an advance warning system during the Mongol invasions of the thirteenth and fourteenth centuries. When Timur the Mongol conquered Iraq in A.D. 1400, he tried to eradicate the pigeon post along with the rest of the Islamic communications network.

The Chinese seem to have learned the art of pigeon training from the Arabs. Strangely, for a civilization with such a well-organized bureauracy, the state never established an intelligence network using carrier pigeons, which were generally used only for commercial purposes. The Arabs also reintroduced the skill to medieval Europe, where it had lapsed after the fall of the Roman Empire in the fifth century A.D. After the collapse of the Roman light telegraph system (see **Telegraphy**), the pigeon post was left as the fastest means of communication in the world. And so it remained until the perfection of the electric telegraph (by Samuel Morse in 1844) and radio (by Guglielmo Marconi in 1895). It was normal practice, even well into this century, for navies, military installations and even businessmen to have pigeons on the payroll. The range of tasks for which pigeons have been employed in the modern world has changed little since ancient times.

But there may also have been a further, far subtler, use that the ancients had for the pigeon post. Surprisingly enough it was a religious one.

Pigeons and Prophecy

Somehow or other the oracle centers of the ancient world seemed to have had genuine knowledge of faraway events. At numerous temples scattered throughout the ancient Mediterranean a priest or priestess would give prophetic utterances on behalf of the local deity—the oracle of Apollo at Delphi, in Greece, was the most famous of all. The answers they gave to individual questions were usually in verse and were so subtly worded that whatever the outcome, the oracle gave the impression of having been right. In some cases, however, they gave quite specific information about remote circumstances that in the normal course of affairs they could not have known about.

King Croesus of Lydia (560–546 B.C.), once a powerful state in eastern Turkey, is best known as a byword for opulence, but he should also be remembered for conducting the first known state-sponsored experiment to investigate claims of the paranormal. According to the Greek historian Herodotus, Croesus set up what he thought would be a conclusive test of the Delphic Oracle's power. He sent his messengers to Delphi with instructions that, on the hundredth day after setting off, they were to ask the oracle what King Croesus of Lydia was doing that day, note the response, and return home. On the appointed day Croesus performed a deliberately pointless and unlikely task—he boiled a tortoise and a lamb together in a bronze caldron. He was most impressed when his messengers returned from Delphi with the following lines:

> The smell has come to me of a hard-shelled tortoise
> Boiling and bubbling with lamb's flesh in a bronze pot:
> The caldron underneath is of bronze, and of bronze the lid.

How could the oracle know what Croesus was doing that day in Lydia, some 350 miles away? Did the priests running the oracles have some secret method by which they could rapidly obtain information over long distances? This is the theory proposed by Robert Temple, a historian of science who has made a penetrating study of ancient oracles and their methods: "I believe that the oracles were fed with information about important events, such as the outcomes of battles, by carrier pigeon. Such news would then be announced as prophecies, sometimes on the very days of the events. Some days or even weeks later, when human messengers were able to arrive, the "news" they brought confirmed the 'prophecies.' "

Temple's pigeon-post theory could well explain the Croesus case. An agent of the Delphic Oracle could have reported Croesus's experiment by pigeon post. A good racing pigeon can achieve speeds of over ninety miles per hour, which means it could have flown from Croesus's court to Delphi in seven hours or so, and certainly within the day allotted for the experiment. It was well worth it for the oracle to go to such efforts—Herodotus provides a long list of the gifts that Croesus lavished on the successful oracle, including his wife's jewelry, solid gold statues and six tons of gold bullion.

Temple's case is based on circumstantial details—as it would have to be, since the priests would have guarded their secret very carefully. But these clues are so persuasive that it can hardly be doubted that he has discovered the truth about an "advance news" system used by the oracles. There is a mass of literary evidence to show that the oracles kept cotes for pigeons, doves or other birds, ostensibly because they were sacred to the divinity who spoke through the oracle. For example at Dodona, in northern Greece, the priests were said to receive messages from the god Zeus by listening to the cooing of doves; the guild of priestesses who transmitted the oracles even called themselves the Rock Pigeons.

Herodotus recounts a telling anecdote about one Aristodicus, who kept receiving advice from the Oracle of Apollo at Miletus that was not to his liking. In retaliation he went into the temple precincts and "stole away the sparrows and all other families of nesting birds that were in it." As he did so, a voice from inside the shrine boomed out calling Aristodicus the "wickedest of men." As Temple notes, the oracle would hardly have been so enraged if these really were just wild birds nesting in its grounds—more likely Aristodicus had guessed their secret and hit the oracle where it hurt by stealing the key to its advance-news system.

This is not to say that oracles never gave advice or utterances on the basis of genuine inspiration. Still, even those who accept the re-

A pigeon or dove approaches a seated goddess or priestess. The *omphalos* ("navel stone") in front of her shows that the scene is an oracle center such as Delphi, where such a stone was a centerpiece of the cult. She clasps a torch, ears of wheat and poppy heads. From a Greek plate of the mid-5th century B.C.

ality of paranormal abilities admit that they are highly unpredictable. Temple's theory that the oracles also used another, eminently practical, way of obtaining advance information seems to be the only way of making any sense of the curious pattern of bird associations and legends that surrounds the oracles.

As a clinching piece of evidence Temple cites a tradition related by Herodotus. He was told the following by the priestesses who delivered the utterances of Zeus at the temple of Dodona:

Two black doves, they say, flew away from Thebes in Egypt, and one of them alighted at Dodona, the other in Libya. The former, perched on an oak, and speaking with a human voice, told them that there, on that very spot, there should be an Oracle of Zeus. Those who heard her understood the words to be a command from heaven, and at once obeyed. Similarly the dove which flew to Libya told the Libyans to found the Oracle of Ammon—which is also an Oracle of Zeus.

This thinly disguised story, involving "talking" doves flying between the Egyptian oracle center of Thebes and those at Dodona, in Greece, and Siwa, in Libya, is almost enough by itself to spill the beans. It naturally raises a wider question that has not as yet even been considered by ancient historians—that the various oracles, such as at Dodona and Siwa, had long been in cahoots by communicating via carrier pigeons or doves. How and when they began their partnership of subterfuge still awaits investigation.

TELEGRAPHY

According to the Greek dramatist Aeschylus, advance news of the capture of Troy by the Greek commander Agamemnon was sent home to Mycenae using beacons. Aeschylus's play *Agamemnon* opens with a watchman who has been patiently awaiting this victory signal for a year. While there is no concrete evidence that beacons were used for communication at the time of the Trojan War (traditionally dated to 1194–1184 B.C.), Aeschylus's play shows that their use was familiar in his own time (about 525–455 B.C.). And with beacons using simple prearranged signals, the prehistory of the telegraph begins.

Not long after Aeschylus's time we find that signaling with torches had been developed into a systematic method of communication. In the fourth century B.C. a Greek military scientist named Aeneas Tacticus (see **Codes and Ciphers**) developed an ingenious method for communicating messages by using the simplest of torch signals. Two armies wishing to communicate would first have to prepare two sets of identical equipment. Each would take a large earthenware jar, drilled at the bottom with a hole of exactly the same size, and a cork float just a little smaller than the neck of the jar. Each cork was pierced with a rod, clearly marked into sections at three-fingerbreadth intervals, and the sections were inscribed with useful military messages—such as "cavalry have arrived," "corn available," and so on.

Having agreed on the most useful messages, the two armies could then separate and use torches to signal each other. Each jar was filled

The decoding jar designed by Aeneas Tacticus in the 4th century B.C. for use in telegraphy. The idea was based on the *clepsydra,* or water clock (see **Clocks** in **High Tech**). The cork is pierced by a rod divided into sections, on which different messages are written. When the plug is removed, water flows out and the cork float sinks, gradually bringing different messages to the level of the jar mouth. If two armies have identical sets of equipment, they can "transmit" messages by signaling that the plugs should be removed, and signaling again when the required message reaches the mouth of the jar.

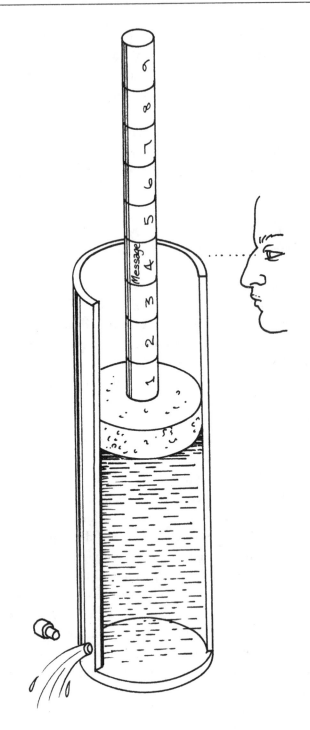

with water, the hole at the bottom plugged and the cork with the rod in it inserted into the mouth. When army A wished to speak to army B, a signalman waved a torch until it was seen by the signal-man of B, who responded by raising a torch. When contact had been made, each signalman unplugged his jar, letting the water drain out. The cork float was allowed to sink until the message that sig-nalman A wished to communicate had reached the level of the jar mouth. He waved his torch again and B read off the message from his own equipment. Since the two sets of jars, holes, and so forth were identical, the floats would sink at the same rate and always pro-vide the same message at any given moment.

The method is a trifle zany, but it would obviously have been completely uncrackable as a code. Nobody but armies A and B could possibly work out what was written on the rods. But the sys-tem was also rather cumbersome. Only one message could be given at a time, after which the equipment would have to be reset by plugging the jars and filling them with water again.

The First Semaphore

Another fault with Aeneas Tacticus's system was that it had no pro-vision for unforeseen circumstances. If the enemy launched a sur-prise attack mounted on elephants, it is unlikely that such a contingency would have been written on the rods. A far more flex-ible system was needed. The Greek soldier and historian Polybius (203–120 B.C.) has left us a detailed account of the new method, the first system of semaphore ever recorded. Polybius describes it with pride. While it was originally devised by two Greek military scien-tists, he stresses that it was "perfected by myself."

First, said Polybius, the alphabet had to be divided into groups of five letters and written onto five tablets. (The last group would only have four, there being twenty-four letters in the Greek alphabet, but this makes no difference.) Communication between the armies be-gan when a signalman raised two torches into the air, to be ac-knowledged by two torches. The signalman raised his torch one to five times to the left, indicating which tablet was to be read, and then the required number of times to the right to indicate the letter.

Clarity could be improved by providing screens, the height of a man, to the left and right of the signalman; the signalman moved his arms behind these, so that the raised torch appeared above the screen and a lowered one was invisible. To narrow the field of vision, Polybius recommended a *diopter*, by which he meant a sighting tube or a device with "sights" similar to that used in surveying (see **Tunneling** in **Working the Land**).

As Polybius explained, messages had to be kept as short and simple as possible to avoid confusion, but with practice a signalman would become extremely fluent with the system. Yet its range was rather limited; too great a distance and it would be impossible to tell the left from the right sides of the signalman. But masters of military organization that they were, the Romans devised a variation of Polybius's system that, with one minor adjustment, was much more effective. The method was recorded for us by Julius Africanus, a Christian philosopher of the early third century A.D.: "The Romans use a system, in my opinion a most remarkable one, to tell each other all kinds of things by means of fire signals."

An ancient system of sempahore, devised by the Greek soldier Polybius in the 2nd century B.C. Two armies could communicate by having agreed upon an arrangement of the letters of the alphabet written on a series of five tablets. The signalman sending the message stood behind a screen and was observed through a sighting tube by his opposite number. He indicated a tablet number by raising his torch a number of times to the left, and a letter number by raising it to the right. The observer could then decipher the message by reading off the letters from the tablets.

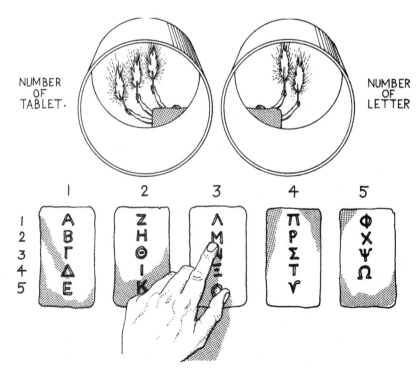

The method described by Africanus required three signalmen, standing a fair distance apart, and the code used was the Greek alphabet divided into three parts. *Alpha* would be shown by the man on the left lifting the torch once, *beta* by his lifting the torch twice, and so on; the second part of the alphabet was indicated by the man in the middle, who raised the torch once for *iota* and so on. If the signalmen stood about ten feet apart, the visibility could be as good as six miles. R. J. Forbes, a leading expert on ancient technology, has concluded that this method was by far the most effective ancient telegraphic system, describing it as "a true precursor of Morse."

Africanus also tells us how the Roman telegraph was used: "Those who receive the signals write down the letters received in the form of fire signals and transmit them to the next station, which then transmits them to the following one and so on until the last of the fire-signaling stations." Thus the distance over which the telegraph was effective could be increased infinitely, through a series of signaling stations or fortresses. Strings of fortresses ideally suited to these purposes were in fact built shortly before Africanus wrote. On the Roman frontiers in both Britain and Germany, barbarian aggression was a long-standing problem, and the great emperor Hadrian (A.D. 117–138) built massive new defense works to hold them off (see **Introduction** to **Military Technology**). Hadrian's Wall in Scotland, an impressive stone structure some 120 miles long, has a pair of stone towers every mile that served as signal stations. A guardpost having trouble at one part of the wall could rapidly telegraph a message for help along the line to the next major garrison.

By the early fifth century A.D. the Romans had pulled out of Britain, and Germanic tribes swarmed into western Europe. The Roman telegraph died, but a similar system was carried on by the Byzantines, the successors to Roman power in the eastern part of the empire. Torches and lanterns were used, and another method (the "heliograph," after the Greek *helios*, or sun) involved mirrors that could flash coded messages by reflecting the sun's rays.

The Byzantine light telegraph, like the Roman, was organized for defensive purposes. During the ninth century A.D. the most vulnerable border of the Byzantine Empire was the province of Cilicia, in southeastern Turkey, where the Christian Byzantines were constantly under threat from the Arab caliphate of Baghdad. So a light tele-

A Roman telegraph system, as depicted on Trajan's Column in Rome (built about A.D. 115). Three signal towers are shown, from the windows of which torches are being waved by the signalmen. Next to the tower on the left are two haystacks and a large platform made of felled trees. In times of emergency the hay would be piled on the platform and ignited as a beacon—the Roman equivalent of a distress call.

graph was developed, running from hilltop to hilltop and eventually to Byzantium (now Istanbul), where the hub of the system was formed by the great lighthouse built on the terrace of the imperial palace. An Arab incursion over the Cilician border, some six hundred miles away as the crow flies, could be known about in the capital within an hour. The system was perfected by Leo "the Mathematician" under the ninth-century emperor Theophilus. The Arabs developed their own systems of fire signals. In the tenth century A.D. the Fatimid caliphs, who ruled all of North Africa, built a string of beacons running 2,200 miles along the coast from Ceuta, in Morocco, to Cairo, in Egypt. A later Egyptian ruler, the thirteenth-century Sultan Baybars, extended the system from Cairo to Baghdad to provide advanced warning against Mongol attacks.

Sadly we know little more about these medieval systems, which may only have involved the simple use of beacons for prearranged emergency signals rather than codes. The secrets of ancient military communications were rarely committed to paper, and it is remarkable as it is that we have so much information about Roman telegraphy, the most sophisticated until the invention of the mechanical telegraph in the eighteenth century.

The Shouting Telegraph

In the New World the North American Indians had their own system of telegraphy, the well-known smoke-signaling method. This was not restricted to codes based on numbers of puffs; different-colored smokes were produced by adding various substances to the fire. On his first voyage to America Columbus noted how the Indians communicated by smoke signals, "like soldiers in wartime." The Amerindian art of smoke signaling probably goes back to time immemorial, but there is simply no way of dating its earliest use by archaeological or written evidence.

Finally, we shouldn't overlook the extraordinary vocal telegraph used by the Gauls of ancient France. It required no technical apparatus whatsoever—just the healthy lungs of farmers—but it was extremely effective for transmitting messages over long distances with amazing speed.

The technically minded Romans were clearly impressed when they came across the system, which required nothing more than simple rustic organization. Julius Caesar described it in the memoirs he wrote of his Gallic campaigns. It was 52 B.C., and the tribe of the Carnutes had just broken into open rebellion:

> The news spread swiftly to all the tribes of Gaul. For when anything specially important or remarkable occurs, the people shout the news from one field to the next, and from farm to farm; step by step the message is received and passed on. In this event, what happened at Cenabum at dawn was known before eight o'clock at night in the country of the Arverni, about a hundred and fifty miles away.

The speed of this ancient French shouting telegraph was nearly twelve and a half miles per hour.

SPORT AND LEISURE

CHAPTER TWELVE

INTRODUCTION

The first recreation was, of course, spontaneous. In the right circumstances gathering, hunting and fishing can also be fun. The first people would have enjoyed these and other simple pleasures, such as feasting, drinking and sex, along with an increasing variety of games and rituals involving horseplay, dance and song. Of these the role of music in the recreation of the earliest human beings may have been paramount—indeed the first tangible evidence of any kind of leisure activity comes from musical instruments.

Not surprisingly, with the wide variety of instruments available in the ancient world, bands and orchestras made an early appearance. Specialized professions developed with the rise of urban civilizations (see **Introduction** to **Urban Life**), and musicians were among the earliest. The world's oldest depiction of a band is on a seal found at the archaeological site of Chogha Mish, in Iran. From the fourth millennium B.C., it shows a small concert with kettledrum, harp and horns. Whether they were professionals can't be said, but the orchestras who played to the Egyptian aristocracy during the New Kingdom (about 1550–1100 B.C.) certainly were. Their main source of musicians was Syria and Palestine, which has been famed for its musicians throughout Near Eastern history. They were so prized that the pharaohs sometimes listed in the official records of their military campaigns the musicians they had captured—in one instance 270 musicians were brought back. With this input from the Levant, Egyptian orchestras became increasingly elaborate under the New Kingdom, the instruments played including harps, lyres, double oboes, flutes, lyres, lutes, tambourines and rattles. The players were often scantily clad young women, highly popular at the private parties of the rich.

We have no ancient equivalent of the "hit parade," but the ancient Chinese invented a curious institution known as the Bureau of Music to regulate styles and standards. Founded in the fourth or third century B.C., it was thought to have become out of touch with popular taste by the second century and was reformed by the energetic emperor Wu-ti (140–87 B.C.). He appointed the musician Li Yannan chief of harmony and ordered him to create a new style of religious music based on the light airs favored at secular occasions. Traditionalists loathed it because they had been brought up with the teaching of the philosopher Confucius that music made purely for pleasure was wrong. Despite their disapproval, the bureau became an important department of state, responsible for both official ceremonies—at which orchestras, choirs and choreographed dancers appeared—and for entertainments at court. It was also charged with the task of collecting popular songs and adapting them to the court style. But the Confucian critics eventually had their way, and the bureau was shut down in 7 B.C. on the grounds that it had encouraged and practiced a corrupt form of music opposed to the morally uplifting type played previously. At the time it was abolished, the bureau had 829 employees, including singers, instrumentalists, craftsmen and musicologists.

Harp player—from an Egyptian tomb of the New Kingdom Period (c. 1550–1070 B.C.).

The central role played by music in the ancient world is well illustrated by a comment that Aristotle made about the ancient Etruscans, whose civilization flourished in the eighth to third centuries in central Italy before the rise of Rome. He stated that the flute was played during their religious rituals, at boxing matches, when slaves were being beaten (to drown out the noise?), when food was being cooked and even during hunts. The Greeks, too, were great lovers of music and considered its enjoyment to be one of the better, more ennobling, recreational pursuits.

The Greeks were also particularly fond of dancing, which played an important role in the theater and religious worship as well as recreation per se. They took it seriously and organized the earliest-known dance schools—here dancing was taught like modern ballet, as an art of movement involving mental as well as physical discipline. The Romans took a very different attitude. While they tolerated some stiff forms of dancing in religious rituals, they were very prudish about dancing for fun, or as an art form, which struck them as a foreign, even dangerous, innovation. Artistic dance was introduced to them early in the second century B.C. by the Greeks and the Etruscans, to the horror of many serious-minded Romans. Around 130 B.C. the politician Scipio Aemilianus could barely believe his eyes when he found dozens of the sons and daughters of well-respected Romans cavorting in a dancing school: "I have been taken to these academies where I swear there were more than five hundred children of both sexes dancing with cymbals and striking attitudes that would dishonor the most worthless slave."

Shortly afterward all dancing schools were closed down, but the trend could not be stopped, and teachers from abroad flooded in.

The Romans of course preferred more "manly" pursuits, a taste shared by many of the ancients and displayed in a wide variety of ball games, combats and races, ranging from the hectic to the extremely violent. After their introduction to Rome by the Etruscans in 264 B.C., the Roman gladiatorial "games" became immensely popular. Elaborate training was provided in special gladiatorial schools, and for the top fighters and all swordsmen, style was as important as victory. They became the "movie stars" of their day, admired by the public for their physique as well as for the speed and cunning of their performance. Painters reproduced their likenesses for the masses, and some had admirers rich enough to commission mosaics, so that we can still see their brutish expressions and overblown muscles today.

Roman gladiators performed in vast arenas, such as the famous Colosseum, which opened in A.D. 80 with a hundred days of spectacle. These arenas were also the scene of the *naumachia*, reenactments of famous naval combats of the past, and of beast fights, in which thousands of wild beasts were killed. At the Colosseum these animals were kept in cages below ground until required, at which time they were herded into elevators worked by counterweights and pulleys, which brought them directly into the arena.

Gladiatorial combats were gradually eclipsed in popularity by chariot races. The earliest account of a chariot race occurs in Homer's *Iliad*, written in the eighth century B.C. where it was the first event in the funeral games of Patroclus, a hero of the Trojan War. By the fifth century B.C. chariot racing had become an event in the ancient Olympics. Pausanias, who wrote a guide to Greece in the second century A.D., described the starting equipment for the race, invented by an architect named Kleoitas. Housed in starting boxes along one side of a triangular enclosure at the beginning of the track, the chariot teams waited until a cable suspended in front of them dropped. At the start of the race the clerk of the course operated a mechanism that made a bronze eagle shoot into the air; simultaneously the cable holding the team farthest from the racetrack

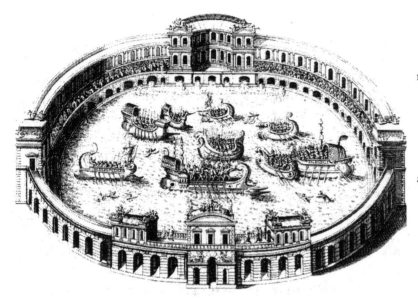

A Renaissance reconstruction of one of the spectacular naval battles held in the Colosseum in A.D. 80, during the celebrations that accompanied its opening. The Colosseum provides a prime example of how the Romans used their technology in the service of leisure.

dropped. As this team passed the second box, the cable for the next team fell, and so on, so that all the teams were in motion by the start of the track.

Like gladiatorial combats, organized chariot racing in Italy began under the Etruscans, but only reached its full height of elaboration under the Romans. They were the first to construct special hippodromes (Greek for "horse courses"), such as the Circus Maximus, built by the Emperor Augustus (31 B.C.–A.D. 14) to hold a quarter of a million spectators. Down the center of the track was a barrier holding statues, columns and sometimes stolen Egyptian obelisks. In the Circus Maximus this barrier also supported a tower from which seven bronze dolphins "dived" one by one into a water channel as each lap was run. Each race consisted of seven laps of about six hundred yards and lasted about twenty minutes. Roman racing chariots used two or four horses, and up to twelve teams competed, racing under four colors—blue, green, red and white—which had organized supporters' clubs or factions. Successful charioteers, like gladiators, became stars, and statues were erected to them by ardent admirers. The greatest charioteers won over three thousand races during their careers. The passionate support of the sporting factions sometimes boiled over into fighting after the races had finished, and in Byzantium during the later Roman Empire there were serious riots between the "blues" and "greens" (which by then had absorbed the other colors) during which many lost their lives.

Horse racing without chariots was already popular in Arabia in pre-Islamic times, and the Arabs were the first to develop it into an organized sport. In the eighth century A.D. the caliph Mahdi of Baghdad laid out the first hippodrome for horse racing per se, as opposed to chariot sports. In the early Arab hippodrome races ten horses were entered, with seven prizes awarded. The first eight horses were allowed into the winners' enclosure, and their owners received robes. The owner of the last horse in was humiliated by seeing a monkey mounted on the back of his pride and joy. Since common people couldn't afford to take part in these races, they imitated them with contests involving camels, donkeys, mules and dogs. In the ninth century A.D. there were also races for homing pigeons (see **Pigeon Post** in **Communications**). This even became something of a social problem in the time of Caliph Harun al-Rashid, when some devotees were squandering their life savings on young birds with a good pedigree or on wagers at the races.

While we often have a wealth of information on the sports enjoyed in the ancient world, the question of how and why they were invented can be more difficult to answer. In some there certainly seems to have been a ritual, even sacrificial element. One can certainly detect this in the dangerous bull-leaping sport of Minoan Crete, the Roman gladiatorial contests and the sacred ball game of ancient Mexico. In each of them death was a frequent outcome—albeit of a bull or an athlete. Other sports, such as the ancient Chinese ancestor of badminton, seem to have evolved from nonviolent rituals. The original Olympic Games seem to have begun as races to accompany a religious festival; still, what religious meaning, if any, the racing itself had is hard to determine. The idea that the ancients were terribly solemn people who were motivated to do everything by religious considerations is dubious, if only because for most ancient civilizations religion was not something separate from everyday life.

Indeed the whole question of the evolution of different sport and leisure activities from religious rituals is a vexed one. We can never be absolutely certain that the activity was not developed in its own right—and just for the fun of it—and then grafted on to a religious

The ancient Egyptians seem to have invented a version of darts. As this scene from a tomb painting shows, boards were placed on the ground, and competitors threw metal points into them; it is thought that the boards were marked with rings for scoring.

festival or ritual at a later date. The controversial origin of Greek theater, as we shall see, is a good case in point.

More sedentary pastimes were so common in ancient times that they pose a recurrent problem for the archaeologist. Gaming pieces and fragments of decorated boards are often found during excavations, but the discoveries generally receive little publicity for the simple reason that, while the excavators are thrilled to find them, the evidence is almost always frustratingly incomplete. Dice, tokens or scraps from an inlaid gaming board are interesting highlights from a site that may otherwise have produced an unrelenting catalog of pottery, but are generally relegated to the back of archaeological reports because little sense can be made of them.

A large percentage of the small, "unidentifiable" objects found during excavations were almost certainly made for games of various kinds. Small balls of clay, stone or glass are a good example. It is difficult to identify them conclusively as marbles, though we know the game existed in classical times. According to the Greek writer Athenaeus (about A.D. 200), a game like marbles was already being played at the time of the Trojan War, traditionally dated to 1200 B.C. While the hero Odysseus was making his long journey back from Troy, the suitors of his queen, Penelope, played the game against each other: one marble represented the queen; the first to hit it had another turn, and if he succeeded again, then he was first in line to propose to her.

Dice at least are easily recognizable, and there are numerous references to dice playing in classical sources. According to the Greek historian Herodotus, writing around 450 B.C., the invention of dice was claimed by the Lydians of western Turkey. In fact along with

EXSI

NON
TRIA DVAS
EST

Roman dicers from a fresco at Pompeii, playing what is thought to be an early version of backgammon. The cartoonlike captions above their heads describe a disputed call: the man on the left cries, "I've won!" while the one on the right objects, claiming, "It's not a three, it's a two." Unfortunately we'll never know what happened next.

dice they supposedly invented knucklebones, ball games and "all games of this sort except checkers," as a way of alleviating the effects of a severe famine they suffered during the thirteenth century B.C.: "The way they used these inventions to help them endure their hunger was to eat and play on alternate days—one day playing so continuously that they had no time to think of food, and eating on the next without playing at all. They managed to live like this for eighteen years."

Despite these diversions, says Herodotus, the Lydians eventually took more drastic measures, drawing lots to divide the population into two parts, one of which was sent off as a colony to Italy, where they became the Etruscans. Not surprisingly this piece of folklore has been contradicted by archaeological discoveries: Dice have been found in the cities of the Indus Valley civilization of India and Pakistan, more than a thousand years older than Herodotus's Lydians. These were six-sided, although later Indian dice have only four faces.

Herodotus's story was probably meant to explain the popularity of dice playing among the Etruscans—something inherited by their successors, the Romans, who were obsessive dice players. There were Roman associations of professional dicers who made a living out of gaming, and the emperor Claudius (A.D. 41–54) even wrote a book on dice play. Dice and shakers are common finds at Roman houses and taverns. That there were unscrupulous dice players in Roman times is shown by finds of loaded dice: they were doctored by the insertion of minute spots of mercury, which would cause the dice to fall with the lowest score downward. The related game of dominoes is commonly supposed to be an Italian invention of the eighteenth century, but the state archives of China record that in A.D. 1120 a set of thirty-two pieces marked with 227 pips was presented to the emperor.

In ancient times as now, children had their own special play equipment. Some of the most intricate children's toys from the ancient world have been found in Egypt, including figures with movable limbs and even simple mechanical devices. In 1887 the British archaeologist Sir Flinders Petrie discovered at Kahun an Egyptian doll-making industry dating to around 1800 B.C. In several houses there were remains of painted wooden dolls with pegged, movable limbs, as well as dolls of blue glazed pottery that were cut off at the knees and held their hands at their sides. These were decorated with tattoo patterns of spots or lines on their thighs and with a girdle line and waist. In one house—thought to be the doll maker's home—

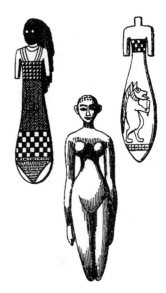

Wooden dolls from ancient Egypt.

Ancient Egyptian mechanical toys. Left, a snapping crocodile. Right, a baker who "kneads dough" when the string is pulled.

Children playing with swing, from a classical Greek vase.

Petrie found a large stock of dolls' hair. In Crete swings seem to have been popular. Pottery models dating to about 1600 B.C. depict goddesses sitting on swings suspended from trees. While these were cult objects, there is no reason to think that children (and adults) did not enjoy the same simple pastime.

The Greeks of classical times were the first to have a developed leisure ethic. Ancient Greek civilization reached its zenith during the fifth and fourth centuries B.C. when the ruling classes (about 20 percent of the overall population) were entirely freed from work by the extensive use of slaves and serfs. With income being provided by the sweated labor on their estates, even everyday business was taken care of by menials. Since their only duties were occasional political and military ones, the citizens of Athens and the other Greek city-states had a considerable amount of free time on their hands. Many of course sat around eating cabbage (thought to stimulate the brain) and engaging in philosophical discourse, including discussion of which activities were suitable as leisure for the rest of the population. Cerebral activities such as music and theater, providing they were the right kind, were favored, and the state invested large sums in building theaters and arranging drama and music festivals. For those who preferred more active pursuits, there were parks and gymnasia for exercise and a wide variety of sports. Modern Westerners would find themselves comfortably entertained if they could

The world's first sword-swallowing act. Hittite acrobats, as depicted on a relief carved about 1300 B.C., from central Turkey. To the left, one swallows a sword, while to the right, another performer wobbles on an unsuspended ladder.

somehow be transported back to Athens in the fourth century
B.C.—given of course that they were in the upper classes.

If one were to ask the question How did people amuse themselves
in ancient times? the answer would be simple: They had developed
practically every form of leisure activity enjoyed by people in the
twentieth century before the advent of radio, the cinema and TV.

BULLFIGHTING

According to Greek legend, Pasiphaë, the wife of King Minos of
Crete, developed a strange desire to be made love to by a particular
bull of extraordinary beauty. She ordered Daedalus, the brilliant
court inventor, to make for her a model of a cow in which she
could hide and present herself to the bull. In due course she was satisfied, and the child of the union was a bull-headed monster that
came to be known as the Minotaur. The monster was confined to
a special labyrinth, also constructed by Daedalus, which was so complex that the exit could not be found. Every year seven youths and
seven maidens were sent as tribute from Athens to King Minos; their
fate was to wander helplessly through the maze of passages until the
Minotaur caught and devoured them.

The monster's reign of terror came to an end with the arrival of
Theseus, son of the king of Athens. Ariadne, the daughter of Minos,
had fallen in love with Theseus, so when it was his turn to enter the
labyrinth, she gave him a ball of thread with which he could leave
a trail through the maze. Theseus entered the labyrinth, killed the
Minotaur, retraced his steps by following the thread and, meeting up
with Ariadne as arranged, sailed home to Athens in triumph.

All this was regarded simply as legend, until the discovery in the
early 1900s of the lost civilization of Crete, dubbed "Minoan" by the
excavators. The enormous palace complex at Knossos, whose construction began around 2000 B.C., seems to have inspired the story
of the labyrinth. Of course archaeologists found no trace of the
Minotaur, but they did uncover a wealth of evidence showing the

importance of bulls to the ancient Cretans. On the walls of the palace are extraordinarily vivid paintings of acrobats engaged in remarkable feats of agility involving bulls. The performer grabbed the bull by the horns, was thrown into the air, and landed on the animal's back, on either the hands or the feet; he or she then leaped off the bull's back, turning another somersault to land with arms outstretched on the ground behind the animal, in exactly the same way a modern gymnast completes a move. One of the Knossos frescoes shows all three stages of the bull leap. These feats of skill took place in front of an aristocratic audience seated on rows of benches, as shown in some of the paintings (see **Theaters**).

Many authorities have argued that the acrobatics depicted were impossible, since a charging bull will toss his head violently in an attempt to wound his opponent. However, Cretan bull leaping was not solely a sporting activity but had strong ritual overtones, so the risk may have been an essential element. It must indeed have been dangerous, even if the bull was partly tamed (or drugged?) and its horns filed down, for one Minoan bronze vessel carries a depiction of an unlucky acrobat impaled on the horns of a charging bull. Deaths caused by the sport would have contributed to the Minotaur legend.

Bullfighting, as practiced today, also seems to be derived from a ritual, but in this case one of bull sacrifice. Classical writers recorded

The (impossibly?) dangerous sport of bull leaping, as reconstructed from a fresco at Knossos by excavator Sir Arthur Evans. The acrobat seizes the charging bull by its horns (1), leaps upward as it tosses its head (2), completes a backward somersault (3) onto the bull's back (4) and then springs to the ground. No modern acrobat has dared to replicate this feat.

that in Thessaly, in northern Greece, a bull was killed at the end of a religious ceremony that included a bullfight. The bullfight was later introduced to the arenas of imperial Rome, where the bulls were roused to fury by red rags, then fought by Thessalonians on foot or on horseback: they leaped onto the bull's back, caught it by the horns and brought the animal to the floor before killing it.

The Spanish bullfighting tradition, dominant today, was not continuous from Roman times. It disappeared as an organized spectacle between the fifth and eleventh centuries after the barbarian invasions, though it was presumably kept alive in the countryside. Bullfights were re-created as part of the celebrations surrounding major events for the Spanish nobility, such as the birth of an heir, marriages or visits by foreign monarchs. El Cid, the famous warrior, who died in A.D. 1099, is said to have been the first man, in the revived sport, to spear a bull in front of an audience.

THE ORIGINAL OLYMPIC GAMES

According to tradition, the Olympic Games were first held at Olympia, in southern Greece, in 776 B.C., although the true date is probably somewhat later. Thereafter they were held every four years, as the modern ones are. Today we have a highly romantic picture of the ancient Games as a true contest of amateur sportsmen. As the Games developed, most of the competitors came to be rich men or top athletes, sponsored by cities, who competed for money; surviving lists of prize money show that, contrary to our image, the violent events were the most popular; and bribery was common, despite the heavy fines imposed on those who were caught.

The original Olympic Games were a one-day event serving a ritual purpose, perhaps on the occasion of a funeral, with sacrifices to Zeus, king of the gods, followed by a footrace. As time went on, however, more events were added, and the religious impulse became overshadowed by the competitive spirit.

The Olympic program, as established by 500 B.C., lasted for three days, with races for horses and chariots, three combat events (boxing, wrestling and an extremely violent free-for-all called the *pankration*), four footraces (two hundred yards, four hundred yards, a long-distance race and a race in armor), and the pentathlon, consisting of throwing the javelin and the discus, the long jump, a two-hundred-yard race and a wrestling bout. There were both men's and boys' competitions for the human part of the program, except for the pentathlon, which was limited to adults for no apparent reason. By this time the Games were attracting crowds of forty to fifty thousand.

Two features of the ancient Olympics would undoubtedly deter modern athletes. The first is the shape of the track. We are used to one made up of two straights joined together by curves, with a large open area in the middle for field events. The Greek stadium was a rectangle about two hundred yards long and twenty-five to forty yards wide with a starting line at each end and a turning post in the

A Greek vase of the 6th century B.C. showing an athlete performing the long jump at a Greek games festival. He is holding two weights shaped like flatirons. Modern trials suggest that, after a short runup, athletes swung the weights behind them and then forward at the moment of takeoff. Here the athlete is just landing.

The three pegs in the ground below him mark previous jumps.

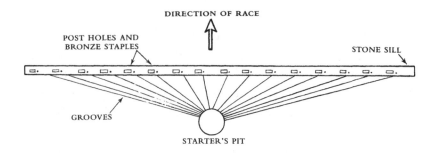

DIRECTION OF RACE

POST HOLES AND
BRONZE STAPLES

STONE SILL

GROOVES

STARTER'S PIT

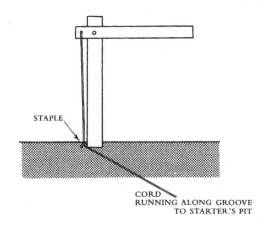

STAPLE

CORD
RUNNING ALONG GROOVE
TO STARTER'S PIT

Diagram of the mechanical starting gate uncovered at Corinth, used for the Isthmian Games, one of the rivals of the Olympics. A similar starting line was employed at Olympia. The runners were held in place by wooden bars until the starting judge tugged on all the cords simultaneously, lifting the bars out of the way.

middle of each line. In longer races the twenty runners had to make for the post and run around it, which must have led to some real scrambles. The second feature is the fact that the athletes competed naked. Tradition has it that in the Athenian Games, a rival to the Olympics, a runner's shorts slipped down; he tripped over them and was killed, with the result that the magistrates banned the wearing of shorts.

Both foot and chariot races had mechanical starting gates after 450 B.C., as we know from excavated examples. In the combat events there was no division into weights, which must have made some of the bouts a foregone conclusion. The wrestling match was won by the best of three falls, but both the boxing and the *pankration* carried on without rounds until one fighter submitted by holding up his hand. Boxing gloves made of hard leather strips were introduced in the fourth century B.C. to stop damage to competitors' hands. The *pankration* was the most popular of all events: it was a vicious and highly dangerous contest in which all parts of the body except the teeth could be used against the opponent. The pentathlon was rather different to today's sport, in which a competitor can come from behind to snatch victory; in its original form only those who had won one of the first three contests (the long jump and the two throws) were allowed to proceed to the fourth event, the two-hundred-yard race. The champion had to win three events out of the five—if he won the first three, that was the end of the contest.

The greatest Olympic victors of all time appear to have been Leonidas of Rhodes, who won twelve victories in the four Olympiads of 164 to 152 B.C., and Milo of Croton, in southern Italy, who won the boys' wrestling in 540 B.C. and then the adult version five times in a row from 536 to 520 B.C. Milo was thirty-four years old at the time of the last of his triumphs—quite an old man by ancient Greek standards.

Significantly, given present-day worries, the original Olympics seem to have died through an excess of commercialism. In the first century B.C. they ran into financial difficulties, since the richer Greek cities on the coast of Turkey were attracting the best athletes to their games. Olympia itself revived slightly through tourism under the Roman Empire, but the Games eventually fell foul of Christian worries about the nudity involved: they were officially closed in A.D. 396 by order of Emperor Theodosius. That was the end of the Olympic idea until it was revived in modern times, with the first Olympiad at Athens in 1896.

BALL GAMES

An ancient Egyptian leather ball, about three inches in diameter.

Ball games have existed throughout recorded history in a vast variety of forms. Simple ball games were played in almost all ancient societies, from Egypt to Australia. Egyptian footballs of four thousand years ago were made of soft leather or fine linen stuffed with cut reeds or straw. They were used for kicking around casually and for games of catch with simple rules. Greek ball games weren't heavily organized either: mostly played by boys, they involved either trying to hit each other with the ball or throwing and catching. One game, however, was a distant precursor of football: The players had to throw or carry the ball, an inflated ox or pig bladder, over a line while their opponents tried to stop them.

Soccer

The clear ancestor of soccer was the Chinese game of *t'su chu*, played by the third century B.C. The ball was made of leather, at first stuffed, then in later times inflated so that it carried farther. The feet and body, but not the hands, were used to propel the ball. In an aristocratic version played in front of the emperor's palace, the opposing teams tried to kick the ball through a tiny hole in a silken net. Even

A medieval picture of Chinese soccer, here a practice session in the garden.

the emperor would occasionally take part. Women also played in a version of the game with eight players called Eight Immortals Crossing the Sea. This Chinese "sport of kings" was, oddly enough, the sport of peasants in England during the fourteenth century A.D.— mobs numbering hundreds played; broken limbs were common and deaths not unknown.

The Amerindian equivalent to English soccer was lacrosse (or *baggataway* in Iroquois), which was played in the eastern half of North America. It was described by European observers in the seventeenth century as an ancient sport. Up to five hundred warriors on each side, often from neighboring villages, took part in this roving battle, with one team painted dark and the other light. The idea was to gather the deerskin ball in a hollowed-out stick (in later times a curved racket with sinew or hemp strings) and run with it toward the goal. Scoring was made even more difficult by the fact that the goalposts were formed by medicine men, who wandered across the field of play as the spirits directed them.

Polo

Back in China the popularity of soccer was overtaken in T'ang Dynasty times (A.D. 618–906) by polo. The game of polo was invented in ancient Persia (Iran), perhaps before 500 B.C. It was even claimed in medieval romances that Alexander the Great (356–323 B.C.) was given a polo stick after his conquest of the Persian Empire. More plausible is the tradition that Ardashir, founder of the Sassanid Dynasty of Iran in A.D. 224, was a great polo enthusiast. The game was played on horseback, and the aim was to hit a leather ball with a long-handled wooden mallet through a goal made of two posts set twenty-four feet apart.

As the only ball game conducted on horseback, polo naturally proved an immense hit with the nomads of Asia, and it was through them that it spread west to Byzantium by A.D. 1100 and east to China by the early eighth century, where princes maintained specially bred herds of horses for the game.

Ladies watch a game of polo from a covered stand— detail from a medieval Persian manuscript.

Golf

Golf is a relative latecomer, and many histories of the game state that its origin lies in the medieval Dutch sport *kolven*, which was played on the ice of frozen lakes or canals or on a paved court called a *kolf bann*. The contestants tried to hit two posts placed at opposite ends of the court with the ball, using the minimum number of strokes with the club (*kolf*). However, the ball used in this game—the size of a grapefruit and weighing some two pounds—would have been pretty hard to shift on land.

In fact the earliest version of golf seems to have been the Chinese game of *chiuwan* ("the ball-hitting game"), first mentioned in A.D. 943. It was very popular during the Sung Dynasty (A.D. 960–1279), particularly among soldiers, and even some emperors would play. *Chiuwan* involved tees, holes in the ground marked by flags, hardwood balls and clubs with bamboo shafts and hardwood heads, so it is a closer match to golf than *kolven*. Chinese historians believe that traders brought *chiuwan* back to Europe, where it was developed into the sport of today. Given the closeness of the words *golf* and *kolf*, however, it may be safer to say that golf was the product of combined influences.

Wherever it came from, golf had spread to Britain by the fourteenth century, when it was depicted in a stained-glass window in Gloucester Cathedral. The first written mention of golf in Britain comes in a law passed by the Scottish Parliament in March 1457, under which it was proposed that "goff be utterly cryit doune and not usit." The reason for this antigolf legislation was that it was interfering with the regular archery practice needed for civil defense. In another hundred years, however, golf had weathered this phase of disapproval to become the national sport of Scotland, and in 1552 Saint Andrews Golf Club, the oldest in the world, was founded.

Badminton and Tennis

Although the modern sport of badminton derives from a game dreamed up by the Duke of Beaufort around 1870 in his stately home of Badminton in Gloucestershire, its roots actually lie in ancient China. Some two thousand years ago a ritual game was played there with a feather "bird." The goal was to keep the bird in the air as long as possible without missing; the number of times it went up indicated the number of years the player would live. This ritual evolved into a game, played by two people, in which the "bird" (a rounded piece of cork with feathers stuck in it) was hit backward and forward, with a point being scored for each miss by the opponent. This game had reached England by the fourteenth century A.D., when it was called shuttlecock, eventually becoming a recognized sport in the late nineteenth century.

The origins of tennis are far more mysterious. "Real" (or "royal") tennis, from which the modern game of lawn tennis was adapted, was played in France in the twelfth century, especially in monasteries. At that time the ball was hit with the palm of the hand, then players started to wear gloves and later still, very peculiar-looking gloves with strings stretched across them. Where the monks got the idea, however, is unknown. It is possible that the game was one of the many new ideas brought back from the East by the Christian Crusaders, but at present there is no proof of this attractive theory.

Shuttlecock—the forerunner of badminton— shown in an English manuscript of the 14th century A.D.

The Mexican Ball Game

A Mexican ball-game player of the 1st millennium B.C.? These massive Olmec culture heads are often shown with a helmet, leading to suggestions that they may have been the first civilization in the Americas to play the sacred ball game.

The most extraordinary sport of the ancient world was without doubt the sacred ball game of Central America and the southern United States. It was first played in about 1000 B.C. by the Olmecs, who lived along the Bay of Mexico, and by all the later great civilizations of the region. From its very start it was played by the most important members of society. The colossal Olmec heads—carved from basalt brought down from mountains fifty miles away and weighing up to forty-four tons—show Olmec rulers wearing head coverings. A plausible explanation is that these are protective helmets (like those of modern football players) worn by the Olmecs when playing their sacred ball game.

The earliest ball courts were simple basins with earthen retaining walls, but by A.D. 1000 they had become far more elaborate. At Chichén Itzá the parallel walls were 283 feet long, 100 feet apart and 27 feet high. In Aztec times ball courts were shaped like a capital I, with temples at both ends and banks of seating along the sides. In the middle of the walls, which were usually twelve feet high, were set stone or wooden rings. The ball was a solid rubber sphere about six inches in diameter. To protect themselves against injuries from the heavy ball, the noble players (and a sprinkling of professionals)

Detail from the 15th-century *Codex Magliabecciano,* showing the Mexican ball game about to start. The rings, here shown outside the court, actually projected into it at right angles to the side walls.

wore protective helmets, wide belts of hard wood and leather, hip pads, knee pads and a single glove.

After the ball was thrown into play, players had to pass it to their teammates using their hips, elbows, or legs, without letting it run into the other side's end of the court, for this counted as a point against them. The excited crowd would bet on the outcome: according to a Spanish chronicler they would wager "gold, turquoises, slaves, rich mantles, even cornfields and houses." Star players were able to hit the ball up through the ring on the side of the court, thereby winning the game. The victorious side had the right to grab the clothes and jewelry of any spectators who couldn't get away fast enough. That the game also had a religious significance is shown by the fact that omens were read from the movement of the ball and the nature of the victory. That the losing team may have paid the ultimate price for defeat is suggested by the sculptural reliefs found next to many ball courts, which show a ball player being decapitated as a sacrifice to the gods.

GARDENS AND GARDENING

The English, whose abilities in almost everything else seem to have slumped so low, are still the world's most dedicated and skilled gardeners. Their gardening abilities are the result of long experience, stretching back to the Middle Ages, when the basic techniques, stock and forms of modern Western gardening were synthesized from a number of traditions. There may have been some input from the Islamic world, via Spain, but the main foundation for western European gardening was laid under the Roman Empire.

The Romans were expert horticulturalists and brought a wide variety of new plant species to western Europe. In Britain they introduced the grapevine; vegetables such as beets, lettuce, radishes and cabbage; trees including the sweet chestnut, mulberry, fig and peach; and new flowers, such as the rose, lily and violet. While the art of gardening, like so much else, declined with the collapse of the Ro-

A window box shown in a detail from a French miniature of the 15th century A.D. Window boxes were already common in Roman houses more than a thousand years before such medieval examples.

man Empire in the fifth century A.D., the plants of course remained. Specialized horticultural techniques, where preserved, were largely in the hands of monks and nuns, who transmitted some of the horticultural writings from Roman times and created new works of their own.

One extremely informative document from the early monasteries is a detailed map drawn up by a French cleric shortly before A.D. 840 showing an idealized plan for the grounds of a monastery. They are divided into three areas—an orchard for fruit trees, doubling as the monks' cemetery; a kitchen garden with plots for onions, garlic, leeks, shallot, lettuce, celery, parsley and other culinary herbs; and an infirmary garden (*herbularius*) divided into sixteen beds for different herbs. The monasteries, which were the hospitals of the medieval world, naturally specialized in the cultivation of medicinal herbs; they were also the leading experts on viticulture (see for example, **Wine, Beer and Brewing** in **Food, Drink and Drugs**).

While the largest gardens of medieval Europe were those in the monasteries, the fourteenth century saw the proliferation of private gardens in towns. Citizens of the later Middle Ages cultivated a wide range of blooms familiar to us today, including marigolds, camomile, wallflowers, delphinium, foxgloves, snowdrops and tulips.

Roman Gardening Technology

Centuries earlier private gardens had been immensely popular among the Romans. They generally placed their gardens at the backs of houses, leaving them open to the air but enclosed on all sides by colonnades and rooms. So keeping a Roman garden must have been like growing houseplants, as the top light that was available would have suited few species. Most planting was done in raised beds held by retaining walls two to three feet high.

Like modern enthusiasts, green-thumbed Romans would buy new stock for their gardens from nurseries, which also sold gardening accessories. A Roman nursery discovered by archaeologists had ranks of painted flowerpots, ready for use in gardens, in roof gardens or on balconies. Roman gardeners also used forcing frames for quicker plant growth. Modern versions usually consist of a box with a lid of glass or plastic, under which the heat generated by the sun's rays is trapped to encourage the growth of particular vegetables. The Roman forcing frame was essentially the same, except that the lid was made of *lapis specularis* ("mirror stone"), a transparent stone thought to be mica—sheet glass was available in Roman times, but it was expensive (see **Glass Windows** in **House and Home**). Under these frames they grew cucumbers in baskets of dung—then as now the best organic fertilizer.

Further clues to the gardening technology of Roman times come from a palace built at Jericho around 15 B.C. by King Herod of Judah. Archaeologists have excavated the palace complex extensively, including parts of the gardens. Set into the ground were large numbers of flowerpots, shaped like upside-down bottles with holes perforated near the necks. Many were broken, but with no sign of having been deliberately smashed, so it seems that they were broken by pressure from the roots of the saplings they contained. The Roman writers Pliny and Cato describe the method of using such pots to transport and propagate new trees. A young branch or shoot of a tree would be slipped into a perforated pot, packed in with soil and left for about two years, until it had rooted. It could then be separated from the parent tree and planted in a new environment. The species grown at Jericho have not yet been established, but likely candidates are the citron, palm tree or Judaean balsam.

Urban Romans living in flats cultivated window boxes with roses, lilies and violets. According to Pliny, writing in the late first century A.D., "the common people of Rome, with their miniature gardens in their windows, [once] offered to the eye a reflection of the country." Unfortunately things had changed for the worse by his day, as he went on to regret the fact that "the vast number of shocking burglaries compel us to shut away such sights from passers-by with bars."

The Romans also seem to have invented the art of topiary around the end of the first century B.C. Pliny describes the elaborate hedges

A Roman-style propagation pot from around 15 B.C., excavated from the Royal Garden at Jericho created by King Herod—exterior (left) and cross-section of contents (right). Such pots were used to transplant and propagate young trees. After planting, the vessel helped to conserve moisture around the sapling, which, if it took well, would eventually break open the pot. In Herod's gardens they were found in rows about five feet apart, with three feet between each pot in the row.

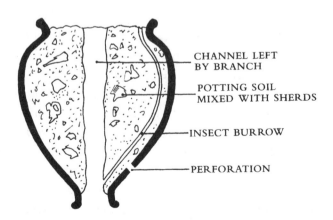

CHANNEL LEFT
BY BRANCH

POTTING SOIL
MIXED WITH SHERDS

INSECT BURROW

PERFORATION

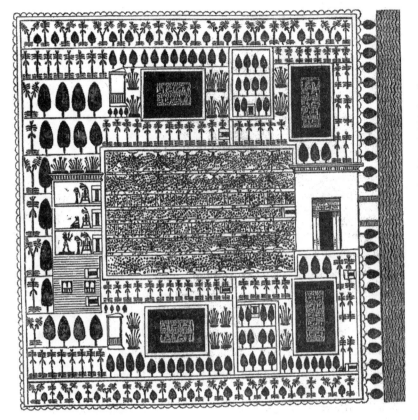

The neatly planned garden of the Egyptian official Sennufer, as painted in his tomb at Thebes about 1400 B.C. Like most Egyptian plans the perspective is curious, for it shows many features in elevation. For example the main gate is depicted (at the center right) as it would look if one were walking through it. Likewise the trees and plants are all upright, but the beds and ponds (containing ducks) are shown in plan. The center of the garden is occupied by vines trained over pergolas. The square plan is typical of Egyptian gardens of this period.

of the rich in his own time, clipped into hunting scenes or fleets of ships. The Romans acquired much of their gardening knowledge from the Greeks, even employing professional Greek landscape gardeners to add an authentic touch to grand pleasure gardens. But both the Greeks and the Romans owed a debt to the much older civilizations of Egypt and the Near East.

The ancient Egyptians were infatuated with flowers, not only for their beauty but also for the symbolic qualities they were thought to possess. Garlands, bouquets and collars of fresh flowers were made for every conceivable occasion, from banquets to funerals, and every large house had a garden. Egyptian gardeners were highly skilled in transplanting fully grown trees: they dug up the trees in a soil ball of roots and transferred them to huge pots, which they carried slung from poles on straps. Judging from tomb paintings, the blue water lily was their favorite flower, partly for its bright color but mainly for its scent. They also grew the red and white lotus, irises, poppies, corn-marigolds, jasmine, delphinium, oleander, convolvulus and chrysanthemum coronarium.

Botanical Gardens

The Egyptians were also the first to make botanical collections, beginning a long tradition in the near East of importing exotic plants from other lands. In the fifteenth century B.C. the female pharaoh Hatshepsut brought back frankincense trees from Somaliland, a move inspired partly by economic motives: Egyptian temples consumed vast amounts of incense (see **Perfume** in **Personal Effects**). But her nephew and successor, Thutmose III, seems to have had a more disinterested curiosity in botanical matters—illustrations of booty carved on the temple walls at Karnak include a huge number of plants and seeds collected during his campaigns in Palestine and Syria. Later pharaohs kept up the tradition. In the early twelfth century B.C. Ramesses III presented over five hundred gardens to temples. In addition to pleasure they provided a valuable source of income to the temples, producing incense, foodstuffs and raw materials that the priests would sell on behalf of the gods who owned them.

A similar fad for collecting new plants developed among the Assyrian rulers of northern Iraq. King Tiglath-pileser I (late 12th century B.C.) was particularly proud of his botanical collections: "Cedars

Pharaoh Thutmose III, who ruled Egypt in the mid-15th century B.C., is the world's first known serious collector of botanical specimens. A relief on the great temple at Karnak depicts in minute detail the plants and seeds he brought back from a military expedition to Palestine and Syria.

and box I have carried off from the countries I conquered," he boasted, "trees that none of the kings my forefathers possessed; these trees have I taken and planted in my own country, in the parks of Assyria." Later Assyrian rulers continued the tradition, including Sennacherib (701–681 B.C.), who gathered together in his gardens at Nineveh aromatic plants from Syria, including myrrh (which he claimed grew better there than in its native land), vines from the hills, and fruit trees from many countries. It even seems as if he obtained cotton bushes from India, since he writes of "trees which produce wool." In one place he cut a five-acre temple garden out of solid rock, by digging out planting holes and five-foot-deep water channels.

In the ancient Near East, it seems, gardening was very much a sport of kings. The Greek historian Xenophon records a fascinating dialogue that took place in 407 B.C. between the Greek ambassador Lysander and Cyrus, son and heir presumptive to the emperor of Persia:

> When Lysander brought gifts to Cyrus from the friendly cities of Greece, Cyrus entertained him, and among other things showed him his garden, called the Paradise of Sardis. Lysander was astonished at the beauty of the trees, their regular planting, the even rows, their positions at regular angles to one another: in a word, the rectangular symmetry of the whole, and the delightful smells they gave out.
>
> He couldn't help extolling the beauty of this fair scene, especially admiring the fine skill of the hand that had arranged it. Cyrus was pleased, and answered: "Actually, I planned, measured and laid out the plantation myself, and some of the trees I even planted with my own hands. . . . I assure you, whenever I'm in good health, I never dine before doing some exercise with weapons or husbandry, till I sweat."

The fabulous gardens built by the Persian emperors gave rise to the word *paradise*, which originally carried the simple meaning of "park" in Persian. The term was adopted by the Greeks, and then the Romans, when they built their own pleasure gardens in imitation of the royal gardens of the Near East, eventually taking on its modern meaning when the Fathers of the Christian Church used it to signify the abode of the blessed dead.

Greek gardening was inspired by models as far east as India. The Greek ambassador Megasthenes describes with wonder the beautiful parks that surrounded the palace of the Indian king Chandragupta Maurya (322–298 B.C.):

> Tame peacocks and pheasants are kept, and they [live] in the cultivated shrubs to which the royal gardeners pay due attention. Moreover there are shady groves and herbage growing among them, and the boughs are interwoven by the woodman's art. The actual trees are of the evergreen type, and their leaves never grow old and fall: some of them are indigenous, others have been imported from abroad.

Many ancient Indian texts describe the wealthy at play in their suburban parks. These contained artificial lakes and pools, often with fountains, and man-made hills, perhaps the earliest examples of landscape gardening. The parks were watered by channels leading from main tanks to the trees and flower beds. Chandragupta II (A.D. 375–415), king of northern India, had palace gardens stocked with brightly colored songbirds and storks chained by one leg. On hot days a water machine, described by the dramatist Kalidasá, would cool the air with a fine haze of moisture from a revolving spray like those used today for watering lawns.

Not all grand gardens were exclusively for the rich; from an early date public parks were created in both India and China. The Indian king Asoka (269–232 B.C.) took great pride in the fact that he had planted groves of trees for public recreation, and later rulers followed his example. During the T'ang Dynasty (A.D. 618–906) of China, public parks were created within towns and cities, playing an important part in urban life. Residents flocked to them on holidays and festivals to climb the hills, go boating, walk on the grass, admire the flowers, lie in the shade and watch acrobats and street musicians. Snacks, drinks, trinkets and toys were sold in the parks, making them a combination of playground and marketplace.

In the Western Hemisphere the kings of the great Mesoamerican civilizations developed gardens quite as sumptuous as those of the Old World emperors. The summer palace of the Aztec city of Texcoco lay at the center of miles of gardens built by King Nezahualcoyotl (A.D. 1440–1472) on a hill at nearby Tetzcotzinco. According to the accounts of the *conquistadores*, there were menageries and aviaries, sweet-scented groves and shrubberies, gardens of medicinal herbs, fountains and ornamental ponds surrounded by marble pavements, all overlooked by pavilions where the emperor and his mistresses would play during the summer. Water was distributed around the gardens in channels running from massive ornamental ponds.

The most developed botanical gardens of the ancient world seem to have been those of pre-Columbian America. Montezuma I (A.D. 1440–1468), emperor of the Aztecs, created a country garden at Huaxtepec, in Mexico, which he filled with tropical flowers, including dahlias, and young trees sent from the coastal provinces under his sway. Imperial officials oversaw the transport of trees and shrubs with their root balls wrapped up in matting. The emperor himself opened the gardens with a religious ceremony. The irrigation methods here were so good that the inhabitants grew tropical species, such as vanilla and cacao, under the supervision of forty gardeners who, with their families, had been brought by the ruler from the "Hot Lands" to the south, where these crops were native.

The seventeenth-century writer Garcilaso de la Vega described the extraordinary botanical and ornamental palace garden at Yucay enjoyed by the Inca of Peru in the days before the Spanish conquest of 1531:

> Here were planted the finest trees and most beautiful flowers and sweet-smelling herbs in the kingdom, while quantities of others were reproduced in gold and silver, at every stage of their growth, from the sprout that hardly shows above the earth, to the full-blown plant, in complete maturity. There were also fields of corn with silver stalks and golden ears, on which the leaves, grains, and even the corn silk were shown. In addition to this there were all kinds of gold and silver animals in these gardens, such as rabbits, mice, lizards, snakes, butterflies, foxes and wildcats; there were birds set in the trees and others bent over the flowers, breathing in their nectar.

The earliest European botanical garden was founded at Padua in 1545, apparently inspired by what the *conquistadores* had seen in Mexico and Peru.

From Flowers to Rocks

The gardeners of ancient China were the first to elevate the breeding of flowers into a fine art. Their favorite bloom was the Chinese chrysanthemum, praised in writings from the fifth century B.C. onward. The yellow-flowered variety was the first to be mentioned, and white and other colors became popular during the T'ang Dynasty (A.D. 618–906). Specialist cultivation began early on: around A.D. 1100 a manual by Liu Meng records thirty-five varieties of chrysanthemum, and a few years later the poet Fan Ch'eng-ta wrote that he had seen a collection of paintings showing more than seventy colors and forms of the flower.

During the Han Dynasty (202 B.C.–A.D. 220) the great gardens of China were deliberately simple and natural, but the simplicity was extremely subtle and the nature highly conventional. Gardening came to be seen as an art form closely allied to painting. Among the gardener-painters of the sixth century A.D., Chang Seng-yu was the finest. He designed many gardens, but in a sense they were all the same, for he worked within very clear limits. A thatched cottage for a resident philosopher-hermit was hidden among trees at the base of a "mountain"; there was a stream and a lily pond, some chrysanthemums, scattered fruit trees grown for their blossom, an old and preferably gnarled pine, perhaps a clump of bamboo and some strangely shaped rocks. Once established, this framework lasted for centuries.

Later on, rocks ("mountains") became ever more important, as can be seen from the work of the major figure in medieval Chinese gardening—Emperor Hui Tsang (A.D. 1100–1125). He was a talented painter and expert cultivator of plants, but above all he loved stones. He collected water-sculpted stones for his gardens at the capital of Hangchou, not only from lakebeds but also from older private gardens. At times barges bearing his stones blocked all other traffic on the canals around the capital. He also sent out a ruthless official, Chu Mien, who confiscated fine stones and old trees from private gardens throughout the land, with the aid of vast armies of peasants employed to scour the countryside.

The most favored rocks were limestones eroded by water into bizarre shapes, the best being those fished up from the bottom of Lake

Tai Hu, in southern China, honeycombed with holes. The largest were over six feet high; small examples were combined with sea-shells and colored gravel to make "tray gardens" for the house. Eventually Tai Hu stones became so expensive that an industry of fakes grew up, with craftsmen carving less worn limestones into fantastic shapes so cunningly that all but the most experienced connoisseurs were fooled.

This gardening with stones, utterly different from anything in the West, was taken to its ultimate extreme by the Japanese. Their great contribution to gardening was the invention of the "dry garden" (or "Zen garden") during the fifteenth century A.D. Developed largely in Buddhist monasteries, these gardens were utterly plantless and waterless, the only elements being rocks and sand. The rocks were arranged in tiers to imitate waterfalls, and the sand was regularly raked to prevent weeds from growing. Miniature versions were set in lacquer trays and displayed in the home.

The Japanese were also pioneering landscape gardening by the tenth century A.D. Gardeners moved earth around on a large scale, creating man-made mountains with the soil dug displaced by digging artificial lakes. The gardens of Kyoto and other centers became admired far beyond Japan. The first tourist party arrived from China in 1402—probably the world's earliest organized international garden-visiting tour. These aristocratic gardens were very different from the "dry gardens" of monasteries. They contained a wealth of flowering trees, pines and bamboos, and although rocks were used, water was just as important.

Floating and Hanging Gardens

Some of the strangest gardens of the ancient world were the floating gardens of Tenochtitlán, in Mexico. In the mid-fourteenth century A.D., when the Aztecs arrived at the site of their future capital on the edge of a large marshy lagoon, they set about creating the farmland they needed. First they cut canals through the reed swamps to reach the clear water below. On the matted floating vegetation left between the waterways, they weaved together the roots of the aquatic plants with small branches. These then formed a platform strong enough to support beds of fertile soil dredged up from the bottom of the lake while deepening the canals.

This was the origin of the *chinampas* that the Spanish *conquistadores* were amazed to find floating on the water when they reached Tenochtitlán in 1519. They were fifteen to thirty feet wide and up to three hundred feet long, with beds of soil up to four feet deep. The largest could even sustain willow trees, and were stable enough to support a cane or reed hut, where the gardener and family lived. To prevent them from floating off around the lake, they were held in place by posts and wicker hurdles. The *chinampas* were a major source of vegetables and flowers, including roses, for the teeming population of the capital.

But without doubt the most famous gardens of the ancient world were those built in the sixth century B.C. in southern Iraq—the "Hanging Gardens of Babylon," now almost a byword for horticultural extravagance. According to classical writers, they were built by Nebuchadnezzar II of Babylon (604–562 B.C.) in imitation of the Persian landscape in order to please his wife, a homesick princess. Their name, which comes to us from classical tradition, suggests something quite extraordinary—gardens suspended in air that one could walk underneath. This is in fact how they are described by the first-century B.C. Greek writer Diodorus Siculus, who is thought to have given the most accurate account of the Hanging Gardens, based on the writings of earlier Greek travelers.

According to Diodorus, from a distance the Hanging Gardens looked like a terraced hillside or the rows of seats in a Greek theater, as "the several parts of the structure rose from one another tier upon tier," the base occupying an area of about one-quarter square mile. Each terrace of this massive edifice was planted with trees and shrubs, the topmost level being a massive roof, so that one could actually walk underneath the gardens. Below it was a system of galleries, made from supporting walls twenty-two feet thick separated by passageways ten feet wide. Inside were the bases for each terrace, made from massive slabs of stone, covered with plants and jutting out so that each level projected over the one below. Holes in the superstructure allowed light into the inside galleries, while the top level, or roof, was strong enough to hold massive trees. It had a base of lead, to prevent water draining to the lower levels, covered with

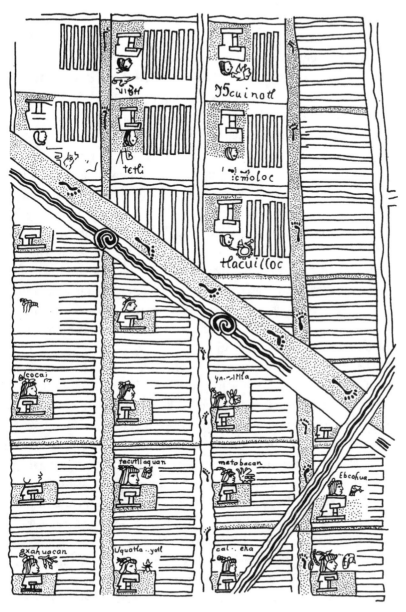

Part of the amazing floating gardens (*chinampas*) at the Aztec capital of Tenochtitlán, as depicted in a Mexican map of the 16th century A.D. The square allotments, separated by small canals, have smaller blocks within them containing the owner's name in both Aztec hieroglyphs and Spanish. The diagonal lines containing curly symbols are abstract representations of the main canals running through the area; one has a path alongside it, indicated by the footprints.

two courses of baked brick, a layer of reed matting set in asphalt and finally a thick covering of soil in which the trees were planted.

An intriguing problem concerns how such raised gardens were kept moist in the arid climate of Iraq. Three of our ancient sources specifically mention "machines" that lifted up water from the river

Euphrates to irrigate the gardens. The Greek geographer Strabo called these machines "screws," but this kind of water-lifting device is usually supposed to have been invented by Archimedes in the late third century B.C. (see **The "Claws" of Archimedes** in **Military Technology**). A recent study by engineer David Stevenson reviewed the other possibilities and suggested that the device used was a waterwheel of the kind called *noria*, which is well known from the Near East. Such wheels have clay or wooden buckets fixed to the inside of the wheel rim. The simplest way of powering them is by a treadmill, operated by a team of men or a donkey. Stevenson envisages a series of such wheels in use at Babylon, each drawing water up to a reservoir at a higher level, until the gardens at the top were reached. But if Stevenson is right, the history of the *noria* will have been pushed back by some five centuries. Wheels that either lifted or were powered by water only begin to appear in the written record in the first century B.C. (see **Watermills and Windmills** in **Working the Land**). Alternatively was Strabo right? Perhaps Archimedes's device was a reinvention.

The machines that watered the Hanging Gardens therefore remain a mystery. We are not helped by the fact that the location of the gardens has never been successfully identified among the extensive archaeological remains of Babylon. Nothing found so far actually matches the classical descriptions. Yet the latter are persistent—and reasonably consistent. It may be that the gardens were dismantled at some stage, the enormous stone slabs that provided the key to the "hanging" effect being robbed for use in other structures. If anything, all we might find today would be the foundations and some of the brickwork.

The ancient city of Babylon is so enormous that it has never been fully excavated, and until recently it was possible that some trace of the Hanging Gardens might yet be found. But in 1987 archaeologists under the direction of Saddam Hussein began to reconstruct Babylon into a theme park dedicated to the glorification of his regime. The price paid has been that large areas of the ancient city have been systematically bulldozed. Such brutal "restoration" techniques may have swept away the last vestiges of the most ingenious experiment in gardening ever undertaken, rated since classical times as one of the Seven Wonders of the Ancient World.

ZOOS

Although Regent's Park Zoo in London rightly sees itself as the world's oldest existing example, zoos were actually surprisingly common in ancient times. Ptolemy II (284–245 B.C.), Hellenistic ruler of Egypt, established a remarkable zoological collection in the gardens adjoining the Museum at Alexandria, the greatest center of learning in the ancient world (see **Introduction** to **High Tech**). It housed lions, leopards, lynxes and other cats, Indian and African buffaloes, wild asses from Moab (Jordan), a forty-five-foot-long python, a giraffe, a rhinoceros, and even a polar bear, as well as parrots, peacocks, guinea fowl and pheasants. Many of the animals in the collection were trapped in the course of elephant hunts. Ptolemy looked on his zoo not as a research institution but as an exotic curiosity of the kind a great king should possess.

King Ptolemy was following a long–established Egyptian tradition of royal animal keeping. In the fifteenth century B.C. Pharaoh Thutmose III made a collection of interesting plants, birds and an-

A scene from the tomb of Rekhmire, vizier to Pharaoh Thutmose III (mid-15th century B.C.). It shows tribute brought to Egypt from Syria, including a collection of animals— horses and, as shown in this detail, a bear and a baby elephant. Elephants (of the Indian variety) were once common in Syria and only became extinct there during the 1st millennium B.C.

imals, including a rhinoceros, that he came across during his military campaigns. Specimens brought back from Syro-Palestine were depicted on the walls of the great Temple at Karnak (see **Gardens and Gardening**), while paintings in the tomb of his vizier Rekhmire show exotic animals from Egypt's subject countries being delivered as tribute.

Similar activities were current on the other side of the world. In China, during the tenth century B.C., King Wen of the Chou Dynasty had a great pleasure garden constructed for his personal use, including a wooded area called the Holy Animal Park, where deer were bred. The great emperors of China revived this tradition, and in the early fourteenth century A.D. Marco Polo visited the pleasure gardens at Hangchou: "Two-thirds of the enclosure were laid out in lakes filled with fish and groves and exquisite gardens, planted with every conceivable variety of fruit tree and stocked with all sorts of animals such as roebuck, harts, stags, hares and rabbits."

The beginnings of a more serious interest in natural history can be seen in the Bible's account of the great Israelite monarch Solomon (10th century B.C.). He was famous for his botanical and zoological learning: "And he spake of trees, from the cedar tree that is in Lebanon even unto the hyssop that springeth out of the wall; he spake also of beasts, and of fowl, and of creeping things, and of

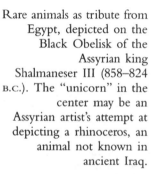

Rare animals as tribute from Egypt, depicted on the Black Obelisk of the Assyrian king Shalmaneser III (858–824 B.C.). The "unicorn" in the center may be an Assyrian artist's attempt at depicting a rhinoceros, an animal not known in ancient Iraq.

fishes" (1 Kings 4:33). It seems likely that Solomon had a zoological collection—the cargo brought by his navy included apes and peacocks as well as gold, silver and ivory (1 Kings 10:22).

Roman Safari Parks

Pleasure in, and scientific curiosity about, the animal kingdom were mixed in the Roman world, where wealthy private individuals and sometimes emperors maintained parks, animal enclosures, aviaries and even aquariums. In the first century B.C. the wealthy Quintus Hortensius had a dining room constructed to overlook his private park at Laurentum, on the coast near Rome, so that his guests could be treated to the spectacle of his animals converging on a servant got up as Orpheus (a mythical character who tamed wild animals with his music) when he blew a horn to summon them. Presumably the horn announced their feeding time.

Around 40 B.C. the Roman landowner and agricultural writer Varro built in the grounds of his villa at Casinum (modern Cassino) a domed aviary with fine hemp netting to keep the birds in. According to Varro, they were "chiefly songsters, such as nightingales and blackbirds."

The grounds of Nero's Golden House, in the heart of Rome, built after the great fire of A.D. 64, were stocked with all kinds of wild and domestic species. According to the Roman "gossip columnist" Suetonius, the Golden House grounds were vast: "There was an artificial lake to represent the sea, and on its shores buildings laid out as cities; and there were stretches of countryside with fields and vineyards, pastures and woodland, and among them herds of domestic animals and all sorts of wild beasts."

An enormous enclosure must have been built to house the animals collected for a triumphal procession by Emperor Gordian III before his sudden death in A.D. 244. They were eventually exhibited by his successor, Philip the Arab, at the celebrations in A.D. 248 marking the one-thousandth birthday of Rome as a city. According to the official *Augustan History* there were thirty-two elephants, ten elks, ten tigers, ninety tame and ten wild lions, ten hyenas, ten giraffes, twenty wild asses, forty wild horses, six hippopotamuses, and a single rhinoceros.

Menageries vanished in the West after the fall of the Roman Empire, only reappearing centuries later. Around A.D. 1100 Henry I of England established a small zoo at Woodstock, near Oxford; many of the animals in his collection had been presented to him by other monarchs. In the thirteenth century Frederick II, king of Sicily and Jerusalem, kept a large menagerie. Many of his animals, including elephants, camels, lions, cheetahs, panthers and monkeys, took part in a grand procession at Worms, in Germany, on the occasion of his marriage in 1235 to Isabella, daughter of King John of England. Her brother, Henry III, moved the English royal collection to the Tower of London Zoo, where it remained until the opening of Regent's Park Zoo in 1828. The most exotic inhabitant was a polar bear, which Londoners regularly saw fishing in the Thames, as the mean king had allowed only 1½ cents a day for its upkeep. Not to be outdone, the fourteenth-century French king Philip VI kept lions and leopards in the Louvre palace in a building called the Hôtel des Lions du Roi. This eventually gave rise to the Royal Menagerie at Versailles, established in 1665 as the first modern zoo, using the animals kept there for research and education.

Aztec Animal Houses

Had the European rulers known of the collections of the Aztec ruler Montezuma II (1503–1520), they would have turned green with envy. Not only did he create a botanical garden in the country, he also kept a zoo and an aviary in his capital, Tenochtitlán. However, the zoo did not meet with the approval of the *conquistador* Bernal Diáz del Castillo:

> [There was a] large house where they kept many idols whom they said were their fierce gods, and with them every kind of beast of prey . . . most of which were bred there. They were fed on deer, fowls, little dogs, and other creatures which they hunt. . . . They also had many vipers in this accursed house— poisonous snakes which have something that sounds like a bell on

the end of their tails. These, which are the deadliest serpents of all, were kept in jars and great pottery vessels full of feathers in which they laid their eggs and reared their young.

Other snakes were kept in long chests or big troughs half filled with muddy water. The carnivores were held in roomy, light, well-ventilated cages made of thick wooden planks securely pegged together. They were fed on turkeys, other poultry, deer, dogs and sometimes people. Whatever we may think of their diet, the fact that these animals bred in captivity and the nature of their cages suggests they were better looked after than the animals in many later zoos. Nearby were large rooms containing deer, llamas, vicuñas and bison.

Above the animals were birds of prey, ranging in size from kestrels to eagles. They were kept in very modern-sounding cages, which had an inner compartment as well as one open to the air, although not to the outside world. Five hundred turkeys a day were needed to feed them. Altogether the animals and birds of prey required a staff of three hundred keepers.

Montezuma's aviary of other birds was equally well appointed. In the center was a freshwater lake supplied by an aqueduct. The birds were looked after by another three hundred keepers, who caught 250 pounds of fish daily and scoured the surrounding countryside for insects. At molting time the feathers were gathered up and taken to a workshop next door, where the emperor's wives adorned coats with them. The most highly valued decoration was carried out with the brilliant green tail feathers of the quetzal, today the national symbol of Guatemala. Aztec interest in the birds did not end even with their death. Montezuma had selected rare specimens stuffed and placed in an ornithological museum—an entirely new concept to the Spanish, who later captured the building.

THEATERS

The theater as we know it in the Western world was almost certainly invented by the ancient Greeks. This is true not only of the theater as a building but of the art form as a whole. But exactly *how* it originated in Greece is another matter.

A major clue is usually thought to come from *The Poetics*, a treatise written in the fourth century B.C. by the great philosopher Aristotle. This states that tragedy (which the Greeks wrote in verse) derived from the custom of singing "dithyrambs," dramatic hymns composed in honor of the god Dionysus at his spring festival. The exchanges between the dithyramb singers evolved into crude plays featuring satyrs (the mythical goat-men who formed the god's entourage) and then solemnized into tragedy. Comedy—or at any rate burlesque theater—evolved from similar rituals, processions in which worshipers of Dionysus donned huge phalluses.

The orthodox theory based on Aristotle also sets great store by the resemblance—though perhaps it is only a superficial one—between the Greek words for "tragedy" (*tragoidia*) and "goat" (*tragos*). Tragedy, according to some scholars, came to be so named because the dithyramb singers dressed up as goat-men (satyrs). But though other sources confirm that, from the sixth century B.C. onward, there were sponsored contests at Athens and elsewhere that awarded prizes for dithyramb singing (usually accompanied by heavy drinking), there is nothing in these accounts about the contestants wearing goatskins. And while some Greek plays did include satyrs as the chorus, the Athenians, who according to Aristotle developed drama, depicted them as men with horse's ears and tails, not goat's legs like the god Pan. An alternative theory is that a goat was offered as the sacrifice or prize at dithyramb-singing contests. Goats were indeed later awarded as prizes in drama contests. But this explanation of the word *tragedy* makes the goat element rather incidental. It leaves little support for Aristotle's claim that tragedy evolved from Dionysiac rituals via the satyr play.

There is actually evidence to contradict the traditional theory at every step. The drama contests at Athens, first organized by the state in the mid-sixth century B.C., were linked with the spring festivals

Athenian theater scene from a vase of about 490 B.C. The chorus of six youths swan about as they sing their part, in front of an actor seated on an altar (*thymele*) to Dionysus, the patron deity of the dramatic arts.

of Dionysus, and there was an altar dedicated to him at the center of the great theater at Athens. But few plays ever featured Dionysus. The Greeks even had a proverb about the theater—"Nothing to do with Dionysus"—which both ancient and modern writers following Aristotle's theory have had trouble explaining. Although Dionysus was later to become its patron deity, this doesn't mean that theater evolved from singing hymns in his honor.

What is more, the introduction of satyr plays to the Athenian Festival is credited by ancient sources to the playwright Pratinas about 500 B.C., half a century after the first prizes had been won there for tragedies. This difficulty has been circumvented by the rather awkward argument that Pratinas was *re*introducing satyr plays. Yet despite these enormous stumbling blocks, most theories still cling to the basic idea that Greek tragedy must have evolved from rituals connected with the god Dionysus, with goats somehow playing an important role. Aristotle's influence has never been fully shaken off.

The notion that tragedy sprang from religious ritual involving goats must surely be taken with a pinch of salt. With regard to comedy, people don't need the structure of a fertility rite in order to start playing the fool. We know that clowns were common as entertainers in the ancient world. Egyptian tomb paintings show buffoons cavorting about in headgear reminiscent of the "dunce cap," or tall hat worn by modern circus clowns. These touring performers may well have put on short mimes and plays; the scripts and plots, however, could have been transmitted orally and never committed to writing.

The same possibility of course applies to Greece, where some interesting traditions exist concerning Thespis, the actor, poet and playwright whose name gave the term *thespian* to the acting profession. He won a prize for a production given at the Athenian Festival about 534 B.C.; he is also said to have been the first to use masks in the theater and to have individual actors speak separately from the general chorus. This much fits with the traditional theory of evolution of theater at the festivals of Dionysus. But another tradition, recorded by Roman poet Horace, is usually glossed over or rejected. This says that Thespis took his plays around the country on a wagon, traveling with a troupe of actors who stained their faces red with wine dregs. The story sounds eminently realistic.

Ancient Egyptian clowns.

Using the example of Thespis, it is logical to imagine that long before the official performances at the festival, itinerant bands of actors, clowns, dancers, singers, and musicians put on performances of different kinds to suit the needs of their customers. The spring festival of Dionysus would have attracted them—he was the god of wine and so a likely patron deity for traveling performers. Once the formal prize giving had begun, members of the profession turned their hand to the kind of heavyweight production that suited a state festival. When classical sources mention the various "firsts" in the development of tragedy and comedy, such as Thespis introducing masks and single actors, they probably meant "first" with respect to the Athenian Festival.

This is not to deny that there was a religious element in Greek drama. It was in fact paramount—the main theme of the early Greek dramatists was mankind's relationship with the Olympian gods as well as with more primeval forces, such as Fate. But this does not mean that classical theater evolved from a specific religious ritual. It surely drew on a wide range of art forms—from the skills of traveling performers to the composition of poetry in general (including dithyrambs). Aeschylus, the first of the great Athenian dramatists (about 525–456 B.C.), described his own work as "slices from the feast" prepared by Homer, the poet who composed the wonderful epics of the Trojan War in the eighth century B.C.

A final nail in the coffin of the "religious ritual" theory for the origin of drama is provided by a recent archaeological find from the

Near East. In 1990 the Russian excavators of the fortress at Armavir-blur, in the ancient kingdom of Urartu (Armenia), published two fragmentary clay tablets from the eighth century B.C., inscribed with a curious text written in Elamite, one of the ancient languages of Iran. It concerns the adventures of the ancient Babylonian hero Gilgamesh, relating the episode when he meets Siduri, the barmaid who lived at the world's end. The story is presented as a dialogue, with parts for Gilgamesh, Siduri and "All," that is, a chorus. In short the tablets give part of a script for a play! The excavators are certainly justified in noting that "everything is amazing" about this discovery. The point is that there was no "religious ritual" involved here—the tale of Gilgamesh had been circulating as a ripping yarn for at least a thousand years by the time the "protoplay" from Urartu was composed. In the view of the excavators, the tablets may reflect a performance put on by a touring company from Iran, perhaps for a royal marriage. In the eighth century the Greeks began to travel widely in the Near East, and it may have been here that they first become acquainted with structured, scripted dramas.

The First Seating Plans

There is no longer any reason to imagine that the theater developed suddenly in the late sixth century B.C. All the same the plays written by the great dramatists of fifth-century Athens remain the world's earliest surviving complete plays. The same century saw the crystallization of tragedy and comedy into polished art forms. The invention of the theater as a formal meeting place for public entertainments must also, as far as we know, be credited to the Greeks. Indeed it may have had a longer history in the Aegean than the traditional theory allows.

The Minoan civilization of Bronze Age Crete, which flourished long before the classical era, had organized spectator sports including boxing and bull leaping (see **Bullfighting**). Just to the north of the great Minoan palace at Knossos is a small open space with a floor about fifty feet square, flanked on two sides by rows of stone steps. Here, it is thought, courtiers would stand or sit to watch dancing, boxing or other performances. Inside the palace a fresco, painted about 1400 B.C., depicts rows of aristocratic ladies sitting in what must have been an enormous grandstand.

Remains of such a grandstand have yet to be found. The architecture of the grandstand in the fresco closely resembles that of the great courtyard inside the palace. However, on the assumption that the audience was watching the bull-leaping game, it has been argued that there was simply not enough room within the palace for such a violent sport—for example dragging the struggling bulls through the narrow corridors would have posed quite a problem. So the excavator of Knossos, Sir Arthur Evans, believed that temporary wooden structures may have been built outside the palace. On the other hand the Minoans depicted in the fresco may have been watching some other kind of entertainment more suitable for the confines of the palace.

In the Minoan grandstands we can see a distant forerunner of the classical theater, the Greeks' major contribution to the physical side of the business. By about 500 B.C. formal theaters on a grand scale were built in Greece. Instead of having grandstands supported by timbers—dangerous when holding the weight of thousands of people—classical Greek theaters were solidly constructed of stone. The oldest known is the theater of Dionysus on the southern slope of the Acropolis, in Athens, built in the fifth century B.C. to house an estimated six thousand spectators. Its earliest remains are obscured by later rebuildings, but it seems to have followed the same plan as slightly later theaters, such as that at Epidaurus.

The "Grandstand Fresco" (detail) from the Minoan palace at Knossos (Crete), painted about 1400 B.C. Row upon row of heads are shown, belonging to the finely dressed ladies of the Minoan court who are watching bull leaping or some other entertainment. (Courtesy of Ashmolean Museum, Oxford, UK)

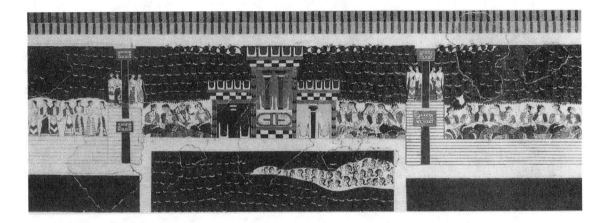

The most impressive feature of the classical Greek theater was of course its huge array of seats (or *theatron*), placed in a gigantic semicircle around a circular *orchestra* ("a place for dancing" in Greek) where the chorus danced as they sang. In the center of the *orchestra* was the *thymele*, an altar to Dionysus. At the rear of the orchestra was the *skene* (from which our word *scene* comes), a long wooden booth with doors in it from which the actors made their entrances and exits. In the earliest theaters the acting took place in the space between the *skene* and the *orchestra*. To bring the action closer to the audience, a platform—the first stage—was later added to the front of the *skene* and, to the sides, the *periaktoi*, or wings, behind which the actors could disappear offstage instead of using the doors in the *skene*.

Sometimes the simplest inventions are the most powerful. With regard to the theater, the Greeks invented the essential ingredients of all later playhouses. Centuries after the collapse of classical civilization their remains continued to inspire theater builders of later cultures. And some Greek theaters were so solidly built that they survive almost intact today. Plays are still performed in many, such as that built in the third century B.C. at Syracuse, a major Greek colony in Sicily. These reenactments provide "living" proof of how the skillful construction of classical theaters allowed several thousand

One of the best preserved of the Greek theaters is the beautiful example at Epidaurus, constructed during the 4th century B.C. The huge arena of seats, with a capacity of twelve thousand, was built to last—after more than two thousand years it is still a working theater, frequently used to stage revivals of classical Greek plays.

people to see and hear the whole performance comfortably, without the use of twentieth-century technology.

The masks that ancient Greek actors habitually wore also helped with the projection of the actors' voices. Modern experiments with replicas of these masks have explained why the mouthpieces were shaped in a rather peculiar fashion. They actually worked as speaking-tubes or megaphones, not only amplifying the voice but improving clarity.

The Roman Theater

The Romans took over the theater, along with so much else, from the Greeks. After an initial ban on building permanent playhouses imposed by the Senate (which deemed the art to be decadent), Rome's first stone theater was built by the dictator Pompey in 52 B.C., and many more followed. The Romans freely modified the Greek theater design to suit their own tastes. Comedy became increasingly popular and drama more secular. These requirements meant that more intimate contact between the audience and actors was in order, and the shape of theaters changed to suit. The circle of the *orchestra* was reduced to a semicircle, and since the traditional full chorus of Greek tragedy was now less popular, the *orchestra* area was reserved as a seating space for the senators and other notables, who relaxed on portable chairs and litters. The stage was broadened and lowered so that the worthies in the orchestra weren't always craning their necks to see.

Roman audiences as a whole were more interested in comfort. While the Greeks seem to have sat in awe during performances (except at comedies, when fruit could be thrown), the Roman public was generally less respectful of the theater as a medium and needed other diversions. Thus Roman theaters, unlike the Greek, were usually surrounded by huge enclosures where food, drink and sideshows were available. Some theaters were also equipped with canvas awnings that could be hoisted over the audience on rope riggings to shield them from the rain and sun.

The art of theater building thrived under the Romans, and the famous architect Vitruvius (1st century B.C.) left detailed instructions on their construction in his handbook. Vitruvius was particularly concerned with the geometry of theaters, in which, as he correctly understood, lay the secret of good acoustics. He also described in some detail how the acoustics of a theater could be improved by the use of strategically placed "sounding vessels." These were made of bronze, and cast in such a way that they resonated to particular frequencies of the musical scale. The musical theory followed by Vitruvius has been heavily criticized by modern acoustic experts, throwing doubt on the value of his sounding vessels. It seems there were also scoffers in Vitruvius's own time, so he can give his own defense:

> Somebody will perhaps say that many theaters are built every year in Rome, and that in them no attention at all is paid to these [acoustic] principles, but he will be in error. . . . If, however, it is asked in what theater these vessels have been employed, we cannot point to any in Rome itself, but only to those in the districts of Italy and in a good many Greek states. We also have the evidence of Lucius Mummius, who, after destroying the theater in Corinth, brought its bronze vessels to Rome and made a dedicatory offering at the Temple of Luna [the moon goddess] with the money obtained from the sale of them. Besides, many skillful architects, in constructing theaters in small towns, have, for lack of means, taken large jars made of clay, but similarly resonant, and have produced very advantageous results by arranging them on the principles described.

"Hearing is believing" was Vitruvius's argument. Some experimental archaeology is needed in order to confirm whether Vitruvius's sounding vessels would really have worked or not. It has already proved Vitruvius correct with regard to the odometer he described (see **Odometers** in **Transportation**).

Special Effects

The theater of classical times was not just an invention in itself. It also housed within it many smaller inventions, for the ancient Greeks developed most of the basic stage contraptions familiar today. As early as the fifth century B.C. theaters had cranes from which actors playing the gods would make their entrances, suspended in midair. In Aristophanes' comedy *Peace*, a character named Trygaeus is whisked off to heaven on the back of a giant dung beetle; the script includes an aside from Trygaeus, screaming out to the crane operator to be careful.

Trapdoors were built through which actors could make surprise appearances or descents into the Underworld. The wings (*periaktoi*) were often made of wooden triangular prisms, which turned on pivots, each side painted with different scenery. Vitruvius described how they were used:

> When the play is to be changed, or when gods enter to the accompaniment of sudden claps of thunder, these may be revolved and present a different decorated face.

Other special effects were developed in the "high-tech" Hellenistic era, which followed Alexander the Great's conquest of the Near East in the late fourth century B.C. Heron, the brilliant engineer from Alexandria, was particularly interested in theatrical gadgetry. His designs included magic four-wheeled stands—automata that ran on stage, performed maneuvers and ran off again—and even miniature mechanical theaters that performed short plays (see **Automata** in **High Tech**).

For his miniature version of the tragedy *Nauplius*, Heron designed a mechanism that dropped a bolt of lightning (painted on a board) onto the stage; this then bounced up again, simultaneously releasing a new backdrop for the stage. Whether this was translated into a full-scale version for a real stage is not known. Another of Heron's special effects, his thunder machine, certainly could have been. Modern theaters (even after the advent of recorded sound) often use enormous metal sheets to simulate the sound of a storm. Heron's device released a stream of bronze balls like an avalanche down a chute containing projections onto a large tin sheet. The noise must have been deafening—and highly effective.

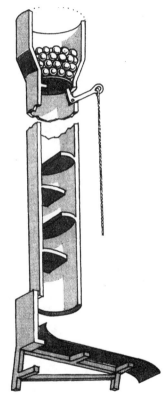

Theatrical thunder machine designed by Heron of Alexandria in the 1st century A.D. When the rope is pulled, the trapdoor falls, releasing the bronze balls from the top compartment. They clatter over the metal projections on the sides of the chute, then crash in rapid succession onto a tin sheet at the bottom.

Classical Survivals

Theater design and effects continued to improve during Hellenistic times and under the Roman Empire. If anything, the Romans, unrivaled as architects in the ancient world, were better at building theaters than the Greeks. But the Roman public was notoriously fickle, and its appetite for novelty and spectacle was insatiable. The expression "bread and circuses" was truly invented for them (by the first-century A.D. satirical poet Juvenal), as a bitter description of what kept the ordinary populace quiet. Within a few generations the large crowds had drifted away to other attractions. The amphitheaters where the Roman "games" were held were putting on greater spectacles than the classical theater could ever hope to stage, such as live re-creations of naval battles in flooded arenas, or the immolation of Christians by the hundred.

The history of the theater under the Romans has been described as one of steady decline. Burlesque pantomime was the only form that remained steadily popular, and it was a long-standing enemy of Christianity—it was vulgar and some plays were written specifically to ridicule the faith. Not surprisingly, then, when the empire became officially Christian, in the fourth century A.D., the theater was banned. It went underground and was preserved by small groups of traveling players.

There are some indications that the classical theater continued to thrive officially in the Eastern Roman Empire, based at Byzantium, but we lack any detailed evidence about the kind of shows performed. Much further east, classical theater may have inspired the Sanskrit drama of ancient India.

Though the evidence is rather patchy, it seems that the Sanskrit theater came into being sometime between 200 and 100 B.C. The earliest known fragments of Sanskrit plays, from the first century A.D., reflect a well-developed dramatic tradition, and a literary reference probably dating to about 140 B.C. obliquely refers to pantomime. Sometime between the second century B.C. and the second century A.D., the *Natya Shastra* ("drama science"), the first major classic of Indian theater, was written. Its composition is attributed to one Bharata, who in the opening chapter describes the mythical origins of the theatrical art. It was devised by the gods as a healthy diversion for humanity at a time when morals were lax: the great god Brahma himself trained the first actors (Bharata and his hundred sons) and commanded the heavenly architect Visvakarma to build the first theater.

The historical origins of Indian theater are less clear, though some input from the classical world seems plausible. In 327 B.C. the armies of Alexander the Great reached northern India, and over the next two centuries Greek adventurers established kingdoms in Bactria (Afghanistan) and northwestern India. It was during this period, it has been argued, that Indians became acquainted with the classical theater. One of the most famous Sanskrit plays, *The Little Clay Cart*, indeed bears a striking resemblance, in plot structure and characterization, to Greek comedies of Alexander's time. *Yavanika*, the word for a stage curtain used in the *Natya Shastra*, may be derived from *yavana*, the Indian word for "Greek."

Whatever the case, the guidelines for Indian drama laid down in the *Natya Shastra* remained dominant there for well over a millennium. The Sanskrit theater declined in the Middle Ages, to be replaced by other forms, but the tradition survives in *Kutiyattam*, a form still performed in the compounds of less than half a dozen temples in Kerala, on the coast of southwestern India. It may well be the oldest art form in the world.

Whereas India has a continuous tradition, Europe had to wait for centuries for a widespread revival of the theater following the Christian ban. Dramatists such as the tenth-century German nun Hrosvitha, who produced Christian versions of the comedies written by the Roman author Terence (2nd century B.C.), were exceptional. But the Church gradually developed its own flair for drama, putting on little plays illustrating the Nativity and the Crucifixion (or Passion—hence "passion plays") to instill the gospel into the illiterate. These performances—the miracle or passion plays—eventually passed into secular hands. Meanwhile survivors of the old tradition of Roman theater—singers, dancers, jugglers and acrobats—struggled on as a wandering fraternity and eventually began to find patronage at courts during the later Middle Ages.

All these strands came together during the Renaissance, when scholars and artists effectively rediscovered the literature of the classical world. Shakespeare was a post-Renaissance man, and he, for one, would have acknowledged the enormous debt he owed to the theater of the ancient world.

MUSICAL INSTRUMENTS

The strikingly early date of many musical instruments makes the kind of Stone Age band portrayed in "The Flintstones" cartoons seem almost feasible.

Wind instruments, in particular, have a venerable antiquity. Cross-blown flutes, similar to pennywhistles, with three to seven finger holes, have been found in Old Stone Age deposits in France, eastern Europe and Russia; they are fifteen to twenty-five thousand years old. The French examples are made from hollow bird bones, while those from eastern Europe and Russia are of reindeer or bear bone. In Georgia a tomb of the thirteenth century B.C. contained a delicately fashioned flute some eight inches long, made from the tibia of a swan. Georgian musicologists have found that simple tunes can still be played on this remarkable instrument. In the late first millennium B.C. the Scythians and their neighboring tribes in southern Russia still played flutes made from eagle and vulture bones. The bones of birds of prey make excellent flutes—they are still used today by Bulgarian folk musicians to make instruments that have a particularly gentle but rich tone.

Roaring Bulls, Thundering Mammoths

After flutes the second musical instrument to be invented may have been the bull-roarer, still widely used by the Aborigines of Australia and other indigenous cultures around the world. It is less familiar in modern Western society—except in the plastic or cardboard versions that occasionally appear as free gifts in boxes of breakfast cereal. The bull-roarer is an ingeniously simple device—almost any material cut into a fish shape of the right size will do. A hole is drilled through one end and a long string attached to whirl it around in the air; the noise created is quite surprising. The Victorian anthropologist Andrew Lang described a performance of bull-roarer playing in a lecture given at the Royal Institution in London in 1885:

> At first it did nothing particular when it was whirled round, and the audience began to fear that the experiment was like those chemical ones often exhibited at institutes in the country, which contribute at most a disagreeable odour to the education of the populace. But when the bull-roarer warmed to its work, it justified its name, producing what may best be described as a mighty rushing noise, as if some supernatural being "fluttered and buzzed his wings with fearful roar." Grown-up people, of course, are satisfied with a very brief experience of this din, but boys have always known the bull-roarer in England as one of the most efficient modes of making the hideous and unearthly noises in which it is the privilege of youth to delight

The troublesome brats about whom Lang wrote were using a musical instrument invented during the last Ice Age. Examples of bull-roarers found in Germany and Denmark date to around 12,000 B.C.

Possible drums are known from an equally early date. It has been suggested that mammoth skulls, which would have made fine percussion instruments, were used as drums at Stone Age sites such as Mezhirich, in Ukraine, where around 12,000 B.C. the inhabitants seem to have been using mammoth bones for everything from house building to mapmaking (see **Mammoth-Bone Houses** in **House and Home**, and **Mapmaking** in **Transportation**). But the earliest clearly identifiable drums, made of clay, come from Germany and Czechoslovakia and date to around 3000 B.C. Drummers are depicted on Near Eastern pots of about the same age. By the third millennium B.C. the Chinese were manufacturing drums from crocodile skin, as we know from grave finds in southern Shensi Province. Around 1400 B.C. they developed bronze drums set on legs, as well as wooden drums suspended in a frame. In the Western Hemisphere the Peruvians were making drums of leather by the late first millennium B.C.

A bull-roarer, about eleven thousand to fifteen thousand years old, discovered at La Roche, in the Dordogne, France. One of the oldest musical instruments in the world, a bull-roarer can be fashioned from almost any material—wood, stone, plastic or even cardboard. Larger ones can be made, but the following dimensions give the required ratio: about ⅙ inches thick, 8 inches long and 3 inches wide. The edges can be left as they are or sharpened, and a hole needs to be bored near one end. A length of string, about 30 inches long, is threaded and knotted through the hole. The bull-roarer is then whirled in the air around the head to produce a deafening noise.

Progressive Percussion

Drums are perhaps the most familiar instruments of the percussion family to be used in the distant past. The range of other percussion instruments invented in the ancient world is vast—a full catalog would run into the thousands, from the simplest (two specially prepared sticks) to the quaintly elaborate, such as the Chinese "tiger box." Popular during the Chou Dynasty (722–480 B.C.), this was a hollowed block of wood, carved into the shape of a tiger with a serrated back. The tiger box was played by brushing its back with a specially made stick split into twelve leaves at the end, which produced a rasping sound.

The favorite percussion instrument of the ancient Egyptians was the sistrum. This was a U-shaped metal hoop pierced with three or four holes, through which thick wires were threaded. The noise it made when shaken was meant to imitate the sound of rustling papyrus reeds. Used as early as 2500 B.C., this sound became the most common musical accompaniment to religious worship in Egypt. As

Aztec band of percussionists and vocalists, from a Mexican Codex of the 16th century A.D.

Egyptian sistrum—from a tomb painting of the New Kingdom (c. 1550–1070 B.C.) at Thebes.

the nineteenth-century Egyptologist John Gardner Wilkinson noted, "The sistrum was the sacred instrument *par excellence*, and belonged as peculiarly to the service of the temple, as the small tinkling bell to that of a Roman Catholic chapel." The comparison is apposite. Sistra are still used in the Ethiopian Church (a descendant of the Egyptian Coptic Church)—a remarkable survival, giving the sistrum more than four thousand years of continuous use since the time of the pyramid builders.

Of all ancient percussion instruments, perhaps the bell has the strangest history. From rather humble beginnings in the Mediterranean it reached a point, in ancient China, where it assumed a status unrivaled by any other musical instrument.

The earliest known bells were made by the Minoan civilization of ancient Crete. Dating to the beginning of the second millennium B.C. they are curious objects, since they are made of clay rather than metal. Bells they are, though, as they have holes in the top through which a string was threaded to hold a wooden clapper. They are thought to have been hung on the branches of trees to act as wind bells, attracting friendly spirits.

Whether the Chinese invention of the bell was independent of that in the Mediterranean is not known. But the first metal bells were certainly manufactured in China, during the Shang Dynasty

Clay bell from Crete, made by the Minoan civilization of the early 2nd millennium B.C. A wooden clapper was hung inside the bell from string threaded through the holes. Another string would be run through the loop on top to suspend the bell, most probably in the branches of trees. Played by the wind, it would attract friendly spirits.

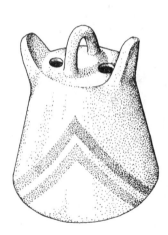

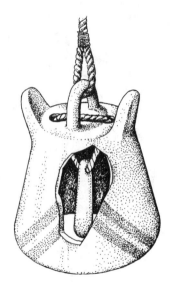

(c. 1500–1000 B.C.). Small bronze examples with a clapper, they mark the beginning of a grand tradition of bell making in the Far East that produced instruments of ever-increasing size and quality. By the fifth century B.C. the Chinese were manufacturing bells of such perfect pitch that they played a key role in the state regulation of weights and measures (see **Written Music**). Ancient Chinese bells were generally struck rather than rung and could emit two distinct notes—one if struck in the center and another if struck near the edge.

The largest bell of ancient times was made by King Hyegong of Korea in A.D. 771. This bronze monster, which still exists, is 9½ feet high, 7½ feet in circumference, and weighs seventy tons. For over nine hundred years it was the heaviest bell ever cast, and even today it is the third largest soundable bell in the world. It was built for primarily religious reasons: Korean Buddhists believed that when the bell was struck, it sent out spiritual waves to the ends of the earth, so the bigger the bell, the taller the waves.

Wind Instruments

The flute kept its popularity throughout prehistoric times. But its history, together with that of other instruments of the wind family, becomes more difficult to trace with the rise of the major urban civilizations of the Near East in the fourth millennium B.C. Strangely this is not due to a lack of evidence but to too much of a certain kind—while excavated examples of musical instruments from such an early date still remain rare, the pictorial record (from sculpture, decorated pottery and wall paintings) suddenly blossoms. Around 3000 B.C. scenes depicting musicians, playing a wide range of new instruments, begin to appear in abundance. But it is often difficult to tell exactly what is being played—particularly where wind instruments are concerned.

For example a carved stone vessel from Bismaya, in Iraq, dating from about 2600 B.C., shows a group of musicians, one of whom is blowing down an instrument that looks like a modern oboe. It has sometimes been held to be the earliest depiction of a reed instrument. While flutes are played by blowing across a hole, reed pipes

Egyptian flute player from
a tomb painting at Thebes.

are played by blowing straight down them, the note being produced
by a reed or other sharp edge near the mouthpiece or within the
tube. But the Bismaya "pipe" could just as easily be a trumpet, a
wind instrument that works on a different principle, since the vibra-
tion is produced by the trumpeter's lips pressed against the mouthpiece.

Even depictions of flutes can be rather confusing. Some flutes can
be played by holding them vertically and blowing right across the
top, so they can look like pipes or trumpets in pictorial evidence.
Fortunately in some cases the evidence is more certain. The Egyptians also had "transverse" flutes, played like the modern orches-
tral instrument by blowing across a hole in the side. These can be
seen in tomb paintings from as early as the Middle Kingdom
(c. 2040–1780 B.C.). Some were well over two feet long and, to
judge from the paintings, involved considerable arm stretching to
play them.

The case of the ocarina is also reasonably clear. Ocarinas are small
pottery flutes, generally oval in shape. Most standard histories of
music credit their invention to the nineteenth-century Italian musician Giuseppe Donati, which is actually quite wrong. Ocarinas were
first made by the Maya civilization of Guatemala and southern Mexico. Mayan examples in the form of ornate human and animal-
shaped figurines date back to at least the ninth century A.D.

Panpipes also belong to the flute family. Made from a row of reed
tubes of graded lengths tied together to produce different notes, they
were the favorite instrument of ancient Greek shepherds and take
their name from the rustic god Pan, who was believed to have invented them. The earliest known examples date to 2000 B.C. and
were found in southern Ukraine. Panpipes seem to have been independently invented in many parts of the globe. They were known to
the Incas of Peru, and are still a popular instrument in South America. In ancient China an enormous version like a giant mouth organ
was developed. The tomb of the marchioness of Dai, who was
buried in 168 B.C., contained a set of bamboo panpipes 2½ feet high
with twenty-two tubes in two rows and twelve pipes of different
lengths to produce the twelve notes of the Chinese scale: the name
of the note is written in ink at the bottom of each pipe.

The earliest-known reed instruments come from Mesopotamia. The first certain examples date to the second millennium B.C., when the reed pipe became the most popular wind instrument throughout the Near East. In the Greco-Roman world it was adapted into a family of reed-blown pipes, known as the *aulos*, which are precursors of the modern clarinet and oboe. The *aulos* is always shown—for example on Greek pottery—being played in pairs, one fingered by each hand. They were usually secured to the player's head by a leather band, which presumably enabled him or her to blow with enormous force without actually spitting out the instruments.

Trumpets have a long record that stretches back into prehistory, since many naturally-occurring shapes, such as conch shells and hollow animal horns, make perfectly good, though rudimentary, instruments. The first metal trumpets may have been manufactured by the Egyptians—two trumpets, one of bronze and gold, and one of silver, were found in Pharaoh Tutankhamen's tomb. The curved trumpet or horn was one of the favorite war instruments of the ancient world. The Roman legions attacked to the sound of the *cornu*, a large horn curved in the form of a letter G. The Romans believed it was invented by the Etruscans, the cultural ancestors of Rome, who flourished in central Italy during the eighth to sixth centuries B.C. But the bronze war horn seems to be an invention of northwestern Europe. Both the Irish and the Scandinavians were making bronze horns by about 1000 B.C. They became known as a characteristic instrument of the Celts; the Greek writer Diodorus Siculus said, "Their horns are of a peculiar barbaric kind; they blow into them and produce a harsh sound which suits the tumult of war."

But of all the wind-instrument family, we think of the bagpipes as the most typically Celtic—as played by the Bretons in western France, the Irish in Ireland and the Scots all over the world at tremendous volume. Strangely enough, however, the skirl of the bagpipes seems to have first been heard in Egypt. Greco-Egyptian sculptures of the first century B.C. showing bagpipe players antedate the first written and pictorial evidence from the British Isles by almost a thousand years. According to contemporary sources, the Emperor Nero (A.D. 54–68) played the bagpipes, at that time a relative novelty in Rome. While the situation is still very uncertain, it seems that bagpipes were introduced to Britain during the Roman occupation (1st–5th centuries A.D.).

The First Strings

The earliest stringed instruments are harps and lyres, almost certainly invented by the Sumerian civilization of ancient Iraq. Some of the earliest written texts in the world come from the Sumerian city of Ur; dating to around 3200 B.C., they include symbols representing simple harps. The shape of these instruments readily suggests how they may have been invented. We can picture some bright spark in the fourth millennium B.C., or even much earlier, twanging a taut bow string and conceiving the idea of adding some further strings to produce different notes. Making one arm of the bow shape into a soundbox to increase resonance and volume was the next step. The lyre was developed from the harp by replacing the single bow shape with two upright arms joined by a crosspiece; and the strings, instead of joining the soundbox directly, were made to run over a bridge attached to the box.

Symbols for the lyre as they appears on early Sumerian texts, c. 3200 B.C.

A measure of the respect in which the Sumerians held the harp is that its invention was credited directly to the great god Enlil. The construction of these instruments could be highly elaborate, even at an early date. Our earliest excavated examples come from the royal cemetery at the great city of Ur, where Sumerian rulers were buried around 2500 B.C. (see **Jewelry** in **Personal Effects**). One is of considerable size (three feet six inches high), with a soundbox shaped like a flat-bottomed boat, and was stood upright to be played like a modern harp. The fittings to secure the strings were of gold and silver. Another burial at the royal cemetery contained three lyres with different animal heads on the soundboxes for different pitches—a bull for the bass, heifer for tenor and stag for alto. All were beautifully decorated with scenes made from shell inlay.

Harps and lyres remained the favorite musical instruments in Mesopotamia for some four thousand years and achieved great popularity elsewhere in the Near East. They took root particularly in ancient Israel, as we can see from the numerous references in the Bible; King David (10th century B.C.), for example, was a noted lyre player.

Guitars and Lutes

We tend to think of the guitar as a modern instrument. Yet a Hittite relief from Alaça Hüyük, in central Turkey, carved about 1300 B.C., shows a musician playing what appears to be a guitar, with a characteristic figure-eight shaped body and even frets. But the definition of ancient musical instruments is a difficult field—particularly when we only have pictorial evidence—and the identification of this Hittite instrument as a guitar has been disputed. Purists insist that it should be defined as a lute, since the curvature of the sides of the instrument is too shallow. Lutes generally have a pear-shaped body, so perhaps "guitar-shaped lute" is the best definition for the Alaça Hüyük instrument.

The argument is a rather semantic one, since the guitar is by definition a short-necked lute—so the instruments share a common history. The lute seems to have developed from the lyre in Mesopotamia and is first depicted on two Babylonian seals of around 2300 B.C. From here it spread throughout the Near East, notably to Egypt, where it became immensely popular in the New Kingdom Period (c. 1550–1070 B.C.). Over the following millennia it gave birth to an enormous family of instruments throughout the Old World. The Greeks introduced it to northern India during the third century B.C.; from there it traveled eastward and spawned the Chinese *yüeh-ch'in* and the Japanese *shamisen*. In India itself the lute became the ancestor of the *sitar*. Versions of the lute adapted for playing with a bow gave rise to the ancestors of the violin family.

The Arabs continued the ancient Near Eastern tradition of lute playing and introduced it to Europe during the period of their conquest and occupation of Spain (A.D. 711-1492). The very name of the lute comes from Arabic *oudh*, via medieval French *l'oudh*. Here we return to the guitar, for which some have argued an Arab origin. Others have suggested that the original guitar was the ancient Greek *kithara*, a kind of lyre with a soundbox made from a tortoiseshell; the names at least are connected, but the instruments are different. Others again have argued that it was a western European invention of the Middle Ages.

Matters are further complicated by the guitar-shaped lutes, such as that from Alaça Hüyük and another shown on a frieze of the first century A.D. at Airtam, in Uzbekistan, which closely resembles the modern instrument. The problem is a complex one, and at present it is fairest to say that lutes through the ages have often been guitar-shaped. Some of these belonged to the family of gitterns, popular in late medieval Europe, and the first modern guitar was derived from the gittern, almost certainly in Spain. Johannes Tinctoris's book *De inventione et usu musicae*, published in 1487, states that the *guitarra*, "which others call the ghiterne," was "invented by the Catalans." Renaissance guitars were like modern Spanish ones in all respects except for the strings. They generally had only four or five, the sixth string only being added in the late eighteenth century.

KEYBOARDS

In the third century B.C. a brilliant young man named Ctesibius was working in his father's barbershop in Alexandria when he made an intriguing discovery. He was experimenting with a contraption—the first in a long career of inventing (see **Introduction** to **High Tech**)—to make it easier to raise and lower a mirror in the shop by counterbalancing it with a lump of lead. Suspended from a cord, the lead weight ran up and down a pipe concealed in a wooden beam behind the mirror. When the lead fell rapidly down the pipe, it compressed the air below, forcing it out with a squeak.

The Mighty Water Organ

This simple device suggested to Ctesibius the principle behind the world's first keyboard instrument—a small organ powered by compressed air. (In effect this was an enlarged and mechanized version of the panpipes, a series of reed pipes set in a row.) As with many of his other achievements, it was improved upon two hundred years later by Heron, also of Alexandria, who manufactured a novelty wind-powered version. A design incorporating further refinements is described by the Roman architect Vitruvius near the end of the first century B.C. This was the Roman *hydraulis*, or water organ, in its final form, which spread throughout the Roman world.

The essence of the instrument was that air from a pump, or in Heron's case from a miniature windmill, was blown into a chamber consisting of an inverted metal bowl immersed in water; the increasing air pressure forced water out of the bowl, thus raising the water level in the cistern surrounding it, while the surplus air was forced up into a pipe chest above the water cistern. Between strokes on the

A mosaic from a Roman villa at Nennig, Germany, shows musicians playing the water organ and *cornu* (Roman horn), probably accompanying a gladiatorial combat.

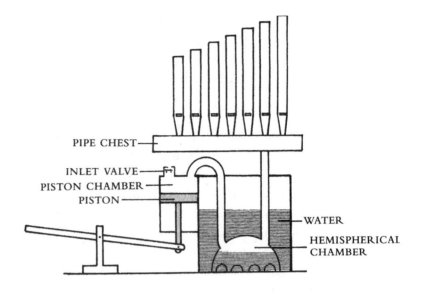

PIPE CHEST

INLET VALVE
PISTON CHAMBER
PISTON

WATER

HEMISPHERICAL
CHAMBER

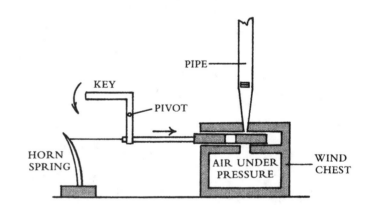

PIPE

KEY

PIVOT

HORN
SPRING

AIR UNDER
PRESSURE

WIND
CHEST

(Above) Diagram of Heron of Alexandria's version of the water-organ. (Below) The key mechanism of the water organ.

pump the air pressure in the metal bowl was kept steady by the water. In the Roman version described by Vitruvius, the flow of air from the pipe chest into the individual pipes was controlled by a valve made up of a square chamber closed by a square wooden slide: each slide had a corresponding key, which the organ player pressed down to allow the air in. As in the modern organ, the notes were created by different-sized pipes—up to nineteen in the Roman examples.

From the second century A.D. there was also a Roman organ powered by a bellows, but it was smaller and less powerful than the *hydraulis*. The superiority of the water organ lay in the fact that the metal bowl inside the cistern acted as a reservoir of air under pressure. No matter how hard the pumper tried, it was impossible to maintain sufficient air pressure to operate a large organ with a bellows mechanism.

The water organ became popular right across the Mediterranean world and eventually throughout the Roman Empire. An inscription of 90 B.C. at Delphi, in Greece, records a man named Antipatros "covering himself with glory" by playing the *hydraulis* for two days straight in a competition. It was also the notorious emperor Nero's favorite instrument. By the second century A.D. it was regularly played to accompany gladiatoral contests. The water organ was also heard in theaters and amphitheaters, at games, circuses and processions, during the banquets and weddings of the wealthy, and even as part of the elaborate ceremonies surrounding the swearing in of new state officials. Its use in almost every conceivable public setting is shown by the remains of a water organ found in the ruins of the headquarters of the Clothworkers' Guild at Budapest, which burned down in the third century A.D.

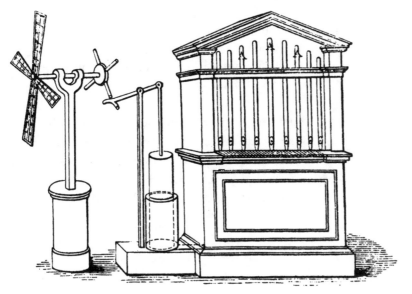

Heron also designed a water organ in which the hand pump was replaced by a small windmill (Renaissance reconstruction).

The Rise of the Keyboard

Yet despite its great popularity the water organ did not survive beyond A.D. 500. The skills required to build it were lost during the barbarian invasions, and it was the simpler bellows-type organ that was preserved by the Eastern Roman Empire. Arab engineers began to produce bellows organs in the ninth century, and soon the imperial courts of Byzantium and Baghdad echoed to the sound of the organ.

Organs were reintroduced to the West by gifts from Byzantium, such as the fine example sent to Pepin, king of the Franks, by the emperor Constantine Copronymous in A.D. 757. The Holy Roman Emperor Louis (son of the famous Charlemagne) sent to Venice in A.D. 826 for a priest named Georgius to come and build him an organ. *The Utrecht Psalter*, an illustrated book of psalms of the early ninth century, shows an instrument similar to that built for Louis.

These early medieval organs were for secular rather than religious use, and it took another century before they became part of church services. Once accepted, though, the use of church organs spread rapidly, and around A.D. 990 the monk Wulfstan composed a poem in praise of the organ at Winchester, in England, ending,

> The sound so clamours—echoing here and there—
> That one closes with his hand the openings of his ears,
> Hardly able to bear the roar in drawing near,
> Which the various sounds render in their clamour,
> And the melody of the muses is heard everywhere in the city,
> And flying fame goes through the whole country.

From here the church organ traveled around the world to become the parent of all keyboard instruments. Although keyboards now work on entirely different (and largely electronic) principles, we can trace their evolution back more than two thousand years to the inspiration of a boy in an ancient Alexandrian barbershop.

WRITTEN MUSIC

In 1893 French archaeologists excavating at Delphi, the sacred city of the Greek god Apollo, uncovered two curious stone slabs. They were inscribed with hymns in praise of Apollo, written in the usual ancient Greek alphabet of the late second century B.C. But what made them different from any other Greek inscription found before were the conspicuous extra characters carved between the lines of the hymns. Much to the delight of the excavators, these proved to be musical scores, which, although fragmentary, are still the most extensive surviving examples of ancient Greek music.

The stone from Delphi inscribed with the oldest-known example of ancient Greek music, written over two thousand years ago.

MUSIC FROM ANCIENT GREECE

A modern copy of the opening lines from the Delphi musical score, with a transcription and translation of the accompanying hymn to the god Apollo. Try playing it. You may not find it very jolly, but it should be remembered that this is a religious piece—it may have sounded archaic even when the inscription was carved in the late 2nd century B.C. Unfortunately, we have just the raw notes, and nothing to indicate the rhythm of the piece.

Since this amazing discovery numerous other fragments of Greek musical scores have turned up, ranging in date from the mid-third century B.C. to the third century A.D. Most are on papyrus, often in frustratingly small scraps, though large enough to see that they are mainly extracts from the music composed to accompany plays, such as the famous *Orestes* written by Euripides about 450 B.C.

These rare archaeological discoveries were easy to identify as ancient Greek musical notation. Conveniently enough, one Alypius had set down a record of the classical Greek system of musical notation in the third or fourth century A.D. He listed the symbols used (a mixture of alphabetic and pseudoalphabetic characters), and provided a detailed explanation of their meanings. Fortunately his work was continuously preserved from classical times by the efforts of the feverish manuscript copiers of the Middle Ages. Knowledge of the ancient Greek system of musical notation was never really lost— indeed our own system of musical notation, evolved through the Middle Ages and crystallized in the seventeenth century, was ultimately inspired by the Greek example.

Music and Mathematics

While the earliest surviving examples of Greek musical notation come from the third century B.C., the system was probably invented long before that. Most likely it was developed in the aftermath of the acoustic discoveries made by the great sage and mathematician Pythagoras in the late sixth century B.C.

Tradition has it that Pythagoras, while pondering how to find a mathematical formulation for the musical scale, happened to walk past a blacksmith's shop. Intrigued by the chimelike sound of hammers ringing out on the anvils, he went in to investigate. At first he wondered whether the strength of the men was determining the pitch of the notes produced; but he excluded this possibility by asking the smiths to exchange their hammers, which continued to produce the same notes. He then weighed the hammers and found that the heaviest produced the lowest sounds.

Pythagoras went away to do further experiments, this time with strings. By varying their length he discovered the mathematical laws governing musical intervals. Expressed simply, the longer the string, the deeper the sound; the shorter the string, the higher the sound. Whatever note a string produces, halving it will produce the same note an octave higher. The other notes of the octave lie at fixed points in between; for example what we call a fifth is produced by using three-quarters of the same string, and a fourth by using two-thirds. Pythagoras expressed the whole as a series of mathematical ratios. Whether he personally took the next step of applying these ratios to cosmology is doubtful. But later tradition claimed that he turned them into a universal principle, in which the arrangement of the planetary orbits followed the same ratio as the musical intervals: as the planets move through the ether, they make it vibrate, producing the mystical "Music of the Spheres," which is only audible to the initiate.

While modern science has never come to grips with these more esoteric ideas, Pythagoras's basic musical discoveries are generally hailed as a landmark in scientific history. Indeed they represent the first clear step toward a mathematical expression of the natural world.

Another fallout, in the atmosphere of serious experimentation generated by Pythagoras's work, was the development of the first Greek musical notation. In the centuries following Pythagoras we find an explosion of interest in musical theory, scales and notation, not only in Greece but in other Old World civilizations. India's earliest surviving work on the theatrical arts, the Sanskrit text *Natya Shastra*, was written sometime between 200 B.C. and A.D. 200 (see **Theaters**). It includes comments on musical theory, describing how Indian scales (with octaves, like the Greek and our own) could be subdivided into twenty-two microtones. Still, despite a keen interest in the more abstruse mathematical and theoretical side of music, the musical tradition in ancient India was basically an oral one, in which written music was of minor importance. (Likewise most Arab musicians prefer improvisation to reliance on other artists' compositions.) The earliest-known score from India, from a rock-cut inscription at Kudumiyamalai, in southern India, dates to the seventh century A.D.

In China evidence for the scientific study of music is almost as ancient as Pythagoras. In 1978 an extraordinary set of musical bells, the finest ever recovered from the ancient world, was found during the excavation of the tomb of the marchioness of Yi, a noblewoman of the Chinese state of Cheng in the late fifth century B.C. The sixty-four-piece graduated set was mounted on wooden beams and was played by five musicians. They are chime bells, struck to produce a note. The surface of each bell is inscribed with a text setting out the note it played, its place in the scale used in Cheng, and the relationship of that scale to those used in other states. The notes inscribed on the bells represent a kind of written music in themselves, and although no scores have survived, it is almost impossible to imagine that the highly literate Chinese of the fifth century B.C. wrote none.

It seems that the Chinese of this period had been making musical investigations similar to those of Pythagoras. This is clear from the fact that their whole system of weights and measures was regulated musically. The standard was the note given by a special *chung* (a downward-facing bell) kept by the state. Measures were transferred from it by using a *chun*, a tuning device seven feet long that held a string. The length of the string would be adjusted until it gave exactly the same note as the official *chung*. The length of the string was the state unit of measurement. The process could be reversed and bells cast to match the pitch of the official *chung*. Why the Chinese came to choose the bell as their standard unit is a mystery, but it clearly gave a spur to the development of the highly sophisticated metallurgical skills needed to produce bells with perfect pitch.

This Oriental interest in the relationship between mathematics and music all sounds very Pythagorean. On the other hand there was a major difference. While the ancient Indian scale had seven notes like the Greek, the Chinese divided the octave into twelve notes, again standardized by official bell sets cast by the state. What connections, if any, existed between the musical researchers of ancient India, China and Greece is difficult to say. The question forms part of a wider problem relating to Pythagoras concerned with the extraordinary similarities between his philosophical teachings and those of his near contemporary the Buddha, the semilegendary Indian teacher (see **Introduction** to **Food, Drink and Drugs**). Cross-fertilization of ideas between the Far East and the Mediterranean area at this early date cannot be ruled out: only a few centuries later strikingly similar developments in engineering (see **Introduction** to

High Tech) and cartography (see **Mapmaking** in **Transportation**) are evident in the Roman and Chinese empires.

On the other hand both the Western and the Eastern musical researchers of the late first millennium may have been inspired by a common, far older, source. Pythagoras was supposed to have spent twelve years of his life studying with the wise men of Babylonia (southern Iraq). Curiously enough, it is in this very region that recent research has opened a completely new chapter in the history of music.

The World's Oldest Song

Literally tens of thousands of tablets, including entire libraries, have been unearthed from the civilization of Babylonia and its northern neighbor Assyria (see **Introduction** to **Communications**). While the wedge-shaped (cuneiform) script and language of the texts is intelligible to the philologists who decipher them, a large percentage of the tablets, concerning more technical subjects, such as mathematics and astronomy, can only be understood with the collaboration of specialists from other disciplines. Music is a good case in point. Some of the musical texts from Babylonia were available as early as 1919, but it took nearly fifty years before the terminology on them was recognized.

The ball was started rolling by English archaeologist Dr. Oliver Gurney. Having a keen interest in music, he was able to recognize that one of the unpublished tablets in his keeping, a text from the later first millennium B.C. found at Ur, in southern Babylonia, gave a list of names for the strings of a musical instrument, apparently a lyre. He gave a copy of the tablet to Anne Kilmer (now professor of Assyriology at the University of California, Berkeley), who was at the time working on a Babylonian mathematical text that also contained musical terms. With the help of the information from the Ur tablet, Kilmer was able to begin the complex task of deciphering the mathematical text. It became clear that while nine strings were referred to, the eighth and ninth strings gave octaves of the first and second. The Babylonians, it appeared, had used a "heptatonic" (seven-noted) scale just like our own.

Other discoveries soon followed. Gurney came across another

tablet, which he worked on with the help of Oxford musicologist David Wulstan. The text turned out to be a remarkably precise set of instructions for the tuning of each string of the Babylonian lyre. Even more surprising was its date—the tablet was written no later than about 1800 B.C. Other musicologists joined the fray, while Kilmer was busy analyzing additional fragmentary musical texts and pulling all the strands together.

By the mid-1970s some certainty was beginning to be possible. The Babylonian terms for the strings, intervals and tuning were all reasonably well understood, and Kilmer turned her attention to a curious tablet that had been excavated at Ugarit, in northern Syria, in the 1950s. It had already been published, but by an Assyriologist who was somewhat baffled by its contents. At the top of the tablet was a hymn addressed to the wife of the moon god and written in Hurrian, one of the ancient languages of Syria. Immediately beneath the hymn were two lines; below the lines Kilmer recognized the by-now familiar musical terms from the Babylonian tablets. Adopting the reasonable assumption that each syllable of the hymn matched a note in the accompanying text, she was able to read the musical notation and reconstruct the song note for note. The Ugaritic tablet dates to around 1400 B.C., making it, in the words of her collaborator, the eminent musicologist Richard Crocker, "the oldest 'sheet music' known to exist."

Crocker's colleague Robert Brown made lyres modeled on ancient Near Eastern examples, and in 1974 Kilmer, Crocker and Brown gave a public performance of the music from the Ugaritic tablet. Two years later they released the song as a record, *Sounds from Silence*, together with a booklet detailing their discoveries. The hymn from Ugarit, they concluded, was in the modern seven-note diatonic scale ("do, re, mi . . ."); further, it was not just a simple melody of single notes but was composed of harmonies. Before Kilmer's work it was believed that ancient music was virtually devoid of harmonies—even in the time of the classical Greeks.

The discoveries from ancient Babylonia have totally revolutionized our thinking on the origin of Western music. As the search continues for further musical texts from the ancient Near East, further surprises can be expected—the handful of tablets deciphered so far are merely the tip of the iceberg.

FIREWORKS

A beneficial spin-off of the Chinese invention of gunpowder for military purposes was its use in fireworks. The firecrackers let off by the thousand in Chinese New Year celebrations today have their origins in the third century A.D., but these early crackers were simply pieces of bamboo thrown into a fire, which exploded when the heat reached the air trapped in the stalk. Around A.D. 1050 the development of gunpowder transformed the firecracker, and fireworks were born (see **From Gunpowder to the Cannon** in **Military Technology**).

The Chinese soon created every imaginable kind of firework, from rockets to the first "Catherine wheels." They were able to create a vast variety of effects using different chemicals to produce explosions in every conceivable color. The brilliant sparkling of their "Roman candles" was achieved by combining minuscule shavings of cast iron or steel dust with the gunpowder.

Of all the products of the Chinese firework industry, the most spectacular were the "water rats" and "ground rats," first seen in the twelfth century, which raced along like rats spewing out flames behind them. Water-rat fireworks were fixed onto tiny skis and sent shooting across ponds and lakes, while ground rats performed on land. These spectacular creations sometimes went out of control, once at a festival in the imperial palace in A.D. 1264. A contemporary account records the pandemonium that ensued:

> When the Emperor Li Tsung retired, he prepared a feast . . . in honor of his mother, the Empress-Mother Kung Sheng. A display of fireworks was given in the courtyard. One of these, of the "ground rat" type, went straight up the steps to the throne of the Empress-Mother, and gave her quite a fright. She stood up in anger, gathered her skirts around her, and stopped the feast. Li, being very worried, arrested the officials who had been responsible for making the arrangements for the occasion, and awaited orders from the Empress-Mother. At dawn the next day he went to apologize to her, saying that the responsible officials had been careless, and took the blame on himself. But the Empress-Mother laughed and said, "That thing seemed to come specially to

An illustration from A.D. 1643 of traditional Chinese fireworks, showing types made from the 13th century onward. Near the bottom of the pole is a Catherine wheel.

frighten me, but probably it was an unintentional mistake, and it can be forgiven." So mother and son were reconciled and just as affectionate as before.

Fireworks probably spread west along with gunpowder, reaching southern India by the thirteenth century. Several Tamil books of this date, written by a fireworks expert named Boganathar, give nearly a hundred formulae for the manufacture of an array of variously colored fireworks. It was not until 1540 that fireworks were mentioned in Europe, in Vanuzzio Biringuccio of Siena's *Pyrotechnia*. He says that they had by then become part of the celebrations surrounding the election of a new pope and a frequent feature of feast days, which they remain today.

MAGIC LANTERNS

The magic lantern, or zoetrope, is little known today, but it amazed and delighted audiences in Victorian drawing rooms. It also led, in the twentieth century, to the invention of the slide projector and the cinema. In its final form the zoetrope was a cylinder-shaped canopy of thin material suspended over a lamp. Vanes placed at the top caught the hot air rising from the lamp and made the cylinder rotate slowly. On the sides of the canopy were thin panes of paper on which were painted pictures. As the cylinder revolved around the lamp, the light shone through the succession of pictures to give the illusion that the painted figures were moving—exactly the same principle as that behind modern motion pictures.

In 1868 Mr. W. B. Carpenter, the vice president of the Royal Society of London, stated with some confidence that the magic lantern had been invented by Michael Faraday, the famous pioneer of electricity, as recently as 1836. He was certainly wrong, as one John Bate had already written about the zoetrope early in the seventeenth century. However, even Bate was merely describing a device long known elsewhere.

The principle of using heat to make small figures rotate is extremely ancient. In the Near East it goes back to Heron of Alexandria, who invented a toy with moving dancers in the first century A.D. (see **Introduction** to **High Tech**). The Chinese version is even older, and far more advanced, as it involved the projection of images. In 121 B.C. a magician named Shao Ong staged a sort of séance for the Emperor Han Wu-ti, using moving images projected onto a screen. An earlier emperor, Han Gaozu, had a lamp in his possession in 207 B.C. that, when lit, showed the sparkling scales of turning dragons. Han Gaozu seems to have owned another magic lantern called "the pipe that makes fantasies appear." This was turned by a small windmill, according to a sixth-century A.D. account of Han Gaozu's treasures: "There was a jade tube two feet three inches long, with twenty-six holes in it. If air was blown through it, one saw chariots, horses, mountains and forests appear in front of a screen, one after another, with a rumbling noise. When the blast stopped, all disappeared."

Around A.D. 180 the inventor Ting Huan (see **Perfume** in **Personal Effects**) created a "nine-storied hill censer," which was apparently an extremely complicated multiple magic lantern. On it

The earliest illustration of a magic lantern, from Athanasius Kircher's *Ars Magna Lucis et Umbrae* of 1671. The more elaborate early Chinese magic lanterns would have produced effects such as these.

were strange birds and unusual animals, which turned around as the lamp burned.

By the twelfth century A.D. the most common form of zoetrope was the "horse-riding" or the "horse-pacing" lamp. After the lamp was lit, a succession of prancing horses was projected onto the walls, moving as if they were alive. More sophisticated examples probably used lenses to produce stronger images. This was the type of magic lantern seen by early European visitors, and it undoubtedly provided the inspiration for the Jesuit priest of the China mission Martin Martini (1614–1661). Martini presented the first lantern slides in Europe at Louvain, Belgium, in 1654, soon followed by other European scientists fascinated by the properties of light.

PLAYING CARDS

When playing cards first arrived in Europe toward the end of the fourteenth century A.D., they caused quite a furor. In 1377 the town council of Florence complained that the playing of "a certain game called *naibbe* has recently been introduced into these parts," and by a vote of 98 to 25 decided to prohibit it. In the same year cards reached Paris, where new city regulations cracked down on working-class cardplayers but apparently left noble devotees alone. The following year, in the Bavarian city of Regensburg, the council tried to limit card games to small stakes. By 1387 cards had arrived in the Spanish kingdom of Castile, where the government tried to ban them.

The killjoys were fighting a losing battle, however, for even at this early stage cards began to acquire royal patrons. In 1379 the prince of Brabant, in Belgium, bought a highly decorated pack of cards, while in 1392 the mad French king Charles VI received three packs of cards painted by artist Jacquemin Gringonneur "for his amusement during the intervals in his sad illness."

Playing cards soon led to the emergence of cardsharps, and the mother of all card swindles is recorded in the Parisian court annals for 1408. Two dubious characters lured a traveling merchant into an inn with talk of a good currency deal. One of them then produced a pack of cards from his pocket and demonstrated an amusing game of guessing the identity of a card while seeing only its back. The astute merchant soon noticed that one of the cards had a slight but distinctive mark on the reverse, so he happily joined in when the betting started. When the marked card turned up, the trader put his shirt on it, only to find that the front of the card was not the same, as it had been switched for another.

The French also made one great contribution to the development of playing cards by inventing, around A.D. 1480, the names and shapes of the four suits (spades, hearts, diamonds and clubs) we still use today. These simple geometrical shapes did much to encourage card playing. By the end of the fifteenth century playing-card manufacture was a major industry, and even Johannes Gutenberg, often claimed to be the inventor of movable type (see **Books and Printing** in **Communications**), became involved. He developed some of the mechanical methods of production and, at a time when his finances were in desperate straits, he used drawings that his artists had prepared for his famous Bible to decorate the back of a deck of playing cards.

From this point the history of Western playing cards is clear. But who brought them to Europe in the first place? The subject is swathed in mystery, and it has at different times been claimed that they were introduced by Marco Polo (1254–1324), the Crusaders, or the Gypsies. The most exotic theories credit the Gypsies with the invention of cards (as a means of divination), and it has therefore been argued that their origins lie in India or even Egypt. The truth is that playing cards are a Chinese invention, but the problem has been that little is known of their transmission from China to the West.

Chinese Gamblers

Playing cards had been invented in China by at least the ninth century A.D., when, according to tradition, a princess and her relatives played the "leaf game," or cards. Women were certainly important in the development of card games, for one apparently wrote the world's first book on the subject (now lost), later in that century. By the eleventh century, cards were printed with woodcut blocks, and in

An early Chinese playing card (of about A.D. 1400) found near Turfan, in western China. The card, once held in the Museum für Völkerunde, Berlin, was lost during the Second World War.

the early Ming Dynasty (A.D. 1368–1644) famous artists were employed to design card backs with portraits of characters from favorite novels, such as *The Water Margin*. Chinese cards were much smaller than ours (about two inches long and one inch wide) and were printed on fairly thick paper, which made them hard-wearing but difficult to shuffle. Chinese "money cards" had four suits: cash, strings (of cash), myriads (of strings) and tens (of myriads), with the numbers 2 through 9 in the first three and 1 through 9 in the fourth.

The Chinese of yesteryear were enthusiastic cardplayers and gamblers, as they are today. Ming Dynasty books on cards praised them as superior to all other amusements, for they "were convenient to carry, could stimulate thinking and could be played by a group of four without annoying conversation, and without the difficulties which accompanied playing chess or meditation." As well as this, "cards could be played in almost any circumstances without restrictions of time, place, weather, or qualification of partners."

But this still leaves us without a link to Europe, for early Western cards don't resemble Chinese ones and have different suits. The missing link appears to be the Islamic world, despite the fact that card playing was frowned on by Muslim clerics. In 1938 Professor L. A. Mayer came across a pack of fifty-two cards while searching through the collections of the famous Topkapi Museum, in Istanbul, Turkey. They had been made in Egypt around A.D. 1400, using designs that closely resemble those of early Italian cards. The Arabic inscriptions on the court cards make clear the origin of the word *naibbe* for cards (used by the Florence council): they are called the *Malik* (King), *Na'ib Malik* (Governor), and *Na'ib Thani* (Deputy Governor). They are in four suits—swords, polo sticks, cups and coins (equivalent to modern clubs, spades, hearts and diamonds).

The only significant difference between these and early Italian cards is that the Egyptian ones are, like the Chinese, long and thin. Even this difficulty seems to have been overcome by the find of a single card with an Arabic inscription made around A.D. 1200: its dimensions are like those of Italian cards, which are still slightly narrower than those made today in the rest of Europe. There can now be little doubt that the Arabs were the intermediaries for the widespread transmission of one of ancient China's most popular inventions.

Islamic playing cards from Egypt, dating to the 15th century A.D. (Left) the *Malik* of Cups (equivalent to the modern King of Hearts); (right) the Seven of Swords (equivalent to the modern Seven of Clubs).

S O U R C E S

To give full references for every source consulted while preparing this book might require another volume! So we have listed here only the most important sources, given
for each section within the chapters, and arranging them in the order in which they
have been used or directly quoted. When several sources were used on a given topic we have
listed the most accessible. Asterisks denote books and articles frequently referred to, full details
of which can be found in the Bibliography which follows the Sources. (Places of publication are
only given for small presses and publishers outside of North America and Britain.)

INTRODUCTION

*Bahn & Vertut 1988, pp. 20–23 [Altamira]
J. Mellaart: "Early Urban Communities in the Near East, c. 9000–3400 B.C.," in P. R. S. Moorey (ed.): *The Origins of Civilization*
    (Clarendon Press, 1979), pp. 22–33 [Çatal Hüyük]
J. Hawkes: *Atlas of Ancient Archaeology* (McGraw-Hill, 1974), p. 139
A. Flinder: *Secrets of the Bible Seas* (Severn House, 1985), pp. 101–112 [Caesarea; Raban quote]
A. Maalouf: *The Crusades Through Arab Eyes* (Al Saqi Books, 1984), p. 54
G. Stuart: *America's Ancient Cities* (National Geographic Society, 1988), p. 180 [Tenochtitlán population comparison]
*Needham IV:2, pp. 427 [paddle-wheelers]; V:2, pp. 192–193 [aluminum]

CHAPTER 1: MEDICINE

INTRODUCTION

E. Trinkaus: *The Shanidar Neandertals* (Academic Press, 1983), pp. 404–405
*Klein 1989, p. 334
D. W. Frayer, W. A. Horton, R. Machiarelli & M. Mussi: "Dwarfism in an Adolescent from the Italian Late Upper Palaeolithic," *Na-
    ture* 330 (1987), pp. 60–62
*Saggs 1989, p. 240 [Egypt]
*Sarton II, pp. 331–383 [Hippocrates], 409–410 [Varro], 129–134 [dissection]
*Needham 1970, pp. 379 [Persia], 276–279 [Chinese medical training; hospitals], 380 [Chinese state medicine]
*Jackson 1988, pp. 65–66 [Roman doctors], 134 [military hospitals]
T. W. Africa: *Science and the State in Greece and Rome* (John Wiley, 1968), p. 71 [Archagathus]
A. J. S. Spawforth: "Roman Medicine from the Sea," *Minerva* 1:6 (1990), pp. 9–10
*Rice 1967, p. 72 [Byzantium]
G. Sonnedecker (ed.): *Kremer's and Urdang's History of Pharmacy* (Lippincott, 1976), p. 28 [Baghdad]
J. Soustelle: *The Daily Life of the Aztecs* (Allen & Unwin, 1961), p. 27
*Majno 1975, pp. 216–221 [medicinal spices]
V. J. Vogel: *American Indian Medicine* (University of Oklahoma Press, 1970), pp. 267–414
*Temple 1986, pp. 123–124 [circulation of blood], 135–137 [smallpox]

1 . S U R G I C A L I N S T R U M E N T S

J. T. Rowling: "The Rise and Decline of Surgery in Dynastic Egypt," *Antiquity* 63 (1989), pp. 312–319

★Saggs 1989, pp. 250 [*Smith Papyrus*]

E. Deilaki, *Archaiologikon Deltion* 28 (1973), Chronikon, pp. 92–93 & Pl. 91

C. Singer & E. A. Underwood: *A Short History of Medicine* (Clarendon Press, 1962), p. 8 [Kom Ombo]

★Jackson 1988, pp. 70–73 [cupping], 92–93 [gynecological instruments], 104 & 125 [catheters]

E. Bennion: *Antique Medical Instruments* (Sotheby's, 1979), pp. 61–65 [scalpels & forceps]

★Majno 1975, pp. 269–275 [India], 304–309 [ants]

2 . E Y E O P E R A T I O N S

★Jackson 1988, pp. 121–123 & 161 [Roman]

★Majno 1975, pp. 378–379 [Indian]

★Saggs 1989, p. 265 [Babylonian]

J. B. Pritchard (ed.): *Ancient Near Eastern Texts Relating to the Old Testament* (Princeton University Press, 1969), p. 175 [Hammurabi's laws]

3 . P L A S T I C S U R G E R Y

★Majno 1975, pp. 285–293, 379–381 & 516

4 . B R A I N S U R G E R Y

T. Wilson Parry: "The Art of Trephining Among Prehistoric and Primitive Peoples," *Journal of the British Archaeological Association* 22 (1916), pp. 33–68 [Broca]

K. M. Kenyon: *Archaeology in the Holy Land* (3rd edn., Ernest Benn, 1970), p. 293

W. J. Bishop: *The Early History of Surgery* (Robert Hale, 1960), p. 24 [19th-century survival rate]

F. P. Lisowski: "Prehistoric and Early Historic Trepanation," in D. Brothwell & A. T. Sandison (eds.), *Diseases in Antiquity* (Charles C. Thomas, 1967), pp. 651–672

P. Bennicke: *Palaeopathology of Danish Skeletons* (Copenhagen: Akademisk Forlag, 1985), pp. 92–101

M. L. Ryder: *Sheep and Man* (Duckworth, 1983), pp. 778–779

E. Tapp: "Disease and the Manchester Mummies—the Pathologist's Role," in ★David and Tapp 1984, pp. 78–95

J. A. Brongers: "Ancient Old-World Trepanning Instruments," *Berichten van der Rijksdienst voor het Oudheidkundig Bodemeonderzoek* 19 (1969), pp. 7–16

F. L. Margetts, "Trepanation by Medicine Men: Present Day East African Practice," in Brothwell & Sandison (eds.) 1967, pp. 673–701

5 . F A L S E T E E T H A N D D E N T I S T R Y

J. Woodforde: *The Strange Story of False Teeth* (Routledge, 1968)

E. Bennion: *Antique Dental Instruments* (Sotheby's, 1986), pp. 10–11, 82–84 [Etruscan goldwork]

F. F. Leek: "Dental Problems During the Old Kingdom—Facts and Legends," in ★David & Tapp (eds.) 1984, pp. 104–131

P. Bennicke: *Palaeopathology of Danish Skeletons* (Copenhagen: Akademisk Forlag, 1985), p. 45

★Needham 1970, p. 372 [China]

6 . F A L S E L I M B S

★Herodotus IX:37

L. J. Bliquez: "Classical Prosthetics," *Archaeology* 36 (1983), pp. 25–29

★Pliny VII:104–106 [Sergius Silus]

Anon.: "Nouvelles archéologiques d'U.R.S.S.," *Archeologia* 42 (1971), p. 89 [Kazakhstan]

W. J. Bishop: *The Early History of Surgery* (Robert Hale, 1960), p. 38 [India]

A. R. David: "Introduction" in ★David & Tapp (eds.) 1984, pp. 3–42 [mummy]

7. ANESTHETICS

*Jackson 1988, pp. 69–70 & 112 [Roman]

*Pliny XXV:150

L. E. Voigts & R. P. Hudson: "A Surgical Anesthetic from Late Medieval England," in S. Campbell, B. Hall & D. Klausner (eds.): *Health, Disease and Healing in Medieval Culture* (Macmillan, 1992), pp. 34–56

*Chinese Academy 1983, pp. 369–370

S. Miyasita: "Mandrake Once Traveled in China as an Anaesthetic," *Japanese Studies in the History of Science* 5 (1966), pp. 189–192

8. ACUPUNCTURE

G. T. Lewith: *Acupuncture: Its Place in Western Medical Science* (Thorsons, 1982) [modern use]

G. D. Lu & J. Needham: *Celestial Lancets* (Cambridge University Press, 1980)

R. Melzack & P. Wall: *The Challenge of Pain* (Penguin, 1982), pp. 319–326

CHAPTER 2: TRANSPORTATION

INTRODUCTION

P. G. Bahn: "The 'unacceptable face' of the West European Palaeolithic," *Antiquity* 52, 1978, pp. 183–192

D. W. Anthony & D. R. Brown: "Origins of Horseback Riding," *Antiquity* 65 (1991), pp. 22–38

S. Piggot: *The Earliest Wheeled Transport* (Thames & Hudson, 1983), pp. 38–50

*Neuburger 1930, pp. 453–460 [Roman roads]

*Needham IV:3, pp. 4–27 [Chinese roads], 269–307 [Chinese canals]

A. Kendall: *Everday Life of the Incas* (Batsford, 1973), pp. 138–139

*Needham 1970, p. 348 [Chinese traffic cops]

O. A. W. Dilke: *The Ancient Romans: How They Lived and Worked* (David & Charles, 1975), pp. 63–64 [Caesar]

*Saggs 1988, p. 40 [Lagash canal]

*Sarton I, p. 294 [Athos Canal]

*Herodotus VII: 22–24, 33–37 & 117

Seutonius: *Life of Caligula* 19

*Landels 1978, pp. 182–183 [Diolkos]

1. MAPMAKING

J. B. Harley & D. Woodward (eds.) 1987: *The History of Cartography*, Vol. 1 (University of Chicago Press)

*Sarton I, pp. 82–84, 184–188, 298–330; II, pp. 99–109 [Babylonia & Greece]

*Herodotus IV:36

C. Gordon: *Riddles in History* (Crown, 1974), pp. 93–106 [Vinland Map]

M. Magnusson & H. Pálsson: *The Vinland Sagas* (Penguin, 1965), pp. 120–121 [Stefansson's map]

C. Hapgood: *Maps of the Ancient Sea Kings* (Chilton, 1966) [portolans]

T. A. Cahill *et al.*: "The Vinland Map, Revisited: New Compositional Evidence on its Inks and Parchment," *Analytical Chemistry* 59:6 (March 15, 1987), pp. 829–833

BOX ON MAPMAKING MARVELS OF THE ANCIENT ORIENT

*Needham III, pp. 497–590 [quote, p. 547]

*Pirazzoli-t'Serstevens 1982, pp. 56 & 209–210 [tomb finds]

2. SKIS AND SKATES

G. M. Burov: "Some Mesolithic Wooden Artifacts from the Site of Vis I in the European North East of the U.S.S.R.," in C. Bonsall (ed.), *The Mesolithic in Europe* (John Donald, 1989), pp. 391–401.
P. Lunn: *The Guinness Book of Skiing* (Guinness, 1983), pp. 3–7 [Rödöy and Haakon]
*Needham IV:3, p. 14 [Chinese accounts]
A. MacGregor: "Bone Skates: a Review of the Evidence," *Archaeological Journal* 133 (1976), pp. 57–74
A. MacGregor: *Bone Antler Ivory and Horn* (Croom Helm, 1985), pp. 142–143 [Fitzstephen]
M. Heller (ed.): *The Illustrated History of Ice Skating* (Paddington Press, 1979), p. 17 [St Liedwi]

3. ODOMETERS

*Needham IV:2, pp. 281–286 [China]
*China Academy 1983, pp. 433–435 [China]
*Vitruvius X:ix
*Drachmann 1963, pp. 157–159
A. Sleeswyk: "Vitruvius' Odometer," *Scientific American* 245:4 (Oct. 1981), pp. 158–171

4. WIND CARS AND ROCKET CARS

*Needham IV:2, pp. 275–281 [wind cars]
*Lindsay 1974, p. 33 [rocket cars]

5. SHIPS AND LINERS

A. J. Parker: "Classical Antiquity: the Maritime Dimension," *Antiquity* 64 (1990), pp. 335–346 [obsidian]
*Klein 1989, pp. 392–395 [Australia]
P. Johnstone: *The Sea-Craft of Prehistory* (Routledge, 1988), pp. 26–28 [Star Car and deep-sea fishing], 46 [Pesse], 85–87 [megaliths], 121–139 [curraghs]
*Caesar III:7–19
T. Severin: *The Brendan Voyage* (Hutchinson, 1978)
B. Landström: *Ships of the Pharaohs* (Allen & Unwin, 1970)
*Needham IV:2, pp. 413–435 [paddle-boats]; IV:3, pp. 440–441 [Wu navy], 484–494 [Cheng Hô and Yung-lo], 627–656 [rudders]
*Chinese Academy 1983, p. 488 [Guangzhou]
*Landels 1978, pp. 153–163 [Hellenistic]
*Sarton II, pp. 120–126 [Athenaeus]

6. THE FIRST SUEZ CANAL

C. W. Hallberg: *The Suez Canal: its History and Diplomatic Importance* (Columbia University Press, 1931)
A. T. E. Olmstead: *History of the Persian Empire* (University of Chicago Press, 1948), pp. 145–147
C. Tuplin: "Darius' Suez Canal and Persian Imperialism," in H. Sancisi-Weerdenburg & A. Kuhrt (eds.): *Achaemenid History VI: Asia Minor and Egypt* (Nederlands Instituut voor Het Nabje Oosten, 1991), pp. 237–283
*Herodotus II:158; IV:40 & 42

7. THE COMPASS

*Chinese Academy 1983, pp. 152–165 & 500–501 [China]
*Needham 1970, pp. 239–249 [China]
*Needham IV:1, pp. 229–314 [China]
J. B. Carlson: "Lodestone Compass: Chinese or Olmec Primacy?" *Science* 189 (Sept. 1975), pp. 753–760

8 . LIGHTHOUSES

D. B. Hague & R. Christie: *Lighthouses: their Architecture, History and Archaeology* (Gomer Press, 1975), pp. 1–13
*Needham IV:3, pp. 660–662 [Chinese and Arab]
K. Sutton-Jones: *Pharos* (Michael Russell, 1985), pp. 11–12 [King Richard]

9 . DIVING GEAR

D. J. A. Ross: *Alexander and the Faithless Lady: a Submarine Adventure* (Birkbeck College, 1967)
Aristotle: *On the Parts of Animals* II:16
*al-Hassan & Hill 1986, pp. 244–245 [al-Biruni]
J. Gimpel: *The Medieval Machine* (Gollancz, 1977), p. 234 [Hussites and Keyser]
J. Crane: *Submarine* (BBC, 1984), p. 107 [Leonardo and Bourne]

10 . MAN-BEARING KITES AND PARACHUTES

*Temple 1986, pp. 175–179 [China]
T. Streeter: *The Art of the Japanese Kite* (Weatherhill, 1974), p. 161
C. Hart: *Kites: an Historical Survey* (Faber & Faber, 1967), pp. 27–28 [Marco Polo]
C. Hart: *The Dream of Flight* (Faber & Faber, 1972), pp. 108–109 [parachute]

BOX ON GLIDERS

B. Laufer: *The Prehistory of Aviation* (Field Museum of Natural History, Chicago, 1928), p. 64 [Archytas]
Aelius Gellius: *Attic Nights* X:12 [Archytas]
I. van Sertima (ed.): *Blacks in Science: Ancient and Modern* (Transaction Books, 1983), pp. 92–99 [Sakkara]

12 . BALLOONING

J. Woodman: *Nazca* (John Murray, 1980)
T. Morrison: *The Mystery of the Nazca Lines* (Woodbridge: Nonesuch Expeditions, 1987), p. 135 [Reiche's reaction]

CHAPTER 3: HIGH TECH

INTRODUCTION

B. Farrington: *Greek Science* (Penguin, 1961), pp. 50–52 [pythagorean experimentation], 171–177 [Strato], 200 [magnetic statues], 301 [quote]
*Sarton II, pp. 3–28 [Ptolemies], pp. 343–350 [Philon & gimbals]
J. Pappademos: "An Outline of Africa's Role in the History of Physics," in I. Van Sertima (ed.): *Blacks in Science: Ancient and Modern* (Transaction Books, 1983), pp. 177–196
*Needham 1970, pp. 58–59, 232 [quote]
D. de S. Price: "An Ancient Greek Computer," *Scientific American* 220:6 (June 1959), pp. 60–67 [quote p. 67]

1 . COMPUTERS

D. de S. Price: *Gears from The Greeks* (American Philosophical Society, 1974)
*Hill 1984, pp. 187–188 [Arab]
D. de S. Price: *Science Since Babylon* (2nd edn., Yale University Press, 1975), p. 48 [quote]

2. CLOCKS

*Saggs 1989, pp. 235–236 [shadow clocks]
B. Cotterell & J. Kamminga: *The Mechanics of Pre-Industrial Technology* (Cambridge University Press, 1990), pp. 59–61 [Amenemhet]
*Brumbaugh 1966, pp. 68–69 [Athens]
*Ronan 1983, pp. 113–115 [Ctesibius and Tower of the Winds]
*Hill 1984, pp. 224–225 [clepsydras]
*al-Hassan & Hill 1986, p. 57 [Islamic]
J. Needham, L. Wang & D. de S. Price: *Heavenly Clockwork* (Cambridge University Press, 1986)

3. COIN-OPERATED SLOT MACHINES

*Lindsay 1974, pp. 331–332 [Heron]
*Drachmann 1963, p. 196 [Philon]

4. AUTOMATIC DOORS

*Lindsay 1974, pp. 330–334

5. THE STEAM ENGINE

*Lindsay 1974, pp. 335–338
*Landels 1978, pp. 28–31
A. Toynbee: "If Alexander the Great Had Lived On," in Toynbee (ed.): *Some Problems in Greek History* (Oxford University Press, 1969), pp. 441–486

6. AUTOMATA

*Hill 1984, pp. 199–222
Homer: *Iliad* XVIII:368–380, 416–422
*Brumbaugh 1966, pp. 30–58 & 98–129 [Heron's theaters]
D. R. Hill (trans.): *"Banu Musa bin Shakir"—The Book of Ingenious Devices* (Dordrecht: Reidel, 1979)
D. R. Hill: "Mechanical Engineering in the Medieval Near East," *Scientific American* 264:5 (May 1991), pp. 64–69
*Needham 1970, pp. 207–208 [Heron], 234–235 [Arabic clocks]
*Needham IV:2, pp. 286–303 [south-pointing carriage]

7. EARTHQUAKE DETECTORS

*Needham III, pp. 633–634 [China]
*Temple 1986, pp. 162–166 [China]
*Hill 1984, p. 206 [Virgil]

8. ELECTRIC BATTERIES

*Lindsay 1974, pp. 270–272 [electric fish]
Lucretius, *On the Nature of the Universe* VI:145–238
P. C. Rice: *Amber: The Golden Gem of the Ages* (Van Nostrand Reinhold, 1980)
*Needham IV:1, pp. 233–238 [China & amber]
W. Winton: "Baghdad Batteries B.C.," *Sumer* 18 (1962), pp. 87–89
A. Al-Haik: "The Rabbou'a galvanic cell," *Sumer* 20 (1964), pp. 103–104
P. T. Keyser: "The Purpose of the Parthian Galvanic Cells: A First-Century A.D. Electric Battery Used for Analgesia," *Journal of Near Eastern Studies* 52 (1993), pp. 81–98
F. Hitching: *The Atlas of World Mysteries* (Collins, 1978), p. 124 [2nd Winton quote—interview]

BOX   ON   ANCIENT   LIGHTNING   RODS

*Neuburger 1930, pp. 348–349 [Edfu]

C. P. Kardara: "The Coming of the Indoeuropeans and the Twin Pillars," in *Acta of the 2nd International Colloquium on Aegean Pre-history: the First Arrival of Indo-European Elements in Greece* (Athens, Ministry of Culture and Sciences General Directorate of Antiquities, 1972), pp. 168–174

9 .   MAGNETS   AND   MAGNETISM

*Lindsay 1974, 245–266

Augustine: *City of God* XXI:5

*Sarton I, pp. 171–172 [Thales]

Plato: *Ion* 533D–534B

Lucretius: *On the Nature of the Universe* VI:906–916, 998–1064

*Chinese Academy 1983, p. 162 [medicine]

L. C. Tai: "The riddle of the magnetic doors," *Nature* 319 (Feb. 13, 1986), p. 544

1 0 .   MAGNIFYING   GLASSES

W. Gasson: "The Oldest Lens in the World: A Critical Study of the Layard Lens," *The Ophthalmic Optician* (December 9, 1972), pp. 1267–1272

S. N. Kramer: "A Sumerian Document with Microscopic Cuneiform," *Expedition* 1:3 (Spring, 1959), pp. 2–3

G. Sines & Y. Sakellarakis: "Lenses in Antiquity," *American Journal of Archaeology* 91 (1987), pp. 191–196

L. Gorelick & A. John Gwinnett: "Close Work Without Magnifying Glasses?" *Expedition* 23:2 (Winter 1981), pp. 27–34

*Pliny XXXVI:62–64

A. Bammer: "Recent Excavations at the Altar of Artemis in Ephesus," *Archaeology* 27 (1974), pp. 202–205

J. M. F. May: "The Alexander Coinage of Nikokles of Paphos," *Numismatic Chronicle* 12 (1952), pp. 1–18 [quote p. 4]

BOX   ON   BURNING   LENSES   AND   MIRRORS

Aristophanes: *The Clouds* 766–772

*Needham IV:1, pp. 111–118 [China]

CHAPTER 4: SEX LIFE

INTRODUCTION

K. J. Dover: *Greek Homsexuality* (Vintage, 1980)

*Needham IV:2, pp. 477–478 [Emperor's sex schedule]

*Tannahill 1989, pp. 182–190 [China], 128–154 [Christian Church], 180 [Mâcon], 91–95 [Greek brothels], 246–248 [eunuchs]

*Herodotus I:196

C. Hibbert: *Cities and Civilizations* (Weidenfeld & Nicolson 1987), p. 52 [Roman brothels]

Anon: " 'Maiouma' laid bare at Aphrodisias," *Minerva* 1:6 (1990), p. 2 [sex shows]

*Bullough 1976, pp. 107–108 [Crete], 232 [Arab transvestites], 302–310 [Chinese homosexuality and eunuchs], 326–330 [Byzantine eunuchs]

S. Runciman: *Byzantine Civilization* (Meridian, 1956), p. 162

Herodian: *History of the Roman Empire* V:v,3–6 & viii,1–2 [Elagabalus]

Dio Cassius: *Roman History* LXXX:11, 13–17 [Elagabalus]

*Temple 1986, pp. 127–131 [Chinese knowledge of hormones]

1. APHRODISIACS

P. & M. Harrison: *Aphrodisiacs* (Jupiter, 1979)

P. V. Taberner: *Aphrodisiacs: the Science and the Myth* (Croom Helm, 1985)

H. E. Wedeck: *Dictionary of Aphrodisiacs* (Peter Owen, 1961)

N. Ohler: *The Medieval Traveller* (Boydell & Brewer, 1989), p. 219 [Ibn Battuta]

E. Chou: *The Dragon and the Phoenix* (Michael Joseph, 1971), p. 148 [China]

★Lindsay 1974, p. 245 [Assyrian iron filings]

Theophrastus: *On Plants* IX:viii,8; IX:ix,1

2. DILDOS

Theophrastus: *The Characters* VI

★Tannahill 1989, pp. 89–90 [Greece]

Al. N. Oikonomides: "The 'Bread-Stick' of Mantios," *Horos* 3 (1985), pp. 130–131

——"*Kollix, olisbos, olisbochollix*," *Horos* 4 (1986), pp. 168–178

J. Muirhead-Gould (ed.): *The Kama Sutra of Vatsyayana*, trans. by R. Burton and F. F. Arbuthnot (Panther Books, 1963), VII:2

★Bullough 1976, pp. 298–299 [China]

3. CONTRACEPTIVES

N. E. Himes: *Medical History of Contraception* (Allen & Unwin, 1936)

★Bullough 1976, pp. 63 [Egypt], 324 [Aëtio], 355 [Church]

★Tannahill 1989, pp. 61–66 [Egypt], 88 [Greece]

★Saggs 1988, p. 158 [Sumer]

J. M. Riddle: "Oral Contraceptives and Early-Term Abortifacients During Classical Antiquity and the Middle Ages," *Past and Present* No. 132 (1991), pp. 3–32

★Jackson 1988, pp. 108–110 [Soranus and Pliny]

M. C. Stopes: "Positive and negative control of conception in its various technical aspects," *Journal of State Medicine* 39 (1931), pp. 354–360

Vogel, V. J.: *American Indian Medicine* (University of Oklahoma Press, 1970), pp. 239–245

G. M. Planas & J. Kuć: "Contraceptive Properties of Stevia Rebaudiana," *Science* 162 (29th Nov., 1968), p. 1007

BOX ON CONDOMS

N. E. Himes: *Medical History of Contraception* (Allen & Unwin, 1936), pp. 187–192

4. PREGNANCY TESTS

★Saggs 1989, p. 247

E. Reiner: "Babylonian Birth Prognoses," *Zeitschrift für Assyriologie* 72 (1982), pp. 124–138

5. SEX MANUALS

J. Muirhead-Gould (ed.): *The Kama Sutra of Vatsyayana*, trans. by R. Burton and F. F. Arbuthnot (Panther Books, 1963)

★Bullough 1976, pp. 245–246 [India], 285–289 [China]

★Tannahill 1989, pp. 191–220 [India], 160–169 [China]

CHAPTER 5: MILITARY TECHNOLOGY

INTRODUCTION

K. Kenyon: *Archaeology in the Holy Land* (3rd edn. Ernest Benn, 1970), pp. 43–44
J. T. Hooker: *Mycenaean Greece* (Routledge & Kegan Paul, 1976), pp. 93–107
S. Johnson: *Hadrian's Wall* (Batsford, 1989)
*Needham IV:3, pp. 47–55 [Great Wall of China]
G. Connah: *African Civilizations* (Cambridge University Press, 1987), p. 79 [Axum]
J. & F. Gies: *Life in a Medieval Castle* (Abelard, 1974), pp. 9–12
*Saggs 1989, pp. 40–41 [Sargon]
*Olivová 1984, p. 24 [Tell Agrab]
V. Karageoghis: *Salamis in Cyprus* (Thames & Hudson, 1969), pp. 67–72 & 78–98
A. L. F. Rivet: "A Note on Scythed Chariots," *Antiquity* 53 (1979), pp. 130–132
A. Ferrill: *The Origins of War* (Thames & Hudson, 1985), pp. 37–38 [Qadesh]
J. B. Pritchard (ed.): *Ancient Near Eastern Texts Relating to the Old Testament* (Princeton University Press, 1969), pp. 278–279 [Qarqar]
A. Pacey: *Technology in World Civilization* (Blackwell, 1990), p. 5 [Sung army]
T. Woody: *Life and Education in Early Societies* (Macmillan, 1949), p. 91 [swimming skins]
*Temple 1986, pp. 235–237 [land mines]
Frontinus: *Strategems* III: Introduction

1.   HUMAN AND ANIMAL ARMOR

A. E. Dien: "Warring States Armor and Pit Three at Qin Shihuangdi's Tomb," *Early China* 5 (1979–80), pp. 46–47 [Chinese lacquer
    & mail armor]
M. E. L. Mallowan: *Early Mesopotamia and Iran* (Thames & Hudson, 1965), p. 109 [Ur]
L. Cottrell: *The Lion Gate* (Evans Brothers, 1963), pp. 110–116 [Dendra]
H. R. Robinson: *The Armour of Imperial Rome* (Arms and Armour Press, 1975), pp. 153–154 [scale], 164 [mail]
*Temple 1986, p. 83 [paper]
*Bray 1968, p. 189 [Aztecs]
R. Humble: *Warfare in the Ancient World* (Cassell, 1980), p. 204 [horses]
M. C. Bishop & J. C. Coulston: *Roman Military Equipment* (Shire, 1989), p. 55 [Dura-Europos]

2.   TANKS

R. Humble: *Warfare in the Ancient World* (Cassell, 1980), p. 32 [Assyrians]
Y. Yap & A. Cotterell: *The Early Civilizations of China* (Weidenfeld & Nicolson, 1975), p. 192 [Sung]
*Needham V:7, p. 421 [Mongols]
H. W. Koch: *Medieval Warfare* (Bison, 1978), pp. 165–171 [Hussites]

3.   CATAPULTS AND CROSSBOWS

W. Soedel & V. Foley: "Ancient Catapults," *Scientific American* 240:3 (March, 1979), pp. 120–128
*Landels 1978, pp. 106 [Syracusa], 123–126 [repeating catapults]
E. W. Marsden: *Greek and Roman Artillery: Historical Development* (Clarendon Press, 1969), p. 96 [Maiden Castle]
*Temple 1986, pp. 218–224 [crossbow]

4. THE "CLAWS" OF ARCHIMEDES

A. W. Lawrence: "Archimedes and the Design of Euryalus Fort," *Journal of Hellenic Studies* 66 (1946), pp. 99–107
*Livy XXIV:34; XXV:23–25
Polybius: *Histories* VIII, 2–7
Plutarch: *Life of Marcellus* 19–23
S. Strandh: *Machines: an Illustrated History* (Mitchell Beazley, 1979), pp. 26–31
*Landels 1978, p. 95–98

5. FLAMETHROWERS

J. J. Norwich: *Byzantium: the Early Centuries* (Viking, 1988), pp. 323–324 [Callanicus]
Anon.: *The Russian Chronicles* (Century, 1990), p. 42
P. Hetherington & W. Forman: *Byzantium: City of Gold, City of Faith* (Orbis, 1983), p. 70 [Anna Commena]
*Thucydides: *History of the Peloponnesian War* IV:100, 114
*Needham V:7, pp. 220–253 [fire-lance], 414–421 [flame-thrower batteries]

6. HAND GRENADES

P. Pentz: "A Medieval Workshop for Producing 'Greek Fire' Grenades," *Antiquity* 62 (1988), pp. 89–93
G-D. Lu, J. Needham & C-H. Phang: "The Oldest Representation of a Bombard," *Technology and Culture* 29 (1988), pp. 594–605 [Ta-tsu]
*Temple 1986, p. 233 [T'ieh Li]

7. FROM GUNPOWDER TO THE CANNON

*Needham V:7 *Military Technology: the Gunpowder Epic* (1986)
G-D. Lu, J. Needham & C-H. Phang: "The Oldest Representation of a Bombard," *Technology and Culture* 29 (1988), pp. 594–605 [Ta-tsu]

8. POISON GAS

J. R. Partington: *A History of Greek Fire and Gunpowder* (Heffers, 1960), p. 284 [Brazil]
*Needham V:7, pp. 117–125 [China]

CHAPTER 6: PERSONAL EFFECTS

INTRODUCTION

Martial *Epigrams* IX:37
*Jackson 1988, p. 54 [Roman]
S. Greenberg & E. L. Ortiz: *The Spice of Life* (Michael Joseph, 1983), p. 53 [cloves]
M. D. Coe: *The Maya* (Thames & Hudson, 1987), p. 158 [Maya]

1. MIRRORS

B. Schweig: "Mirrors," *Antiquity* 15 (1941), pp. 257–268
*Mellaart 1967, p. 211 [Çatal Hüyük]
G. C. Vaillant, *Aztecs of Mexico* (2nd edn., Penguin, 1965), pp. 151–152, 236
Z. Y. Saad: *The Excavations at Helwan* (University of Oklahoma Press, 1969), p. 54 [earliest Egyptian]
*Corson 1972, p. 32 [Indus Valley]

M. Sullivan: *The Arts of China* (University of California Press), p. 19 [Shang]
★Chinese Academy, pp. 171–172 [periscope]
Exodus 38:8
A. Fox: "The Holcombe Mirror," *Antiquity* 46 (1972), pp. 293–296
★Macnamara 1973, p. 115 [Etruscan]
★Forbes V, pp. 187–189 [Rome]
Suetonius: *Life of Domitian* 14 & 17
★Neuburger 1930, p. 163 [Roman glass mirrors]

BOX ON MAGIC MIRRORS

★Temple 1986, pp. 66–67
★Needham IV:1, pp. 94–97 [quote, p. 97]

2 . MAKEUP

★Tannahill 1989, pp. 104 [Sumer, Greece], 140 [St. Jerome]
★Manniche 1989, p. 47 [Egypt]
H. Wallnöfer & A. von Rottauscher: *Chinese Folk Medicine* (Crown, 1965), p. 164 [nails]
★Needham V:3, p. 16 [lead]
★Corson 1972, pp. 38–41 [Greece], 54 [Poppaea]
T. McLaughlin: *The Gilded Lily* (Cassell, 1972), pp. 42–43 [Elagabalus]
★al-Hassan & Hill 1986, p. 144 ["Blackbird"]
★Bray 1968, p. 30 [Aztecs]

3 . TATTOOING

K. Spindler: *The Man in the Ice* (Weidenfeld & Nicolson, 1994), pp. 167–173
R. S. Bianchi: "Tätowierung" in *Lexikon der Ägyptologie* VI (Otto Harrassowitz, 1986), p. 146 [Nubia and Egypt]
★Rudenko 1970, pp. 109–114 [Pazyryk]
★Herodotus V:6
E. Kidder: *Ancient Japan* (Phaidon, 1977), p. 43
★Caesar V:14
C. P. Jones: "*Stigma*: Tattooing and Branding in Graeco-Roman Antiquity," *Journal of Roman Studies* 77 (1987), pp. 139–155
H. Ebenstein: *Pierced Hearts and True Love* (Derek Verschoyle, 1953), pp. 13–14 [Christianity]

4 . SOAP

★Pliny XXVIII:51
★Forbes III, p. 187 [Mesopotamia]
★Herodotus IV:73
★al-Hassan & Hill 1986, 150 [Arabs]
W. A. Hassall: *How They Lived* (Blackwell, 1962), p. 113 [Britain]
★Needham 1970, pp. 368–372 [soap-bean]

5 . RAZORS

★Bray 1968, p. 149 [Aztecs]
M. Stead: *Egyptian Life* (British Museum, 1990), p. 50
M. Dayagi-Mendels: *Perfumes and Cosmetics in the Ancient World* (Israel Museum, 1989), pp. 72 [Mesopotamia], 78 [Roman first shave]
P. V. Glob: *The Mound People* (Faber & Faber 1973), p. 116
★Caesar V:14

*Corson 1972, pp. 33–34 [India]

J. Carcopino: *Daily Life in Ancient Rome* (Routledge 1941), pp. 159–164 [Greece & Rome]

Martial *Epigrams* XI:84

*Pliny XXIX:114

L. Dunkling & J. Foley: *The Guinness Book of Beards and Moustaches* (1990), p. 42 [Louis VII]

6. PERFUME

J. Hawkes: *The First Great Civilizations* (Hutchinson, 1973), 359 [Hatshepsut]

R. Genders: *A History of Scent* (Hamish Hamilton, 1972), p. 26 [Edfu]

*Manniche 1989, pp. 48 [balanos], 51–52 [perfume cones]

*al-Hassan & Hill 1986, pp. 142–144 [distillation]

*Rice 1967, p. 125 [Zoë]

J. Trueman: *The Romantic Story of Scent* (Aldus, 1975), p. 92 [Greece]

*Jackson 1988, p. 55 [Rome]

*Needham IV:1, p. 233 [gimbals]

7. WIGS

John Woodforde: *The Strange Story of False Hair* (Routledge & Kegan Paul, 1971)

J. Stevens Cox: *The Story of Wigs Through the Ages* (Guernsey: Toucan Press, 1983)

*Wilkinson 1854, ii, pp. 325–329 [Egypt]

Xenophon: *Cyropaedia* I,iii,2

Ovid: *The Art of Love* III:163, 245

BOX ON FALSE BEARDS

*Wilkinson 1854, ii, pp. 325–327, 329

A. Gardiner: *Egypt of the Pharaohs* (Oxford University Press, 1961), pp. 183–184

*Rudenko 1970, p. 105 [Pazyryk]

L. Dunkling & J. Foley: *The Guinness Book of Beards and Moustaches* (1990), p. 43 [Europe]

8. CLOTHING AND SHOES

*Klein 1989, pp. 370–371 [Sunghir]

G. Clark & S. Piggott: *Prehistoric Societies* (Penguin, 1965), 89–91 [Buret]

*Mellaart 1967, pp. 219–220 [Çatal Hüyük]

E. J. W. Barber: "Neolithic textiles in Europe and the Near East," *Archeomaterials* 4 (1990), pp. 63–68

*Needham V:9, pp. 58 [cotton], 272–419 [silk], 23–39 & 226 [ramie & hemp]

*Kramer 1963, p. 104 [Sumerian weaving]

*Wilkinson 1854, ii, pp. 72–80 [Egyptian linen]

*al-Hassan & Hill 1986, pp. 181–182 [Arab silk]

*Ronan 1983, p. 51 [rubber]

*Temple 1986, p. 82 [paper-mulberry]

*Needham III, p. 656 [asbestos]

J. P. V. D. Balsdon: *Life and Leisure in Ancient Rome* (Bodley Head, 1969), p. 34 [costume changes]

D. J. Symons: *Costume of Ancient Rome* (Batsford, 1987), pp. 22 [trousers], 27 [bikini]

M. Sichel: *Costume of the Classical World* (Batsford, 1980), pp. 51–52 [shoes]

R. Brain: *The Decorated Body* (Hutchinson, 1979), p. 88 [poulaines]

BOX ON SEWING AND KNITTING

*Klein 1989, p. 370 [Palaeolithic needles]
W. H. Manning: *Catalogue of the Romano-British Iron Tools, Fittings and Weapons in the British Musuem* (1985), p. 35 [Manching]
E. F. Holmes: *A History of Thimbles* (Cornwall, 1985)
R. Rutt: *A History of Hand Knitting* (Batsford, 1987)

9 .   JEWELRY

J. Ogden: *Ancient Jewellery* (British Museum, 1992)
*Mellaart 1967, p. 211 [Çatal Hüyük]
H. Tait: *Seven Thousand Years of Jewellery* (British Museum 1986), pp. 11 [Arpachiyah], 23–24 [Ur]
C. Renfrew: "Varna and the Social Context of Early Metallurgy," *Antiquity* 52 (1978), pp. 199–203
*Hawkes & Woolley 1963, pp. 567–568 [Ur; granulation]
Plutarch: *The Life of Artaxerxes* XXIV:6
D. Keys: "Jade Finds Re-write Chinese History," *Minerva* 2:1 (1991), p. 3
*Saggs 1989, p. 129 [lapis]
*Forbes 1966, V, pp. 122–123, 133–145 [paste]
K. R. Maxwell-Hyslop: *Western Asiatic Jewellery c. 3000–612 BC* (Methuen, 1971), pp. 36–37 [granulation]

1 0 .   SPECTACLES

E. Rosen: "The Invention of Eyeglasses," *Journal of the History of Medicine* 11 (Jan. 1956), pp. 13–46; April 1956, pp. 183–218
*Needham IV:1, pp. 118–122.

BOX ON SNOW GOGGLES

R. McGhee: *Canadian Arctic Prehistory* (Ottawa: National Museum of Man, 1978), pp. 79 & 95

1 1 .   UMBRELLAS

T. S. Crawford: *A History of the Umbrella* (David & Charles, 1970)
M. C. Miller: "The Parasol: an Oriental Status-symbol in Late Archaic and Classical Athens," *Journal of Hellenic Studies* 112 (1992), pp. 91–105
*Temple 1986, pp. 96–99 [China]
D. Morgan: *The Mongols* (Blackwell, 1986), p. 177 [India]

CHAPTER 7: FOOD, DRINK AND DRUGS

INTRODUCTION

*Klein 1989, pp. 171, 218, 221–222 [Choukoutien]
J. A. J. Gowlett: *Ascent to Civilization* (Collins, 1984), p. 164 [pottery]
*Tannahill 1988, pp. 54 [preserving], 27–29 [fermentation], 14 [prehistoric ovens], 29, 49–51 [sweeteners], 110, 142, 144 [sugar], 86–91 [Roman spice trade], 131 [chopsticks], 187–188 [forks], 69 [Athenian *hors d'oeuvres*]
*Woolley 1976, p. 7
*Lindsay 1974, p. 370 [natural gas]
N. Liphschitz, R. Gophna, M. Hartman & G. Biger: "The Beginning of Olive *(Olea europaea)* Cultivation in the Old World: a Re-assessment," *Journal of Archaeological Science* 18 (1991), pp. 441–453
S. J. Keay: *Roman Spain* (British Museum, 1988), p. 103 [Rome dump]
*Pliny XII:42
*Chang 1977, pp. 55–57 [tomb with ginger]

J. I. Miller: *The Spice Trade of the Roman Empire* (Oxford University Press, 1969), pp. 54–55 [Fa-Hsien], 82 [Hippocrates]
*Manniche 1989, p. 136 [Ramesses' peppercorns]
H. W. Allen: *A History of Wine* (Faber & Faber, 1961), p. 45 [Philoxenus]
*Rice 1967, p. 170 [forks]
Suetonius: *Life of Vitellius* 13
*Chang 1977, pp. 1–22 [Chinese court]
J. Barkas: *The Vegetable Passion* (Routledge, 1975)

1. RESTAURANTS AND SNACK BARS

*Chang 1977, pp. 81–82, 85–140, 156–163 [Chinese restaurants]
C. Hibbert: *Cities and Civilizations* (Weidenfeld & Nicolson 1987), p. 71 [Hangchow]
*Tannahill 1988, pp. 234–237 [pasta debate]
G. Wiet: *Cairo: City of Art and Commerce* (University of Oklahoma Press, 1964), p. 95
*al-Hassan & Hill 1986, pp. 228–230 [censor of morals]
H. Johnson: *The Story of Wine* (Mitchell Beazley, 1989), p. 64 [Pompeii]
C. Blegen: *Troy and the Trojans* (Thames & Hudson, 1963), pp. 153–154
*Woolley 1976, pp. 25, 32–33, 153–154

2. COOKBOOKS

Anon.: "Some Tasty Recipes for a Babylonian Feast," *National Geographic* 178:6 (December, 1990)
J. Edwards: *The Roman Cookery of Apicius* (Ryder, 1984)

BOX ON RECIPES FROM A ROMAN COOKBOOK

*Tannahill 1988, pp. 82–84 [liquamen]
J. Edwards: *The Roman Cookery of Apicius* (Ryder, 1984), pp. xxii-xxiii [liquamen], 65, 175, 287 [recipes]

3. REFRIGERATION

*Forbes VI, pp. 104–121 [Rome, Egypt, India, Estonia]
Seneca: *Natural Questions* IVB:13
Scriptores Historiae Augustae XXIII:8 [Elagabalus]
Pliny the Younger: *Letters* I:15
J. M. Sasson: "Ice at Mari," *Biblical Archaeologist* 47:2 (June, 1984), pp. 115–116
A. Cotterell: *The First Emperor of China* (Macmillan, 1981), p. 74 [Shih Huang Ti]
*Majno 1975, pp. 525–526 [Chinese court]
V. Liščák: "Excavations at Yongcheng Site, 1959–1986," *Early China* 13 (1988), pp. 274–287

4. CHEWING GUM

*Tannahill 1989, p. 299 [prostitutes]
M. Hooper: *Everyday Inventions* (Angus & Robertson, 1972), p. 39 [Maya]
D. Adamson: *The Ruins of Time* (Allen & Unwin, 1975), p. 191 [chicle hunters]

5. TEA, COFFEE AND CHOCOLATE

J. Blofeld: *The Chinese Art of Tea* (Allen & Unwin, 1985)
P. Griffiths: *The History of the Indian Tea Industry* (Weidenfeld & Nicolson, 1967), p. 5 [Lu Yü]
R. Whitlock: *Everyday Life of the Maya* (Batsford, 1976), pp. 36 & 61
R. E. W. Adams: "Rió Azul," *National Geographic* 169:4 (April 1986), pp. 420–451
C. Roden: *Coffee* (Faber & Faber, 1977), pp. 19–20

6. WINE, BEER AND BREWING

H. Johnson: *The Story of Wine* (Mitchell Beazley, 1989), pp. 17–18 [Georgia], 44 [kottabos], 40 [Vix]

Genesis 9:20

M. Stead: *Egyptian Life* (British Museum, 1986), pp. 30–31

*Forbes III, pp. 74–78 [Egyptian wine], 70–72 [Egyptian beer], 130 [Julian], 132–133 [Bavaria]

*Diodorus V:xxvi,3

R. Rolle: *The World of the Scythians* (Batsford, 1989), p. 8

M. Yazici & C. S. Lightfoot: "Two Roman Samovars *(authepsae)* from Caesarea in Cappadocia," *Antiquity* 63 (1989), pp. 343–349

M. Grant: *The Visible Past* (Weidenfeld & Nicolson, 1990), p. 90 [Settefinestre]

*Jackson 1988, pp. 36–37 & 178 [lead]

*Tannahill 1988, pp. 48–52 [beer in Iraq & Egypt]

K. Gill: "Brewers Seek Ancient Secret of Pharaoh," *Times* 23rd Feb. 1990

D. Samuel: "Ancient Egyptian Cereal Processing: Beyond the Artistic Record," *Cambridge Archaeological Journal* 3:2 (1993), pp. 276–283

*Temple 1986, pp. 77–78 [sake]

Anon.: "The World's Oldest Brew", *Archaeology Today* 9 (1988), pp. 20–21; "A Drop of the Hard Stuff," *Glasgow Herald* 13th Nov. 1987 [Rhum]

N. Hammond: *Ancient Maya Civilization* (Cambridge University Press, 1982), p. 285

BOX ON DISTILLING

*Needham V:4, pp. 122 [Europe], 108–109 [Mexico]

*Temple 1986, pp. 101–103 [China]

7. DRUGS

R. Lewin: "Stone Age Psychedelia," *New Scientist* No. 1772 (8th June 1991), pp. 30–34

U. Jones: "Metates and Hallucinogens in Costa Rica," *Papers from the Institute of Archaeology* 2 (1991), pp. 29–34

R. E. Schultes: "Antiquity of the Use of New World Hallucinogens," *Archeomaterials* 2 (1987), pp. 59–72

J. Soustelle: *The Daily Life of the Aztecs* (Allen & Unwin, 1961), p. 155 [de Sahagún]

I. Cameron: *Kingdom of the Sun God* (Random Century, 1990), p. 206 [coca]

*Coe, Snow & Benson 1986, pp. 158 & 187–189 [coca]

S. Connor: "Cocaine 'first used 2,000 years ago,' " *Independent on Sunday* 16th June, 1991

G. Andrews & S. Vinkenoog (eds.): *The Book of Grass* (Penguin, 1972), pp. 75–95 [Amerindians], 59 [Washington], 17–27 [India], 43–45 [assassins]

*Herodotus IV:74–75

*Rudenko 1970, p. 285 [Pazyryk]

*Needham V:2, pp. 150–153 [taoists]

R. E. M. Hedges, R. A. Housley, I. A. Law, C. Perry & J. A. J. Gowlett: "Radiocarbon Dates from the Oxford AMS System: Archaeometry Datelist 6," *Archaeometry* 29 (1987), pp. 289–306 [Lalibela Cave]

R. Merrillees: "Opium Trade in the Late Bronze Age Levant," *Antiquity* 36 (1962), pp. 287–292

V. Karageorghis: "A Twelfth-century Opium Pipe from Kition," *Antiquity* 50 (1976), pp. 125–129

J. Renfrew: *Palaeoethnobotany* (Methuen, 1973), pp. 161–162 [Swiss poppies]

M. Grant: *The Visible Past* (Weidenfeld & Nicolson, 1990), p. 114 [Aphrodisias]

Homer: *Odyssey* IV:219–234

*Temple 1984, pp. 49–57 [Greek oracles]

*Pliny XXV:35–37 & 106

8. TOBACCO AND PIPES

J. Wilbert: *Tobacco and Shamanism in South America* (Yale University Press, 1987), pp. 9–10 [de las Casas], xvii [Marajó]

R. Whitlock: *Everyday Life of the Maya* (Batsford, 1976), p. 76

*Bray 1968, p. 42 [Aztecs]

*Coe, Snow & Benson 1986, pp. 50–53 [Hopewell]

L. Balout & C. Roubet (eds.): *La Momie de Ramsès II* (Éditions Recherche sur les Civilizations, 1985), pp. 182–191, 383

CHAPTER 8: URBAN LIFE

INTRODUCTION

J. Mellaart: "Early Urban Communities in the Near East c. 9000–3400 B.C.," in P. R. S. Moorey (ed.): *The Origins of Civilization* (Clarendon Press, 1979), pp. 22–33

*Mellaart 1967 [Çatal Hüyük]

H. Frankfort: *The Art and Architecture of the Ancient Orient* (Penguin, 1969), pp. 6, 51–52 [ziggurats]

*Herodotus I:178

V. Liščák: "Excavations at the Yongchang site, 1959–1986," *Early China* 13 (1988), pp. 274–287

A. Badawy: *A History of Egyptian Architecture* (University of California Press, 1966), p. 38 [El Lahun]

B. & R. Allchin: *The Rise of Civilization in India and Pakistan* (Cambridge University Press, 1982), p. 150

J. Cramer: *The World's Police* (Cassell, 1964), pp. 5–9 [Egypt and Rome]

A. L. Basham: *The Wonder that was India* (Sidgwick & Jackson, 1967), p. 105

*Neuburger 1930, pp. 244–245 [street lighting]

*Temple 1986, p. 81 [street lighting]

1. SEWERS

*Childe 1931, pp. 18–19 [Skara Brae]

*Jansen 1989 [Indus Valley]

*Wright 1980, pp. 6 [Knossos], 31 [monasteries]

*Forbes I, pp. 77–80 [Mesopotamia]

E. J. Owens: "The Koprologoi at Athens in the Fifth and Fourth Centuries B.C.," *The Classical Quarterly* 33 (1983), pp. 44–50

*Neuburger 1930, p. 444 [*Cloaca Maxima*]

2. PIPES AND PLUMBING

*Jansen 1989 [Indus Valley]

*Forbes I, p. 153 [Abusir]

*Wright 1980, p. 6 [Knossos]

*White 1984, p. 162 [Pergamum]

*Landels 1971, pp. 42–43 [Roman piping]

*Jackson 1988, pp. 44–45 [Frontinus]

A. H. M. Jones: *The Later Roman Empire* (Blackwell, 1964), p. 696 [Constantinople]

3. APARTMENT BUILDINGS

*Vitruvius II:viii & xvii

*White 1984, p. 83 [*pozzolana*]

A. Boëthius & J. B. Ward-Perkins: *Etruscan and Roman Architecture* (Harmondsworth, 1979), pp. 113–114

*Livy XXI:62

J. J. Deiss: *Herculaneum* (Thames & Hudson, 1985), pp. 114–115

S. Arenson: *The Encircled Sea* (Constable, 1990), p. 71 [Acre]

*Coe, Snow & Benson 1986, p. 79 [Pueblo Bonito]

4. FIRE ENGINES

*Landels 1978, pp. 75–77 [Heron]
S. Goodenough: *Fire! The Story of the Fire Engine* (Orbis, 1978), pp. 23–29 [Rome]
Pliny the Younger: *Letters* X:33–34

5. BANKS

G. Roux: *Ancient Iraq* (Penguin, 1966), p. 194 [prostitutes]
F. J. Frost: *Greek Society* (Heath, 1980), pp. 75–76
A. T. Olmstead: *History of the Persian Empire* (University of Chicago Press, 1948), pp. 83–85, 192, 299, 358 [Babylonians, Jews]
A. K. Bowman: *Egypt After the Pharaohs* (British Museum, 1986), p. 113
*al-Hassan & Hill 1986, pp. 17–18 [Islamic]

6. COINS AND PAPER MONEY

K-C. Chang: *Shang Civilization* (Yale University Press, 1980), pp. 153–154
R. A. G. Carson: *Coins* (Hutchinson, 1970) pp. 539 [Tsu], 5–7 [Lydia]
J. Dayton: "Money in the East Before Coinage," *Berytus* 23 (1974), pp. 41–52 [money rings]
*Sarton I, p. 79 [Mesopotamia]
A. L. Basham: *The Wonder that was India* (Sidgwick & Jackson, 1967), p. 223
*Needham I, pp. 246–247 [China]
B. Hobson & R. Obojski: *Illustrated Encyclopaedia of World Coins* (Robert Hale, 1970), p. 110 [Aegina]
Y. Yap & A. Cotterell: *The Early Civilizations of China* (Weidenfeld & Nicolson, 1975), p. 86 [Wu-ti]
*Temple 1986, pp. 117–119 [flying money]
W. Dodsworth: *History of Banking in All the Leading Countries* IV (The Journal of Commerce and Commercial Bulletin, 1876), pp. 153–154 [Venice]

CHAPTER 9: WORKING THE LAND

INTRODUCTION

*Klein 1989, pp. 395 [Australian boomerang], 375–377 [fishing hooks]
P. Valde-Nowak, A. Nadachówski & M. Wodan: "Upper Palaeolithic boomerang made of a mammoth tusk in south Poland," *Nature* 329 (1987), pp. 436–438
T. Champion, C. Gamble, S. Shennan & A. Whittle: *Prehistoric Europe* (Academic Press, 1984), p. 39 [Parpalló]
D. Collon: "Ivory," *Iraq* 39 (1977), pp. 219–222 [elephant parks in Syria]
D. D. Luckenbill: *Ancient Records of Assyria and Babylonia* I (Chicago University Press, 1926), p. 86
C. Scarre (ed.): *Past Worlds: the Times Atlas of Archaeology* (1988), p. 80 [first farming]
A. G. Sherratt: "Plough and Pastoralism: Aspects of the Secondary Products Revolution," in I. Hodder, G. Isaac & N. Hammond (eds.): *Pattern of the Past* (Cambridge University Press, 1981), pp. 261–305
*Wilkinson 1854, i, p. 44 [monkeys]
*Sarton I, pp. 83–85
J. A. J. Gowlett: *Ascent to Civilization* (Collins, 1984), p. 175 [Geokysur]
*Hill 1984, pp. 50 ["Dam of the Pagans"], 28 [Sennacherib], 52 [Nabataeans], 53 [Subiaco]
J. Hawkes: *The First Great Civilizations* (Hutchinson, 1973), p. 93 [al-Gharrif]
J. Merson: *Roads to Xanadu* (Weidenfeld & Nicolson, 1989), p. 20 [Zheng Gou]
R. W. Clark: *Works of Man* (Century, 1985), p. 28 [*Piscina Mirabilis*]
Zeylanicus: *Ceylon* (Elek, 1970), pp. 44–54
J. Ellis Jones: "The Silver Mines of Athens," in B. Cunliffe (ed.): *Origins* (BBC, 1987), pp. 108–120
*Needham IV:3, p. 671 [Mêng Ching]

1. THE REAPING MACHINE

*Pliny XVIII:296

*White 1984, pp. 29–30 [Roman reaper]

*Needham VI:2, pp. 342–343 [push-sycthe]

E. A. Thompson, *A Roman Reformer and Inventor* (Clarendon Press, 1952), p. 80 [Palladius, Loudon & Ridley]

2. WATERMILLS AND WINDMILLS

*Vitruvius X:v

T. S. Reynolds: *Stronger Than a Hundred Men* (Johns Hopkins University Press, 1983), pp. 36 [Venafro], 18 [Denmark], 39 [Barbégal], 32 [Belisarius]

Straabo: *Geography* XII:556

*Needham IV:2, pp. 366–370 [China]

*al-Hassan & Hill 1986, p. 54 [Mosul]

D. R. Hill: "Mechanical Engineering in the Medieval Near East," *Scientific American* 264:5 (1991), pp. 64–69

E. J. Kealey: *Harvesting the Air* (University of California Press, 1987), p. 51 [England]

3. PESTICIDES

*Needham VI:1, pp. 473–548 [China]

*Pliny XVII:45 & 74; XIX: 34 & 58

*Temple 1986, pp. 94–96 [ants]

4. BEEKEEPING

E. Crane: *The Archaeology of Beekeeping* (Duckworth, 1983), pp. 21 [Spain], 34–42 [Egypt], 63 [Maya]

*Saggs 1989, pp. 215–216 [Near East]

5. FISH AND OYSTER FARMS

*Wilkinson i, 238 [Egypt]

*Diodorus XI:xxv,4 [Gelon]

*Pliny IX:59, 106–123, 167–173

N. Kokkinos: *Antonia Augusta: Portrait of a Great Roman Lady* (Routledge, 1992), p. 153

Columella: *On Agriculture* VIII:xvi-xvii

H. Higginbotham: "Roman Fishtanks of the Late Republic and Early Empire in Italy," *American Journal of Archaeology* 94 (1990), p. 304

C. Davaras: "Rock-cut Fish Tanks in Eastern Crete," *Annual of the British School at Athens* 69 (1974), pp. 87–93

K. Nicolaou & A. Flinder: "Ancient Fish-tanks at Lapithos, Cyprus," *International Journal of Nautical Archaeology and Underwater Exploration* 5:2 (1976), pp. 133–141

A. Flinder: *Secrets of the Bible Seas* (Severn House: London 1985), pp. 113–123 [Caesarea]

J. G. D. Clark: *Symbols of Excellence* (Cambridge University Press, 1986), p. 80 [Caesar]

*Needham IV:3, pp. 674–676 [eastern pearl cultivation]

6. DRILLING AND MINING

*al-Hassan & Hill 1986, p. 145 [Baku]

H. U. Vogel: "The Great Well of China," *Scientific American* 268:6 (June, 1993), pp. 86–91

C. Scarre (ed.): *Past Worlds: the Times Atlas of Archaeology* (1988), p. 68 [Koonalda]

R. Shepherd: *Prehistoric Mining and Allied Industries* (Academic Press, 1980), pp. 23–107 [European flint]

R. F. Tylecote: *The Early History of Metallurgy in Europe* (Longman, 1987), p. 33 [Rudna Glava]

B. Rothenberg: "Timna," in M. Avi-Yonah & E. Stern (eds.): *Encyclopedia of Archaeological Excavations in the Holy Land* IV (Oxford University Press, 1978), pp. 1184–1203 [quote p. 1202]

*James *et al.* 1991, pp. 201–203 [Solomon's mines]

A. R. Millard: "Does the Bible Exaggerate King Solomon's Wealth?" *Biblical Archaeology Review* May/June 1989, pp. 20–34 [Egyptian gold]

*Diodorus III:xii–xiv

*Chinese Academy 1983, pp. 268–269 [coal]

S. S. Frere: *Britannia* (Routledge, 1987), p. 288 [Roman coal]

J. Ellis Jones: "The Silver Mines of Athens," in B. Cunliffe (ed.): *Origins* (BBC, 1987), pp. 108–120

S. J. Keay: *Roman Spain* (British Museum, 1988), pp. 63–64

*Sarton II, p. 377

G. E. Sandström: *The History of Tunnelling* (Barrie & Rockliff, 1963), pp. 41 [Hadrian], 28–30 [Hannibal]

*Neuburger 1930, pp. 461–462 [Hannibal]

*Livy XXI:37

*Pliny XXXIII:21

*Vitruvius VIII:iii,19

J. F. Healy: *Mining and Metallurgy in the Greek and Roman World* (Thames & Hudson, 1978), pp. 84–85

I. Beit-Areih: "Serabit el-Khadim: New Metallurgical and Chronological Aspects," *Levant* 17 (1985), pp. 89–116 [mirrors, quote p. 103]

I. Beit-Areih: "Mirrors as Mining Tools: A Response," *Levant* 22 (1990), p. 163

7.   TUNNELING

P. Beaver: *A History of Tunnels* (Peter Davies, 1972), pp. 22–29 [Rome, Egypt, India]

J. C. Harle: *The Art and Architecture of the Indian Subcontinent* (Penguin, 1986), pp. 43–57, 118–135

P. Brown: *Indian Architecture* Vol. 1 (2nd edn., Bombay, 1942), pp. 23–32 [quote p. 25]

H. E. Wulff: "The Qanats of Iran," *Scientific American* 218:4 (April 1968), pp. 94–105

*Herodotus III:60

J. Goodfield: "The Tunnel of Eupalinus," *Scientific American* 210:6 (June 1961), pp. 104–112

*Sarton I, p. 192

*Neuburger 1930, pp. 420–422 [Eupalinus], 396 [diopter]

Virgil: *Aeneid* VI

R. F. Paget: *In the Footsteps of Orpheus: The Discovery of the Ancient Greek Underworld* (Robert Hale, 1967)

*Temple 1984, pp. 3–31

*Vitruvius VIII:i

CHAPTER 10: HOUSE AND HOME

INTRODUCTION

*Klein 1989, pp. 126–127 [earliest fire]

L. Wright: *Home Fires Burning* (Routledge, 1964), pp. 19 & 62 [chimneys]

*Needham IV:2, pp. 150–151 [Ting Huan]

S. A. de Beaune & R. White: "Ice Age Lamps," *Scientific American* 266:3 (March 1993), pp. 74–79

*Manniche 1989, pp. 34–36 [Egypt]

*Pliny XXXI:39

*Needham V:7, p. 75 [China]

*Macnamara 1973, p. 111 [Etruscans]

*Temple 1986, pp. 119–120 [Chinese lamps], 78–81 [gas], 98–99 [matches]

P. V. Glob: *The Mound People* (Faber & Faber, 1974), pp. 134 [fire-making kit], 93 [stools]

*Childe 1931 [Skara Brae]

M. Stead: *Egyptian Life* (British Museum, 1986), pp. 14–15 [furniture]

T. B. L. Webster: *Everyday Life in Classical Athens* (Batsford, 1969), p. 36

J. M. Cook: *The Persian Empire* (Dent, 1983), p. 74 [Artystone]

Xenophon: *Cyropaedia* VIII:16

L. Harrow: *From the Lands of Sultan and Shah* (Scorpion, 1987), p. 13 [earlier carpets]

★Rudenko 1970, pp. 298–304 [Pazyryk]

S. Dalley: "Ancient Assyrian Textiles and the Origin of Carpet Design," *Iran* 29 (1991), pp. 117–135

F. R. Cowell: *The Garden as a Fine Art* (Weidenfeld & Nicolson, 1978), p. 67 [Chosroes]

★Needham V:1, pp. 118–120 [wallpaper]

R. L. Hills: *Papermaking in Britain 1488-1988* (Athlone Press, 1988), p. 81 [Louis XI]

J. Simpson: *The Viking World* (Batsford, 1980), p. 55

1. MAMMOTH-BONE HOUSES

J.-P. Lhomme & S. Maury: "Les battiseurs de la préhistoire," *Archéologia* 250 (1989), pp. 34–43

J. Jelínek: *The Evolution of Man* (Hamlyn, 1975), pp. 248, 256

2. CATS AND DOGS

S. J. M. Davis: *The Archaeology of Animals* (Batsford, 1987), pp. 137–148 [Palestine], 133–134 [Khirokitia]

J. Clutton-Brock & N. Noe-Nygaard: "New Osteological and C-Isotope Evidence on Mesolithic Dogs," *Journal of Archaeological Science* 17 (1990), pp. 643–663 [Seamer Carr]

B. Fagan 1985: *New Treasures of the Past* (Windward, 1985), p. 66 [Koster]

R. & J. Janssen: *Egyptian Household Animals* (Shire, 1989), pp. 9–19

J. M. C. Toynbee: *Animals in Roman Life and Art* (Thames & Hudson, 1973), pp. 103–107

J. Clutton-Brock: *The British Museum Book of Cats* (British Museum, 1988)

★Diodorus I:1xxxiii,8–9

★Herodotus II:66

3. LAVATORIES

★Childe 1931, p. 18 [Skara Brae]

★Jansen 1989 [Mohenjo-Daro]

★Saggs 1988, p. 169 [Eshnunna]

★Wright 1980, pp. 11 [El-Amarna], 6, 14 [Rome]

T. G. H. James: *Pharaoh's People* (Oxford University Press, 1985), p. 227 [Kha]

A. Evans: *Palace of Minos* I (Macmillan, 1921), p. 230

★Jackson 1988, pp. 50–53 [Rome]

BOX ON TOILET PAPER

★Temple 1986, pp. 83–84

4. SAUNAS

A. S. Ingstad: *The Discovery of a Norse Settlement in America: Excavations at L'Anse aux Meadows, Newfoundland, 1961–1968* (Oslo: Universitetsförlaget), pp. 217–219

L. Barfield & M. Hodder: "Burnt Mounds as Saunas, and the Prehistory of Bathing," *Antiquity* 61 (1987), pp. 370–379

★Herodotus IV:75

V. Buckley (ed.), 1990: *Burnt Offerings: International Contributions to Burnt Mound Archaeology* (Dublin: Wordwell)

D. A. O'Driscoeil: "Burnt Mounds: Cooking or Bathing?" *Antiquity* 62 (1988), pp. 671–680

J. W. Hedges: *Isbister Chambered Tomb and Liddle Burnt Mound* (Orkney: The Isbister and Liddle Trust, 1985)

Linton Sattherthwaite Jr., 1952. *Piedras Negras Archaeology: Architecture. Part V: Sweathouses* (Pennsylvania University Museum) [Mexico; Clavijero]

5. BATHS

Plato: *Critias* 117B
B. & R. Allchin: *The Rise of Civilization in India and Pakistan* (Cambridge University Press, 1982), p. 176–181
T. G. H. James: *Pharaoh's People* (Oxford University Press, 1985), pp. 225–227
C. Singer, E. J. Holmyard & A. R. Hall: *A History of Technology* I (Clarendon Press, 1954), p. 255 [Mesopotamia]
J. W. Graham: *The Palaces of Crete* (Princeton University Press, 1962), pp. 99–108, 214, 220
*Neuburger 1930, pp. 440–443 [Greeks], 259–260, 366–376 [Romans]
*Vitruvius V:x [Romans]
L. Barfield & M. Hodder: "Burnt Mounds as Saunas, and the Prehistory of Bathing," *Antiquity* 61 (1987), p. 375 [Byzantium]
*Daumas 1969, pp. 343, 371 [Arabs]
M. Magnusson, 1972: *Viking Exploration Westwards* (Bodley Head), pp. 109–110 [Snorri]

BOX ON SHOWERS

*Neuburger 1930, pp. 441–442

6. CENTRAL HEATING

*Needham IV:3, pp. 134–135 [China]
J. DeLaine: "Recent Research on Roman Baths," *Journal of Roman Archaeology* 1 (1988), pp. 11–32 [Orata]
*Neuburger 1930, pp. 259–266 [Rome]
*Forbes I, pp. 83–85 [Constantinople]

7. GLASS WINDOWS

S. H. Vose: *Glass* (Collins, 1980), pp. 26–27 [early glass]
S. Frank: *Glass and Archaeology* (Academic Press, 1982), pp. 19–20 [Jerusalem & northern Europe], 22 [stained glass]
*Forbes V, pp. 181–183 [Pompeii]
A. K. Bowman: *Egypt After the Pharaohs* (British Museum, 1986), p. 39 [Alexandria]
D. B. Harden: "Domestic Window Glass, Roman, Saxon, Medieval," in E. M. Jope (ed.): *Studies in Building History* (Odhams Press, 1961), pp. 39–63 [crown glass]
T. S. Reynolds: *Stronger Than a Hundred Men* (Johns Hopkins University Press, 1983), p. 34 [unbreakable glass]

8. KEYS AND LOCKS

Homer: *Odyssey* XXI:5–7 & 46–49
*Neuburger 1930, pp. 336–339 [Greece & Rome]
*Daumas 1969, pp. 186–187 [early tumblers], 243 [Roman]
*Needham IV:2, pp. 237–242 [China]
G. de la Bédoyère: *The Finds of Roman Britain* (Batsford, 1989), pp. 113–114 [rotary lock]
W. H. Manning: *Catalogue of the Romano-British Iron Tools, Fittings and Weapons in the British Museum* (1985), p. 95 [padlocks]

BOX ON THE MYSTERIOUS EGYPTIAN KEY

A. R. David: *The Pyramid Builders of Ancient Egypt* (Routledge, 1986), pp. 150–151
H. Anagnostou: "The Mysterious Artifacts of Molino della Badia," *Journal of Mediterranean Anthropology and Archaeology* 1 (1981), pp. 157–166

CHAPTER 11: COMMUNICATIONS

INTRODUCTION

*Bahn & Vertut 1988, p. 78 [Azilian pebbles]
D. Schmandt-Besserat: *Before Writing. Vol 1: From Counting to Cuneiform* (University of Texas Press, 1992) [tokens]
R. Harris: *The Origin of Writing* (Duckworth, 1986)
*Kramer 1963, pp. 302–307 [cuneiform writing]
*Saggs 1988, p. 50 [Enheduanna]
*Daumas 1969, pp. 176–178 [papyrus]
*Bray 1968, pp. 42 [Aztec paper], 62 [Aztec schools]
*Temple 1986, pp. 81–84 [Chinese paper]
*Saggs 1989, p. 101 [scribes and Shulgi]
*Woolley 1976, p. 33 [Sumerian school]
*Ronan 1983, pp. 95–96 [Athenian Academy]
*Needham 1970, p. 382 [Imperial University]
M. Edwardes: *Everyday Life in Early India* (Batsford, 1969), p. 137 [Nalanda]
*Sarton I, pp. 157–158 [Assurbanipal's library]
L. Canfora: *The Vanished Library* (Hutchinson Radius, 1989) [Alexandria]

1. CALENDARS

A. Aveni: *Empires of Time* (I. B. Tauris, 1990)
A. Marshack: *The Roots of Civilization* (Weidenfeld & Nicolson, 1972)
*Kramer 1963, p. 91 [Sumerians]
D. C. Heggie: *Megalithic Science* (Thames & Hudson, 1981)
A. Cotterell: *The Encyclopedia of Ancient Civilizations* (Rainbird, 1980), p. 309 [oracle bones]
*Ronan 1983, pp. 52 [Zapotecs], 54–55 [Maya]
*James *et al.* 1991, pp. 225–228, 378 [Egyptian calendar]
J. V. P. D. Balsdon: *Life and Leisure in Ancient Rome* (Bodley Head, 1964, p. 58 [Rome]
*Sarton II, pp. 320–326 [Julian calendar]

BOX ON DAYS OF THE WEEK

*Sarton II, pp. 326–335
A. Aveni: *Empires of Time* (I. B. Tauris, 1990), pp. 100–101 [biorhythms]

2. THE ALPHABET

R. Harris: *The Origin of Writing* (Duckworth, 1986), p. 29 [quote from Moorhouse]
J. Naveh: *Early History of the Alphabet* (The Hebrew University, Jerusalem, 1982)
*James *et al.* 1991, pp. 81–85 [origins of Greek alphabet]
I. Beit-Areih: "Serabit el-Khadim: New Metallurgical and Chronological Aspects," *Levant* 17 (1985), pp. 89–116 [Sinai inscriptions]
C. H. Gordon: "The Accidental Invention of the Phonemic Alphabet," *Journal of Near Eastern Studies* 29 (1970), pp. 193–197
H. A. Moran & D. H. Kelley: *The Alphabet and the Ancient Calendar Signs* (Palo Alto, California: Daily Press, 1969)

3. CODES AND CIPHERS

C. H. Gordon: *Riddles in History* (Arthur Baker, 1974), pp. 53–69 [Near East]
S. L. MacGregor Mathers: *The Kabbalah Unveiled* (Routledge & Kegan Paul, 1926), pp. 9–11 [*temurah*]
Plutarch: *Life of Lysander* 19

D. Kahn: *The Codebreakers* (Weidenfeld & Nicolson, 1967), p. 71 [China]
P. Way: *Codes and Ciphers* (Aldus, 1977), pp. 18–19 [Alberti], 138–140 [Voynich]

BOX ON SHORTHAND

F. G. Kenyon: "Tachygraphy," in *Oxford Classical Dictionary* (1949), p. 876
*Needham 1970, p. 204 ["grass writing"]

4. BOOKS AND PRINTING

T. McArthur: *Worlds of Reference* (Cambridge University Press, 1986), p. 29 [Rome]
A. J. Mills: *A Penguin in the Sahara* (Royal Ontario Museum, 1990) [Dakhleh]
*Needham V:1, pp. 29 [bamboo], 149 [7th-century A.D. India]
*Bray 1968, p. 92 [Mexico]
M. Sugimoto & D. L. Swain: *Science and Culture in Traditional Japan* (Charles E. Tuttle, 1989), p. 184
S. Jeon: *Science and Technology in Korea* (MIT Press, 1974), p. 168
*Temple 1986, pp. 110–116 [China]

BOX ON THE EARLIEST MOVABLE TYPE

R. W. Hutchinson: *Prehistoric Crete* (Penguin, 1962), pp. 66–67

5. ENCYCLOPEDIAS

R. Collison: *Encyclopaedias: Their History throughout the Ages* (Hafner, 1966), pp. 22–26 [Speusippus]
*Needham VI:1, p. 191 [*Er Ya*]; III, pp. 193–196 [Spring and Autumn Annals]
T. McArthur: *Worlds of Reference* (Cambridge University Press, 1986), pp. 41–43 [Pliny]
J. Fryer: *The Great Wall of China* (New English Library, 1975), p. 140 [Chu Ti]

6. POSTAL SYSTEMS

C. H. Scheele: *A Short History of the Mail Service* (Smithsonian Institution Press, 1970), pp. 7–23
C. B. F. Walker: *Cuneiform* (British Museum Press, 1987), pp. 26–27
*Hill 1984, pp. 77, 86–87
*Herodotus VIII:98

7. PIGEON POST

B. Laufer: *The Prehistory of Aviation* (Field Museum of Natural History, Chicago, 1928), pp. 71–87
*Pliny X:52 & 110
*Hill 1984, p. 87
*Temple 1984, pp. 32–38 [quote p. 32]
*Herodotus I:47, 50–51 [Croesus], 159 [Aristodicus]; II:55 [doves story]

8. TELEGRAPHY

G. Wilson: *The Old Telegraphs* (Phillimore, 1976), pp. 1–4
Aeschylus: *Agamemnon* 1–30
*Forbes VI, pp. 171–183 [Greek & Roman systems]
Polybius: *Histories* X:42–47
G. Webster: *The Roman Imperial Army (of the First and Second Centuries A.D.)* (Adam & Charles Black, 1969), pp. 246–248
*Daumas 1969, pp. 383 [Byzantium], 242 [shouting telegraph]

N. Ohler, 1989. *The Medieval Traveller* (The Boydell Press), pp. 99–100 [smoke signals and Arab beacons]
★Caesar VII:1

CHAPTER 12: SPORT AND LEISURE

INTRODUCTION

M. Duchesne-Guillemin: "Music in Ancient Mesopotamia and Egypt," *World Archaeology* 12:3 (1981), pp. 287–297
★Pirazzoli-t'Serstevens 1982, pp. 106 & 132 [Bureau of Music]
★Olivová 1984, p. 156 [Etruscans]
F. R. Cowell: *Everyday Life in Ancient Rome* (Batsford, 1961), p. 158 [dancing]
J. Pearson: *Arena* (Thames & Hudson, 1973) [gladiators & beast fights]
Homer: *Iliad* XXIII
Pausanias: *Guide to Greece* VI:xx,10–14
★Harris 1972, pp. 167–189 [chariot racing]
A. Cameron: *Circus Factions* (Clarendon Press, 1976)
M. M. Ahsan: *Social Life Under the Abbasids* (Longman, 1979), pp. 243–250 [horse racing]
Athenaeus: *Deipnosophistai* 15:2–7
★Herodotus I:94
A. L. Basham: *The Wonder That Was India* (Sidgwick & Jackson, 1967), pp. 39 & 209 [dice]
K. Branigan: *Roman Britain* (Reader's Digest, 1980), p. 107 [dice]
★Needham IV:1, p. 329 [dominoes]
A. R. David: *The Pyramid Builders of Ancient Egypt* (Routledge, 1986), p. 163 [Kahun]
R. Castleden: *Minoans: Life in Bronze Age Crete* (Routledge, 1990), pp. 131 & 191–192 [swings]
M. & H. R. Chubb: *One Third of Our Time: An Introduction to Recreation Behavior and Resources* (John Wiley & Sons, 1981), p. 15 [Greek leisure]

1. BULLFIGHTING

R. Whitlock: *Bulls Through the Ages* (Lutterworth Press, 1977), pp. 50–56 [Crete]
A. Evans: *The Palace of Minos* III (Macmillan, 1930), pp. 209–232
★Olivová 1984, p. 180 [Rome]
G. Martin: *Bullfight* (Blackwell, 1988), pp. 52–54 [Spain]

2. THE ORIGINAL OLYMPIC GAMES

★Harris 1972, pp. 18–42 [origin, events & decline]
★Olivová 1984, pp. 153 [Leonidas], 119 [Milo]

3. BALL GAMES

★Wilkinson i, p. 200 [Egypt]
★Harris 1972, pp. 77–87 [Greece]
A. Cotterell: *China: a Concise Cultural History* (John Murray, 1988), p. 165 [soccer]
R. Brasch: *How Did Sports Begin?* (Longmans, 1972), pp. 159–164 [lacrosse], 171 [polo], 97–98 [golf], 15–16 [badminton], 226 [tennis]
J. Schauble: "Worrying News from China: It may Have Invented Golf," *London Independent on Sunday*, 4th Aug. 1991
★Coe, Snow & Benson 1986, pp. 100 & 108 [sacred ballgame]
J. A. Sabloff: *The Cities of Ancient Mexico* (Thames & Hudson, 1989), p. 13

4. GARDENS AND GARDENING

J. Harvey: *Medieval Gardens* (Collins, 1985) [English gardening]

D. Clifford: *A History of Garden Design* (Faber & Faber, 1966), pp. 24 [Rome], 46 [Tiglath-pileser]

E. Hyams: *A History of Gardens and Gardening* (Dent, 1971), pp. 95 [forcing frames], 16 [Hatshepsut & Ramesses III], 23 [Chang Seng-yu], 25–27 [Hui Tsang], 66–69 [Japan]

K. L. Gleason: "Garden Excavations at the Herodian Winter Palace in Jericho, 1985–7," *Bulletin of the Anglo-Israel Society* 7 (1987), pp. 21–39

★Pliny XII:13 [topiary], XII:5 [potting trees], XII:71 [window boxes]

★Manniche 1989, pp. 7–32 [Egypt]

R. & J. Janssen: *Egyptian Household Animals* (Shire, 1989), p. 57 [Thutmose III]

I. Finkel: "The Hanging Gardens of Babylon," in P. A. Clayton & M. J. Price (eds.): *The Seven Wonders of the Ancient World* (Routledge, 1988), pp. 38–58 [Sennacherib & Hanging Gardens]

Xenophon: *The Economist* IV:20–24

C. Thacker: *The History of Gardens* (Croom Helm, 1979), pp. 16 [Megasthenes], 55 [chrysanthemum]

M. Edwardes: *Everyday Life in Early India* (Batsford, 1969), p. 116 [Candragupta II]

A. L. Basham: *The Wonder That Was India* (Sidgwick & Jackson, 1967), p. 204 [Asoka]

H. Yang: *The Classical Gardens of China* (Van Nostrand Reinhold, 1982), p. 92 [parks]

★Bray 1968, pp. 106 [Texcotzinco], 114 [Huaxtepec & *chinampas*]

A. Kendall: *Everyday Life of the Incas* (Batsford, 1973), p. 133

★Diodorus II:x

D. W. W. Stevenson: "A Proposal for the Irrigation of the Hanging Gardens of Babylon," *Iraq* 54 (1992), pp. 35–55

5. ZOOS

J. Marlowe: *The Golden Age of Alexandria* (Gollancz, 1971), pp. 79–80

R. & J. Janssen: *Egyptian Household Animals* (Shire, 1989), p. 57 [Thutmose III]

D. Rybot: *It Began Before Noah* (Michael Joseph, 1972), pp. 20 [King Wen]

C. Thacker: *The History of Gardens* (Croom Helm, 1979), p. 48 [Hangchou]

J. M. C. Toynbee: *Animals in Roman Life and Art* (Thames & Hudson, 1973), p. 16

Suetonius: *Life of Nero* 31

S. Zuckerman: "The Rise of Zoos and Zoological Societies," in S. Zuckerman (ed.): *Great Zoos of the World* (Weidenfeld & Nicolson, 1980), pp. 3–26 [Europe]

★Bray 1968, p. 106 [Aztecs]

6. THEATERS

A. Pickard-Cambridge: *Dithyramb, Tragedy and Comedy* (Oxford University Press, 1962)

J. M. Walton: *Greek Theatre Practice* (Greenwood Press, 1980)

N. G. L. Hammond: *The Classical Age of Greece* (Weidenfeld & Nicolson, 1975), pp. 98–102

P. Montet: *Everyday Life in Egypt* (Edward Arnold, 1958), pp. 294–297

Horace: *Ars Poetica* 275–278

I. M. Diakonoff & N. B. Jankowska: "An Elamite Gilgameš Text from Argistihenele, Urartu (Armarvir-blur), 8th century B.C.," *Zeitschrift für Assyriologie* 80:1 (1990), pp. 102–193

J. W. Graham: *The Palaces of Minoan Crete* (Princeton University Press, 1969), pp. 27, 36, 73–83

★Neuburger 1930, pp. 352–361 [classical theaters]

★Vitruvius V:v-viii

★Brumbaugh 1966, pp. 122–123, 127–129 [Greek stage-gear]

F. P. Richmond, D. L. Swann & P. B. Zarrilli: *Indian Theatre* (University of Hawaii Press, 1990), pp. 25–88

7. MUSICAL INSTRUMENTS

*Bahn & Vertut 1988, pp. 68–69 [flutes & bullroarers]

C. Burney & D. Lang: *The Peoples of the Hills* (Weidenfeld & Nicolson, 1971), p. 263 [Georgia]

*Rolle 1989, p. 95 [Scythians]

A. Lang: *Custom and Myth* (2nd edn., Longmans Green, 1885), pp. 29–44

J. Blades: *Percussion Instruments and Their History* (Faber & Faber, 1975)

*Wilkinson i, 131 [sistrum], 108–126 [Egyptian harps & lyres], 127–128 [Egyptian flutes]

P. Price: *Bells and Man* (Oxford University Press, 1983), pp. 73 [Minoan], 2 [China], 36 [Korea]

H. Frankfort: *The Art and Architecture of the Ancient Orient* (Penguin, 1969). p. 19 & Pl. 11(A) [Bismaya]

S. Sadie (ed.): *New Grove Dictionary of Music* (Macmillan, 1980) Vol. 6, pp. 669 [Egyptian flutes]; Vol. 7, pp 663 [*aulos*]

P. F. Healy: "Music of the Maya", *Archaeology* 41 (1988), pp. 24–31 [ocarina]

B. Cotterell & J. Kamminga: *Mechanics of Pre-Industrial Technology* (Cambridge University Press, 1990), p. 283 [panpipes]

*Pirazzolli-t'Serstevens 1982, p. 54 [mouth organ]

P. Holmes & J. M. Coles: "Prehistoric brass instruments," *World Archaeology* 12 (1981), pp. 280–286 [trumpets]

*Macnamara 1973, p. 135 [*cornu*]

*Diodorus V:xxx,3

F. Collinson: *The Bagpipe* (Routledge, 1975)

T. C. Mitchell: "The Music of the Old Testament Reconsidered," *Palestine Excavation Quarterly* July-December 1992, pp. 124–143 [harps, lyres]

New Grove Dictionary of Music Vol. 12, pp. 196–198 [Mesopotamian harps & lyres], Vol. 7, pp. 825–835 [Alaça Hüyük, guitars], Vol. 11, pp. 342–351 [lutes]

8. KEYBOARDS

P. Williams: *A New History of the Organ* (Faber & Faber, 1980), pp. 23–54

*Vitruvius IX:viii; X:viii

9. WRITTEN MUSIC

I. Henderson & D. Wulstan: "Introduction: Ancient Greece," in F. W. Sternfeld (ed.): *Music from the Middle Ages to the Renaissance* (Weidenfeld & Nicolson, 1973), pp. 27–58 [Delphi and Greek notation]

B. Farrington: *Greek Science* (Penguin, 1969), pp. 50–52 [Pythagoras]

*Temple 1986, pp. 199–201 [China]

A. D. Kilmer, R. L. Crocker & R. R. Brown: *Sounds from Silence: Recent Discoveries in Ancient Near Eastern Music* (Berkeley: Bit Enki Publications, 1976)

M. Duchesne-Guillemin: "Music in Ancient Mesopotamia and Egypt," *World Archaeology* 12:3 (1981), pp. 287–297

R. Fink: "The Oldest Song in the World," *Archaeologia Musicalis* 2 (1988), pp. 98–100

10. FIREWORKS

*Temple 1986, pp. 232–238

B. V. Subbarayappa: "Chemical Practice and Alchemy," in D. M. Bose, S. Sen & B. V. Subbarayappa (eds.): *A Concise History of Science in India* (Indian National Science Academy, 1971), pp. 274–349

A. StH. Brock: *A History of Fireworks* (Harrap, 1949), p. 29 [Biringuccio]

11.   MAGIC LANTERNS

*Temple 1986, pp. 87–88
*Needham IV:1, pp. 122–125

12.   PLAYING CARDS

D. Parlett: *The Oxford Guide to Card Games* (Oxford University Press, 1990), pp. 36–43 [Europe], 72 [swindle]
*Temple 1986, pp. 116–117 [China]
G. Beal: *Playing-cards and Their Story* (David & Charles, 1975), p. 10 [Arabs]

al-Hassan, A. & Hill, D. R., *Islamic Technology*. Cambridge University Press, 1986.

Bahn P. & Vertut, J., *Images of the Ice Age*. Facts on File, 1988.

Bray, W., *Everyday Life of the Aztecs*. Batsford, 1968.

Brumbaugh, R. S., *Ancient Greek Gadgets and Machines*. Greenwood Press, 1966.

Bullough, V. L., *Sexual Variance in Society and History*. University of Chicago Press, 1976.

Caesar, J., *The Gallic War*.

Chang, K. C. (ed.), *Food in Chinese Culture*. Yale University Press, 1977.

Childe, V. G., *Skara Brae*. Kegan Paul, 1931.

Chinese Academy of Sciences (Institute for the History of Natural Sciences), *Ancient China's Technology and Science*. Peking: Foreign Languages Press, 1983.

Coe, M. D., Snow, D. & Benson, *Atlas of Ancient America*. Facts on File, 1986.

Corson, R., *Fashions in Makeup*. Peter Owen, 1972.

Daumas, M., *A History of Technology and Invention*. John Murray, 1969.

David, A. R. & Tapp E. (eds.), *Evidence Embalmed*. Manchester University Press, 1984.

Diodorus Siculus: *Library of History*.

Drachmann, A. G., *The Mechanical Technology of Greek and Roman Antiquity*. Copenhagen: Munksgaard, 1963.

R. G. Forbes, *Studies in Ancient Technology*. 2nd ed., Leiden: E. J. Brill, 9 volumes, 1964–1972.

Harris, H. A., *Sport in Greece and Rome*. Thames & Hudson, 1972.

Hawkes, J. & Woolley, L., *Prehistory and the Beginnings of Civilization*. George Allen & Unwin, 1963.

Herodotus: *The Histories*.

Hill, D. R., *A History of Engineering in Classical and Medieval Times*. Croom Helm, 1984.

Jackson, R., *Doctors and Diseases in the Roman Empire*. British Museum, 1988.

James, P., Thorpe, I. J., Kokkinos, N., Morkot, R. & Frankish, J., 1991. *Centuries of Darkness*. Jonathan Cape; Rutgers University Press, 1993.

Jansen, M., "Water Supply and Sewage Disposal at Mohenjo-Daro," *World Archaeology* 21 (1989), pp. 177–192.

Klein, R. G., *The Human Career: Human Biological and Cultural Origins*. University of Chicago Press, 1989.

Kramer, S. N., *The Sumerians*. Chicago University Press, 1963.

Landels. J. G., *Engineering in the Ancient World*. Chatto & Windus, 1978.

Lindsay, J., *Blast-Power and Ballistics*. Frederick Muller, 1974.

Livy: *History of Rome*.

Macnamara, E., *Everyday Life of the Etruscans*. Batsford, 1973.

Majno, G., *The Healing Hand*. Harvard University Press, 1975.

Manniche, L., *An Ancient Egyptian Herbal*. British Museum Press, 1989.

Mellaart, J., *Çatal Hüyük: A Neolithic Town in Anatolia*. Thames & Hudson, 1967.

Needham, J., *Science and Civilization in China*. Cambridge University Press, 6 volumes, 1954 onward.

Needham, J., *Clerks and Craftsmen in China and the West*. Cambridge University Press, 1970.

Neuburger, A., *The Technical Arts and Sciences of the Ancients*. Methuen, 1930.

Olivová, V., *Sports and Games in the Ancient World*. Orbis, 1984.

Pirazzoli-t'Serstevens, M., *The Han Civilization of China*. Phaidon, 1982.

Pliny: *Natural History*.

Rice, T. T., *Everyday Life in Byzantium*. Batsford, 1967.

Rolle, R., *The World of the Scythians*. Batsford, 1989.

Ronan, C., *The Cambridge Illustrated History of the World's Science*. Cambridge University Press, 1983.

Rudenko, S. I., *Frozen Tombs of Siberia*. Dent, 1970.

Saggs, H. W. F., *Civilization Before Greece and Rome*. Batsford, 1989.

Sarton, G., *A History of Science*. The Norton Library, 2 volumes, 1970.

Tannahill, R., *Food in History*. Penguin, 1988.

———. *Sex in History*. Cardinal, 1989.

Temple, R., *Conversations with Eternity*. Rider, 1984.

————. *China: Land of Discovery and Invention*. Patrick Stephens, 1986.

Vitruvius: *On Architecture*.

White, K. D., *Greek and Roman Technology*. Thames & Hudson, 1984.

Wilkinson, J. G., *The Ancient Egyptians*. John Murray, 2 volumes, 1854.

Woolley, L., *Ur Excavations, Vol. VII: The Old Babylonian Period*. British Museum, 1976.

Wright, L., *Clean and Decent*. Routledge, 1980.

Hussein, Saddam, 574
Hussite Wars, 103, *103, 215*
Hyegong, king of Korea, 597
hygiene, 247, 261-63, *262*

Ibn-al-Haytham, 252
Ibn Battuta, 176, 179
Ibn Ben Zara, 69-70
Ibn-Sina, 189-90
icehouses, 321-22, 523
Iceland, 461
"Iceman," 258
Ida, Mount, 159
Igor, prince of Kiev, 227-28
I-Hsing, 126
Ikhnaton, pharaoh of Egypt, 335
Iliad (Homer), 86, 135, 419, 542
Illinois, 437
Incas, 110, 247, 248, 339,
    340-41, 598
  gardens of, 569
  medicine of, 25, 26
  roads of, 52-53
India, 99, 108, 321-22, 394, 438,
    567
  baths in, 460
  currency in, 374
  drugs in, 341-42
  education in, 481
  fireworks in, 615-16
  food in, 303-4, 305, 307-8
  gardens in, 568
  lavatories in, 442
  lighting in, 425
  medicine in, 7, 16-17, 20, *20,*
    21-23, *22,* 23, 28, 37
  military technology in, 212
  music in, 601, 610, 611
  paper in, 479
  personal appearance in, 247,
    248, 256, 264, 265
  printing in, 516
  refrigeration in, 323-24
  sexual matters in, 181, 184,
    189, 193-97
  theater in, 590-91
  tunneling in, 415-16, *415*
  umbrellas in, 296, 297
  see also Indus Valley
    civilization

Indonesia, 305
Indus Valley civilization, 248,
    256, 275
  baths in, *454,* 455
  games in, 546
  lavatories in, 442, *443*
  urban life in, 356-57, *357,*
    359, 361-62
ink wells, 118, 268
inoculations, 11
International Explorers Society
    of Miami, 110-11
Iran, *see* Persia
Iraq and Mesopotamia, 2, *2,*
    58-60, 63, 264, 283, *283,*
    296, 303, 314, 322, 333, 382
  agriculture in, 384
  baths in, 456, 460
  lavatories in, 442-43
  libraries in, 481
  music in, 597, 599, 600-601
  postal system in, *521,* 522,
    527
  transportation in, *51*
  urban life in, 354-55, 360-61,
    373
  writing in, 477-78, 506-7
  see also Assyria; Babylon,
    Babylonia; Baghdad; Sumer,
    Sumeria; Ur
Ireland, 80, 81-82, 336, 449, 599
irrigation, 383-86, 573-74
Isabella, 578
Isaiah, 471
Ishango, 484
Ishigoya Cave, *301*
Isidore, bishop of Seville, 252
Islamic and Arab worlds, 63, 69,
    76, 93, 99, 123, 305, 432, 510,
    535-36
  aphrodisiacs in, 178-79, 181
  automata in, 138-40
  banks in, 372
  baths in, 461
  clocks in, 125-26
  clothing in, 276, 280
  coinage in, 161-62
  drink in, 336
  eunuchs in, 173
  gardens in, 561

hashish in, 343
heating and lighting in, 425
libraries in, 484
mapmaking in, 63, 66
medicine in, 7, 12, 23
military affairs in, 204,
    226-27, 231-33, *232*
music in, 601, 606, 610
oil drilling in, 405
paper in, 479
perfumes in, 267
personal appearance and
    hygiene in, 257-58, 263
playing cards in, 620, *621*
postal system in, 523, 524,
    526-27
sexual matters in, 168, 173,
    186, 189-90, 196
sports and games in, 544, 620,
    621
tunneling in, 416-17
watermills in, 392
windmills in, 392-94, *393*
writing in, 501, 503
Isocrates of Athens, 513
Israel, xviii, 384, 403, 576-77,
    601
  see also Jews, Judaism
Istanbul, *see* Byzantium,
    Byzantine Empire
Istemkheb, 269, *269*
Italy, 69, 106, *106,* 228, 258,
    288, 311, *320,* 372, *496,* 497,
    508, 511
  Stone Age in, 2-3
  see also Etruscans; Rome,
    Roman world

Jackson, Andrew, 206
Jainists, 308
James I, king of England, 245
Japan, xiv, 104-5
  food in, 308, *310*
  gardens in, 571
  music in, 601
  personal appearance in, 247,
    254, 260
  pottery in, 300, *301*
  sake in, 336
  sexual manuals in, 196

PICTURE CREDITS

The authors gratefully acknowledge permission to use illustration material (listed by page number) provided by the following individuals and institutions:

CHAPTER 1: MEDICINE
2 Prof Erik Trinkaus (Dept of Anthropology, University of New Mexico); 4 Museo Ostiense, Rome, Italy [#B-450]; 6 Rheinisches Landesmuseum, Bonn, Germany; 12 Rosalind & Jac Janssen (Dept of Egyptology, University College London, UK) and Juanita Homan; 14 The Wellcome Trust, London, UK; 17 Römisch-Germanisches Zentralmuseum, Mainz, Germany; 18 Musée Barrois, Bar-le-Duc, France [Inv. 850.20.1]; 19 Römisch-Germanisches Zentralmuseum, Mainz, Germany; 25 G. J. Owen, Faculty of Archeology and Anthropology, Cambridge University, UK [Duckworth Collection]; 33 Jack Ogden, Cambridge Centre for Precious Metal Research, Cambridge, UK; 43 *China Reconstructs.*

CHAPTER 2: TRANSPORTATION
59 Courtesy British Museum; 67 By permission of The British Library [Royal 14.C.ix, ff. lv-2r]; 68 Biblioteca Comunale e dell'Accademia Etrusca, Cortona, Italy; 93 Ontario Science Centre, Toronto, Canada; 99 Dr Nikos Kokkinos (London, UK); 102 By permission of The British Library [Royal 20, A.v. f. 71v]; 107 By permission of The British Library [Add. 34113, f.200v]; 108 Francis Hitching (Swalcliffe, Oxfordshire, UK); 111 The International Explorers Society.

CHAPTER 3: HIGH TECH
122 American Philosophical Society (Philadelphia); 127 Needham Research Institute, Cambridge, UK; 141 Trustees of the Science Museum, London, UK; 147 Søren Bo Andersen/Fur Museum, Denmark; 158 Courtesy British Museum.

CHAPTER 4: SEX LIFE
174 Courtesy British Museum; 194 © Ann & Bury Peerless, Slide Resources and Picture Library (Birchington-on-Sea, Kent, UK).

CHAPTER 5: MILITARY TECHNOLOGY
201 Jericho Excavation Fund; 215 Bodleian Library, Oxford, UK [*Mittelalterliches Hausbuch,* p. 53, Shelfmark = *247139,c,4*]; 218 The Dorset County Museum, Dorchester, Dorset, UK; 227 Biblioteca Nacional, Madrid, Spain; 232 Réunion des Musées Nationaux, Paris, France [Musée Nationale du Louvre]; 233 Needham Research Institute, Cambridge, UK; 235 The Governing Body of Christ Church, Oxford, UK [Christ Church Library, Oxford MS. 92, Folio 70 v (bottom), *De Nobilitatibus, Sapientiis et Prudentiis Regum* by Walter de Milemete]; 237 Needham Research Institute, Cambridge, UK.

CHAPTER 6: PERSONAL EFFECTS
246 Museum of London, UK; 249 Copyright British Museum; 251 Courtesy British Museum; 259 Professor S. I. Rudenko; 262 Museum of London, UK; 269 Joann Fletcher (Manchester University, UK); 274 Novosti (London, UK); 279 Museum of London, UK; 281 Hamburger Kunsthalle, Hamburg, Germany; 283 Copyright British Museum; 287 Réunion des Musées Nationaux, Paris, France [Musée Nationale du Louvre].

CHAPTER 7: FOOD, DRINK AND DRUGS
327 National Geographic Society, Washington.

CHAPTER 8: URBAN LIFE
357 Dr. S. R. Rao (Marine Archaeology Centre, National Institute of Oceanography, Goa, India); 360 Illustrated London News, London, UK; 375 Copyright British Museum.

CHAPTER 9: WORKING THE LAND
385 Konstantinos D. Politis (Institute of Archaeology, University College London, UK); 394 By permission of the Syndics of Cambridge University Library, UK [f. 130r from Ee. 2.31]; 406 Courtesy British Museum.

CHAPTER 10: HOUSE AND HOME
426 The Egyptian Museum, Cairo; 429 John Frankish (Cardiff, UK); 431 Professor Karl Jettmar, University of Heidelberg [*Art of the Steppes,* Methuen, 1967]; 432 Historical Museum, University of Bergen, Norway; 441 Copyright British Museum; 444 The Committee of the Egypt Exploration Society, London, UK; 445 Trustees of the British Museum; 452 University Museum, Pennsylvania; 454 Illustrated London News, London, UK; 465 Copyright British Museum; 467 Bedes World, © Jarrow, Tyne & Wear, UK.

CHAPTER 11: COMMUNICATIONS
485 Musée des Antiquités Nationales, Saint-Germain-en-Laye, France. Copyright © Photo MAN.; 486 Musée d'Aquitaine, Bordeaux, France; 495 Archivio di Stato di Siena, Italy [Docc. no. 214 = Biccherna no 117: "La riforma del Calendrio"]; 505 Dr Hugh A. Moran & Professor David Kelley (University of Calgary, Alberta); 511 Beinecke Rare Book and Manuscript Library, Yale University Library; 514 Bodleian Library, Oxford, UK; 517 By permission of The British Library [Or 8210/P2]; 519 Courtesy of the Board of Trustees of the Victoria & Albert Museum, London, UK; 521 Courtesy British Museum.

CHAPTER 12: SPORT AND LEISURE
526 Courtesy of The Oriental Institute of The University of Chicago; 552 Copyright British Museum; 560 Francis Hitching (Swalcliffe, Oxfordshire, UK); 562 Bibliothèque Nationale, Paris, France [Térence des Ducs, Paris MS. Arsenal 664, f47]; 581 Antikenmuseum, Basel, Switzerland; 585 Ashmolean Museum, Oxford, UK [detail; fresco is a copy of that in the Heraklion Museum, Crete, Greece]; 594 © Photo Musée des Antiquités Nationales de Saint-Germain-en-Laye; 603 Rheinisches Landesmuseum, Trier, Germany; 607 Ecole Française d'Archéologie, Athens, Greece; 621 Topkapi Sarayi Museum, Istanbul, Turkey.

PETER JAMES is a professional writer on ancient history and archaeology. He studied at Birmingham and London Universities and describes himself as a "generalist" in the ancient Near East and Mediterranean. He has published numerous articles on ancient technology, chronology, and the history of science, and is the principal author of the highly controversial *Centuries of Darkness*.

DR. NICK THORPE, an archaeologist specializing in prehistory, studied at Reading and London Universities, and is presently Lecturer in Archaeology at King Alfred's College, Winchester. A director of research projects in Britain and Denmark, he has contributed articles on agriculture, chronology, metalworking, astronomy, and prehistoric society to numerous books and journals, and is a co-author of *Centuries of Darkness*.